AF813160

Design Science and Innovation

Series Editor

Amaresh Chakrabarti, Centre for Product Design and Manufacturing, Indian Institute of Science, Bengaluru, India

The book series is intended to provide a platform for disseminating knowledge in all areas of design science and innovation, and is intended for all stakeholders in design and innovation, e.g. educators, researchers, practitioners, policy makers and students of design and innovation. With leading international experts as members of its editorial board, the series aims to disseminate knowledge that combines academic rigour and practical relevance in this area of crucial importance to the society.

Sananda Das · Wasim Akram · Vivek Khanzode ·
Rauf Iqbal
Editors

Human-Centered Digitalization, Volume 3

Proceedings of 4th ACED, HWWE-2023 and BRICSplus HFE 2023

Springer

Editors
Sananda Das
Department of Operations and Supply Chain Management
Indian Institute of Management Mumbai
Mumbai, Maharashtra, India

Wasim Akram
Department of Operations and Supply Chain Management
Indian Institute of Management Mumbai
Mumbai, Maharashtra, India

Vivek Khanzode
Department of Operations and Supply Chain Management
Indian Institute of Management Mumbai
Mumbai, Maharashtra, India

Rauf Iqbal
Department of Operations and Supply Chain Management
Indian Institute of Management Mumbai
Mumbai, Maharashtra, India

ISSN 2509-5986 ISSN 2509-5994 (electronic)
Design Science and Innovation
ISBN 978-981-96-4382-0 ISBN 978-981-96-4383-7 (eBook)
https://doi.org/10.1007/978-981-96-4383-7

© The Editor(s) (if applicable) and The Author(s), under exclusive license to Springer Nature Singapore Pte Ltd. 2025

This work is subject to copyright. All rights are solely and exclusively licensed by the Publisher, whether the whole or part of the material is concerned, specifically the rights of translation, reprinting, reuse of illustrations, recitation, broadcasting, reproduction on microfilms or in any other physical way, and transmission or information storage and retrieval, electronic adaptation, computer software, or by similar or dissimilar methodology now known or hereafter developed.
The use of general descriptive names, registered names, trademarks, service marks, etc. in this publication does not imply, even in the absence of a specific statement, that such names are exempt from the relevant protective laws and regulations and therefore free for general use.
The publisher, the authors and the editors are safe to assume that the advice and information in this book are believed to be true and accurate at the date of publication. Neither the publisher nor the authors or the editors give a warranty, expressed or implied, with respect to the material contained herein or for any errors or omissions that may have been made. The publisher remains neutral with regard to jurisdictional claims in published maps and institutional affiliations.

This Springer imprint is published by the registered company Springer Nature Singapore Pte Ltd.
The registered company address is: 152 Beach Road, #21-01/04 Gateway East, Singapore 189721, Singapore

If disposing of this product, please recycle the paper.

Contents

About the Editors

Sananda Das is Doctoral Research Scholar (Ph.D.) at the Indian Institute of Management (IIM), Mumbai, India, working in the Operations and Supply Chain Management (O&SCM) domain. She completed her M.Sc. in Human Physiology Ergonomics Specialization. Her research interests are work system design, engineering anthropometry, biomechanics of human movement, occupational ergonomics, smart retailing, consumer psychology, digitalization in retailing, supply chain management, industrial engineering, AR-VR technology, and artificial intelligence. Currently, she is working on digital shopping consumer behavior in retail for her thesis work.

Wasim Akram is Doctoral Research Fellow (Ph.D.) in area of Operation and Supply Chain Management at Indian Institute of Management, Mumbai, and his focused area is Ergonomics and Human Factor Engineering. He is working in the area of Operation and Supply Chain Management, Specialization in Human Factor Ergonomics. His research area is "cognitive ergonomics" and thesis topics—"Cognitive Workload Measurement and Its Impact on Network Operation Centre (NOC) Operator". He has experienced in teaching of Master's Student in the area of Ergonomics and also expertise in fieldwork management, ergonomics assessment, occupational safety, and ergonomics tools and techniques handling experimental work from ergonomics perspective.

Vivek Khanzode is Professor at the Indian Institute of Management, Mumbai, India (formerly NITIE, Mumbai). He has completed his M.Tech. and Ph.D. from the Department of Industrial and Systems Engineering at the Indian Institute of Technology (IIT), Kharagpur. His research interests are lean manufacturing, manufacturing strategy, work system design, supply chain management, and simulation modeling. He has published over 40 articles in international journals like *IEEE Transactions on Engineering Management*, *Experts Systems with Applications*, *Safety Science*, *Rapid Prototyping Journal*, *Journal of Enterprise Information Management*, *Process Safety Progress*, and *Benchmarking*.

Rauf Iqbal is Professor of Ergonomics and Human Factors Engineering at the Indian Institute of Management, Mumbai, India (formerly NITIE, India). He presently holds the position of Dean-Student Affairs at NITIE. His research interests are work system design, engineering anthropometry, biomechanics of human movement, occupational ergonomics, supply chain management, and safety management. He has published 125+ journal articles, edited books, and book chapters. He has guided 12 Ph.D. scholars. He is Member of various committees like the Foundation of Professional Ergonomics, USA, BRICS Plus Executive Committee for Human Factors and Ergonomics, the Ergonomics Sectional Committee for Bureau of Indian Standards, and the General Secretary of Indian Society of Ergonomics.

Towards a Flood-Resilient Mumbai: The Case of Mithi River Rejuvenation

Shankar Deshpande and Soudip Kundu

1 Introduction: Urbanization, Climate Change, and Flooding and the Case of Mumbai

The world is rapidly urbanizing, with India experiencing significant economic growth driven by its urban areas, which now house around 34% of the population [1]. This urbanization, combined with the forces of climate change, poses unprecedented threats to quality of life, economic stability, and social well-being. As cities rapidly grow, especially in developing countries with informal settlements and slums, they face challenges addressing climate change and other issues due to governance, infrastructure, and equity deficits. Climate change has far-reaching impacts on our planet and societies. Here are some key impacts:

- **Rising temperatures**: Global warming leads to heatwaves, droughts, and heat-related illnesses. It disrupts ecosystems and affects biodiversity.
- **Melting ice and rising sea levels**: Melting glaciers and polar ice caps raise sea levels, threatening coastal communities with erosion and flooding.
- **Changes in precipitation patterns**: Altered rainfall patterns result in more intense rain in some areas and droughts in others. This affects agriculture, water availability, and increases the risk of wildfires and landslides.
- **Impact on ecosystems**: Climate change disrupts ecosystems, leading to habitat loss, species extinction, and imbalances. This affects vital ecosystem services like pollination, water purification, and carbon sequestration.
- **Health risks**: Global warming poses health risks such as heat-related illnesses and changes in disease distribution and pest patterns.

S. Deshpande (✉) · S. Kundu
Town and Country Planning Division, Mumbai Metropolitan Region Development Authority, Mumbai, India
e-mail: shankardc1@gmail.com

© The Author(s), under exclusive license to Springer Nature Singapore Pte Ltd. 2025

S. Das et al. (eds.), *Human-Centered Digitalization, Volume 3*, Design Science and Innovation, https://doi.org/10.1007/978-981-96-4383-7_1

- **Socio-economic impacts**: Climate change causes economic damage, increased healthcare costs, and disruptions in sectors like agriculture, fisheries, and tourism. Vulnerable communities, especially in developing countries, suffer the most.
- **Climate migration**: Climate change can displace populations due to rising sea levels, extreme weather events, and reduced agricultural productivity, leading to social, economic, and political challenges.

Addressing and mitigating the impacts of global warming and climate change require concerted efforts at the global, national, and local levels. It involves reducing greenhouse gas emissions, transitioning to renewable energy sources, promoting sustainable land and resource management practices, and adopting resilient and adaptive strategies to protect vulnerable communities and ecosystems.

Mumbai, India's commercial capital, has witnessed extensive urbanization. Originally a cluster of seven islands, reclamation projects have expanded the city's land area to accommodate its growing population. The urban sprawl in the Mumbai Metropolitan Region (MMR) is evident, with the built-up area increasing significantly. The city's high population density, about 28,500 persons per sq.km, has led to traffic and transportation challenges. Reduced reliance on public transport has increased traffic congestion and greenhouse gas emissions. Addressing these issues in a sustainable manner is essential for Mumbai's future (Fig. 1).

Climate change's impact on Mumbai is evident in Sea Surface Temperature (SST) patterns [2]. According to the IPCC Fifth Assessment Report, Asia will experience more frequent and intense heatwaves, along with an intensified heat island effect. These factors pose long-term risks of increased heat-related fatalities and illnesses.

This emphasizes the urgent need for effective measures to combat rising temperatures. Proactive strategies, such as heatwave preparedness plans, improved urban planning to mitigate the heat island effect, and awareness campaigns, are crucial

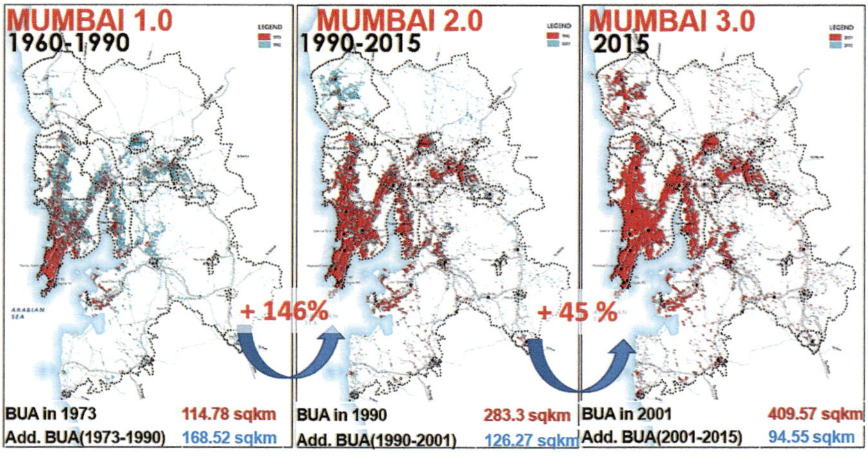

Fig. 1 Urban sprawl in MMR

to protect vulnerable populations and reduce the health consequences of higher temperatures.

2 Global Warming, Climate Change, and the 26th July 2005 Floods

Human-induced global warming has altered worldwide rainfall patterns and intensified heat island effects in urban areas. These changes in precipitation lead to water stagnation in coastal cities, causing socio-economic impacts like traffic congestion and disruptions to city life. The stability of Indian monsoon rainfall remains a puzzle amid rising global temperatures [3, 4].

Climate change intensifies threats to urban areas, including sea-level rise, ice shelf collapse, and warming seas. Coral reefs, vital for coastal protection, are endangered. These changes impact urban residents' health and well-being. Rising sea levels in low-lying coastal zones lead to inundation, flooding, saltwater intrusion, erosion, and loss of habitable land. Intensified storms worsen hazards for these populations. Flooding results from population growth, urbanization, altered river channels, and heavy rainfall, influenced by urbanization, climate change, and local flood risks. Coastal cities experience socio-economic repercussions, exacerbated by climate-induced rainfall pattern shifts. South Asia faces extreme humid heat due to climate change, disproportionately affecting vulnerable communities. Intense rainfall events, regardless of location, can lead to flooding, with climate change increasing their frequency and intensity [5, 6].

In the dynamic urban landscape, the ancient Mithi River has silently witnessed Mumbai's growth from barren lands to a thriving metropolis. It observes the evolving daily life and the constant movement of its connected residents. Surrounding Mumbai are the Arabian Sea and various creeks, including Malad, Mahim, Mahul, and Thane, impacting its drainage system. Additionally, the city is crossed by three other rivers—Poisar, Dahisar, and Oshiwara—originating from hilly regions and merging with the Arabian Sea through creeks [7–9]. Mumbai's stormwater system was designed for 25 mm/hr precipitation with 50% runoff initially. But extensive development increased runoff to nearly 100%, making the old design insufficient. The century-old stormwater and sewerage systems, including open drains, rivers, and box drains, deteriorated, reducing their capacities [10].

Heat islands contribute to urban river flood risks, as seen in the devastating 26 July 2005, Mumbai flood linked to climate change. Mumbai came to a standstill with 944 mm of rainfall in 24 h, coinciding with a 4.48 m high tide, resulting in 698 fatalities and extensive property damage. The Mithi River gained global attention. The 26th–27th July 2005 flood, exacerbated by the heat island effect, was similar to the 1974 event when Colaba received 570 mm of rainfall. The situation worsened due to encroachments, culvert obstructions, altered stream flows, crowded structures, and silting, reducing waterway capacity. The Fact-Finding Committee Report identified

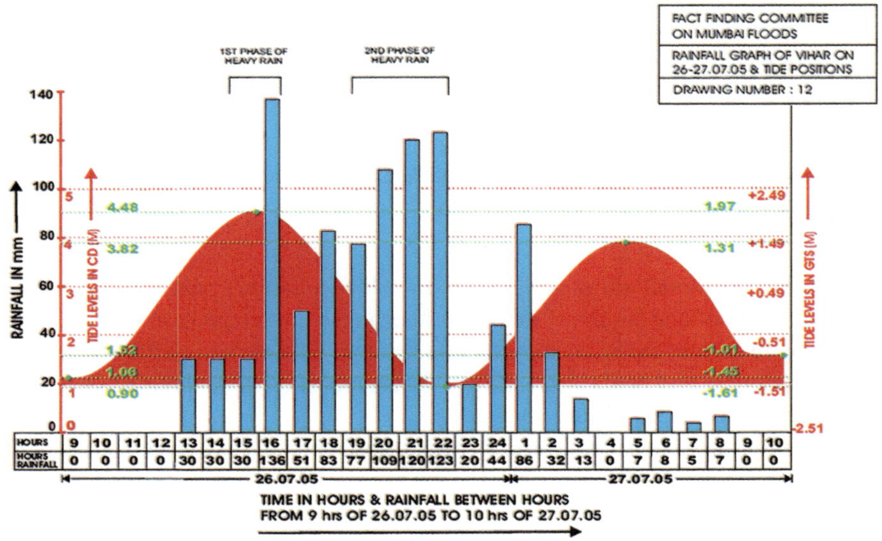

Fig. 2 Rainfall graph of Vihar on 26th–27th July 2005 [10]

two flood waves. The first wave, during high tide from 14:30 to 20:30, caused a deluge. The second wave, from 20:00 to 22:00, should have passed during the ebb period, but a clogged drainage system hindered effective water flushing [10] (Fig. 2).

The excess rainfall, coupled with encroachments and the improper maintenance of the river, led to its rapid overflow and caused severe inundation in many parts of the city. The 2005 Mumbai flood highlighted the vulnerability of urban areas to the impacts of climate change, particularly in relation to urban rivers. The event served as a wake-up call, prompting the recognition of the need for improved urban planning, infrastructure, and disaster management strategies to mitigate flood risks.

3 Rainfall and Characteristics of Mithi River and Vakola Nalla

Mumbai experiences an average annual rainfall of 2129.8 mm, predominantly occurring between June and September. July is the wettest month, typically receiving 742.2 mm of rainfall. During the monsoon season, it is not unusual to encounter heavy downpours exceeding 250 mm in a single instance [10]. Mithi River with catchment area of 7295 ha has its origin at 246 m. above mean sea level and has a total length of 17.84 km. The Mithi River originates from spillovers of Vihar and Powai Lake traverses through Mumbai's suburban areas, viz. SEEPZ, Marol, Andheri and then flows below the runway of International Airport and then meanders through areas of Bail Bazar, Kurla, BandraKurla Complex and meets Arabian sea at Mahim

Table 1 Administrative limits of Mithi River and Vakola Nalla

	MMRDA (Special planning authority for BKC)	**MCGM** (Planning authority as local authority)
Mithi River—17.84 km Catchment: 7295 Ha	6.0 km (CST Bridge on SCLR to Mahim Creek)	11.84 km (Vihar Lake to CST Bridge on SCLR)
Vakola Nalla—3.8 km Catchment: 907 Ha	1.8 km (Vakolla Bridge on SCLR to BKC)	2.0 km (Airport to Vakola Bridge on SCLR)
Adopted Width and Depth of Mithi River	***Tidal portion*** As per CWPRS for Return Period 1 in 100 years	***Non-tidal portion*** As per FFC Report for Return Period 1 in 10 years

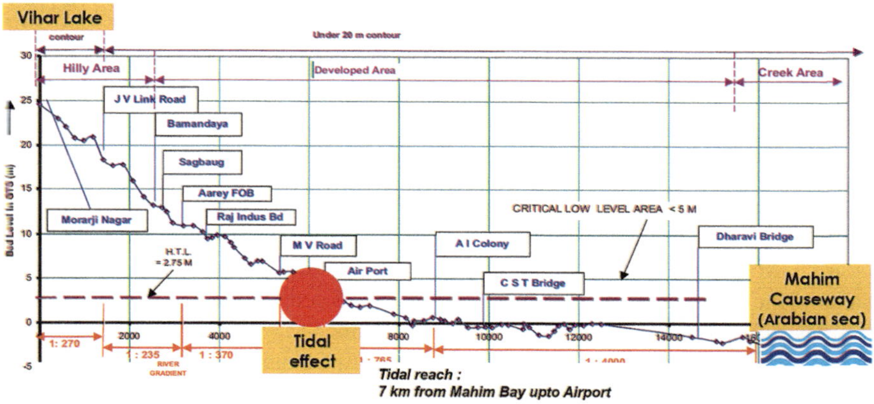

Fig. 3 Characteristics—Mithi River

Bay after flowing below 15 bridges for a length of 17.84 km (Table 1 tells about the administrative limits of Mithi River and Vakola Nalla). Figure 3 shows the characteristics of the Mithi River with a river gradient varying from 1:4000 near Mahim Causeway to 1:270 near the Vihar Lake. The Tidal Effect is observed for a length of 7 km, till the airport, and the river remains perineal only in this portion. The remaining part of the river, till its origin, gets water only during monsoon season.

4 Immediate Actions by the State Government

The Fact-Finding Committee, appointed by the Government of Maharashtra (GoM) after the August 2005 flood, identified flood causes and recommended mitigation measures. Their report was submitted in 2006. The GoM further established the Mithi River Development and Protection Authority (MRDPA) under the Hon. Chief Minister's leadership to prevent future disasters. MRDPA coordinated with the

Mumbai Metropolitan Region Development Authority (MMRDA) and the Municipal Corporation of Greater Mumbai (MCGM) for the Mithi River and Vakola Nalla projects. The Empowered Committee, headed by the Hon. Chief Secretary, conducted meetings since 2005, with the last in November 2013. These meetings addressed various aspects of the Mithi River initiatives, involving technical experts, stakeholders, and the public in decision-making and implementation.

The Central Water and Power Research Station (CWPRS) and the Indian Institute of Technology Bombay (IIT B) provided expert input on deepening and widening the river and environmental improvement. Pollution studies by the Maharashtra Pollution Control Board (MPCB) in 2006 and an Integrated Impact Assessment Study by IIT B and NEERI in 2013 were also conducted.

5 Approval to Implement the River Development Works in Two Phases

In the Writ Petition No. 2116 of 2005 (Kirit Somaiya vs. State of Maharashtra and others), the Hon. Bombay High Court issued an order in August 2005, directing the MCGM and MPCB to identify and regulate polluting industrial units along the banks of the Mithi River. Furthermore, the court ordered that no new encroachments or industrial activities be permitted on the riverbank.

In a subsequent order in March 2006, the Hon. High Court accepted the proposal of the MRDPA to carry out the river development works in two phases. Phase I focused on deepening, widening, and desilting of the Mithi River, as well as the resettlement and rehabilitation of project-affected persons (PAPs). The MRDPA was tasked with monitoring and overseeing the execution of the deepening, widening, and desilting works by the MMRDA and the MCGM. Phase II included additional works such as pitching, construction of retaining walls, service roads, landscape development, beautification, and further widening and deepening of the river.

6 Approval of Development Plan of Mithi River and Vakola Nalla and the Status of the Mitigation Works Carried Out

During the 6th Authority meeting of the MRDPA, chaired by the Hon. Chief Minister on 25 May 2010, the Mithi River's widths were finalized based on recommendations from the Fact-Finding Committee, CWPRS, and IIT B. The development plan, including deepening and widening, was largely accomplished by MCGM and MMRDA, following the approved widths. In the MMRDA section, deepening is complete, while widening reached 90%. Rock excavation is at 58%, retaining walls at 91%, service roads at 73%, and beautification work is finished. In the MCGM section, deepening is 90%, widening is 84%, rock excavation is 85%, retaining walls are 20%, and service roads are at 5%.

These efforts have incurred an expenditure of Rs. 1300 Crores, with about Rs. 530 Crores remaining for pending work. Approximately 10,000 encroachments have been successfully removed along the river, and over 5000 project-affected persons (PAPs) have been rehabilitated. Mumbai International Airport executed several projects along the Mithi River, including an additional channel to enhance discharge capacity, a retaining wall, and a GSE bridge for efficient and safe aircraft operations.

7 Various Adaptation Measures Led by Mumbai Metro

Following the 2005 flood, Mumbai took measures to enhance its resilience and mitigate climate-related challenges. Efforts focused on flood mitigation, transportation infrastructure improvement, and sustainability. The Municipal Corporation of Greater Mumbai (MCGM) plays a key role in maintaining river cleanliness and controlling pollution. The Pollution Control Project involves sewer construction and infrastructure development, with progress on sewer tunnels and packages at various stages. MCGM annually removes on an average more than 2.5 lakhs MT silts every year from the entire length of the river. These ongoing efforts by the MCGM demonstrate the city's commitment to improving the sewer network and preventing pollution in the Mithi River. The progress of these works, including desilting, is regularly reviewed during the annual pre-monsoon preparation meeting, which is chaired by the Honourable Chief Minister. This meeting serves as a platform for discussing the status of the various projects related to the Mithi River and ensuring coordination among the concerned agencies and stakeholders involved in the river's development and maintenance. The Mumbai Metropolitan Region Development Authority (MMRDA) has undertaken construction of more than 350 km of Metro Rail in the metropolitan region and introduced green fuel and electric buses to reduce emissions and improve air quality. The Mono Rail, another important addition to Mumbai's transportation infrastructure, offers a convenient, sustainable and eco-friendly mode of transport for commuters (Fig. 4).

These climate-resilient measures adopted by Mumbai demonstrate a holistic approach to address both transportation needs and environmental concerns. By investing in sustainable and low-carbon transportation infrastructure, the city aims to mitigate the impacts of climate change, improve air quality, and enhance the overall resilience of its urban systems. These initiatives not only contribute to reducing the carbon footprint but also create a more sustainable and liveable city for its residents.

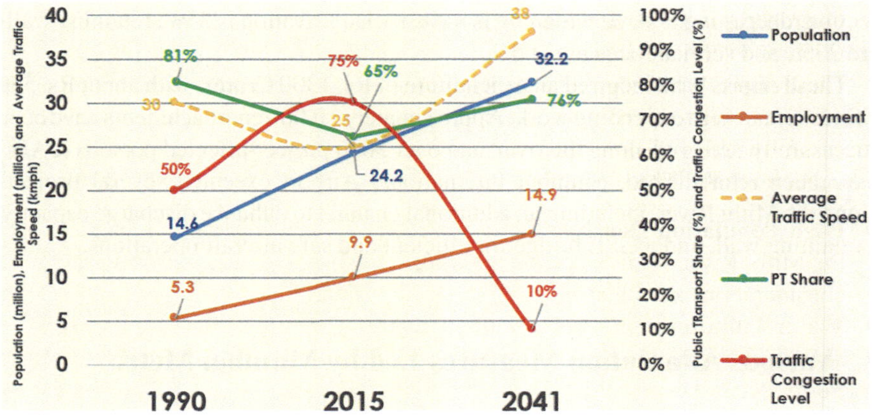

Fig. 4 Positive impacts of Metro Rail project towards climate-resilient adaptation measures

8 Achievement and Challenges and Road Ahead

Achievements: The works carried out by MMRDA, MCGM, and MIAL following the 2005 floods in Mumbai have acted as flood mitigation measures and have prevented fresh disasters that also contribute to climate change adaptation and resiliency. The achievements are summarized below:

- **Enhanced River Capacity**: The Mithi River has been widened by an average of 30 m and deepened by 2 m doubling its water-holding capacity and tripling its discharge capacity, as confirmed by CWPRS in 2011. These changes help reduce flood risks during heavy rains.
- **Flood Risk Mitigation**: The combined measures of deepening, widening, and building retaining walls have reduced the likelihood of floods in vulnerable areas of Mumbai, protecting the region's economy by preventing significant losses and interruptions.
- **Economic Protection**: Flood mitigation efforts have safeguarded the local economy, avoiding productivity losses estimated at Rs. 13,500 Crores since 2005, according to MMRDA's analysis in 2016.
- **Improved Water Quality**: Bed slope adjustments have enhanced flushing activity, leading to better water quality with higher levels of dissolved oxygen and lower biochemical and chemical oxygen demand.
- **Encroachment Protection**: Retaining walls have been built to prevent encroachments, maintaining the river's flow capacity and protecting its dimensions.
- **Service Roads**: The construction of service roads along the river facilitates maintenance and ensures the river's proper functioning.
- **Ecological Restoration**: Efforts have helped restore the river's marine ecology, improve habitats for flora and fauna, and support aquatic life.
- **Bridge Widening**: Eleven bridges have been widened, improving traffic flow, connectivity, and infrastructure resilience.

Overall, the work since the 2005 floods has not only mitigated flood risks but also contributed to climate change adaptation and enhanced Mumbai's resilience.

Challenges: While significant progress has been made in the rejuvenation of the Mithi River, there are still several challenges that need to be addressed in the future to ensure the long-term success of flood mitigation measures and climate change adaptation efforts. These challenges include:

- **High Population Density**: The dense residential and commercial areas around the Mithi River strain its resources. Balancing development with environmental considerations is crucial.
- **Water Pollution**: Ongoing pollution from untreated discharges requires better sewage infrastructure, stricter pollution controls, and public awareness efforts.
- **Solid Waste Disposal**: Effective waste management systems, including collection, segregation, and disposal, need to be strengthened to prevent waste accumulation along the river.
- **Land Constraints**: Limited space for utilities and service roads along the river poses **logistical** challenges that require innovative solutions and careful planning.
- **Encroachment Prevention**: Continuous efforts are needed to prevent new encroachments despite existing retaining walls. Strict enforcement and community engagement are key.

Addressing these challenges will require continued collaboration among government agencies, local communities, and other stakeholders. Long-term planning, effective governance, and sustained efforts are necessary to ensure the success and sustainability of the Mithi River rejuvenation project and its contribution to flood mitigation and climate change adaptation in Mumbai.

References

1 Census (2011) Census data
2 Mumbai Climate Action Plan (2022)
3 GoI MO (2015) Smart cities mission statement & guidelines
4 Mann R, Gupta A (2022) Assessment of vulnerability to urban floods in Greater Mumbai, India using geospatial techniques. Research Sqaure
5 Cities and Climate Change (2011) Global report on human settlements. United Nations Human Settlements Programme, UN Habitat, London, Washington, DC
6. Extreme humid heat in South Asia in April 2023, largely driven by climate change, detrimental to vulnerable and disadvantaged communities. (2023, May 17). Retrieved from World Weather Attribution: https://www.worldweatherattribution.org/
7. Khan M, Vinay N, Gupta K (2014) Flood mitigation on mithi river using detention pond. IOSR Journal Of Environmental Science, Toxicology and Food Technology (IOSR-JESTFT), Mumbai, India, pp 34–39
8. Kirtane G (2020) Why Mumbai must reclaim its Mithi. Observer Research Foundation, Mumbai
9. Singh A, Taulton T, Wu W (2019) Mithi River Action Plan: changing Mumbai's paradoxical equation with its river. Carnegie Mellon University Spring 2019
10. Fact Finding Committee (2006) Mumbai: Government of Maharashtra

Media and Communication? Reflections

Deepshikha and Avinav Krishna

1 Introduction

The paper comprises literature study conducted on subjects such as communication strategies, cognitive load, silent marketing, social media and 'less is more' principle in the following sections. Further, a user study has been conducted with 25 participants belonging to the age group 25–40 years, to understand their social media usage, their idea of utopic communication, cyber law and its features, how they would promote a brand if they owned one and if silent marketing is known to them. The response of the user study conducted has been presented in tabular format. The conclusion section links literature review with user study finding for key insights to summarize the paper.

2 Literature Study

2.1 Communication

Communication is important as it reduces ambiguity and mistrust. Researchers have studies how designers and clients interact to exploit or bridge the gap between two parties. Technology, viewpoints of the client, stakeholders, design requirements,

Deepshikha (✉)
Design and Innovation, Jamia Millia Islamia, Delhi, India
e-mail: emaildeepshikhajha@gmail.com

A. Krishna
International Relations, South Asian University, Delhi, India

© The Author(s), under exclusive license to Springer Nature Singapore Pte Ltd. 2025
S. Das et al. (eds.), *Human-Centered Digitalization, Volume 3*, Design Science and Innovation, https://doi.org/10.1007/978-981-96-4383-7_2

needs of client and users, feedback, values, design requirements and human characteristics have to be managed to a great extent in present communication channels [1]. Knowledge acquisition and knowledge negotiation are also very important for effective communication. Social relationships, collaborative design, problem solving, creativity, influencing relationships and context also play an important role in effective communication of function and behavior [1]. Communication is dynamic to quite an extent as it requires a balance between ideas, opinions, goals, mutual beliefs and assumptions. It must also emerge from credible sources, heard and shared by right people and generate appropriate response in the audience [1]. Communication may be verbal, non-verbal, physical or digital. Hence strategies may also differ according to the medium the information is being transmitted or communication being shared. Few critical elements of information factors may include semantics, emotions, technicalities, tools, messages and relationships [1]. Most of the graphic, product and UI design exercises begin with client brief converted to technical specifications with great detail and precision to avoid high cost of sampling [2]. Communication takes place between different cultures, people, groups, companies, designers, manufacturers, suppliers and the end customers. Communication may be synchronous or asynchronous [3]. Characteristics of communication include—Profane, Universal, Ephemeral, Relational, Multi-causal, ReflexiveTemporal, Social and Factual [3]. Communication may be information-centered, interaction-centered or situation-centered. Design at present does not have a defined taxonomy of tasks or process since it is ever evolving and creative freedom is immense [3]. Strategic communication is practicing persuasive and purposive methods to reach set goals [4]. Communication strategies provide information, help in increasing awareness, may encourage action, build consensus, may change behavior, promote community participation, may resolve conflict, need inputs, help, care, harm, establish relationships or generate business. Visual communication has played an important role in strategizing communications in present times [5]. Strategic communication helps identify the right objectives and channels for propagation.

2.2 Cognitive Load

The cognitive load theory was initially derived out of educational psychology [6]. The three types of cognitive load areIntrinsic cognitive load (where content is to be learnt); Extraneous cognitive load (where learning is not involved) and Germane cognitive load (where learning and performance is monitored) [6]. The purpose of cognitive load theory is to provide information to be processed in such a manner that it reduces cognitive load significantly [7]. It is important to understand cognitive load architecture, the learner, the task at hand and the learning environment to be able to reduce the cognitive load [7]. It is also believed that cognitive load theory is built on two ideas—one, what amount of new information can be processed by human brain at a given time and second, that, there is no limit to how much stored information can be processed at one time. It is also important to understand working memory (where small amounts of information are stored for shorter duration) and long-term memory (where larger amounts of information are stored for longer durations semi-permanently) [8]. Well-being can significantly help in reducing cognitive load by inducing presence of positive emotions, reduced negative emotions and cognitive regulation [9]. This happens due to reduction in load on working memory as the information stored in long-term memory is recalled according to the environmental organization and linking principle and results in thought-action processes emanating from positive emotions [9].

2.3 *Silent Marketing*

It is past when a good product would speak for itself since today anyone can manufacture anything and be ready to sell. As markets mature faster, advertising turns symbolical, brands become iconic and companies become corporate brands, events happening at a very quick pace and brand building has evolved from what it was formerly [10]. According to research and practice there are 5 types of interactions between company and customers—service messages, product messages, planned marketing, unplanned marketing and absence of communication [11, 12]. Present communication between services, marketers and customers are increasingly optimistic about their impact on the customers and overlook the possibilities of silent or invisible communication which communication may not happen from marketer's viewpoint but from customers' viewpoint [13]. Four kinds of silent communication from marketers' perspective include—marketer monologue, miscommunication, marketer non-response and tacit/no communication [13].

2.4 *Social Media*

Social media has had immense positive and negative impact and leads to distress, objectification and change in behavior [14]. People who are more depressed and anxious use social media extensively to escape foul moods; however, it may affect negatively in most cases as there is lack of constructive advice [14]. Facebook has been designed to be addictive, to attune to users feeling so that they want to come back more and more and spend more time on it [15]. A lot of advertising which may not have positive social benefits gets rampant leverage such as alcohol, violence, substance abuse among others and is difficult to retract [16]. Uncontrolled and unadministered usage may lead to supplementing of negative impacts of social media and psychological well-being [17].

2.5 Less Is More

Minimalism involves incomplete stimulus for a sophisticated and innovative understanding and representation in art, architecture or design [18–20]. The minimalist design or advertising strategy does not have creatively divergent or superfluous tropes, is capable of eliciting more surprise and positive response and attitudes [21–24]. Metaphoric representations on the other hand may provide incongruent and incomplete stimulus and there are greater chances of not being understood or completely incomprehensible response in viewers [25–27]. Few studies have been conducted on minimal visual rhetorics and its impact on consumer preferences [18]. It has been also noticed that users or consumers get greater satisfaction upon solving complex modern stimulus [18]. Minimalist designs or art have also been accused of portraying no essence or message which leads to failure of comprehension in viewers [22]. Contrarily, minimalist methods also generate a qualitative silence that leads to generation of metaphoric stimulus in its viewers and increase comprehension [18]. Researchers in an in-depth study found that minimalism does have a significant positive influence on attitude toward the brand, attitude toward the advertisement and subjective message comprehension [18].

3 User Study Findings

The study was conducted with 25 respondents between the age group 25–40 years of which 48% were male respondents and 52% were female respondents. They are young working professionals in the domains of design, finance and management. 30% of the respondents are very active and 40% are active users of social media while 17% frequently use social media. 13% also mentioned that they rarely use social media presently. 75% of the respondents own 2–3 profiles on the social media while 3 respondents own more than four profiles. 85% agreed that social media should not be banned, while 10% said it should be banned. One respondent did not have any opinion on social media being banned at all. Other data collected has been summarized in Table 1.

4 Conclusion

Social media has several disadvantages and they outgrow the number of advantages [28]. To reduce cognitive load and addiction, the author proposes that marketing and promotion should be separated from connecting with friends, family and colleagues and be treated as a separate entity. A disruptive innovation is sought where all promotions take place either online or physically and sharing information should be a separate channel of communication. At present, Facebook, Instagram, Tiktok, etc., have

Table 1 Findings of the user study conducted

S. No.	Question	Response
1	Ways in which social media has benefited YOU–	Important information and announcements; Helped with good connections; Networking and knowledge; Freelancing; Helped me to increase my spiritual horizon; Keeps me updated; Made me aware of current technology and techniques; Improvising my skill set; A lot of inspiration for creatives and some motivation when needed; right profiles on social media give correct information; Exposure; entertainment; experience; paid promotions; motivational videos; time pass; human connection; Diverse information, breaking news, memes (entertainment); trends and new tech launch; new things to know and explore; Staying positive
2	Ways in which social media has harmed YOU–	Exposed sometimes; Spamming; Disturbed Sleep timeline; Time-consuming; Overthinking; Time taking; Mobile phone addiction; spent more time in front of a screen rather than reading a book or going out for that relaxing walk; No peace; Distraction, time waste; mindlessly browsing social media sites; show off by others sometimes leads to introspection; FOMO; Negatively impacts self-worth; kills time; Losing focus on day-to-day activities; eye strain; Privacy gets compromised easily; false information; false connections; Too much of information; continuous scrolling wastes a lot of time
3	What are few most annoying aspects of social media?	Ads and too much irrelevant information; Frequent notifications that seem unavoidable and leads to distraction; A negative feeling due to comparative environment; unnecessary nudity; trolling and unnecessary follow requests from creeps; Pop ups; Sometimes see trends even if not interested; Poorly designed content; Nonsense videos; falsehood; unrealistic standards; Self-aggrandisement; superficial; addictive; Posting unnecessary and irreverent information; The way it circulates news without any check
4	What's best about social media?	Reach people and make connections; updates; Platform to share your talent and in some cases even make a living out of it; Connect with new people; e-learning; What's happening around the world; getting inspired; Let's you interact; lot of opportunities; shared lives and knowledge; Entertainment; New forms of expression; Fast information

(continued)

Table 1 (continued)

S. No.	Question	Response
5	Can you imagine a future without social media? Would it be sad, difficult, boring or would you be relived, happy and comfortable?	26% sad; 27% neutral; 40% happy; 7% didn't say anything about it May be boring; Hard to say, sometimes it acts as a breather in life; Relieved and happy but practically it's difficult due to the technological advancement today in every field; I think there will be some alternative to it because the main purpose of the social media is to connect, share and gain knowledge. If this purpose is fulfilled by any other means then there is no need of social media. There might be some other platform like maybe VR where you will be able to connect to your friends while sitting in your bedroom. So things will change and it will have its own advantages, people will adapt and I think it will be exciting; No, definitely it will be sad, difficult and extremely boring too; Sad; Difficult; I will be neutral about it. I will find something else to explore and spend time learning some other activities that I don't get to do now; Difficult; It will be both. But yes it's a habit nowadays. In its absence after a while people will find new ways of sparing their time. So yes it will be a sad difficult phase for a short while but afterwards we will feel relieved; Neither, I am okay if it is not there; it would be happy and comfortable without social media though; I will be very happy; I think social media will continue to exist. It will probably end when there are severe restrictions on communication and internet itself, which will be very, very sad; I think it would be bit unusual for few days because I will look for other ways for exploring things which I used to explore via social media and with time I will definitely adopt alternate ways. At the end I would be comfortable, happy and relieved; I will adapt to any kind of future. People will too, with time; Difficult; I will be neutral to it

(continued)

Table 1 (continued)

S. No.	Question	Response
6	What would be the most idealist or utopic way of communicating, if you were to design a one and only channel for communication? What would be your strategy?	Not sure; Never thought; I wouldn't design, sorry; One nation one communication; No creepy ads, just free, secure and effective communication channel; A chatbox is quick and easier but on a formal note it should be email; The utopic way would be my channel will work in any situation and is not dependent on the internet; Confession without letting the identity revealed; Should have security; An app that would work on AI so that you can schedule your social app usage and saves you from that addiction; I think all social media platforms are an ideal way in itself. But I think verbal communication on call should be something an ideal way; Calls; Human interaction at physical level without using any particular medium just one to another; Web-conferencing (such as Zoom) is the best way to communicate, because you can communicate with multiple people in real-time through voice and video; I will use call or mail for crisp and concise communication. Other than that, meetups are good; Face to face; Telepathy, maybe advanced versions of neuralink will enable that; I would prefer to have a platform which can be used for both social or professional purposes because both aspects of the life is important. So, that one and only platform should have two spaces one for social and another for professional; I think only one way of communication won't serve the purpose of communication because in different spaces different ways of communication serve the purpose in best way. So, feasibility of communication in different way is to be there; Communication in person is the most ideal because there are more ways to communicate besides language and images; One where one can also share emotions; To encourage people in constructive discussion only; Probably thinking and getting to the place where the other person would be willing to meet or just a call away

(continued)

Table 1 (continued)

S. No.	Question	Response
7	Do you think cyber laws are helpful?	65% said yes Haven't tried or explored; Sometimes because it is controlled by the government hence it can be manipulated by the people in charge; Yes but less awareness; Yes but the system needs to work better; Yes, very much. I have some close influencer friends who received very quick help from cyber security; Yes but they are just laws without getting practiced; Yes, They help protect interests of citizens in the new online world; Yes, but needs to adapt with time; Yeah it is definitely helpful. As social media has cyber space as medium for communication. So, obviously that space need to be secure and safe. They benefit businesses and governments mostly. They can be helpful to the common man but that is just a by-product; To some extent; To an extent; Not really sure; I am not aware; Not in India
8	What would be the most important clauses you would certainly want if you were to amend cyber laws?	Making user aware of what is a crime and when to report; Quick and responsive; Freedom of speechthere are loopholes; Action to be taken within 24 h of complaint lodged; How one portrays the other person on social media that need to be tackled so as to avoid false narratives; I think the security should also be provided quickly even if you are not a celebrity or an influencer. Common people shouldn't think twice before reaching out for help. Cyber laws should be portrayed and even help in such a way that it doesn't seem like a hassle; Content to be verified and validated before going to social media; To take down all fake profiles, the accounts should be verified; 1 h a day policy; I will be allowing only 18 + people on Social media; High levels of control over one's data, and easy to understand Terms and Conditions from the digital platforms; Data protection; Responsibility for generating fake news, clarifying ambits of online freedom of speech and expression, more control of the user over personal data, more stringent privacy laws; Better privacy and companies unable to access and store the information beyond a certain point; Protection of personal information; More stringent laws against cybercrime; To notify people before they click on dangerous links and track criminals quickly before they are able to cause damage. If one can send a message in a second through social media then crime should also be stopped at the same pace; Prohibition of any destructive criticism that may affect mental health; Privacy protection; Not sure; Not aware about it exactly

(continued)

Table 1 (continued)

S. No.	Question	Response
9	Do you believe 'less is more'? How have you contributed to it?	10% said 'no'. 90% said 'yes' Simple and intuitive information; If you can manage things well nothing is less. The Covid has taught us on how to spend less and still live a decent lifestyle; Yes I do believe 'less is more' because less is clearer than being confused with more; Yes, because as a designer my approach toward every product is dependent on efficiency and ergonomics for creating balance; I have donated half of my wardrobe in 2022 and have so much more peace; Less is more. It will always be. It's a human behavior but we cannot change that; Being aware that social media is not reality; Yes, I post less and I try to spend less hours; I do believe that less is more because more you show the less important you become; I think it is a good idea in many scenarios. However, I don't think I have actively contributed to it; I prefer to direct get into call and solve their queries rather than overflowing them with lot of external sources; I try and live frugally; Less social media = more time for self; Less talk = more observe and explore; Less things = managed things; I do post rarely in social media; Yes, trying to filter out information every now and then

(continued)

Table 1 (continued)

S. No.	Question	Response
10	If you were to own a brand, how would you promote it? Please ideate creatively	Social media would be used by 53% of the respondents All modes including social media; Advertisement through social media; Word of mouth works best. Release my offering to a selected group and let them speak about the brand; Initially through my friend circle and society giving some simpler products to try for free. That's the first step before approaching through social media; I would have promoted on any platform which is about to come in the near future; Like Metaverse is new, I would have promoted my brand through Metaverse; I would follow the latest trends going on the social media; As of today I would say for my brand promotion I will firstly connect with the current influencers for promotion and secondly I would work on the presentation of my brand because these were the major factors that will help in connecting with the audience; Through social media and offline marketing. And by doing small collaborations with other similar brands in market for more visibility; If I am owning an apparel brand, social media will be one aspect. But also promoting it in different areas and collaboration with people related to that field is very necessary. People look upon celebrities and personalities. So yes collaboration and promotion in different cities, areas and countries by arranging campaigns and then sharing glimpses on social media will be very helpful; Flood people on WhatsApp, Instagram and Twitter about it; Since these days people visit malls more than any other place, I would have placed a flyer or a poster in malls and at areas with heavy footfalls; I would use the old ways the traditional ways of promoting a brand and I would rather use cinema and films medium to promote it; Many people want to signal a way of life through the choice of products they own/use. If I own a brand, I make that brand stay true to lifestyle/philosophy. I will also make that philosophy well-known through advertising/interviews, etc. E.g., Outdoor gear/clothing brand Patagonia has made conservation of nature as its aim. This is the big draw for people who believe in this cause; Organize meetups, for marketing can connect with known brands like Swiggy, Blinkit, etc., for side promotions; Stalls in open markets; Multi-level marketing; Since my sense of style is quirk, I would definitely follow something in reference to that; I will use both social media and physical ways because there are huge people in both the segments; Ways on social media: short videos, posts, posters, demo, etc.; Physical: pamphlets, posters etc.; Slow, Word of mouth like Muji; I'll create events which are fun for people to get to know about the brand
11	Have you heard of silent marketing? Do you think it can be successful?	No 40% Depends on product; Yes it can be certainly; Only if your product is actually useful for the common people; It might be a successful approach but they are targeted at the human senses and are surprisingly have been proven effective too. Buy I am not a fan of it because these brands are playing with my mind and I do not even know it and I end up buying more expensive and useless stuff than I need to; Could be risky; It seems like a good concept, where customer is not manipulated into buying a product. However, many products (such as a soft drink, or new clothes) may not really be needed by most people; Yeah it can be successful. Having surrounded by lots of hyped commitments from number of brands, everyone looks for genuine and committed brand. One who don't hype much about false promises and delivers better to customers may always have fair chance to become successful; If the product/service is actually good and meaningful; Yes it is very much important in this generation where people are always in a rush

mixed all the purposes for connecting with people, entertainment and marketing and promotions. 53% of the respondents have also agreed that they may use social media for promotion but should social media not have distinctive roles to play? Can we not define what its roles should be? Some existing ways of branding and promotion may include—celebrations, through special events, emailers, websites, focus groups, public meetings, schools, colleges, exhibits, presentations, fairs, etc. All the five kinds of communication strategies [11, 12], can easily be employed for promotions—planned marketing which also includes traditional marketing (brochure, advertising, personal selling, sales promotion); Product messages (perceived when customer uses the product such as pricing, raw materials, design and usefulness); Service messages (logistical interactions); Unplanned communication (news in media, word of mouth, gossip); and silent communication. It is a need to reduce dependence on social media, innovate specific channels for promotion and marketing and streamline marketing activities for less distraction and cognitive load. Also, let us bring back focus to less is more. Let us produce less, only required quantitiesthe best example for this would be fashion and textiles and the detrimental impact that fast fashion has to work environment and natural environment. And the amount of aggressive promotions across social media, online and print media that takes place to promote fast fashion. On the other hand, there are sustainable brands trying to promote themselves as well, but do they have the same reach, following and revenue? Absolutely not. Similarly, food industry, the retail outlets and chains and the amount of apps and publicity and promotions they are involved in, amount of revenue generated and the impact on environment is quite startling. The need is to reduce our needs or have lesser options to fulfill our needs. To tame our desires into more ethical and sustainable contexts to harm people and environment less. Lesser needs, lesser chaos of communication. Does this impact employment generation? Yes, it does. But can we have more ethical and environmentally friendly engagements for our fellow unemployed populations. Yes, we can? So why not shift the focus to tackling social, environmental- and employment-related issues instead of creating more fuss by developing new products in existing competitive environment compete with marketing. Rather design, develop and create disruptive ways of tackling issues and providing better life to the have-nots. Instead of increasing the gap between rich and poor, reduce and bridge the gaps. Is communication a way to promote this? Yes, it is. Can there be new brands, for new products, systems and services and new channels or methods to promote them? Can social media be tamed? Can marketing and promotion be separated from social media? Can silent marketing be employed aggressively? A very good example of that would be Vipassna Meditation Centre, where the publicity happens only by the word of mouth and yet results and revenue generated have are very rewarding which is very refined and deep and meaningful in the most utopic way. Can we not learn from such existing empires of wisdom and make our world a better place. Communication—verbal or non-verbal—can we not refine and make it impactful yet subtle and ideal and prove to the world, yes, we can be the change.

References

1. Norouzi N, Shabak M, Embi MR, Khan TH (2015) The architect, the client and effective communication in architectural design practice. Procedia Soc Behav Sci 172:635–642
2. Stylidis K, Rossi M, Wickman C, Söderberg R (2016) The communication strategies and customer's requirements definition at the early design stages: an empirical study on Italian luxury automotive brands. Procedia CIRP 50:553–558
3. Eckert, C, Maier A, McMahon C (2005) Communication in design. Communication in design. In: Design process improvement, pp 232–261. https://doi.org/10.1007/978-1-84628-061-0_10
4. Frandsen F, Johansen W (2017). Strategic Commun. https://doi.org/10.1002/9781118955567.wbieoc194
5. Goransson K, Fagerholm A (2018) Towards visual strategic communications: an innovative interdisciplinary perspective on visual dimensions within the strategic communications field. J Commun Manag 22(1):46–66. https://doi.org/10.1108/JCOM-12-2016-0098
6. Leppink J (2017) Cognitive load theory: practical implications and an important challenge. Elsevier.
7. Paas F, Merrienboer JJ (2020) Cognitive-load theory: methods to manage working memory load in the learning of complex tasks. Curr Dir Psychol Sci 29(4):394–398
8. Centre for Education Statistics and Evaluation (2017) Cognitive load theory: research that teachers really need to understand
9. Hawthorne BS, Vella-Brodrick DA, Hattie J (2019) Well-being as a cognitive load reducing agent: a review of the literature. Front Educ 4:121. https://doi.org/10.3389/feduc.2019.00121
10. Strannegård MSL (2004) Silence of the brands. Eur J Mark 38(1/2):224–238. https://doi.org/10.1108/03090560410511203
11. Grönroos C (2000a) Service management and marketing. A customer relationship management approach. London, John Wiley & CO
12. Grönroos C (2000b) Creating a relationship dialogue: communication, interaction and value. Market Rev 1(1)
13. Finne A, Strandvik T (2009) Silent communication—A challenge to established marketing communication practice. Hanken School of Economics, Helsinki, Finland
14. Wei R (2021) Analysis of negative effects on social media. J Sociol Ethnol 3(5):179–181. Clausius Scientific Press, Canada.https://doi.org/10.23977/jsoce.2021.030532
15. Baron J (2019) Stressed out people with no offline support more likely to overuse Facebook', Consumer Tech. Available at http://www.forbes.com/sites/jessicabaron/2019/05/28/stressed-out-people-with-no-offline-support-more-likely-to-overuse-facebook/. Last accessed 5 Dec 2019
16. Uhls YT, Ellison NB, Subrahmanyam K (2017) Benefits and costs of social media in adolescence'. Pediatrics 140(November):68–70. https://doi.org/10.1542/peds.2016-1758E.LastAccessed:5/12/2019
17. Ostic D, Qalati SA, Barbosa B, Shah SMM, Galvan Vela E, Herzallah AM, Liu F (2021) Effects of social media use on psychological well-being: a mediated model. Front Psychol 12:678766. https://doi.org/10.3389/fpsyg.2021.678766
18. Margariti K, Boutsouki C, Hatzithomas L (2018) Less is more or less is a bore? In: Cauberghe V et al (eds) Advances in advertising research IX. European Advertising Academy, Springer. https://doi.org/10.1007/978-3-658-22681-7_10
19. Abadi-Nagy Z (2001) Minimalism versus postmodernism in contemporary American fiction. AkademiaiKiado XXVIII/1:129–143
20. Courbet D, Fourquet-Courbet MP, Kazan R, Intartaglia J (2013) The long-term effects of E-advertising the influence of internet pop-ups viewed at a low level of attention in implicit memory. J Comput-Mediat Commun 19(1):274–293
21. Garaszczuk M (2015) Attentional-capture efficacy and brand qualities of minimalist packaging design. Undergraduate Honors Theses, Paper 21
22. Gressel JW, Carlston DF, Oakley J (2013) In advertising, a double negative does not equal a positive. J Promot Manag 19(4):470–487

23. Hu Y, Yin H, Han, D., Yu, F (2014) The application of similar image retrieval in electronic commerce. Sci World J 2014:7
24. Kim J, Baek Y, Choi YH (2012) The structural effects of metaphor- elicited cognitive and affective elaboration levels on attitude toward the Ad. J Advert 41(2):77–96
25. Phillips B, McQuarrie EF (2002) The development, change, and transformation of rhetorical style in magazine advertisements 1954–1999. J Advert 31(4):1–13
26. McQuarrie EF, Mick DG (1999) Visual rhetoric in advertising: text- interpretive, experimental, and reader-response analyses. J Consumer Res 26(1):37–54
27. Phillips B (1997) Thinking into it: consumer interpretation of complex advertising images. J Advert 26(2):77–87
28. Lelisho ME, Pandey D, Alemu BD, Pandey BK, Tareke SA (2022) The negative impact of social media during COVID-19 pandemic. Trends Psychol. Springer. https://doi.org/10.1007/s43076-022-00192-5

The Utilisation of Commercial Art Aboard Boats as a Means of Achieving Efficient Visual Communication

M. Pavan Kumar

1 Introduction

> Visual culture uses and interprets visual imagery to express culture.
>
> Cultural studies, art history, critical theory, philosophy, media studies, deaf studies, and anthropology study this topic.[1]

Images are used to display popular visual culture. The subject matter can encompass religious, political, societal, or commercial aspects. It caters to the preferences of the general populace within a given society. Living art is a form of artistic expression that is prevalent within society. This visual artwork is intended for a wide audience. The visual representation or label should possess an inherent ability to captivate the viewer's attention. It is commonly associated with the urban street culture in contrast to classical or traditional civilisations. The realm of popular art frequently finds itself in competition with elite art, which thrives within exclusive cultural spaces. Indian popular visual cultures are frequently regarded as a broad and homogeneous category, thinking that mass-produced images are popular only in terms of quantity.

The post-independence era saw a widespread rise in the veneration of images among common people. In a parallel vein, fishermen began incorporating depictions of fishing related imagery onto many surfaces, ranging from the walls of their residences to the walls of sacred edifices, as well as their modes of transportation. The prevalence of painted visual culture has become a customary tradition in the region of Andhra Pradesh India.

[1] Heywood, I., & Sandwell, B. (Eds.). (2011). *The Handbook of Visual Culture*. Bloomsbury Publishing.

M. P. Kumar (✉)
Faculty of Fine Arts, Amity University, Noida, UP, India
e-mail: Pavan.2521@gmail.com

© The Author(s), under exclusive license to Springer Nature Singapore Pte Ltd. 2025

S. Das et al. (eds.), *Human-Centered Digitalization, Volume 3*, Design Science and Innovation, https://doi.org/10.1007/978-981-96-4383-7_3

In addition to the fishing boats, we can see a few images of other parts of the community. During the 1970s and 1980s, calendar prints emerged as a prominent medium for featuring renowned paintings and writings from diverse locations.

This research aims to analyse the impact of living traditions on commercial art in the coastal region of Andhra Pradesh. This study places particular emphasis on the fisherman community and the significance of visual culture within the context of fishing boats.

The fishing population residing in the coastal region of Andhra Pradesh engages in the generation of deities associated with water. The community religious practises in question are deeply rooted in longstanding traditions. These kinds of cultural practises have been observed and upheld for an extensive period. The depictions of deities intended for religious veneration were rendered in the form of painted or sculpted idols inside the confines of their respective communities. Prior to engaging in fishing activities, individuals partake in a ritualistic practise of offering prayers to their kuladivam (Caste-based god), which refers to the deity associated with their community or household. Within the fisherman's community, male individuals engage in the act of capturing fish from the sea, while their female counterparts undertake the task of marketing the acquired fish. Additionally, older members of the family contribute by assisting in the repair of fishing nets or any other activities associated with the fishing endeavour. The following tasks are usually distributed throughout the fishing family. Numerous fishing communities are dispersed throughout the coastal villages, with a few having relocated to different coastal regions within Andhra Pradesh.

The principal objective of this form of painting is to embellish the boats, thereby serving as a representation of their community and culture. In India, numerous communities and individuals belonging to the working class engage in the celebration and depiction of various deities, primarily symbolising their respective occupations. During the Dussehra festival, numerous working-class groups engage in the practise of offering prayers for the equipment and machines utilised within their respective workplaces. Regional cultures in India have significantly influenced the way of life in various regions. India is characterised by the presence of diverse regional cultures, which collectively contribute to its distinctiveness as a nation.

The locations of Lingasamudram, Eatamukkala, Pakal, Vllapalem, Ramayapatnam, and ummalapenta are connected to Tada. Various coastal villages engage in the act of venerating these entities during distinct periods of celebration. Each state's folk deity is distinct from one another. The many fishing groups share a common belief in some water deities, such as Gangamma, Pollamma, and Nookalamma. The term "Yellamma" refers to a deity in Hindu mythology. Numerous coastal settlements are characterised by the presence of little shrines near the sea. Prior to embarking on their fishing expeditions, fishermen engage in the ritualistic practise of offering prayers to their ancestral deities. Various deities have been fashioned to serve distinct functions, including those associated with natural elements such as wind, water, sun, and rain. In each successive year, the local community commemorates the festivals dedicated to their respective village deities. The deities depicted on the boats were rendered by signboard artists who adhered to the preferences of the respective boat owners in the area. The boats exhibit a prominent display of vibrant hues, accompanied with ornamental motifs.

1.1 *Background Study*

Visual culture is important to society because it can change how people communicate, express themselves, understand the world, and connect with each other. The thing we're talking about affects many different parts of our daily lives, from how we feel about ourselves to how we think about the world and what we buy. To understand the world, we live in and our place in it, it's important to learn about visual culture and how it affects how we navigate and how social structures are made.

2 North Coastal Andhra Region

The northern border of Orissa is next to the coastal region, where several landscapes depict the scenic beauty of the Andhra Pradesh side. The prevalence of cross-cultural imagery is observable in numerous areas within the region of Konda, owing to its proximity to the border. Furthermore, individuals partake in the festivities of both state festivals without encountering any limitations imposed by state boundaries. Likewise, a limited number of religious depictions were rendered and affixed onto the door facades of the coastal settlement, predominantly Kavita and Bharuva. The images in Canada had a cross-cultural characteristic.

The worship of folk deities, such as the revered deity Puri Jagannath, was widespread among the community. Several families of fishermen from Orissa residing in Baruva, Kanada, actively engage in and partake in festivals celebrated in the state of Andhra Pradesh.

Likewise, those from Andhra Pradesh also partake in the festivities of the Orissa festival. Cross-cultural practises also manifest themselves in the visual cultures observed within their professional environments and domestic spaces.

Certain wooden boats were adorned with odia text and religious imagery from Andhra. The Konark temple figures were sparsely depicted on the walls. The wooden boats are painted by Andhra's regional commercial artists in accordance with the instructions provided by the boat owner. At the outset, the artists apply flat colours as foundational hues. In addition, painters employ several forms of imagery. The painting typically occurs when their return from fishing during the scheduled week-long hiatus. Occasionally, when a boat necessitates repairs, it is customary to document the vessel's condition post-repair using photographs.

Srikakulam is among the nine districts located in the coastal region of Andhra Pradesh. Within this district, one may observe several cross-cultural images that have been depicted through the medium of painting. The art and craft culture of Orissa also serves as a reflection of the visual culture seen in border areas. However, it can be observed that the regions of Vizianagaram and Visakhapatnam do not exhibit significant traces of odia cultural impact.

The mural portrays a depiction of a fisherman returning from a fishing expedition, with his wife assisting with the removal of fish from the boat along the Seashore. The fisherman is observed adjusting the boat to aid his wife in unloading the catch, which is then placed in a bucket and bowl for further examination. The artwork depicted in this context serves as a representation of the various fishing groups inhabiting the area. Numerous locations can be discerned by the analysis of their respective visual characteristics. In numerous rural settlements adjacent to fishing communities, such visual phenomena can be observed.

A diminutive shrine was constructed near the shoreline at Mangamaripeta, a coastal community renowned for its fishing activities, located in the region of Visakhapatnam, inside the state of Andhra Pradesh. Coastal cultures engage in the practise of depicting grama devata through the artistic representation of water-associated fauna, including but not limited to crocodiles, fish,

and tortoises. These animals also serve as symbolic representations of fortune and vitality, while also serving as a vital source of sustenance for these people. This tradition of visuals taken from Hindu puranas. The local fishing community successfully raised funds from the local populace and engaged the services of an artist to create paintings for their boats. If there is a need for a proficient artist to do a painting project, it is imperative that they contact him from the urban area. The local artist exhibits a limited range of skills, and their artwork displays an unsophisticated aesthetic. However, it possesses its distinct impression.

The fees charged by a professional artist are typically more in comparison to those charged by a local artist. These urban artists possess a high level of awareness of contemporary trends and patterns. The artists employ advanced techniques in conjunction with visual technology to create their paintings. Visakhapatnam fishing boats exhibits a higher concentration of skilled artists compared to the regions of Srikakulam and Vizianagaram.

3 Delta Region

Currently, the primary mode of transportation on this waterway consists predominantly of motorised boats. There is a limited number of individuals who engage in fishing activities within this canal. In addition to the presence of speedboats, one can see vibrant paintings and textual elements in certain coastal locations frequented by both Muslim and Christian communities for the purpose of fishing. However, because of prevailing economic difficulties and high rates of unemployment, young individuals from various communities frequently opt for engaging in fishing activities as a means of sustenance. A small number of migrants' families also engage in the occupation of fishing, independent of the local community. Kakinada is home to a significant number of Christian families, although the coastal villages spanning from Kavali to Nellore are predominantly inhabited by the Muslim community.

In the present era, numerous local commercial artists engage in the creation of artwork on various mediums in the region of Andhra. According to a commercial artist from Kottapatnam, painting on fibre boats is comparatively simpler than painting on wooden ships, mostly owing to the supportive surface it provides. This medium offers a high level of comfort and allows artists to have more space for their artistic expression. Numerous images have been replicated from prints found in calendars. There is currently little evidence to suggest that the advent of digital photography has had any discernible impact on the operations or practises of fishing vessels. There is a scarcity of locations were diminutive scarecrow depictions are additionally adorned with paint.

There are numerous fishing villages spanning from Kavita Beach in north Andhra to Tada Nellore Beach in south Andhra. Colourful painted boats can be found. The term "kanlinga" refers to a concept or entity that requires further clarification or the locations mentioned include Patanam Baruva, Konada, Yanam, and Antravedi in Pallepalem.Pakala, Etamukkala, Ullapalem, Ramaya Patnam, and Tummalapenta are coastal settlements characterised by their predominant fishermen communities. Maintaining the boat's colour integrity is of utmost importance to them as well. During the process of data collecting, a significant number of individuals engaged in fishing activities provided valuable insights and information. Visual culture plays a significant role in the lives of individuals, as exemplified by their practise of maintaining and embellishing puja halls and displaying religious imagery within their households.

Furthermore, the living traditions observed in coastal villages are same. Religious practises encompass several forms of worship, including home-based devotion to deities and communal worship. In the lives of individuals, community worship holds more significance, with specific focus on venerating deities such as Gangamma (the Water goddess), Nookalamma, and Ammoru. The deities known as Polamamba were widely worshipped. Parental deity is also considered as one of the prominent regional gods in the northern areas of Andhra Pradesh. A limited selection of deities enjoys representation in artistic depictions.

A small proportion of them are housed within residential dwellings. A subset of these structures can be classified as little shrines. Kuladivam plays a significant role in the preservation and continuation of their living traditions.

The phenomenon of popularity is accompanied by a sense of excitement, and the ceremonial practises within the fishermen community are characterised by their simplicity and distinctiveness. Based on personal experience and observations, it is evident that the entire community actively engages in and commemorates various familial events. In contemporary times, there is a strong adherence and profound dedication towards the observance of ceremonial processions. This is the way they perceive the perpetuation of their historic traditions. A significant number of individuals exhibit a refined attire and provide offerings to the divine being, which consist of various household items and handcrafted delicacies like as sweets, bananas, sugar cane, rice, and newly acquired saris. The belief in the practise of rituals for deities is a prominent aspect of their religious faith. The crucial factor in the context of.

rapidly evolving relationships and cultures is in the absence of significant influence on these local cultural contexts.

4 South Coastal Region

The central region of the coastal Andhra Pradesh, namely the Kakinada port, holds significant historical and cultural significance. During the British time, it gained renown for its trading activities. There were two distinct categories of vessels employed for the purpose of transportation cargo ships, which were utilised for the conveyance of goods, and transport ships, which were designated for the movement of individuals between different locations.

It is mandatory for all vessels to display the national flag of India by the application of paint. Due to a deep sense of patriotism and the imperative of safeguarding national security. The location in question pertains to the boundary of Tamil Nadu, a state in India. Occasionally, adverse weather conditions result in the unintended drifting of boats towards Sri Lanka, serving the purpose of identification. The Indian flag was commonly depicted on ships, alongside the exclusion of any visible representation of their home god's name. This practise served a dual purpose: firstly, as a form of ornamentation, and secondly, as a means of expressing self- interest. Regardless of the surface on which it was displayed, the Indian flag served as a visual medium through which their interests were conveyed. Numerous fishing vessels are adorned with vibrant hues. The colours red, yellow, blue, and green are being discussed. Commonly used colours are frequently employed. The vessel is coated with enamel pigments. The fish motifs are also discernible on the surface. A rudimentary method for applying colour is by visual means.

Tamil devotees adorned the fisherman's boat near Thada (Nellore), which is situated on the boundary of Tamil Nadu and Andhra Pradesh. A limited number of families relocated from Tamil Nadu to coastal communities as a means of ensuring their survival. In the province of Tamil Nadu, a novel artistic style evolved because of the amalgamation of the Tamil and Andhra cultures. Collaborative artistic practises were also observed in the northern Andhra region, namely in certain locations along the odia border. Mainstream deities exert influence on mainstreams through their iconography, wherein the colour scheme adheres to a customary local palette that primarily features natural hues. These paintings have the potential to introduce novel artistic trends. Undoubtedly, the inclusion of such qualities will contribute to the inherent simplicity often found in popular artistic expressions.

The portraits of village goddesses in Tamil Nadu are widely circulated in the coastal areas. The deity grama devata holds significant relevance in several locations as a representation of their religious beliefs. Various surfaces, such as temple walls, interior spaces, and storefronts, are commonly encountered in different settings. Painted on the entrances of commercial establishments are depictions of many deities. Furthermore, adherents of this faith hold the belief that grama devata serves as a safeguard against all malevolent forces.

4.1 *Scope of Study*

Hand-painted coastal Andhra artworks are said to require attentive inspection. Using legends and facts, the goal is to fill knowledge gaps and illuminate the region's culture and people. Thus, coastal Andhra hand-painted visual practises have been extensively studied. The goal of this study is to uncover new insights and reinterpret historical truths.

5 Conclusion

The present study investigated contemporary commercial art practises within the domain of fishing boat painting. The preservation of cultural traditions is heavily reliant on the practise of signboard art. Over the course of time, painted images gradually superimpose upon one another, so rendering photography a captivating method for the preservation and archiving of visual content. Coastal villages undergo a slow transformation, resulting in the emergence of a novel landscape that mirrors their religious beliefs. This study sheds light on the intricate interplay between tradition and modernity within the visual cultures of Andhra fishermen. It demonstrates how these cultures evolve in response to societal changes, while concurrently safeguarding their distinct identities and traditional practises.

Bibliography

1. Lovegrove K (2003) Graphicwallah, graphics in India. Laurence king Publishing Ltd. (P.No. 12)
2. Skoda U, Lettmann B (2018) Edited book, India, and its Visual Cultures. Sage publication, (P.No:xxii) .2018.
3. Gram Devata. https://www.britannica.com/topic/gramadevata
4. Misra B (1969) "Sitala:" The small-pox goddess of India, vol 28, no 2. Asian Folklore Studies, Published By: Nanzan University, pp 133–142 (10 pages)
5. Skoda U, Lettmann B (2018) Edited book, India, and its visual cultures. Sage Publication, (P.No:xvii)
6. Bhattacharyya, Chandra D (1980) Iconology of composite images. Munshiram Manoharlal Publications Pvt. Ltd.
7. Banerji A (1993) Images, attributes, and motifs. Sundeep Kashan Publication, New Delhi
8. Coomaraswamy AK (1956) The transformation of nature in art. Dover Publication, New York
9. Sudha RB (2009) Naturalism in Andhra art. Historical developments of contemporary Indian art. Lalit Kala Academy
10. Schonberger C, Neumayr E. Bharat Mata: India's freedom movement in popular art. Oxford University Press, New Delhi
11. Pinney C (2002) A secret of their own country. In: Contributions to Indian sociology, vol 36, nos 1 and 2. Sage Publication, Delhi
12. MacDowall DW (1992) Indian numismatics, history, art, and culture: essays in honour of Dr. P.L. Gupta, vol II. Agam Kala Prakashan, Delhi

13. Poster for the film Eeswer (Telugu language film). Vijaya Publications, Hyderabad
14. Fischer, E, Pathy D (1996) Murals for Goddesses and Gods. Indira Gandhi National Centre for the Arts, New Delhi
15. Neumayr E. Popular Indian art: Raja Ravi Varma and the printed Gods of India. Oxford University Press, New Delhi
16. Neumayr E. Bharat Mata: India's freedom movement in popular art. Oxford University Press. ISBN 978-0-19-569-518-3
17. Kerlogue F (2004) Arts of Southeast Asia (World of art). Thames & Hudson
18. Geertz C (1973) The interpretation of cultures. New York 1973.
19. Jain J (2007) India's popular culture: iconic spaces, and fluid images. Marg Publications
20. Jain J. Reconfiguring the divine and the political. Essay on Shekhawati Collages
21. Hodin JP, Herbert Read's philosophy of art, the dilemma of being modem. Routledge & Kegan Paul, London, p 244
22. El Mansouri SM (1959) Art culture of India & Egypt. Ghosh Printing House, Calcutta, 1959, p xi
23. Chattopadhyay K (1976) The glory of Indian handicrafts. Indian Book Company, New Delhi, p 22
24. Roget's B, Thesaurus (1990) New York
25. Mahajan VD (1983) Ancient India. S Chand & Company Ltd., New Delhi, p 45
26. Hirayama M (1997) Art experience. Indira Gandhi National Centre for the Arts, New Delhi, p 31
27. Langer SK (1957) Problems of art. Charles Scribner's Sons New York, USA, p 110
28. Berndtson A (1968) Art expression, and beauty. Holt, Rinehart, and Winston, USA, p 145
29. Read H (1979) Art, and industry, principles of industrial design. Faber and Faber Limited, London, p 196
30. Scarfe L (1954) Alphabets. Ipswich Suffolk Publishers, London, p 13
31. Richter IA, Paragone A Comparison of the Arts by Leonardo Da Vinci, Oxford
32. University Press, London, 1957, p 77
33. Ramaswamy S (2010) The Goddess and the nation. London, p 34
34. Mukherjee RK (1964) The flowering of Indian art. The growth and spread of a civilization. Asia Publishing House, Bombay, p 246
35. Nisbet R (1977) Sociology as an art form. Oxford University Press, USA, p 29
36. Rodman S (1949) Artist in tune with the World, masters of popular art in the Americas & their relation to the folk tradition. Simon and Schuster, New York, p 10

Geography, Politics, and the Future of Work: Challenges for Human Factors and Ergonomics from Around the World

Andrew Thatcher

1 Introduction

The International Ergonomics Association (IEA) established their Future of Work Task Force in 2018 (led by Juan Carlos Hiba and Klaus Zink). The Task Force was created partly in response to the International Labor Organization's Future of Work Centenary Initiative which later published a centenary declaration on the future of work [1]. The ILO [1] declaration recognized technological innovations, demographic shifts, environmental and climate change, and globalization as the global drivers behind the future of work. The declaration called for a just transition to decent work for all that leverages off technological progress and skill acquisition. However, the mechanisms through which this just transition could be achieved were opaque, stated in policy terms which were short on detailed interventions. In their state of science paper on the challenges of the future of work for human factors and ergonomics (HFE), Bentley et al. [1] echoed the ILO's [2] global drivers (i.e., technological enhancements, demographic shifts, environmental pressures, and globalization). Bentley et al. [2] correctly noted that as a discipline that is focused on the study of work, it was imperative that HFE keep abreast of the challenges arising from the future of work, but also noted that apart from the challenges arising from technological innovations, HFE had made few contributions to understanding the impact of these global trends on HFE research and practice.

The technological innovations which characterize the future of work included aspects familiar to Industry 4.0 such as the automation of work and autonomous systems, robotics, big data analytics, and 3-D printing [2], and more recently, the challenges arising from artificial intelligence (AI). Demographic changes were identified as an aging population (and workforce) and an intergenerational workforce, more women participating in the workforce, and where there were several forms of

A. Thatcher (✉)
Department of Psychology, University of the Witwatersrand, Johannesburg, South Africa
e-mail: Andrew.Thatcher@wits.ac.za

© The Author(s), under exclusive license to Springer Nature Singapore Pte Ltd. 2025

S. Das et al. (eds.), *Human-Centered Digitalization, Volume 3*, Design Science and Innovation, https://doi.org/10.1007/978-981-96-4383-7_4

greater diversity in the workplace [2]. Globalization referred to the challenges associated with labor migration (immigration in some parts of the world and emigration in others), manufacturing and manual labor shifting to low-income countries, and a downward pressure on working conditions [2]. The challenges from environmental pressures included resource scarcities (e.g., energy, water, food, etc.), renewable and low-energy systems, recycling and circularity, and social sustainability [2]. The extent to which HFE has addressed environmental challenges has also been covered in Thatcher et al. [3] and other reviews since then such as Bolis et al. [4]. Bentley et al. [2] also identified two other general challenges for the future of work: new forms of working and new organizational forms. New forms of working referred to flexible working arrangements, activity-based working, remote and digital working (e.g., telehealth), and co-working hubs. The Covid-19 pandemic has accelerated the growth in some of these new working arrangements [5]. New organizational forms referred to the growing casualization of work particularly through forms such as Gig work, virtual organizations, and distributed organizations [2].

It should be evident however, that the types of challenges and the extent to which these challenges will have an impact is highly dependent on where in the world one is situated. For example, an aging workforce is only really an issue in Global North countries (e.g., countries in Europe, North America, and high industrialized Asia such as Japan and South Korea) whereas in many Global South countries (e.g., countries in most of Asia, Africa, and South America) the population and workforce are getting younger on average [6]. In urging HFE to think about addressing the challenges of the future, Moray [7] noted that we will need to move beyond insular HFE disciplinary thinking and embrace the cultural, political, and geographical differences that emerge when considering differential impacts on regions and countries. This paper makes a small step in that direction as we consider the HFE challenges of the future of work from around the world.

2 Methods

From 2018 to 2021 the Future of Work Task Force concentrated on popularizing the need for more concerted efforts to understand the challenges of the future of work. This culminated in a survey among IEA federated societies, the results of which were presented at the IEA Congress in 2021. It should be noted that the survey concentrated on identifying the extent to which the ILO's centenary initiative trends were present in member societies. Thirty-four federated societies responded, representing 68% of the IEA's 50 federated societies (at that time). This type of analysis is fairly reductive and doesn't demonstrate the significantly regional and national variabilities. From 2021, the Task Force has concentrated on gathering more detailed inputs from federated societies around the world through a Future of Work webinar series (each of the webinars can be viewed on the IEA's YouTube channel: https://www.youtube.com/@iea-internationalergonomic9925/videos). The webinars were structured in such a way that there were two speakers. For each webinar we tried to find one speaker who was a leading or established HFE professional or researcher and another speaker who was a young or emerging HFE professional or researcher. The principle behind this choice was to identify someone who had considerable experience in HFE and the trends in work in their region/country and another person who could identify emerging trends and would be working for longer into the future. In most instances the speakers were identified by the IEA federated societies in the country or region. This paper presents the consolidated outcomes from the first eight webinars presented between November 2022 and September 2023.

3 Results

3.1 Antipodean Lifestyle (Australia and New Zealand)

Australia and New Zealand represent the future of work challenges that are typical of many of the highly industrialized countries/regions of the Global North. The first challenge emerges from increased automation of work roles and workplaces such that repetitive, manual labor will be increasingly replaced by automated technologies. Those people who depend on this type of work will either need to be reskilled or may drop out of labor market. The second challenge emerges from an aging workforce. While healthcare advances have meant that their aging populations will be able to continue working for longer, there are still both deteriorating physical and cognitive risks that cannot yet be completely removed. For HFE this means identifying relevant interventions and recommendations such as work schedule design, workplace accommodations, and removing hazards in order to reduce the impact of these risks. The third challenge emerges from changing patterns of work. These trends have allowed people to work remotely (avoiding commuting costs and office space) with greater flexibility and control over their work pace and work schedule. Some of the negative consequences of these flexible work patterns are starting to emerge including psychosocial risks from working in isolation, fewer labor protections (e.g., leave, medical support), and the need for psychological resilience due to reduced social support. The fourth challenge relates to the need to provide healthcare services from a distance. Telehealth technology enables a range of healthcare services that were previously unavailable including psychological support and remote surgery. Given the enormous geographical distances between major centers in this part of the world, changing patterns of work and telehealth technology may combine to allow an aging workforce to work remotely while still receiving appropriate healthcare. However, HFE research and design is needed to support these transitions.

3.2 "Hidden" Dragon (China)

Given China's enormous population (more than 1.4 billion people at the time of writing), China's future of work issues center on how to logistically coordinate this gigantic workforce for service provision and delivery in four primary areas: agriculture, e-commerce, healthcare, and Gig work. China expects to support the provision and logistics in these areas through automation and AI. First, in agriculture the rural population is shrinking due to urbanization and aging. This produces a significant challenge for food security. Currently agriculture often requires labor-intensive work at varying elevations. One approach is to automate agriculture in greenhouses closer to urban centers, but this will significantly increase the complexity, intensity, and skill-level required to farm and maintain the automation. Second, China already has one of the largest and most sophisticated e-commerce markets in the World. However, the logistics of ensuring hundreds of millions of on-time deliveries per day is increasing bringing AI and possibly even automated transport solutions into the mix to optimize delivery systems. Third, healthcare challenges from an increasingly aging population with more lifestyle diseases (e.g., obesity, diabetes, high blood pressure) will require advanced and expanded healthcare systems. AI will be required to help healthcare practitioners keep data on hundreds of millions of patients and to help identify a complex array of possible treatments. China also expects the number of people employed in Gig work to grow (already more than 200 million people work in some form of flexible employment in China). It is expected that human-led HR systems will be unable to cope with the logistics to assign people work and AI technology will be needed to help manage these processes. For HFE, these four areas raise challenges for human-AI teaming, and human-automation optimization. In particular, the requirement to integrate changing human needs and preferences into AI algorithms for optimal human well-being and safety will be extremely challenging. For example, AI algorithms could be used to plot optimal paths for food deliveries raising customers' expectations but will struggle to address chaotic changes from traffic and weather.

3.3 K-pop Generation (South Korea)

South Korea is a country with an aging population and the world's lowest average birth-rate. Currently this workforce issue is being addressed through encouraging more women to enter the workforce, but this may only be exacerbating the birth-rate problem. Importing foreign workers to support the workforce (especially in manufacturing, agriculture, fishing, and homecare) is happening, but is politically controversial and difficult given language and cultural barriers. For these reasons South Korea has invested heavily in robotics and automation to deal with anticipated labor shortfalls and exoskeletons to assist aging workers with physical tasks in instances where physical attributes are required but are deteriorating. South Korea already has the highest proportion of robots per employee and this proportion is expected to grow. Automation and robotization will involve re-skilling and upskilling the workforce, in many cases making the work more cognitively demanding. Exoskeleton development is intended to keep the workforce as physically involved in work for as long as possible, but there are many issues with exoskeleton design that require HFE, including how to make it easy for exoskeletons to be bespoke for each worker and how to make it easy to put on the exoskeleton (at present it is not a simple task like putting on clothes).

3.4 Asian Tigers (Singapore and the Philippines)

Singapore and the Philippines present two sides of the same coin when it comes to future of work challenges. The Philippines has a large population, but with generally a low-skilled workforce. Singapore, on the other hand, has a relatively small population of highly skilled workers. Both countries have a high demand for skilled labor such as healthcare and engineering, but while the Philippines invests heavily in developing these skills, many skilled workers emigrate from the Philippines (to countries like Singapore to find better paying jobs and higher standards of living), leaving a significant skills gap that remains unfilled despite significant investment. Singapore also invests heavily in skills development, but despite these investments its population is not large enough to meet the work demands and therefore has a large migrant population from across the skills spectrum. To deal with potential skills shortages, Singapore has also invested heavily in developing automated and AI systems. The key HFE issue in Singapore's implementation of automation and AI has been to determine the appropriate places for humans to fit into these systems. A further aspect that bears mentioning for the Philippines is the various challenges associated with disaster management. The Philippines is a nation of more than 7500 islands sitting directly on the ring of fire. Not only must the Philippines deal with continuing threats from earthquakes, but climate change has made the country more susceptible to powerful hurricanes and sea-level rises. For the future of work this means developing disaster management systems that can reduce exposure to these risks and deal with/adapt to the aftermath of extreme weather events.

3.5 Nordic Innovation (Finland)

Finland is known for its excellent education systems and innovative manufacturing systems. Given its relatively small and distributed population, Finland has an outsized influence on global smart manufacturing systems. For Finland, the challenges for the future of work are less about new and innovative work systems (many of these technologies and work systems are already in place or are evolving rapidly), but how HFE can re-imagine itself to be a facilitator of innovation, rather than being perceived as an obstacle to change. In Finland, HFE is associated primarily with HR systems and OHS with a focus on standardization, well-being, and safety rather than system optimization. As a consequence, HFE is very often viewed as a barrier to innovation and productivity. The challenge for HFE is therefore to demonstrate its value in linking manufacturing competitiveness and technology to human well-being. This is a significant challenge given that in Finland, HFE is associated with holding back technological maturity.

3.6 Engineering Excellence (Germany)

Like many other highly industrialized countries, Germany expects the future of work to involve a large proportion of its workforce being displaced by automation, AI, or other forms of digitalization. While they expect that many jobs will not be entirely replaced, they do expect that many will be fundamentally altered, especially due to a significant rise in remote working which has a number of related problems that need to be resolved. Similarly to Australia and New Zealand, these issues include as enabling social cohesion, social support, and exercising managerial oversight/support remotely. Similarly to places like Singapore, China, and Australia, an aging workforce and obesity in the workforce have been identified as similar problems to places. Like Singapore, and many other highly industrialized countries in the Global North, there are numerous concerns arising from large migrant populations, fleeing a range of adversities in their "home" countries including from war, oppression, climate-related issues, and poverty. A large proportion of those fleeing have low skill levels or have skills which are difficult to transfer into the German context. The challenge becomes how to train and integrate migrants with low skills into an increasingly technology-reliant workforce without creating a two-tier system of employment.

3.7 Post-Brexit (United Kingdom)

The UK's primary concern is with filling skills gaps in several key industries, most notably in the healthcare, transport and storage, education, and construction sectors. Some of these skills shortages (i.e., in the transport, education, and construction sectors) are self-inflicted from the ill-conceived Brexit decision. So, while they share a similar problem with South Korea and Singapore regarding skills shortages, the source of those skills shortages differs. Similarly to China, one of the possible ways of addressing the skills shortages being explored in the UK is to use automation, robotics, and AI to manage some of the complexities of the work and work scheduling. In the healthcare sector this could involve AI assisting with tasks like screening and diagnosis support, triage (to reduce waiting lists), and drug-interaction support. In the transport sector this could include autonomous delivery trucks and trains. In the construction sector, modular, off-site construction, and on-site assembly through robotic systems may become prevalent. For HFE this raises issues about how we design automation, robotics, and AI systems that both increase trust while still making their limitations transparent. It also raises classic function-allocation processes which seem to have fallen out of favor in HFE.

3.8 *"Dark" Continent (South Africa and Ghana)*

The last region to be covered in this paper provides a fundamentally different view of the future of work. Unlike South Korea where birth rates are declining and population sizes are shrinking, across Africa the population is growing. If it continues at the current rate, Africa's population may be the same size as India and China combined by 2050. While unemployment rates in African countries generally remain similar to those in the Global North (South Africa's unemployment rate is one of the highest in the world at more than 34%), a significant proportion of the working population participates solely in the informal sector. Informal work is work which is predominantly manual, family-centered, artisanal, without a regular income, and without formal labor protections (i.e., health and safety regulations, protections against unfair dismissal) that are normally found when people are employed by large organizations and receive a regular salary or wage. Within Africa there is also a large amount of internal and cross-border migration as people move to find work (many national borders are porous and considered by locals to be social constructions from the era of colonialism). Technological skill levels remain low, although manual skills may actually be quite high. Due to unreliable energy supplies, reliance on technology (other than simple information and communication technologies) operated by electricity is low (hence the "dark" continent). Despite this rather bleak outlook, there are also many opportunities. African countries are experienced in surviving in low-resource contexts and therefore there are opportunities to leap-frog to sustainable energy systems. Opportunities for transitions to a green economy remain high (think of solar energy farms in the Sahel) if political instability can be contained. Given the huge population growth, Africa also has the potential to supply countries whose birth rates remain low with labor, provided that these countries do not make full transition to high-technology, automated societies and that educational and training facilities can be fostered and grown.

4 Discussion

While there are some similarities between countries (e.g., migration, skills shortages, and a dependence on technology to solve problems), the sources of these problems often have different causes, demanding different solutions, for different sectors in the workforce. Often, the issues that need addressing have complex socio-historical and geo-political stories that are not only difficult to resolve, but also require bespoke interventions. This makes it difficult to extract a single narrative on the future of work that the entire HFE community can rally behind. Instead, there are unique narratives that require careful unpacking and unique solutions that address specific geographical, political, historical, sociological, and anthropological issues (see [7]). To conclude, it is also worthwhile making some observations on the limitations of this analysis. First, the future is notoriously difficult to predict, particularly the specifics and the timing. We (the speakers and I) struggled to identify which future; a near future (a few years), a mid-term future (five to ten years), or a long-term future (2050 and beyond) to contemplate. The best that one can hope to achieve is to extrapolate current trends and then hope that significant global events will not significantly disrupt the interconnected complexities. As a global pandemic such as Covid-19 has shown us, even though a global health pandemic was perfectly predictable, the timing, the exact severity, and the impacts on societies were not predictable at all. The Covid-19 pandemic accelerated some future of work trends (e.g., remote working and automated systems) while stalling others (e.g., recycling work and activity-based workplaces). The work presented here is far from complete. It represents a few voices in a crowded space, with many countries without a voice (so far). There may be biases introduced by who was "chosen" to present in the webinar series or by myself as the primary analyst of these webinars. It is intended that this paper will open, rather than close, this conversation.

Acknowledgements The following webinar speakers are thanked for their inputs during the Future of Work webinar series: Tim Bentley (Australia), Nicola Green (New Zealand), Wei Zhang (China), Liang Ma (China), Rosemary Seva (the Philippines), Shanqing Yin (Singapore), Arto Reiman (Finland), Jari Kaivo-Oja (Finland), Anita Tisch (Germany), Verena Nitsch (Germany), Andrew Todd (South Africa), Augustine Acquah (Ghana), Nigel Heaton (UK), Rob Becker (UK), Taezoon Park (South Korea), and Hong-In Cheng (South Korea).

References

1. International Labour Organization (2019) ILO centenary declaration for the future of work. ILO. Retrieved from the WWW, 26 September 2023. https://www.ilo.org/wcmsp5/groups/public/---ed_norm/---relconf/documents/meetingdocument/wcms_711674.pdf
2. Bentley T, Green N, Tappin D, Haslam R (2021) State of science: the future of work—Ergonomics and human factors contributions to the field. Ergonomics 64(4):427–439
3. Thatcher A, Waterson P, Todd A, Moray N (2018) State of Science: ergonomics and global issues. Ergonomics 61(2):197–213
4. Bolis I, Sigahi TFAC, Thatcher A, Saltorato P, Morioka SN (2023) Contribution of ergonomics and human factors to sustainable development: a systematic literature review. Ergonomics 66(3):303–321
5. Sigahi TF, Kawasaki BC, Bolis I, Morioka SN (2021) A systematic review on the impacts of Covid-19 on work: contributions and a path forward from the perspectives of ergonomics and psychodynamics of work. Human Fact Ergonom Manufact Service Ind 31(4):375–388

6. Kühn S, Milasi S, Yoon S (2018) Population ageing and future labour market challenges. World Employ Soc Outlook 2018(1):45–50
7. Moray N (2000) Culture, politics and ergonomics. Ergonomics 43(7):858–868

Elevating Resilient Acid Attack Survivors: Unlocking Paths to Skill Enhancement and Career Readiness

Preksha Gupta, Pranita Ranade, and Tanmoy Goswami

1 Introduction

In India, acid attacks have grown alarmingly over the past ten years, mainly on women. Any act of "throwing acid in any form on the victim with the intention of or with the knowledge that such person is likely to cause to the other person permanent or partial damage or deformity or disfiguration to any part of such person's body" is considered an acid attack, according to the National Commission of India [1]. UNICEF mentions, "Acid attacks are a severe problem worldwide". In many cases, children are victims of acid attacks. Girls under the age of 18 who have turned down sexual advances or marriage proposals make up the majority of the victims. According to a survey, 78% of reported attack causes are due to people rejecting romance or marriage [2]. Although acid attack violence occurs in many nations, it is most common in Pakistan, Bangladesh, India, and Cambodia. India witnesses 250–500 attacks yearly, the highest worldwide [3]. Victims devastating impacts can be broadly divided into physical, psychological, and economical. Due to the pain, disfigurement, and physical disability, victims also experience tremendous psychological anguish [4]. As the attack obliterates their chances of getting married or finding a job, victims also start to rely on the necessities of daily living physically. Victims frequently attract glares, which causes them to give up on their education [5]. NGOs such as the 'Chhanv Foundation' help by offering legal counsel, medical care, assistance with higher education, and even employment through 'Sheroes Hangout' cafes, but they struggle with funding. However, India now intensified its focus on diversity and inclusion. Acid attack survivors have been hired by organizations such as 'Sodexo' and 'Lemon Tree Hotels' for front-line and back-end positions. Renowned

P. Gupta · P. Ranade (✉)
Symbiosis International (Deemed University), Symbiosis Institute of Design, Pune, India
e-mail: pranitaranade@gmail.com

T. Goswami
Principle Global Services, Pune, India

© The Author(s), under exclusive license to Springer Nature Singapore Pte Ltd. 2025

S. Das et al. (eds.), *Human-Centered Digitalization, Volume 3*, Design Science and Innovation, https://doi.org/10.1007/978-981-96-4383-7_5

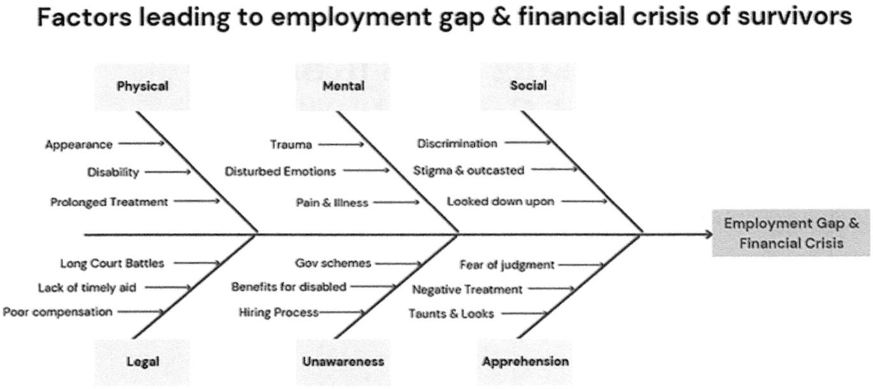

Fig. 1 Fishbone diagram—cause and effect relation of acid attack survivors

companies like TATA Group, Radisson, and Shoppers Stop started hiring acid attack victims under the activity of corporate social responsibility (CSR), providing training and upskilling programs for survivors [6]. The proposed design concept in this paper aims to bridge the digital divide of equal opportunities for acid attack survivors, making growth accessible.

2 Research and Design Methodology

The design thinking strategy was utilized to analyze the current situation and iterate on a proposed design concept. Secondary research was initiated through literature review analysis, competitor studies, and surveys. Primary research was conducted via in-depth user interviews and field studies. The contextual inquiry was carried out to obtain a deeper understanding of stakeholders, which was conducted across the Delhi NCR (India) area with acid attack survivors, hiring professionals and mentors to gain insights (Fig. 1).

3 Literature Review

Various touch points are examined in the literature review, which includes the aftermath of the attack, social stigma, initiatives, economic effects, etc. Springer, Scopus, Google Scholar, and Research Gate were used to access the research papers.

3.1 Acid Attacks Reasons and *MentalPhysical Trauma*

Even though they are horrifying and immoral, acid assaults can have a variety of causes. Relationship conflicts frequently result from refusals to get married or turn down sexual approaches, as well as domestic violence in which acid is used as a tool of intimidation or control [7]. Such attacks may also be motivated by feelings of vengeance or retaliation, with offenders going after individuals they perceive to have wronged them. Gender disparity can be sustained by social structures, cultural values, and lack of legal enforcement. Untreated mental health problems may play a role in a few instances. These crimes highlight the urgent need for judicial action, public education campaigns, victim assistance, and initiatives to combat negative gender stereotypes and strengthen vulnerable populations [8].

As acids are corrosive, they enter deep within the body and burn away the skin, resulting in physical deformities and damaged emotional well-being. Numerous corrective actions must be made, including operations to rebuild the body and medical treatment. Survivors experience deep-seated traumatic concerns that have an impact on their social interactions [9]. They suffer the most from family members' sympathetic looks and ignorance, which are bad for their mental health. People nearby make them feel unwanted and pitiful about their future, resulting in an inferiority complex. These are the main causes for why survivors stop leading a goal-oriented life and quit education and employment [10].

3.2 Financial Struggle, Unemployment, and Legal Challenges

Along with the physical and mental suffering, the survivors also experience extreme financial hardship due to the hospital bills and extensive medical care, which burn a hole in their purses. The victims require quick assistance with medical care, compensation, and justice—all of which are highly expensive, take a long time, and are not accessible to everyone in India [11]. In India, legal disputes spiral out of control, eventually depleting the family's financial resources. Attacks result in a variety of physical abnormalities and disabilities for the survivors, such as blindness, which is a problem for those who are already employed because they are unable to cope and often cannot continue in their current state. People don't want to provide opportunities, forcing them to fend for their food and money [2].

Several laws and penalties have been implemented in India to address crime. The penalties for acid assaults are harsh, with offender's subject to a fine and a sentence of imprisonment of at least 10 years that may be extended to life. The victim may also be granted compensation by the court, which may be used to pay for related costs such as medical bills and lost wages [12]. Although the law imposes severe penalties on offenders, it is nevertheless difficult to put these rules into practice in India. The victims frequently have significant challenges in receiving compensation, and the court system is frequently delayed. The victims of acid assaults sometimes cannot find closure since the attackers are not brought to justice [13].

3.3 Support Initiatives by Government and Private Sector Companies

The National Digital Database of Acid Attacks was constructed by the National Commission of Women (NCW) to facilitate quick compensation, redress, and skill-upgrading training. Many government initiatives, including the 'Manodhairya' Scheme, aim to improve access to justice and quality healthcare [14]. The Indian Government allocates compensation of Rs. 1 lakh for acid attack survivors. Additionally, facilities are made available for their advancement in obtaining remuneration and benefits. The Disability Act of 2016 now includes 14 new categories of physical and mental disabilities, including survivors of acid attacks. With the aid of NGOs, rehabilitation services and medical facilities have been established to raise public awareness and provide facilities to survivors. They also aid in implementing government policies by setting up official training camps, programs, and events [15].

Initiatives in corporate social responsibility (CSR) can help acid attack victims by providing possibilities for professional development. Businesses have also started to provide survivors career training programs that are customized to their interests and talents, allowing them to develop marketable skills. Within the organization, mentorship and internship programs help them advance their careers. Initiatives that promote entrepreneurship and education might also benefit from financial backing. Businesses also focus on developing comprehensive programs that support economic independence and professional progression for acid attack survivors by collaborating with NGOs and vocational training facilities. This promotes inclusion, diversity, and social impact while boosting their own brand recognition [16].

4 Research Methodology and Findings

4.1 Contextual Inquiry

Survivors' real-life problems were discovered through contextual inquiry. In the cultural as projected in Fig. 2, various stakeholders such as acid attack survivors, society, employers, NGO, family and friends, counselors, employers, etc., are mapped according to their correlations.

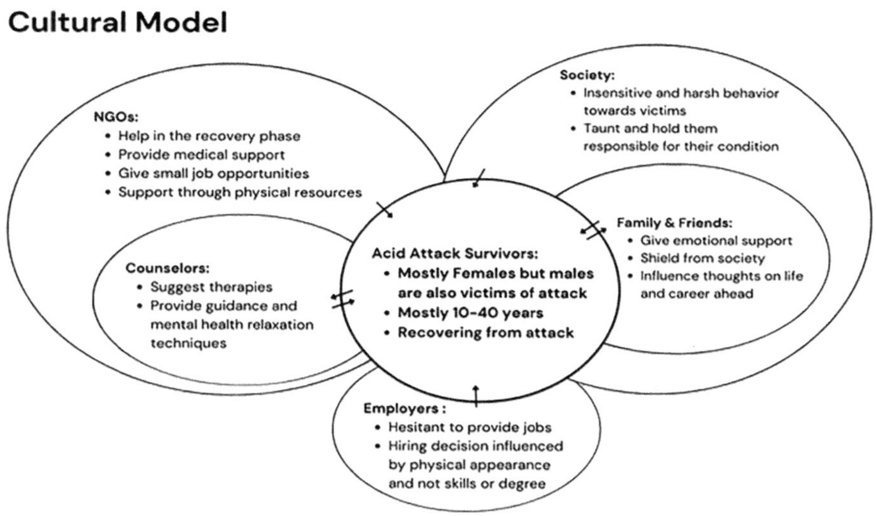

Fig. 2 Model exploring the interrelationship between different stakeholders

4.2 Field Study

A total of 5 NGOs were visited in Delhi NCR (India) to understand the aftermath of acid attack survivors. Field research enhanced understanding of users, including how they act, what they want, and what their attitudes, perceptions, and behaviors are. Users were observed and recorded in their natural environment for the study (Table 1).

5 Design Ideation and Functioning

The GROW application designed for acid attack survivors emphasizes growth opportunities. User journey mapping, in-depth task analysis, and prioritization paved the way for feature ideation.

Features of 'GROW'A proposed Mobile Application

- **Job Opportunities**—Multiple job application processes are simplified through a resume builder. The application encourages the assessment of the mental and physical health of candidates through a short form. AI is proposed for video interviews of candidates to maintain unbiased selection.
- **Upskilling Opportunities**—The focus is to provide opportunities to upskill and enroll in training programs and certifications through corporate social responsibility engagement. Mentorship is provided through verified counselors to broaden skill horizon and become financially independent.

Table 1 Interview insights from acid attack survivors

Acid attack survivor	Age	NGO	Profession	Interview insights
Lady 1	25	Meer Foundation	Call center worker	Application tracking
Lady 2	28	Meer Foundation	Works in Sheroes Café	Hardly finds support. Limited help from NGO
Lady 3	32	Atijveen	Looking for job	No one values skills, degrees, and capabilities in front of disfigurement
Lady 4	47	Meer Foundation	Works in boutique	Forced to settle for low paying jobs even after education
Lady 5	34	Chhaanv Foundation	Wants to study	It is very difficult to approach big companies even after being skillful
Lady 6	21	Atijveen	Education stopped	Rejections based on our appearance. Everyone shuns us away
Lady 7	43	Chhaanv Foundation	Teacher	Unable to find bigger and better opportunities for myself
Lady 8	18	Meer Foundation	Education stopped	No one to guide ahead. NGO doesn't have resources to support
Lady 9	22	Meer Foundation	Mehandi artist	Difficult times for family financially with meager jobs
Lady 10	19	Atijveen	Education stopped	There is no mentorship available for people like us
Lady 11	26	Chhaanv Foundation	Dependent on family	Fear of judgment pulls me back from making a career
Lady 12	36	Chhaanv Foundation	Cook	People are hesitant to even give us a chance at qualified jobs
Lady 13	17	Atijveen	Education stopped	Want to make a bright future ahead but can't find an opportunity

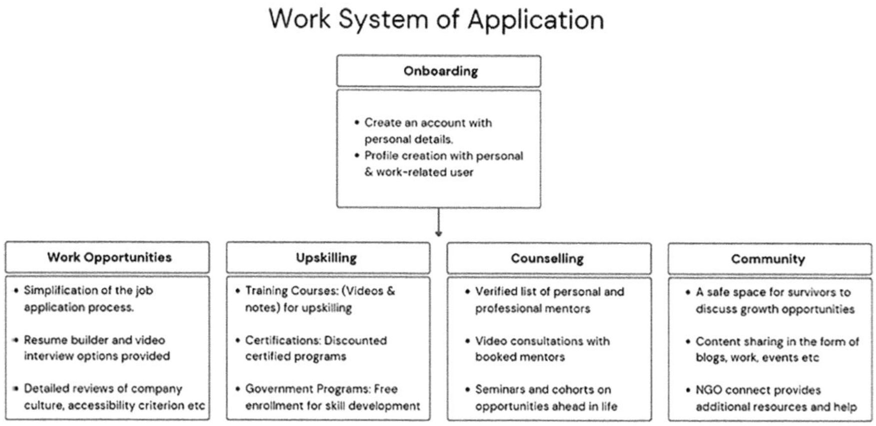

Fig. 3 GROW—Mobile application work system with features

- **Community**—The platform is created to engage with NGOs and the government to support survivor's growth comprehensively (Fig. 3).

5.1 *Role of Ergonomics*

When proposing a mobile application design concept for acid attack survivors, ergonomics is really important. The concept focuses on developing a user interface that meets their unique requirements and difficulties. This includes making font sizes scalable, developing bigger, evenly spaced buttons for users with limited dexterity, and making sure there is strong contrast for those who have visual impairments. User interfaces also support customization so they may adjust to different skill levels. Additionally, the design of the app is sensitive, with simple access to resources that support mental health and content, considering the psychological effects of acid attacks.

5.2 *Accessible Design*

Keeping in mind survivors with visual impairments, cognitive impairments, and mobility, many additional features were curated. Considerations for accessibility were included like captioning and transcripts for multimedia material, utilizing appropriate heading structures, guaranteeing keyboard navigation, testing app with assistive tools like screen readers, and offering alternate text for pictures. Acid attack victims with various disfigurements were also made to test navigation, gesture controls, haptics, and most of them responded positively to the design. Their feedback was incorporated and an iterative design methodology was followed to improve

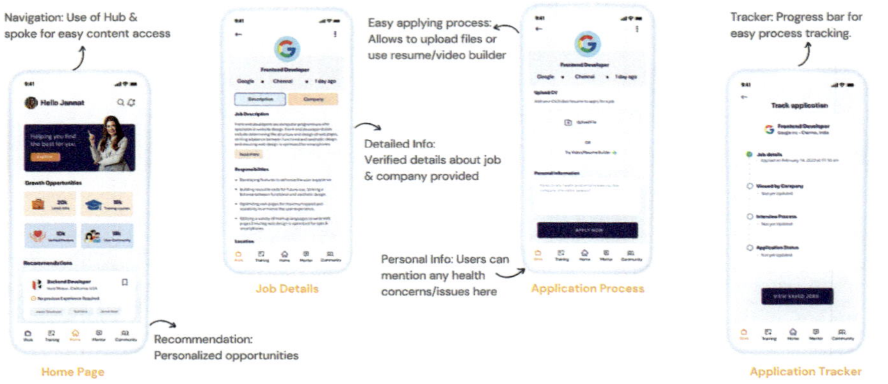

Fig. 4 High fidelity prototype screens for GROW application

user experience. The design concept aims to expand usage to diverse user base of survivors to help them up skill and grow (Fig. 4).

6 Discussion

The research sheds light on the evil after effects in the lives of acid attack survivors. Literature review discusses the mental, physical, societal trauma accompanied by financial struggle of survivors after the incident which hinders their growth ahead in life. Financial freedom and independence are compromised, and survivors often shun themselves from opportunities in life. There is an urgent need to empower them with the help of technology. The concept is discussed for the lack of opportunities & upliftment for acid attack survivors initiating gaps in the employment process. Today, the survivors, by choice, seek education, growth, and empowerment. Currently, NGOs are responsible for most of the help provided to the survivors. With changing times, many corporations and MNCs support acid attack survivors through corporate social responsibility (CSR) funding initiatives. Through corporate social responsibility, acid attack survivors can be helped tremendously as the objective is to develop skills and facilitate employment and future growth opportunities. The main issue with the current scenario is that it relies heavily on physical resources and word of mouth for opportunities resulting in limited access. Hence, the problem is addressed through an accessible digital concept that bridges the gap between survivors and growth opportunities, enabling independent survival. A comprehensive application is required as survivors are left with damaged motor skills, blindness, paralysis, disability, etc. Support is provided through voice-based interactions, screen readers, haptic feedback, and an intuitive navigation model to enhance the digital experience. A user-centered solution would increase the engagement and awareness of digital growth exploration.

7 Conclusion and Scope

Acid attacks impact the physiological, sociological, and psychological behavior of survivors. They face challenges in growth and employability after the attack. There is an opportunity for digital intervention, as most current solutions only have an offline presence scope for providing a platform for survivors which caters to their needs from growth to employment. A digital platform would improve growth outcomes by targeting upskilling, work management, and community culture. The application can be implemented at the national level to increase awareness around acid attacks which would also boost the digitization of government schemes. The financial pay gap between rural and urban landscapes can be bridged through digital intervention. This would empower survivors, increase their economic status, and make them self-reliant.

References

1. Das A (2019) A study on acid attack in India and its impact
2. Chanderia K. Acid attack: discovering a lost identity. J Contemp Issues Law
3. Binoy SM (2019) A critical analysis on acid attacks in India. J Emerg Technol. Innov Res
4. Hameed S (2022) Scarred for life: thoughts on legal perspectives of acid attacks in selected countries with a focus on India
5. Kumar V (2019) Acid attacks in India: a socio-legal report. J Anal Exploit Violence
6. A. Venugopalan, “How finding jobs becomes difficult for acid attack survivors,” Economic Times, 2020.
7. Byard RW (2023) Potential mechanisms of death in cases of fatal acid attacks
8. Tripathi SM (2020) Acid attacks & impact of crimes against the dignity of women in India: an analytical study
9. Mittal S (2021) Exploring the trauma of acid attack victims: a qualitative enquiry
10. Kapur R (2020) Acid attacks: crime having unfavourable effects upon the entire lives of women. Research Gate
11. Chauhan A (2021) Research report on acid attacks in India
12. Bose R (2022) Low compensation and convictions, lax implementation: why acid attack laws fail survivors
13. Khaleelullah A (2020) Challenges faced by acid attack survivors in India
14. Algoworks, Metaverse calls attention to acid attack survivors and handicaps. Algoworks
15. Goswami S (2020) The peril of acid attacks in India and susceptibility of women. Victimol Victim Justice
16. Dutta D (2023) How to ensure inclusive employment for burn survivors

Empirical Analysis in Manufacturing Industry: An Ergo-economical Approach

Sarvesh Saravanan, Pugazhendhi Smruthi, Vignesh Raja, Vigneswaran Chidambaram, and Brajesh Kumar Kanchan

1 Introduction

The concept of 'industry' refers to a collective group of manufacturers or businesses that share the common goal of producing a wide range of goods and services. These entities engage in various economic activities, including manufacturing, production, services, and technology-driven ventures. The categorization of industries depends on the types of products and services they provide and the specific processes they use in their operations. Industries hold immense importance in the global economy, serving as fundamental contributors to economic growth and progress. This discussion focuses on the manufacturing sector, which creates physical goods through intricate processes. Manufacturing plays a pivotal role in driving economic development, fostering innovation, advancing technology, ensuring national security, creating job opportunities, and promoting societal well-being. Workers dedicate their time and effort to delivering high-quality products, making significant contributions to their organization's mission. In return, industries have a solemn responsibility to prioritize the well-being of their workforce. However, workers often engage in repetitive tasks, intricate procedures, and physical exertion during various processes within industrial settings. While these actions are essential for production, they can lead to fatigue, muscle discomfort, and inflammation, ultimately restricting a worker's mobility and causing discomfort. These work-related challenges give rise to musculoskeletal disorders (MSDs), a significant concern for both workers and industries. MSDs encompass a range of issues affecting muscles, tendons, ligaments, nerves, and soft tissues,

S. Saravanan · P. Smruthi (✉) · V. Raja · V. Chidambaram · B. K. Kanchan
Ergonomics Laboratory, PSG College of Technology, Coimbatore, India
e-mail: 20p625@psgtech.ac.in

V. Chidambaram
e-mail: cvn.prod@psgtech.ac.in

B. K. Kanchan
e-mail: brajeshlean@gmail.com

© The Author(s), under exclusive license to Springer Nature Singapore Pte Ltd. 2025

S. Das et al. (eds.), *Human-Centered Digitalization, Volume 3*, Design Science and Innovation, https://doi.org/10.1007/978-981-96-4383-7_6

resulting in symptoms like pain, stiffness, weakness, numbness, tingling sensations, and limited movement. If left unaddressed or managed inadequately, MSDs can lead to long-term impairments or disabilities. Consequently, industries must prioritize worker safety and comfort. By improving workplace conditions and ergonomics, not only can employee well-being be protected, but overall workplace efficiency can also be greatly enhanced. This underscores the crucial need for industries to embrace the principles of ergonomics, a scientific discipline focused on understanding the complex interaction between workers and their workplace systems. The ultimate goal of ergonomics is to optimize the synergy among employees, tools, equipment, and tasks. Ergonomics encompasses a wide range of applications, including physical ergonomics, cognitive ergonomics, organizational ergonomics, and environmental ergonomics. Its utility extends beyond process optimization, as it also plays a pivotal role in designing and improving various elements, such as products, organizational workspaces, digital interfaces, environmental layouts, educational methods, training programs, and even sports equipment. Integrating ergonomics into industry practices yields numerous tangible benefits, including increased productivity, higher work quality, improved employee retention, reduced absenteeism, and lower healthcare costs. Ergonomics is relevant across industries, with a steadfast commitment to enhancing human well-being, safety, and performance, resulting in user satisfaction, comfort, and operational efficiency. However, it's important to acknowledge that the adoption of ergonomic principles and practices may present challenges for small-scale manufacturing industries. Ergonomic solutions are sometimes seen as financially burdensome, which can deter their implementation. Additionally, the widespread use of lean techniques in industries, which aim to eliminate waste, including over-processing, may lead to hesitation in embracing ergonomic alternatives. Therefore, before recommending ergonomic solutions, careful consideration of factors such as initial costs, complexity, resistance to change, and potential increases in time consumption is essential. The intersection of ergonomics and economics is a reality that cannot be ignored. In the industrial environment, balancing the interests of workers and management is essential for the industry's successful operation. Consequently, finding cost-effective ergonomic solutions to address workplace challenges becomes crucial. These alternatives should undergo a comprehensive cost–benefit analysis to ensure they meet both cost-effectiveness and ergonomic effectiveness requirements. To illustrate this concept, a case study is conducted, examining the processes performed within an industry. The goal of this study is to gain a deeper understanding of the processes, the sequences followed by workers, frequently repeated activities, the strain experienced by workers, and the impact of these tasks on muscle health. Through this analysis, suitable alternatives can be identified that not only alleviate ergonomic concerns but also align with the economic interests of the industry. The literature survey provides valuable insights into the significance of ergonomics in the contemporary industrial landscape, especially within the context of Industry 4.0. It highlights how ergonomics plays a crucial role in bridging the gap between the manufacturing sector and human factors [1]. Existing studies reveal a critical finding a significant proportion of workers, approximately 86%, report experiencing upper body pain, with 14% reporting lower body pain, particularly among those aged 18–45 years [2]. The prevalence of musculoskeletal disorders (MSDs) is a global concern, resulting in substantial costs for both employees and organizations. For example, in Great Britain, an estimated 470,000 workers were affected by MSDs in the year 2020–2021, contributing to 28% of work-related ill health [3]. Similarly, in Europe, a significant portion of employees, 5 out of every 28, is affected by work-related musculoskeletal disorders. Australia also faces a high incidence of musculoskeletal disorders, with approximately 87% of workers' claims related to such issues in 2019–2020.The literature underscores the need to educate workers about the importance of ergonomics in preventing work-related musculoskeletal disorders [4]. It also emphasizes that implementing ergonomic interventions tailored to specific workplace needs can lead to a significant reduction in work-related musculoskeletal disorders. However, it acknowledges the financial implications of creating an ergonomic workplace, highlighting the necessity of developing economically feasible strategies that enhance worker efficiency and overall industry productivity. One key concept discussed is break-even analysis, which is instrumental in determining the point at which the revenue generated equals the cost of operations [5]. This analysis helps identify the break-even quantity, representing the production volume required to recoup all manufacturing costs. This mathematical relationship between output units and costs, typically depicted graphically [6], provides valuable insights for decision-makers. Furthermore, the literature suggests that implementing feasible ergonomic interventions can facilitate the redesign of plant layouts in an optimized manner, ultimately resulting in enhanced workplace efficiency and productivity [7]. This highlights the potential for ergonomics to not only improve worker well-being but also yield tangible benefits for the entire industry.

2 Methods

During this study, a structured methodology was devised to comprehensively address the various aspects related to ergonomic interventions in the selected industrial setting. The first step in this work methodology was the selection of the specific work activity to be studied. This process encompassed a series of considerations, including the choice of an industry, the selection of a particular workplace within that industry, and a meticulous analysis of the processes carried out in that chosen workplace. Additionally, an evaluation of the workers involved in these processes was conducted. The ultimate criterion for the selection of the activity under investigation was twofold: it had to be not only a task performed significantly within the workplace but also one that imposed a considerable degree of strain on the workers. Following the selection of the target activity, the next phase involved the design of the experiment. In this context, the study employed a key performance indicator widely utilized in various industries—the learning ratio. This metric served as a valuable tool for assessing and managing the productivity and effectiveness of the workforce. Specifically, the study calculated the learning ratio of the workers involved in the selected activity. This calculation not only allowed us to gauge the learning efficiency of these workers but also laid the foundation for subsequent endeavors aimed at enhancing their performance. The insights derived from this calculation were instrumental in performance improvement assessments, resource allocation for the processes, continuous improvement initiatives, and talent development within the organization.

To gain a deeper understanding of the work environment and the challenges workers face while performing the chosen activity repeatedly, the study embarked on modeling the workplace. This involved the creation of a three-dimensional workplace environment, which facilitated a detailed analysis of the work execution process. It enabled us to pinpoint the specific challenges and ergonomic issues encountered by the workers during their tasks. The study employed the Rapid Upper Limb Assessment (RULA) method to assess the workers' postures and movements [8]. The RULA score, calculated based on this assessment, played a pivotal role in determining the necessity for ergonomic intervention. Subsequently, various alternatives were proposed to address the ergonomic concerns identified in the chosen area of study. Design considerations for ergonomic alternatives hinged on several critical factors, including the nature of the task being performed, the time required for task completion, and the frequency with which the task was repeated to produce the desired output. These considerations informed the development of ergonomic solutions tailored to the specific needs of the workers and the production process [9]. The economic perspective was of paramount importance throughout this study, as it is a key concern for the industry. Therefore, all proposed ergonomic interventions were assessed from an economic standpoint. To ascertain the economic viability of these interventions, the study employed break-even analysis, which allowed us to delve into the cost structure of each intervention and determine the number of units that needed to be sold to recover the investment costs. At the break-even point, the company neither incurs a profit nor a loss, and all the funds invested are recuperated. This analysis served as a crucial tool for identifying ergonomically and economically viable alternatives. Furthermore, the study established an empirical relationship between the RULA scores obtained and the learning ratios of the workers. This relationship informed us about the areas within the study that required ergonomic intervention. Subsequently, the study conducted a comprehensive cost–benefit analysis, considering the models suggested in light of this empirical relationship.

The culmination of this work methodology involved the digital modeling of the identified activity for ergonomic intervention. Workers were engaged in performing the task within this digital model, and subsequent RULA assessments were conducted to calculate the RULA scores. Multiple design iterations were executed to reduce the RULA action levels, and the learning ratio of the workers was recalculated to quantify the improvements in efficiency and productivity resulting from the ergonomic interventions. Ultimately, through a rigorous break-even analysis, the study estimated the costs associated with each ergonomic action model. The break-even analysis was predicated on the established relationship between RULA scores and learning ratios, enabling the identification of the break-even quantity required for each economically feasible action.

2.1 Experimental Study

In this experimental study, the primary objective is to achieve a relative increase in the learning ratio through a decrease in the Rapid Upper Limb Assessment (RULA) score for the ergonomic intervention. The assessed RULA score obtained post-ergonomic intervention serves as a key indicator of the significance and effectiveness of the ergonomic measures implemented. The learning ratio, a pivotal metric in this study, quantifies the time required for workers to accomplish tasks efficiently and productively. The calculation of the learning ratio provides insights into the time taken by workers to achieve specific performance levels. Before the implementation of ergonomic interventions, the learning ratio among workers was notably low. This metric holds paramount importance because the increase in the learning ratio is directly associated with a decrease in ergonomic risk factors within the workplace. The need for ergonomic intervention was identified through several assessment criteria, which led to the development of ergonomic models based on the relationship between RULA scores and learning ratios. Subsequent cost–benefit analysis facilitated the selection of an economically feasible model, which aimed to increase the learning ratio and reduce the associated ergonomic risks.

To conduct the study, a sample of workers was selected from the industry, and their learning ratios were calculated. A subgroup of workers with the same degree of learning ratios was chosen as the focus of the study. The workers are made to perform the process in different postures and an empirical relationship between RULA scores and learning ratios was established.

While the need for ergonomic improvements was substantiated through the study, the feasibility of these alternatives remains a critical consideration. To address this, the recommended RULA action plans were assessed for feasibility using break-even analysis, a tool that identifies the optimum point where neither profit nor loss condition is incurred known as the break-even quantity. A 3D workspace was created to execute feasibility analyses, aiding in the identification of which ergonomic interventions were economically viable. The break-even quantity for each suggested ergonomic model was meticulously calculated. Ultimately, this comprehensive study culminated in the proposal of an ergonomic alternative, guided by empirical relationships, cost–benefit analysis, and RULA scores. The aim is to not only enhance the well-being and comfort of workers but also to drive increased efficiency and productivity within the workplace.

3 Results and Discussion

The worker's learning ratio describes how fast a worker can learn things, perform processes more efficiently, and achieve significant results in a shorter period. A higher worker-learning ratio significantly shows the faster grasping capacity, understanding of new information, and faster learning qualities exhibited by the worker. Calculation of the learning ratio helps to identify the ability of the worker when employed in a new environment which includes productivity and efficiency of the worker. The learning ratio can be calculated by conducting performance assessments, time taken by the worker to complete an intended task, error rates in the activity performed, and self-reporting methods. The suggestion of an ergo-economical alternative involves three stages: the selection stage, the design and economics stage, and the implementation stage. The empirical relationship between worker-learning ratio and RULA assessment is established to prove the need for ergonomic improvement. This approach involving the generation of empirical relationships helps in decision-making, justifying the investments made, ergonomic improvements, healthier work environment, and cost-effectiveness. The industry considered for the study is selected based on the task demands, worker characteristics, and potential risks in the industry. The worker-learning ratio is calculated to assess the capabilities of the workers in the industry. Five different workplaces with maximum worker strain are identified in the industry [10], the process sequence, worker efficiency, worker strain due to improper work posture which may lead to musculoskeletal disorders, duration of the work, and quantity produced are noted and the requirement for ergonomic improvement is determined by giving scores as per the criteria like redesign feasibility and work in orders hand.

Among processes, the part or assembly that involved maximum worker strain through repetitive actions, and improper work posture was chosen for study. The processes are painting, Flange assembly, Stator Winding, Capacitor fixation to the motor body, and Pump Assembly. The chosen processes are observed based on a few criteria, tabulated in Table 1 which are considered for further study and investigated for worker strain (Fig. 1).

Among the processes, the process with a high weighted score depicts high ergonomic danger levels and the need for ergonomic improvement. As observed from Table 1, the Flange assembly process is chosen for ergonomic improvement, and RULA actions are suggested.

Table 1 Workplace selection

Criteria	Painting	Flange assembly	Stator assembly	Capacitor assembly	Pump assembly
RULA action score	4 (3)	5 (4)	4 (3)	3 (2)	3 (2)
Redesign feasibility	8 (4)	9 (5)	8 (4)	7 (4)	7 (3)
Work orders in hand	30 (3)	50 (5)	30 (3)	25 (3)	15 (2)
Weighted score	10	14	10	9	7
Workplace selected		Selected			

where ( ) represents the weighted score ranging from 1 to 5

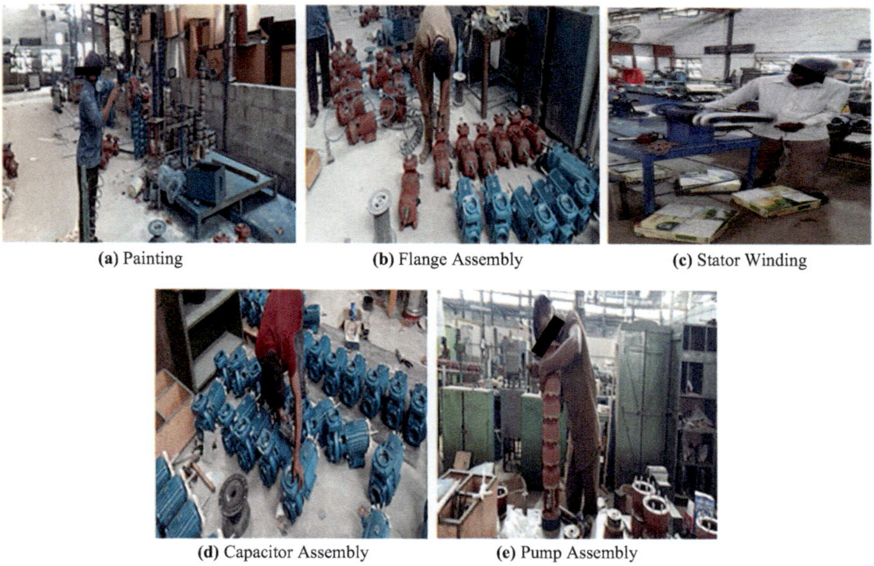

Fig. 1 **a** Painting, **b** flange assembly, **c** stator winding, **d** capacitor assembly, **e** pump assembly

A group of 10 workers are chosen from the industry and their worker-learning ratios are calculated by taking the ratio of time taken to produce n jobs to the time taken to produce $3n$ jobs. From the obtained WLR values the worker efficiency is obtained.

$$\text{Working Learning Ratio} = \frac{t \text{ for } n \text{ jobs}}{t \text{ for } 3n \text{ jobs}} \tag{1}$$

where 't' is the time taken by the worker in minutes.

As observed from Table 2, three workers with the same degree of learning ratio are chosen for further study. Ergonomic interventions positively impact the health of the workers, reducing worker fatigue, and increasing efficiency and productivity. The physical work environment is converted into a digital model, with the workstations, equipment design, and proper lighting conditions.

RULA Action 1: Work table set up for the worker to perform the assembly of flange housing designed based on optimal height suitable for any worker (Fig. 2).

RULA Action 2: A chair setup for the workers to perform the seated assembly of flange housing.

RULA Action 3: A table and chair setup for the workers for the assembly of the flange housing.

The redesigning phase is started by decreasing the RULA Action level, a manikin of the 50th percentile which matches and represents the current industrial anthropometric data is selected. It is incorporated into the designed workplace in a position relative to the working table. Then RULA assessment is conducted for the redesigned

Table 2 Selection of workers for evaluation

Worker	Time taken for 'n' jobs (minutes)	Time taken for '$3n$' jobs (minutes)	Worker-learning ratio
1	25	76	0.3289
2	22	67	0.3283
3	27	82	0.3292
4	23	68	0.3382
5	24	74	0.3243
6	22	67	0.3283
7	26	79	0.3291
8	27	85	0.3176
9	25	77	0.3246
10	23	71	0.3239

workplace. A similar RULA assessment is carried out for all the ergonomic suggestions, and further redesigning is carried out to significantly reduce the RULA assessment scores to an acceptable value.

The chosen set of workers is evaluated with different working postures for different RULA action levels while performing the process at the chosen workplace and the observations are tabulated in Table 3.

The RULA Action with a higher worker-learning ratio is chosen for ergonomic improvement for the final compliance from Table 3.

The cost estimation for the RULA actions with acceptable RULA scores is tabulated in Table 4.

Then break-even analysis was conducted to find the designs with minimal feasibility and guaranteeing the company's economic returns. The break-even quantity must be less than the order quantity. Based on the interpretation from graphs in Fig. 3a–c the ergo-economic alternative is chosen.

The implementation phase consists of introducing and integrating ergonomic interventions into the workplace environment. The implementation phase involves various steps including developing a plan, ensuring the employers are given necessary awareness about the upcoming ergonomic advancement in the workplace, redesigning the workplace according to the ergonomic improvement, and most important of all the feedback mechanism.

The calculated worker's learning ratio is tabulated in Table 1, from which three workers with the same degree of learning ratio are chosen. This set of identified workers has a higher degree of worker-learning ratio among the observed 10 workers, which represents a higher grasping and learning ability of the worker. These workers are made to perform the process in different positions in the suggested RULA Action levels stated in Fig. 3a–c and are tabulated in Table 3. The RULA Action level gives the maximum worker-learning ratio and is considered the most efficient design. However, the economical view is crucial for the industry break-even analysis to

(a) RULA Action 1

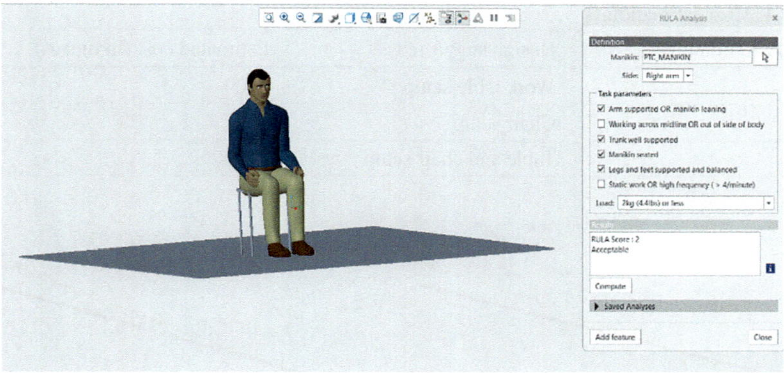

(b) RULA Action 2

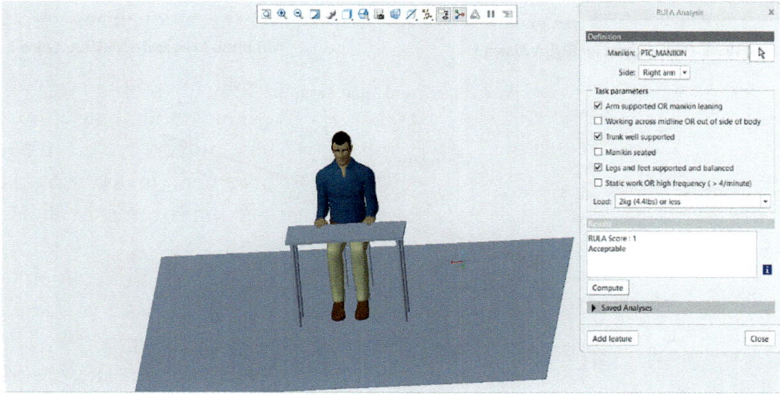

(c) RULA Action 3

Fig. 2 **a** RULA Action 1, **b** RULA Action 2, **c** RULA Action 3

Table 3 Evaluation of learning ratio in different RULA Action levels

RULA Action levels	Worker	Time taken for '*n*' jobs (minutes)	Time taken for '3*n*' jobs (minutes)	Worker-learning ratio
1	1	18	50	0.36
	2	20	54	0.3703
	3	19	55	0.3454
2	1	22	65	0.3384
	2	20	60	0.3333
	3	20	58	0.3448
3	1	21	60	0.35
	2	20	58	0.3448
	3	21	59	0.3559

Table 4 Cost estimation

RULA Action level	Design suggested	Estimated cost (in rupees)
1	Work table setup	4900
2	Chair setup	700
3	Table and chair setup	2000

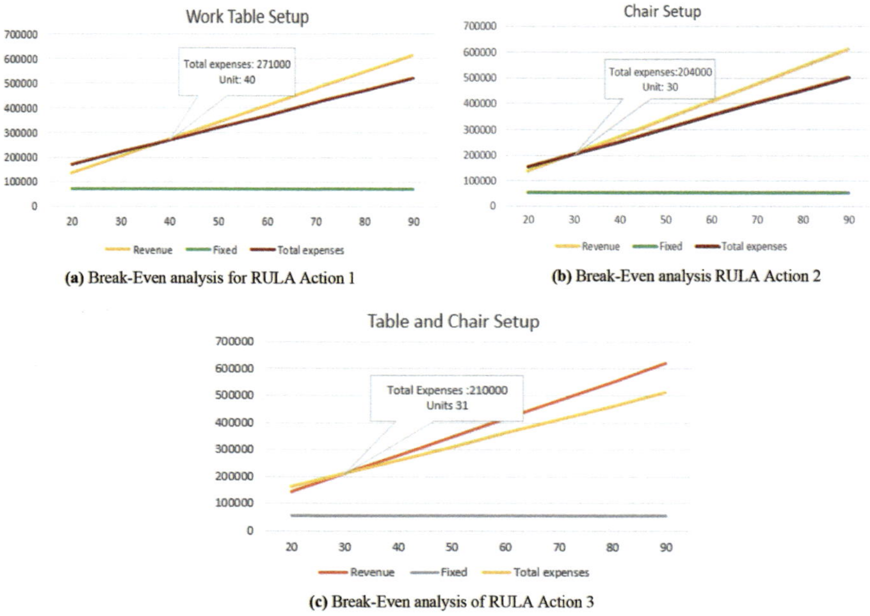

Fig. 3 **a** Break-even analysis for RULA Action 1, **b** break-even analysis RULA Action 2, **c** break-even analysis of RULA Action 3

be performed on all the RULA Action levels, from which the break-even point is identified which shows the optimum number of quantities to be produced for each alternative suggested to attain neither profit nor loss condition. So the appropriate RULA Action can be chosen according to the number of quantities to be produced in the industry.

4 Conclusion

From this research, workers from all industries irrespective of the domain have a higher probability of experiencing medium to high risks of work-related musculoskeletal disorders due to bad working postures and improper working conditions hinging to the need for an ergonomic environment for worker health and safety and also for the productivity of the industry. Often ergonomics involves higher investments which may not be beneficial to the industry, it is important to analyze or study whether the ergonomic solution is beneficial for the industry guaranteeing returns. This research proves the cruciality of cost–benefit analysis in ergonomics by providing an ergo-economic alternative that has a greater significance in worker health as well as guarantees returns for the industry.

Any industry considering opting for an ergonomic environment or an ergonomic improvement can follow this ergo-economic approach to know the feasibility of an ergonomic intervention to be brought into play in terms of safety and tangible investments for the industry.

References

1. Reiman A, Kaivo-oja J, Parviainen E, Takala E-P, Lauraeus T (2021) Human factors and ergonomics in manufacturing in the industry 4.0 context–A scoping review. Technol Soc 65:101572
2. Ayub Y, Shah ZA (2018) Assessment of work-related musculoskeletal disorders in manufacturing industry. J Ergon 8(3):1–5
3. Anugraha RA, Sutan W, Mufidah I (2015) The design of batik stamp tool scraping working table using ergonomics principles. Procedia Manuf 4:543–551
4. Ndaliman MB, Suleiman UY (2011) An economic model for break-even analysis. In: Proceedings of the 2nd international conference on mechanical and manufacturing engineering (ICME), PICC, Putrajaya, Malaysia, pp 23–25
5. Naik SB, Kallurkar S (2016) A literature review on efficient plant layout design. Int J Ind Eng 7(2):43–51
6. Morano P, Tajani F (2017) The break-even analysis applied to urban renewal investments: a model to evaluate the share of social housing financially sustainable for private investors. Habitat Int 59:10–20
7. Chidambaram V, Gopalsamy MM, Kanchan BK, Mouleeswaran S (2023) A holistic methodology for mitigating awkward postural risks: evidence from South Indian small-scale industries. Work (Preprint):1–15

8. Chidambaram V, Gopalsamy MM, Kanchan BK (2023) Ergonomic investigations on novel dynamic postural estimator using blaze pose and transfer learning. Ergonomics, 1–17
9. Vigneswaran C, Madhan Mohan G, Vignesh Raja M, Selva Vignesh E (2022) Computer-aided ergonomic design and assembly of a domestic string hopper machine. In: Industrial automation and robotics. CRC Press, pp 33–42
10. Mohan, GM, Elangovan S, Prasad PSS, Rama Krishna P, Mokkapati AK (2008) Prevalence of job strain among Indian foundry shop floor workers. Work 30(4):353–357

Exploring Opportunities in Sustainable Packaging Design Industry: A Review Study

Taruna Singh, Ranganath M. Sinagri, and Sumer Singh

1 Introduction

The packaging industry is the fifth-largest sector in India's economy, experiencing some of the highest growth rates, with annual increases ranging between 24 and 27%. The food processing industry, holding over 50% of the market share, is a leading consumer of flexible packaging. According to the report *Packaging Industry in India Market Share | Industry Trends 2021 to 2026*, Mordor Intelligence in 2021, the Indian food packaging market is projected to grow at a compound annual growth rate (CAGR) of 10.3%, reaching a market size of USD 108.2 billion by 2027. According to http://ficci.in/spdocument/, in the food sector to increase the level of sanitation, plastics are mostly utilized as it also enhances the shelf life of products. Plastics are frequently favored and employed due to their qualities and versatile in applications. They may be molded into beautiful shapes and are lightweight, corrosion-resistant, moisture-proof, very flexible, and adaptable.

The branding strategy of a company determines its success in the contemporary marketing environment, where comparable companies are vying for market share within comparatively analogous marketplaces [1]. Packaging represents a company's branding, notably for manufacturers and merchants, and has even been referred to as the "silent salesperson" [2]. The path from first brand exposure to product acquisition comes to a finish here. According to this, packaging is crucial to the sales and decision-making processes. For many years, packaging has been a concern for the environment. In society, it has mostly been discussed and contested as an environmental concern. Production, distribution, and consumption of food and beverages

T. Singh (✉) · R. M. Sinagri
Delhi Technological University, New Delhi, India
e-mail: tarunasingh@dtu.ac.in

S. Singh
Indian Institute of Technology, New Delhi, India

© The Author(s), under exclusive license to Springer Nature Singapore Pte Ltd. 2025

S. Das et al. (eds.), *Human-Centered Digitalization, Volume 3*, Design Science and Innovation, https://doi.org/10.1007/978-981-96-4383-7_7

are causing the major negative impact on the environment. An important but sometimes ignored environmental concern is how packaging helps to reduce food losses. After accomplishing its main objective, which is to protect and advertise the goods it contains, packaging is typically not considered helpful. Packaging is therefore viewed as a disruptive waste for customers and a huge burden for the environment. The service viewpoint is employed to better understand how customers interact with packages. Because they are inexpensive and convenient only in the short term, disposable packaging is frequently used. Plastic food packaging box usage has grown to such worrisome levels that it is gravely damaging our ecosystem. The rise in plastic waste has been attributed to the proliferation of online food delivery services, with approximately 22,000 tons of plastic waste generated per month in India. Of this total, only about 60% of the material is subject to recycling, while approximately 43% is non-recyclable and is designed for single-use only [3].

"Packaging and packaging design have grown to be important marketing components for diverse consumer goods and plays an important role in conveying product advantages to the consumer" [4]. Especially in a developing country like in India, solving the problems with sustainable packaging means figuring out a number of complex obstacles. The amount of plastic garbage has increased dramatically due to the growth of e-commerce and meal delivery services. A significant amount of packaging materials are either non-recyclable or made for single use, which contributes to persistent environmental issues even in the face of growing awareness and legislative attempts. The absence of uniform standards and financial limitations also make it more difficult for sustainable packaging alternatives to be widely adopted. A comprehensive strategy is needed to address these issues, one that includes promoting cutting-edge packaging materials, enforcing stricter rules, and enhancing waste management systems. Addressing the issues related to sustainable packaging in India requires a multi-pronged approach.

There are connections between product packaging and other marketing mix components. According to Packaging Federation, these changes occur in a number of areas, such as new technology, the development of materials, logistical requirements, ecological concerns, and consumer preferences, all of which are crucial aspects when choosing a marketing strategy [5]. Government regulations and recommendations for sustainable packaging have been implemented as public awareness of the environment increases. The growing regulatory requirements and market pressures surrounding packaging sustainability have prompted extensive research into both its economic and environmental implications. This research has led to the development of a wide array of sustainability standards, theoretical frameworks, methodologies, and practical tools aimed at promoting more sustainable packaging solutions. These resources have been disseminated across a diverse spectrum of stakeholders engaged in various stages of the packaging lifecycle, including production, transportation, and distribution. Among the key users of these tools are designers, engineers, technologists, marketers, and environmental managers, all of whom play critical roles in ensuring that sustainability is integrated into packaging processes. By providing these stakeholders with access to well-defined sustainability guidelines and instruments, the industry is better equipped to align packaging practices with environmental objectives, while also addressing economic and logistical considerations.

The initial point is that packages should be designed to have as lesser impact on the environment as possible. Tools like lifecycle valuations and inventories management that reduce carbon footprint and environmental effect should be used to help the package design process. Packaging must be designed at a fair economic cost and fulfill market norms. It must be socially and culturally acceptable (e.g., mirror the lifestyle of your target audience) and include social enablers that inspire users to come up with new applications for it. Customers must be facilitated to dispose of products in an ecologically acceptable manner if the packaging design doesn't offer an alternative use [6]. The authors' research is grounded in the premise that the package design system offers substantial opportunities for exploration and advancement that might result in sustainable packaging. This will provide experts and academics a holistic perspective and lay the groundwork for creating new kinds of tools and systems that can be employed when designing sustainable packaging. Concentrating on the attributes of packaging traits may be used to efficiently reduce resource consumption and the environmental impact of the packaging system.

2 Methodology

A decent number of articles were considered for the literature review in this paper to understand what's happening in the field of packaging design. The bibliographic references and content extraction from the reviewed articles were systematically managed using Medley software. All identified documents were subsequently analyzed based on the extracted information as mentioned in Table 1:

In phase 1 (Collection): Data was collected from various journals and publishers in context of packing, sustainable packaging, packaging design. The largest number of data were found using a limited keyword search like *Packaging design*, *Pack* and *sustainable packaging design* from sources including Springer, SCI, SCEI, Emerald, Springer, Elsevier, Science direct, online science, and many more journals. The research was split up into the sub categories mentioned above because package design is a bigger field as mentioned in Fig. 1.

In phase 2 (categorization): The publications identified were thoroughly scrutinized and analyzed by categorizing various facets of packaging design. After being published in 20 separate digital libraries, journals were also sorted into 7 areas of categorization. The data collection for each aspect was analyzed by the literature review which provided a qualitative data.

Table 1 Analyzing criteria for the articles selected

Journal	Author/s	Year	Introduction	Objective	Methods/Tools	Findings	Limitation

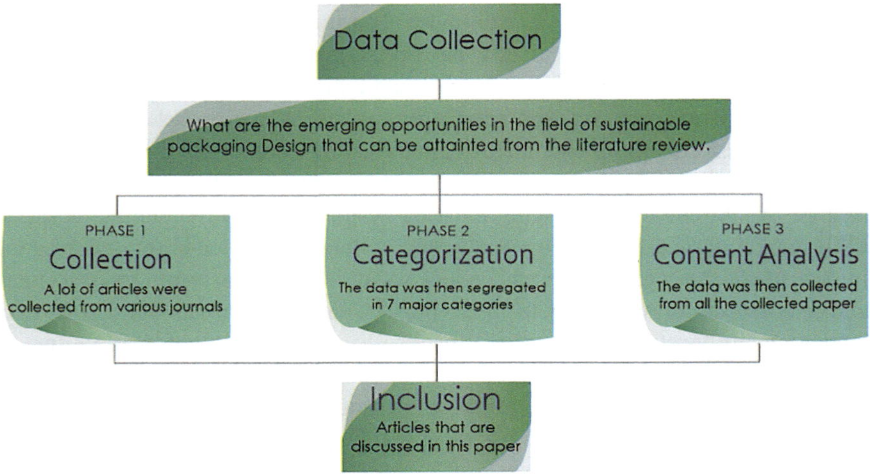

Fig. 1 Methodology flowchart

In phase 3 (content analysis): To facilitate data exploration and mitigate subjectivity in the qualitative study, an analysis of a select number of papers has been incorporated into this research.

3 Results

The fundamental notion is that designers and other stakeholders should research what sustainability means in their specific scenario rather than utilizing a tool or method that was not intended for and may not match their own situation. Similar to how designers typically begin with a concept phase where opportunity of exploration is high and concept of sustainability is followed.

About a third of the studies included experimental data, whereas the majority focused on appraising aesthetics and discussing various elements of design which contributes to the packaging. Thereafter sustainability in second position, where the physical and chemical properties of materials have been discussed along with a trivial discussion about new innovation in materials, as well as their biodegradability and other characteristics.

3.1 Packaging Design Focusing on Aesthetics

The graphic and structural components of design, such as its form, color, size, and typography are referred to as its packaging design [7]. These components could act as subliminal cues that might attract attention and imply product qualities through associations. Compared to more explicit factors such as price, brand, or product promises, these implicit design signals are often not immediately recognized or consciously processed by consumers as valuable sources of information. While clear cues like pricing and branding directly capture attention and inform purchasing decisions, the subtler aspects of design often operate beneath the surface of conscious awareness. Despite their understated nature, these design elements can still exert a significant influence on consumer perceptions and behavior, shaping decisions in ways that are not always readily acknowledged or consciously identified by the client. Therefore, the subconscious aspect of decision-making is more strongly correlated with packaging design [8].

Colors are an important aspect of a packaging design [9–11]. They are the package aspects that respond the quickest, and they have grown to be one of the most crucial visual design components [12]. The ability to distinguish the backdrop color from the other package components researchers will look solely at the impact of color [13].

In the food sector, color has a significant role [11, 14, 15]. This is the reason why producers should use the food packaging to grab customers' attention and pique their desire to buy to create sensory and hedonic expectations for the products that are in line with its actual performance features [10]. Design styles have a great evocative effect for customers, and companies frequently employ them to express certain connections of concepts as color frequently enters human decision-making and discourse. Due to the various interpretations given to colors in various circumstances, distinct connotations of color may surface. Red, for instance, connotes a negative conclusion when competence is being assessed and both good and negative outcomes are conceivable.

There are many different color associations that explain the influence of color, in which a certain color is interpreted as a symbol or a sign. "In addition to shape, color is the most crucial component of a package and is a key component of logos and product packaging for example red for Coca-Cola red, blue for IBM blue, purple for Cadbury, and so on" [16]. According to various studies found during the literature review, colour and form have a substantial impact on customers' associations, predicted preferences, and desire to buy. According to a recent study on the effect of product package colors on customers' purchasing decisions under time constraints, a customer's purchasing decision is largely influenced by the color scheme rather than the amount of time available [17]. To entice customers in their decision-making, however, all package components—including color and shape—must be blended. "When applied to a product's packaging, the appropriate choice of package color, backdrop picture, wrapper design, and unique concepts will foster a cheerful sensation in customers' minds" said Ghosh B. in 2016 [18].

3.2 Packaging Design Focusing on Material and Innovation

With the widespread use of packaged goods and the rapid growth of online delivery services, it is imperative to address the environmental issues associated with single-use, disposable plastics. Industries must assess the suitability of different packaging materials for food products by weighing their respective advantages and disadvantages. Additionally, it is necessary to consider the inclusion of specific attributes in the packaging materials that align with the end-use properties of the food products. For example: researcher explored eco-friendly techniques that were employed to create biodegradable food packaging sheets from the leaves and sheath of banana (Musa spp.) [3]. To maximize its utility, xylanase enzyme was immobilized in calcium alginate beads. This straightforward eco-friendly packaging sheet is suitable for use in food packaging because no hazardous chemicals are used during the manufacturing process. The bio plastic layer acts as a waterproof barrier to prevent spills while being transported.

By comparison to alternative options, cellulose-based compounds present in both woody and non-woody plants have a very low environmental effect, as demonstrated by the body of available studies [19]. One of the most prevalent and major naturally occurring materials/resource is cellulose. While from a merely structural perspective packaging may seem straightforward, from a purely systemic one it is really rather complicated. The process of creating eco-friendly packaging is complex and time-consuming. Finding instances of concept duplication was also simplified by the new format. Both steps (raw material processing) and steps (waste disposal) get a disproportionate amount of attention from researchers.

3.3 Packaging as Marketing and Management Tool

"Today, packaging design is a key component in the marketing of a wide range of "consumer goods" and is essential in helping consumers understand the benefits of the product. As a result, there is a relationship between product packaging and other elements of the marketing mix" [4]. These are strategies that the business can use to adjust to alterations in the marketplace. However, the business environment has a complex set of effects on packaging design. Key factors influencing management decisions regarding marketing strategy encompass advancements in technology, developments in materials, logistical considerations, environmental concerns, consumer preferences, and marketing variables [20, 21].

Because they may be utilized to create and strengthen brand equity, the four Ps of the marketing mix—product, promotion, pricing, and place—have a significant impact on the establishment of a brand. One of the pillars in the whole marketing approach of numerous companies for a long time has been the marketing mix [22]. In the traditional marketing mix, one of the four core elements is the product itself. For fast-moving consumer goods (FMCGs), the most critical aspect of the product—particularly from the customer's perspective at the point of sale—is its external appearance, including its shape and design. This visual and physical presentation plays a pivotal role in influencing consumer choices and perceptions, often serving as the primary point of interaction between the customer and the product [23].

3.4 Packaging Design Focusing on Sustainability

Two organizations, the Sustainable Packaging Alliance (SPA) in Australia and the Sustainable Packaging Coalition (SPC) in the US, have made an effort to define "sustainable packaging" by creating collections of guidelines or tactics that might be used as a foundation for decision-making. As a result of social pressure, marketing strategies, or genuine attempts to address the financial, social, and environmental problems associated with packaging, some industry associations and companies are also adopting the language of sustainable packaging. While the global warming debate has encouraged some firms to focus on the "carbon footprint" of packaging as a relevant and significant subject, new methods have also been developed to analyze the effects of packaging on the environment over its lifespan. However, increasing prices to achieve better sustainability was the major worry. Furthermore, they noted that certain technologies hadn't developed sufficiently to enable SPD [24].

By examining the various problems with modern food packaging and how they influence people's health and the environment paint a picture of the leadership's goals. While food packaging receives more attention in section, other types of packaging, such as those used for drinks, pharmaceuticals, and medical devices, are subject to regular regulation. In light of the prevalence of "post-use" articles, it's plausible that scientists were reticent to offer a detailed summary of their findings. Instead, researchers in this field use systems dynamics, a number of life cycle assessment techniques (LCC, LCS), various optimization algorithms, etc., to acquire a fuller view of the problem.

3.5 Packaging Design Life Cycle Assessment

The use of context units simplifies and puts information regarding eco-friendly packaging into perspective. As a researcher or expert in the food, chemical, raw material, or energy industries, one may find the SLR's life cycle phases useful as a heuristic for navigating the data [25]. While from a merely structural perspective packaging may seem straightforward, from a purely systemic one it is really rather complicated. The process of creating eco-friendly packaging is complex and time-consuming [26].

A wide range of methodologies, such as design for environment techniques like design for recycling and source reduction, are currently employed to assist sustainable design. Corporate methods to package design include creating sustainable design standards and incorporating them into the processes for designing new products, such as the SPC's Design Guidelines for Sustainable Packaging. The Sustainable Packaging Coalition® (SPC), a GreenBlue® project, imagines a world in which all packaging is produced using only renewable energy, produced with moral sourcing, designed to be safe and effective throughout its life cycle, and recycled after use to provide a valuable resource for future generations. The goal of the Sustainable Packaging Coalition is to disseminate a strong, pro-environment message.

3.6 Packaging Design Distribution and Supply Chain

Packaging is often regarded as having a significant negative impact on both the environment and the supply chain. Packaging is frequently viewed as unnecessary or excessive despite the crucial collaborative function it plays in a supply chain [27]. This is brought on by an excessive emphasis on packing's function in a product's life cycle's post-use stages. Brand owners, packaging-developing companies, and regulatory organizations—the major players who have a significant impact on how product-packaging combinations are developed concurrently—want to emphasize "sustainability" more [28]. There is, however, a limited application of sustainability-related aspects as development-directing or criteria for decision-making in the context of product development and innovation.

Furthermore, current business practice and legal requirements permit solutions where sustainable development is subordinate to other determining variables [29]. The problems stem from the inherent complexity of product-packaging development since "sustainable development" has lost much of its original meaning and connotation, and "sustainable development" and "sustainability" are no longer linked together [30]. The integration of sustainability-related considerations into product-packaging design frequently underscores the gap between strategic sustainability goals and their practical realization at the operational level [29]. Although the pivotal role of packaging in influencing the environmental sustainability of products and supply chains is widely recognized, and packaging is increasingly addressed as a separate entity in legislation and industry standards, efforts focused exclusively on optimizing "just" packaging definitions that are unlikely to be sufficient. Such measures may not effectively advance the comprehensive alignment of product-packaging combinations and their associated life cycles with the principles of absolute sustainability. At most, these initiatives will lead to modest increases in local output. How much consideration is given to design and engineering in regard to society, business, and the environment a more rigorous road to progression is provided by sustainability levels that are business-focused? These actions will at most result in slight boosts in local production. On the other hand, the links between the planetary, social, and environmental sciences and the design and engineering.

3.7 Packaging Design as Communication Tool

Sustainability is a multifaceted and often ambiguous term, subject to a wide range of interpretations. While there is frequent acknowledgment of the environmental damage caused by human activities—manifested in issues such as climate change, land degradation, and declining water quality and availability—there remains considerable uncertainty regarding the most appropriate actions for individuals, businesses, and government entities.

In particular, the packaging sector has been under continuous pressure to address these concerns by reducing excessive packaging, minimizing waste, and enhancing the use of recyclable materials. These objectives are complicated by the complex role that packaging plays within supply chain systems, which can create substantial challenges. As a result, it is frequently unclear whether individual decisions and efforts lead to a net positive impact on environmental performance, given the intricate interactions and trade-offs involved.

The Sustainable Packaging Alliance (SPA) in Australia and the Sustainable Packaging Coalition (SPC) in the United States are two prominent organizations that have endeavored to define "sustainable packaging" by formulating sets of principles or strategies intended to inform decision-making processes. In response to social pressures, marketing strategies, and earnest efforts to address the financial, social, and environmental challenges associated with packaging, various industry associations and companies have begun to adopt the concept of sustainable packaging. The ongoing debate surrounding global warming has also led some firms to focus on the "carbon footprint" of packaging as a critical issue. This focus has catalyzed the development of new approaches for evaluating the lifecycle environmental impacts of packaging.

4 Discussion and Conclusion

Sustainable innovation is "a process in which sustainability aspects (environmental, social, and financial) are integrated into design process from idea generation through research and development and commercialization" [31]. The strategy for using product-packaging materials has altered as a result of recent advancements and worldwide development. Novel packaging results from innovative thinking that goes beyond the framework of traditional thinking. It is difficult to strike a balance between sustainability and its demands, particularly when cost becomes a major factor in SPD process. There is a conflict between cost and sustainability in current literature, but there are limited concrete solutions that are available. The principal obstacle to the integration of SPD into product development and design processes is the associated cost. A prevalent perception among users is that clients will be unwilling to incur the additional expenses for SPD technologies, thereby creating a substantial impediment to its ado [24]. Moreover, the numerous factors that influence how consumers use packaging, combined with the rapidly changing nature of packaging technology, make it increasingly challenging to achieve sustainable package design. Consumers demonstrate a strong interest in environmental sustainability, particularly regarding material selection and recycling. However, current research frequently overlooks the comprehensive integration of these attributes into economically viable SPD practices.

Thus, there is a critical need to identify the key attributes that lead to an effective SPD solution. Examining sustainable packaging attributes is essential for addressing this need, as these attributes are pivotal in minimizing environmental impact, aligning with consumer preferences, ensuring economic viability, adhering to regulatory standards, and promoting innovation. For sustainable packaging to achieve broader adoption in the future, a thorough understanding of these attributes is essential. Addressing these gaps in existing research and implementing the proposed strategies will enable the packaging sector to progress toward a more sustainable future, balancing environmental concerns with economic considerations. Due to the complexity of the packaging design process and the need for thorough evaluation of a number of factors that give a 360-degree approach for SPD, more research into this process is necessary. This cannot be accomplished by a single organization, and hence greater collaboration throughout the entire packaging industry is required to make this happen.

References

1. Brown M (2002) Keeping up appearance. Financ Manag Jul/Aug 2001, pp 38–39
2. Pilditch J (1973) The silent salesman: how to develop packaging that sells
3. Jeenusha KS, Amritkumar P (2020) Production of biodegradable food packaging material from Musa (banana plant) leaves by ecofriendly methods. IOSR J Environ Sci Toxicol Food Technol 13(4):1–5. J Market 59(3):16–30
4. Czinkota MR, Ronkainen IA (2007) International marketing. Thomson, London
5. Rundh B (2005) The multi-faceted dimension of packaging: marketing logistic or marketing tool? Br Food J 107(9):670–684
6. van Ooijen I (2016) Packaging design as communicator of product attributes: effects on consumers' attribute inferences
7. Underwood RL (2003) The communicative power of product packaging: creating brand identity via lived and mediated experience. J Market Theory Pract 11:62–76
8. Silayoi P, Speece M (2004) Packaging and purchase decisions: an exploratory study on the impact of involvement level and time pressure. Br Food J 106(8):607–628
9. Ampuero O, Vila N (2006) Consumer perceptions of product packaging. J Consum Mark 23(2):100–112
10. Ares G, Deliza R (2010) Studying the influence of package shape and colour on consumer expectations of milk desserts using word association and conjoint analysis. Food Qual Prefer 21(8):930–937
11. Hutchings JB (2003) Expectations and the food industry: the impact of color and appearance, vol 11. Kluwer Academic/Plenum Publishers, New York
12. Swientek B (2001) Uncanny developments. Beverage Ind 92:38–39
13. Marshall D, Stuart M, Bell R (2006) Examining the relationship between product package colour and product selection in preschoolers. Food Qual Prefer 17(7):615–621
14. Deliza R, Macfie H, Hedderley D (2003) Use of computer-generated images and conjoint analysis to investigate sensory expectations. J Sens Stud 18(6):465–486

15. Piqueras-Fiszman B, Spence C (2011) Crossmodal correspondences in product packaging. Assessing color–flavor correspondences for potato chips (crisps). Appetite 57(3):753–757
16. Hynes N (2009) Colour and meaning in corporate logos: an empirical study. J Brand Manag 16(8):545–555
17. Javed SA, Javed S (2015) The impact of product's packaging color on customers' buying preferences under time pressure. Market Branding Res 2(1):4–14
18. Ghosh B (2016) Impact of packaging on consumers' buying behaviour: a case study of Mother Dairy, Kolkata. Parikalpana: KIIT J Manag 12(1):63
19. Gupta VK, Carrott PJM, Singh R, Chaudhary M, Kushwaha S (2016) Cellulose: a review as natural, modified and activated carbon adsorbent. Biores Technol 216:1066–1076
20. (The) Packaging Federation (2004a), Packaging in the 3rd Millennium. Competitiveness Study for the Packaging Industry in UK, The Packaging Federation, London
21. (The) Packaging Federation (2004b), Packaging's Place in Society. Resource Efficiency of Packaging in the Supply Chain for Fast Moving Consumer Goods, London
22. Nandan S (2005) An exploration of the brand identity-brand image linkage: a communications perspective. J Brand Manag 12(4):264–278
23. Bloch PH (1995) Seeking the ideal form: product design and consumer response
24. Ma X, Moultrie J (2018) Understand sustainable packaging design in practice. In: DS 92, pp 2693–2704
25. Azeredo (2009) Nanocomposites for food packaging applications 42:1240–1253
26. Haanaes K (2013) Sustainable packaging design 54(2)
27. Luttikhuis EO, de Lange J, Lutters E, Ten Klooster R (2014) Using actor networks in decision making during content-packaging development. Procedia CIRP 15:419–424
28. Niero M, Hauschild MZ, Hoffmeyer SB, Olsen SI (2017) Combining ecoefficiency and eco-effectiveness for continuous loop beverage packaging systems: lessons from the Carlsberg Circular Community. J Ind Ecol (21):742–753
29. De Koeijer B, de Lange J, Wever R (2017) Desired, perceived, and achieved sustainability: trade-offs in strategic and operational packaging development. Sustainability (9):1923
30 Olawumi TO, Chan DWM (2018) A scientometric review of global research on sustainability and sustainable development. J Clean Prod (183):231–250
31. Clark T, Charter M (2007) Sustainable innovation: key conclusions from sustainable innovation conferences 2003–2006 organised by the centre for sustainable design

"Sustainability in Fashion: An Exploratory Study into the Upcycling of Rugs into Wearable Textiles"

Priyanka Sharma and Nilakshi Yein

1 Introduction

The textile industry in India is renowned for its constant commitment to sustainability, employing regionally accessible resources and environmentally conscious methodologies throughout its varied terrain. The commitment exhibited here encompasses not alone the selection of primary materials, but also encompasses the utilization of many surface embellishment methods in the production of extraordinary textiles and apparel. Cotton, silk, and jute are the primary raw materials utilized in the textile production process [1]. Cotton is often utilized as the predominant foundation for numerous conventional textiles manufactured within the nation. In addition, several surface improvement techniques, including block printing, dyeing, embroidery, and painting, include the utilization of natural raw materials. These processes not only contribute to the distinctive and complex designs observed in Indian textiles but also adhere to sustainable and recycled practices, thereby highlighting India's extensive legacy of environmentally responsible textile manufacturing.

The Khatri community, known for their expertise in the craft of Ajrakh printing, remains dedicated to the use of natural raw materials in their intricate printing processes. One commonly employed natural material is Harda [2], also known as myrobalan, which plays a pivotal role in fabric preparation for Ajrakh and block printing. When exposed to sunlight, Harda imparts a subtle yellowish tint, serving as the foundation for the exquisite Ajrakh patterns. In addition to Harda, the art of Ajrakh printing continues to rely on natural dyes extracted from sources such as tamarind and turmeric [3]. These traditional techniques not only preserve the authenticity and

P. Sharma · N. Yein (✉)
University of Petroleum and Energy Studies, Dehradun, Uttarakhand, India
e-mail: yeinnilakshi@gmail.com

P. Sharma
e-mail: priyankasharma142000@gmail.com

© The Author(s), under exclusive license to Springer Nature Singapore Pte Ltd. 2025
S. Das et al. (eds.), *Human-Centered Digitalization, Volume 3*, Design Science and Innovation, https://doi.org/10.1007/978-981-96-4383-7_8

vibrancy of Ajrakh prints but also underscore the Khatri community's commitment to sustainable and eco-friendly practices in their craft.

Assam is renowned for its traditional practice of harnessing naturally extracted dyes sourced from plant stems, flowers, and roots. These exquisite dyes are employed in the art of dyeing handcrafted textiles woven from a variety of fine materials such as cotton, silk, and woolen yarns [4]. Similarly, the painted textiles of India, particularly Kalamkari, are renowned for their use of natural dyes and mordants in crafting intricate designs on fabrics. This practice not only preserves the artistic heritage but also demonstrates a deep commitment to environmental eco-sensitivity [5]. Kantha embroidery, a traditional textile craft, serves as an emblematic illustration of sustainable textile upcycling. This art form employs the running stitch to seamlessly amalgamate aged, weathered, and used saris and fabric remnants, ingeniously repurposing them into quilts [6].

The Kashmir Valley produces a diverse range of sustainable rugs is skillfully crafted, including the renowned Namda and Gabba carpets. These carpets represent a cost-effective alternative to traditional rugs, yet they maintain their distinctive charm and utility. Originally born from the innovative repurposing of discarded woolen clothing, these rugs are meticulously fashioned by sewing together these materials to form durable and practical floor mats. This unique woolen felt base is further adorned with intricate chain stitch embroidery, enhancing both their aesthetic appeal and functionality [7]. A similar variety of rugs is also prevalent in the region of Rajasthan, where coarse wool that might not be suitable for apparel finds new purpose. In the Tonk village of Rajasthan, artisans skillfully embroider these rugs with intricate floral patterns. As the demand for these rugs continues to grow, artisans have innovatively added value through techniques such as appliqué, hand embroidery, or machine embroidery, enhancing both their aesthetic allure and market appeal [8]. India boasts a rich tapestry of crafts that embrace the use of organic materials and champion the principles of recycling and reusing textiles and resources. These sustainable practices are not confined solely to the industrial sector but are also deeply ingrained in households across the nation [9]. The textile waste is divided into three main categories namely pre-consumer waste which is the waste that is created due to the surplus orders, sales, the garments that could not be sold. This is a new category of waste that is created by fast fashion products. The second category of waste is the industrial waste which is created by the production units in cutting and sampling department. The third category of waste is the post-consumer waste that is created by the consumers as discarded garments, bedsheets, and other household items. The waste textile material received of all kinds is majorly processed in Panipat where the textile is shredded in fibers and blanket are made which are exported to countries like Africa and distributed in relief camps [10]. In parts of Haryana waste fabric is cut in strips and used to make Panja Dari which got its name from the claw-like tool used in weaving process. The durries adorn geometric patterns and motifs of human and birds [11]. Furthermore, Rajasthan and Madhya Pradesh also produce Panja Dari from cotton, jute, and silk yarns [12] Most of the post-consumer waste is collected from household in exchange of steel utensil and sold in markets as second hand clothing [9]. Coming to Uttarakhand recycled rugs "Chindi Dari" (as shown in

Fig. 1 Original Chindi Dari made with thick fabric yarns

Fig. 1) make use of the worn-out clothes, old saris, and stoles to produce bedsheets, bedspreads, and even floor mats [13]. The fabric waste is increasing tremendously and ending into the landfill, most of the waste produced can be recycled but cannot be recycled due to lack of resources. Therefore, recycling needs to be introduced at every state and village level so that the waste is reduced and creates employment opportunity for the people. It is found that this repurposed fabric is highly used in making interior products. With the help of this paper, we examine the same fabric can be utilized for garments. This paper further discusses the methodology to examine Chindi Dari as a wearable textile, which leads to the design process in manufacturing the textile and finally evaluation of the textile.

2 Methodology

To achieve the specified objectives, a mixed methods approach was utilized. The first step was acquiring data from a range of primary and secondary sources. To gather primary data, a series of visits were organized in the district of Dehradun to observe the weaving methods employed in the production of Chindi Dari. The literature review for secondary data was conducted using the Science Direct and Web of Science databases. The data collection process encompassed interviews, group discussions, and telephonic interviews. The process of weaving was recorded in the village of Timli, located in the Dehradun region. Voluntary response non-probability sampling was done to test the wearability of the garments. The garments constructed were given to 15 volunteers. The female volunteers were selected from the courses conducted in UPES within an age group of 18–35 years ($n = 15$: mean age: 23.8 ± 3.95, 100% female). The volunteers were later asked to fill a feedback form about the fabric, design, and feel of the garment. This is further evaluated and described.

Fig. 2 Spinning of yarn with charkha on bobbin

3 Design Process

3.1 Preparation of Raw Material

The weavers inventory served as the source for the raw materials. The author also sourced some of the raw materials, which consisted primarily of scraps of sari and dupatta fabric. This raw material was segregated by color and fiber type. Embroidery of any type was discouraged since it slowed down the process of preparing the yarn tearing process. Fabric options mostly consisted of polyester georgette, cotton, and chiffon. The fabric was cut an inch wide through slash and tear method along the longest possible length. The border was taken off the saris and stoles as that could not be woven along with the other yarn. Bobbins were filled with this yarn once it was wound into a ball. Using a charkha (Fig. 2) for the yarn spinning process [14], the cloth ball was methodically rolled onto a plastic bobbin. To speed up the weaving process, multiple bobbins were loaded and arranged near the loom. In this case, the weaver chooses the palette, keeping the complementary colors in mind.

3.2 Production of Fabric

The dobby loom [15] is prepared by arranging the warp yarn, while the bobbin is inserted into the shuttle to serve as the weft yarn. The shuttle's back-and-forth movement results in the creation of a plain weave fabric (Fig. 3). The weaver must take into consideration any occurrence of thread breakage in the loom, since it has the potential to result in defects in the cloth. The process of weaving 3 m of fabric requires a minimum time of 3 h. The width of the fabric is 1.5 m.

Fig. 3 Weaving process on loom

3.3 Cutting

Upon receiving the fabric from the loom, a quality inspection is conducted on the fabric. The fabric is divided into predetermined patterns by the process of cutting. The designs have been deliberately maintained in classic style for the purpose of facilitating a more comprehensive evaluation of the finished garment. An extra allowance is maintained in the cutting process due to unraveling of the fabric.

3.4 Garment Construction

During the cutting phase of the garment, the edges were promptly overlocked to secure any fraying edges, after which they were sent for subsequent construction. The single-needle stitch is employed for the purpose of creating seams, which are subsequently reinforced with topstitching. To achieve a polished appearance, the hem is skillfully adorned with a bias fabric tape.

4 Evaluation of the Garments

Based on the questionnaire, it was found that the participants are aware about the base fabric (Fig. 4) that is used in making the garments as 60% of the participants identified it as "recycled handmade fabric" (Fig. 5). Although many recognized it as "handmade fabric."

The selection of raw materials was dependent upon the weaver's access to the raw material, the color combinations chosen by the artist are widely admired by individuals of various age groups. The silhouette of the garment was liked by 86.7% respondents (Fig. 6).

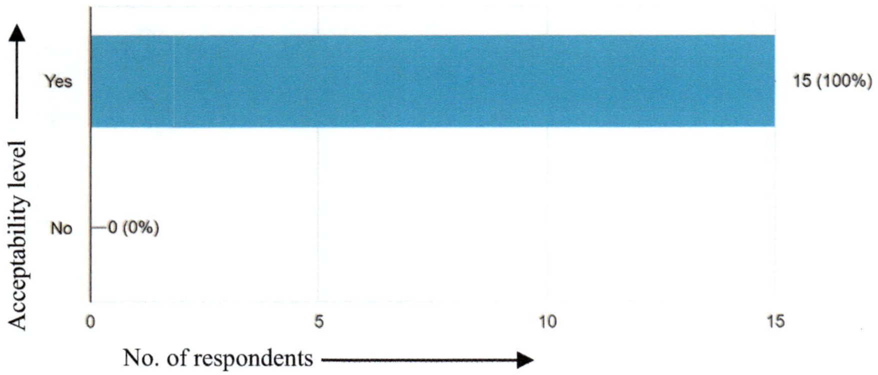

Fig. 4 Identification of Chindi Dari fabric

Fig. 5 Type of fabric as identified by the respondents

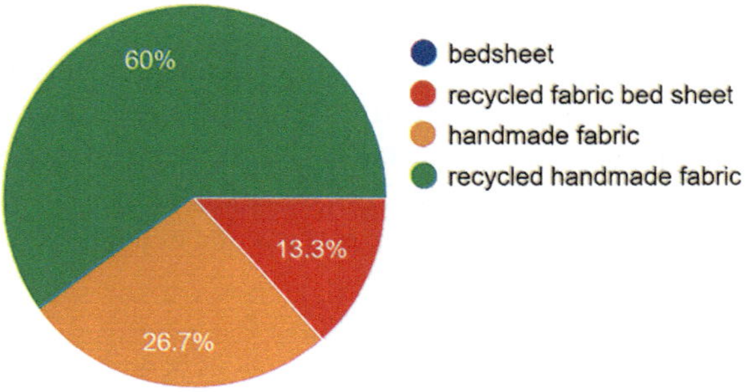

Fig. 6 Silhouette appreciation by the respondents

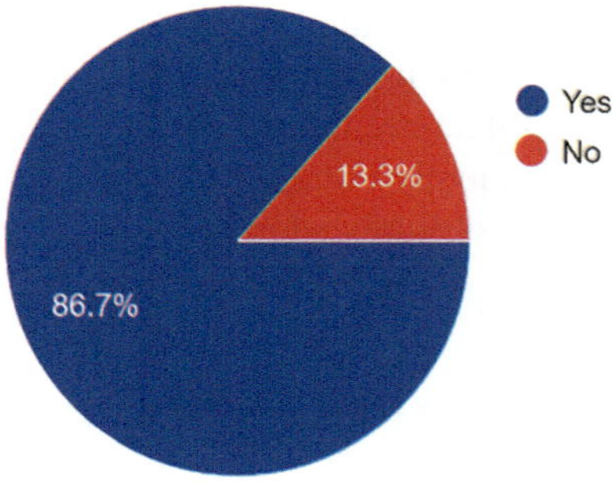

The primary raw materials employed in the production of the textile were predominantly synthetic fibers such as polyester chiffon and georgettes. Due to the coarse texture of the fabric, it was partially accepted by the participants in terms of its tactile qualities. Consequently, the level of comfort experienced was impacted. The feedback received was influenced by the type of item worn, with the jacket being perceived as more comfortable in comparison with the skirt and dress (Fig. 7).

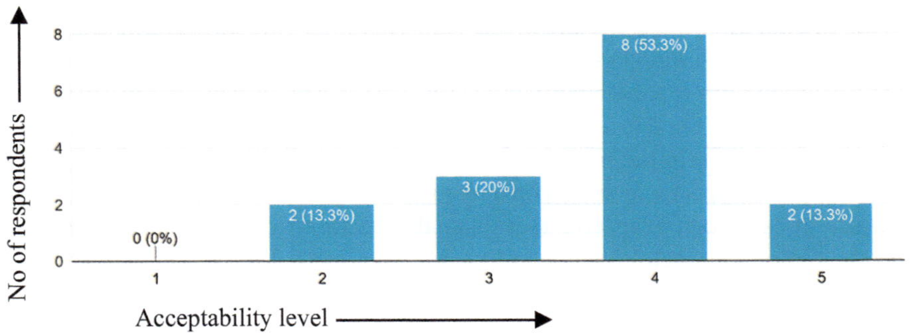

Fig. 7 Level of acceptability based on comfort

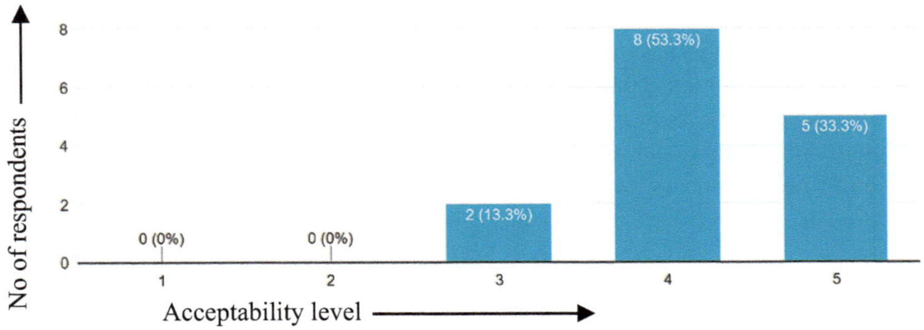

Fig. 8 Level of acceptability based on durability

When it comes to its visual appeal, the garment's design has been unanimously praised at 80% for its style and 86% for its silhouette, earning it the label of "classic" that will never go out of style.

The durability of the garment was highly appreciated by the participants (Fig. 8) which is also an important factor for consumers to buy a product ([16]). The fabric of the garment was widely recognized and valued for its distinctive qualities. The commendable attributes of the fabric included its sustainability, durability, and conformity with current fashion trends.

5 Conclusion

The garments produced from the fabric were highly appreciated by the participants. It was identified as a sustainable product which was unique and highly durable. The emphasis on the garment's comfort level was centered on its silhouette, as the individual who wore the sleeveless jacket did not express any dissatisfaction regarding its texture and tactile experience. The individuals who wore the skirt reported that it possessed a substantial thickness and suggested it as suitable attire for the winter season. Individuals who wore the bodycon dress reported experiencing discomfort when wearing ti for extended periods of time, despite the presence of a lining within the dress. It is suggested to avoid direct contact with the skin for those who have skin irritation issues. From the literature review we found that garment fit and comfort is an important aspect which are considered during the purchase process [17] which also serves as the one of criteria to evaluate the wearability of a garment. Therefore, the most important factor being comfort was examined. Another study reveals that quality is the most important factor followed by comfort, cost, and brand factor [16]. From the survey we found that the garments produced were durable. It is one of kind in the market which has classic styles and some of the silhouettes were highly comfortable. Although it is suggested to avoid silhouettes that are body hugging as it may cause irritation to the wearer. The fabric that was manufactured exhibited a lower density and thickness in comparison with the originally produced Chindi Dari. Therefore, it is suggested to be worn in winters. The outfits' color scheme received significant appreciation. The study was limited to the use of nylon warp threads in the weaving process, as these materials were only available at the weaver's workplace. If cotton strands had been used as the warp yarn, the fabric's behavior during the clothing-making process might have been different.

6 Discussion

The utilization of Chindi Dari, a formerly prominent component of conventional floor coverings, has progressively declined in prominence because of the emergence of alternative materials within the realm of floor covering. To revive Chindi Dari, this article examines its possible application in the field of trendy apparel and conducts a thorough evaluation of its viability as such. By doing thorough analysis, our objective of determining the feasibility of this fashion product is accomplished. This investigation provides insights into its practicality, visual appeal, and general acceptability in the realm of modern fashion. The cloth demonstrates enhanced suitability for colder winter weather conditions due to its considerable thickness. The reduction of fabric width can be achieved through the utilization of cotton fabric strips as the principal raw material. This alteration will improve the tactile comfort of the fabric, making it more appropriate for hot summer environments. Potential areas of focus for future academics could involve the exploration of strategies to enhance the scalability of garment manufacturing within the industrial sector for mass production. Moreover, it is recommended to do additional experiments into diverse weaving processes to enhance the adaptability and excellence of these fabrics. This article demonstrates our dedication to advancing sustainability as a communal endeavor within households, with the objective of reducing landfill waste as "Sustainability is not an end goal, it's a journey"—Eileen Fisher.

References

1. Meena B, Chitra A (2014) A Glorious Journey of Roghan Painting from Kutch, India [Online]. Available: www.isca.me
2. Karolia A, Buch H (2008) Ajarkh, the resist printed fabric of Gujarat
3. Batham M (2014) Bagru, the Traditional Hand Printed Textile of Jaipur [Online]. Available: www.indiancraftsmen.com
4. Kar A, Borthakur SK (2008) Dye yielding plants of Assam for dyeing handloom textile products
5. Divakala M, Muthian V (2017) Temple cloth to textile craft: the progression of Kalamkari (Vraatapani) of Srikalahasti. Craft Res 8(1):79–99. https://doi.org/10.1386/crre.8.1.79_1
6. Mandal R, Ornamental surface of Nakshi Kantha. An Int Refereed Res J 4(7)
7. Akhter M (2017) The importance of Kashmiri art and craft [Online]. Available: www.academicsjournal.com
8. Karolia A, Sardiwal S (2014) Namda-The traditional felted craft of Rajasthan
9. Jain P, Gupta C (2016) Textile recycling practices in India: a review, pp 21–36 [Online]. Available: www.tjprc.org
10. Sumita PB (2017) Exploring the potential of Chindi (textile waste) for producing quality products for women empowerment. Peer Rev J, Refereed Journal, Indexed Journal ISSN Number, 3(3):2395–7476 [Online]. Available: www.homesciencejournal.com
11. Singh S, Rani J (2021) Textile & Leather Review Traditional Indian Textile Techniques Used to Upcycle and Recycle Textile Waste Traditional Indian Textile Techniques Used to Upcycle and Recycle Textile Waste. Online Journal homepage: www.tlr-journal.com Journal, pp 2623–6281. https://doi.org/10.31881/TLR
12. PT & Sinha Meghna IK (2022) Caraivéti Démarche de sagesse Peer Reviewed and Referred Biannual International Journal 5(2):63–73 [Online]. Available: www.caraiveti.com
13. Tolia P (2018) the crafts of Uttarakhand 1:300–303
14. Mehta R, Sharif S (2019) A study on wool weaving & trade of Harsil-Uttarakhand; looming into promoting livelihood [Online]. Available: www.tjprc.org
15. Stankard S (2015) Yarn to fabric: weaving. In: Textiles and fashion: materials, design and technology, Elsevier Inc., pp 255–287. https://doi.org/10.1016/B978-1-84569-931-4.00011-8
16. Pandit P, Nadathur GT, Jose S (2018) Upcycled and low-cost sustainable business for value-added textiles and fashion. In: Circular economy in textiles and apparel: processing, manufacturing, and design, Elsevier, pp 95–122. https://doi.org/10.1016/B978-0-08-102630-4.00005-4
17. Kaplan S, Ays A, Okur A (2008) The meaning and importance of clothing comfort: a case study for Turkey

Enhancing AFC Gates and Implementing a User-Centric Navigation System for Ahmedabad Metro Stations

Riddhi Katariya, Sanika Deshpande, Amaan Dola, Devanshi Pandya, Swadha Mahajan, and Agnivesh Sharma

1 Introduction

The Ahmedabad Metro Project is a major infrastructure development initiative aimed at improving public transport in the city of Ahmedabad in the state of Gujarat, India. The strategy is designed to meet the growing transportation needs of the city's growing population while reducing traffic congestion. The governments of Gujarat and India established the Metro Link Express for Gandhinagar and Ahmedabad (MEGA) Company Limited as a special purpose body to implement the Ahmedabad Metro. The initiative aims to provide a reliable, effective, and eco-friendly mode of transport to the citizens of Ahmedabad. The subway network will consist of two corridors: the north–south corridor and the east–west corridor. The north–south corridor will stretch for about 18.87 kms and connect APMC Vasna with Motera Stadium. The east–west corridor will cover around 20.73 kms of distance and connect Thaltej Gam with Vastral Gam. These corridors will intersect at Ashram Road area

R. Katariya · S. Deshpande · A. Dola · D. Pandya · S. Mahajan
Karnavati University, Gandhinagar, Gujarat, India
e-mail: ua2125501@karnavatiuniversity.edu.in

S. Deshpande
e-mail: ua2125553@karnavatiuniversity.edu.in

A. Dola
e-mail: ua2125054@karnavatiuniversity.edu.in

D. Pandya
e-mail: ua2125160@karnavatiuniversity.edu.in

S. Mahajan
e-mail: ua2125652@karnavatiuniversity.edu.in

A. Sharma (✉)
Indian Institute of Technology, Bombay, India
e-mail: agnivesh@iitb.ac.in

© The Author(s), under exclusive license to Springer Nature Singapore Pte Ltd. 2025
S. Das et al. (eds.), *Human-Centered Digitalization, Volume 3*, Design Science and Innovation, https://doi.org/10.1007/978-981-96-4383-7_9

and form an interchange station (See Fig. 2). Depending on the terrain and urban environment, the Ahmedabad Metro will have elevated and underground sections and will run on standard gauge tracks. The metro system would utilize state-of-the-art rolling stock with facilities for passenger comfort and superior safety features. The aim of the project is to improve transportation options for locals, visitors, and tourists by reducing travel times and improving connectivity within Ahmedabad and neighbouring areas. As a result, a more sustainable urban environment should be promoted by dramatically reducing traffic congestion, air pollution, and carbon emissions [1].

Automatic Fare Collection (AFC) using smart cards installed at the entry and exit points of subway stations in various countries that provide public transportation services allows urban researchers to collect all passenger data, including their boarding and alighting times [2]. The latest ridership data for specific stations is not readily shared by metro authorities, making it challenging to collect this data in developing nations [3]. Furthermore, the statistics from smart cards cannot reveal the form of transportation that a transit user frequently uses or the difficulties they have while trying to enter the station where they frequently board. Analysing the Ahmedabad Metro Station is crucial for maximizing transportation effectiveness, directing urban planning and development, improving passenger experience and safety, assessing economic impacts, and promoting environmental sustainability.

2 Literature Review

The AFC improvement is relevant to understanding the importance of the AFC Gate globally, the AFC Gate (Automatic Fare Collection) plays an important role in the metro system worldwide, contributing to the efficiency, safety, and overall effectiveness of urban public transport networks. AFC gates automate the fare collection process and allow passengers to enter and exit subway stations quickly and conveniently. This reduces long queues. And ticket counters minimize waiting time, resulting in a smoother and more efficient passenger flow. AFC gates are critical for accurate revenue collection. They track and log the boarding and disembarking of each passenger, ensuring that the appropriate fare is deducted and collected. This helps metro authorities maintain financial sustainability and plan future investments. Automation through AFC gates reduce the need for manual ticket inspection and fair collection, resulting in lower labour costs for subway operators. It also minimizes the risk of human error associated with manual ticket creation. The system generates valuable data on passenger flows, ticket sales, and usage patterns. This data can be analysed to optimize service planning, station operations, and resource allocation, resulting in a more responsive and customer-focused subway system. Many modern AFC gate systems support contactless payment methods. This encourages the use of cashless transactions. The system can be equipped with accessibility features such as wider gates for passengers with disabilities, wheelchair ramps, and textile flooring to ensure subway stations are inclusive and accommodating to all passengers.

Understanding relevance and reliability also through visualization of AFC data: Shanghai Metro, the article explores the use of AFC data to gain insights into passenger flow characteristics and assess the reliability of travel times within the Shanghai subway system. It uses visualization techniques to analyse this data and provides valuable information to improve the efficiency and effectiveness of the subway system. The article addresses a relevant problem in the urban transportation system, namely understanding passenger flow and travel time reliability. Using AFC data for analysis is a unique approach because this data is readily available and provides real-time information about passenger movement. The use of visualization is commendable, as it can make complex data more accessible and help communicate results effectively. Focusing on the Shanghai Metro as a case study allows the findings to be applied to a real-world scenario and may provide insights that can be transferred to other subway stations [4].

A study of optimization schemes in subway stations using arbitrary logic, this article is about how to reduce passenger crowds in subway stations. The study considered three optimization schemes and simulated them using any logic software. The results show that, compared to corner optimization and escalator expansion measures, widening the existing stairwell in the station is the better choice to reduce the rush when changing trains. The study also made two suggestions for the design of new subway stations: use round corners if possible, and only use stairs if the lounge areas in front of the stairs or escalator zones are small.

3 Objective

The objective of a study for enhancing the navigation system and redesigning the Automatic Fare Collection (AFC) gate at an Ahmedabad Metro Station is to create a more efficient, passenger-friendly, and secure environment within the station. This study aims to improve the passenger experience by reducing congestion, enhancing safety, promoting accessibility, and incorporating modern technology.

Considering the case study of Ahmedabad Metro Station Phases I and II, the objectives for the present research are formulated as given below:

1. **Improve Passenger Experience**: Enhance the overall passenger experience by reducing congestion and waiting times and providing a user-friendly environment.
2. **Efficiency and Flow Optimization**: Streamline passenger flow within the station to minimize bottlenecks and congestion, especially near the AFC gates.
3. **Enhanced Safety and Accessibility**: Ensure a safe and accessible station environment with clear evacuation routes, accessibility features, and security measures.
4. **Wayfinding and Signage**: Enhance wayfinding with clear and effective signage, maps, and information displays to help passengers find their way within the station easily.

Fig. 1 Methodology

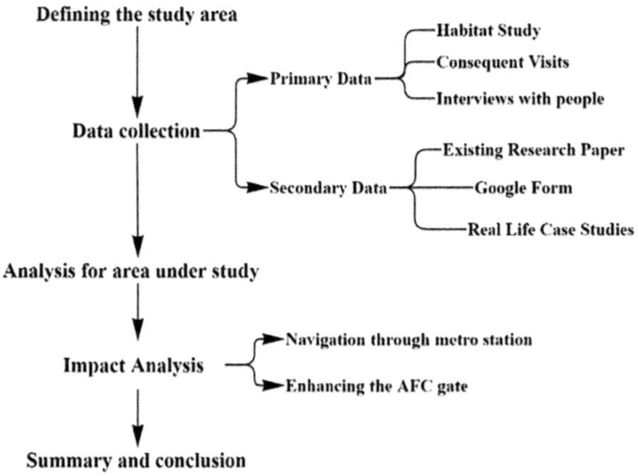

5. **User-Friendly Design**: Redesign AFC gates and navigation elements to be user-friendly, ensuring that passengers can easily understand and use the systems without confusion.

4 Methodology

The objectives of the study were set, and data have been collected from primary sources. Surveys have also been carried out to measure quantitatively and qualitatively how frequently people are mentally impaired while using the Automatic Fare Collection System, AFC, as well as on their journey through metro stations. These interviews gave us the opportunity to look at both individual experiences and daily challenges from people's point of view (Fig. 1).

5 Data Collection and Analysis

5.1 Defining the Study Area

The Ahmedabad Metro train system in Ahmedabad, Gujarat, India, includes the Ahmedabad Metro Station. The Ahmedabad Metro is an efficient means of travel for both residents and visitors. To make it easier for passengers to board and exit metro trains, a metro station is being built. It has been equipped with ticket stations, platforms, escalators, elevators, and additional facilities to provide passengers with an enjoyable and convenient journey. The metro station in Ahmedabad is located between APMC and Motera Stadium, part of the North South Line Phase 1, which begins at APMC. The line will cover Sabarmati, Gandhigram, the old High Court, and several important parts of Ahmedabad. Construction is now underway for Phase

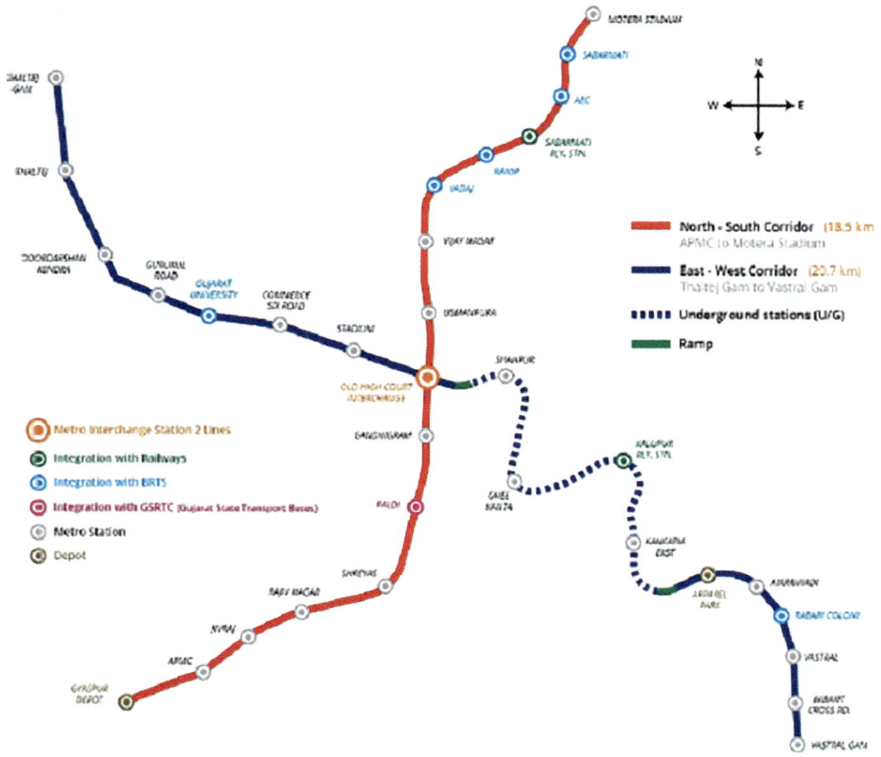

Fig. 2 Ahmedabad Metro routes

2 of the EastWest Line, which will connect Thaltej Gam and Vastral Gam. Changes to the precise route and stations can be made as construction progresses (see Fig. 2).

5.2 *Data Collection*

The metro system presents several challenges [problems, difficulties, and issues] for passengers that impact and influence their overall experience. These issues include difficulties [complexities, complications] in navigating ticket gates, language barriers and communication hindrances at ticket counters, and the struggle to find a seat during peak hours. To overcome these difficulties, it is necessary to provide real-time seat availability information, reorganize ticket office layouts, and provide better language support. Additionally, favouring smart cards over traditional tokens for payment options can significantly increase convenience and optimize the travel experience. However, the lack of a dedicated subway app increases the inconvenience for passengers, as it prevents them from receiving real-time information about train schedules and prevents them from topping up their smart cards online. The introduction of

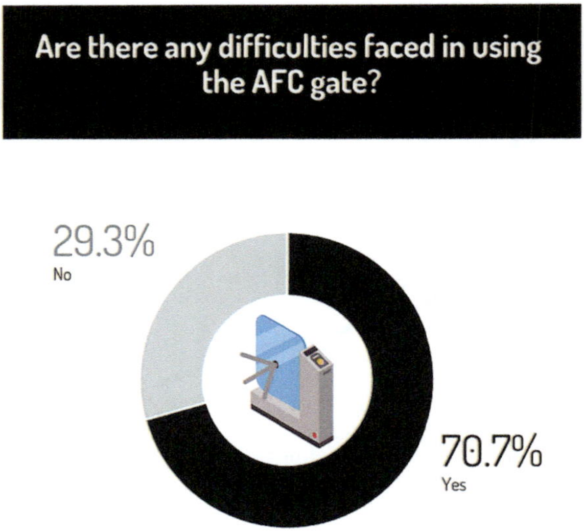

Fig. 3 Pie chart for the difficulties faced in using the AFC gate

an official Metro app could address these issues by providing a sleeker and more user-friendly system.

Additionally, passengers, especially those who are not tech-savvy, face challenges with ticket sales and card reloading machines. For some passengers, these machines may not be intuitive, and the requirement to have the exact fare may cause difficulties. To improve the experience, it is crucial to consider usability and accessibility in the design and operation of these machines. Finally, the design and clarity of signage in subway stations play a critical role in passengers' ability to navigate effectively. Inconsistent signage, complex floor plans, and other factors can make it difficult for passengers to find their way around. Metro authorities must prioritize clear, consistent, and easy-to-understand signage to improve the overall user experience, regardless of how familiar passengers are with the station (see Fig. 3).

In underground subway stations, the lack of reliable network connectivity is a common problem, hindering passengers who rely on their smartphones for various purposes. To address this issue, significant network infrastructure improvements are required to ensure a seamless and connected experience for all subway passengers. In summary, focusing on these aspects can improve the overall subway experience for all passengers and improve comfort, communication, and accessibility (see Fig. 4).

5.3 Analysis for Area Under Study

The data analysis of passengers' feedback reveals key problems in the Ahmedabad Metro system. These problems include difficulty in placing Automatic Fare Collection gates within the Automatic Fare Collection zone, communication difficulties at ticket counters caused by language barriers and loud windows, uneven availability

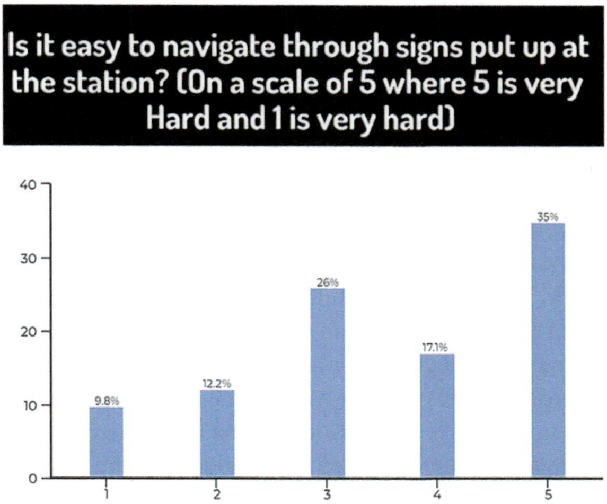

Fig. 4 Bar graph to understand how easy people find navigation in metro station

of spaces during periods of peak traffic, as well as preference for smart AFC cards over tokens. These problems include difficulty in placing Automatic Fare Collection gates within the Automatic Fare Collection zone, communication difficulties at ticket counters caused by language barriers and loud windows, uneven availability of spaces during periods of peak traffic, as well as preference for smart AFC cards over tokens. These issues include the difficulty of placing automated gates to collect fares in Automatic Fare Collection Zones, communication problems at ticketing establishments due to language barriers and loud windows, uneven availability of spaces over periods of high demand, as well as a preference for smart AFC cards instead of tokens. The investigation suggests several remedies to address these problems, e.g. the construction of an app on metros for real-time information and smart card recharging, optimization of gate positions at AFC facilities offering language comfort assistance, and enhanced passenger mobility in stations, compartments, or anywhere else. These proposed improvements aim to increase passenger convenience and overall efficiency within the metro system while considering cost effectiveness and continuous monitoring of performance.

5.4 *Impact Analysis*

Several key stakeholders have been identified in the impact analysis of the proposed improvements to the metro system, with passengers at the forefront. These enhancements are expected to make travelling more comfortable, reduce wait times, and provide a better metro experience for passengers. Nevertheless, to help passengers make the most of the changes successfully, efficient communication, and support strategies are essential.

The improvements are expected to bring significant benefits for metro authorities, such as an increase in revenues and more efficient management of traffic flows, which will make the metro system better organized and easier to manage. To ensure that these improvements remain viable for a longer period, however, they must be careful about the upfront investment costs and continued maintenance expenditures.

Businesses in these towns are to be affected both positively and negatively. While they might expect to see an increased number of visitors due to better metro accessibility soon, they need to be ready for a period of temporarily interrupted construction work. There are opportunities for suppliers to supply equipment, but they should be prepared to deal with potential disruption during installation. These improvements are expected to enhance revenue generation and increase safety by means of improved AFC gates, as well as improve passenger satisfaction. The implementation of a user-centric navigation scheme is intended to reduce traffic congestion, increase accessibility, and enhance the efficiency of the metro system in terms of overall operational excellence. Installation costs and ongoing maintenance costs are part of the project, which requires careful financial planning. However, the project is faced with several technical difficulties, including resistance to change, potential construction delays, and possible technical errors. To be successful in projects, a commitment to local or national legislation and the active involvement of the public and stakeholders are essential so that they have access to regular feedback collection and can address problems consistent with community and stakeholder expectations.

6 Result

Significant problems related to overcrowding, confusing signs, and a high cognitive ergonomic load have been identified in the research paper on improving metro stations in Ahmedabad. The study is likely to have identified existing station layout and signs as contributing factors to overcrowding, particularly during peak periods, which results in long waiting times and passenger dissatisfaction. Moreover, the study showed that confused graphics and an excessive number of screens create detrimental effects on cognitive ergonomics, which lead to a reduction in functionality for passengers. A thorough assessment of the research itself is needed to fully understand this paper's specific findings and propose solutions.

The lack of information on how the AFC system works is an important concern in the current design. The lack of guidance may lead to confusion among passengers who wish to benefit from the system. Due to a confusing user interface with unclear instructions, coin handling problems caused by a small coin slot and lack of sensor technology, and a lack of user education, travellers are experiencing cognitive difficulties at AFC gates. These factors all add to confusion and delays in fare collection. AFC machines now have animated displays, which make it easier for users to understand how to operate them. This is just one of several enhancements made to improve the user experience. Furthermore, in order to minimize any potential confusion for travellers, larger signage was placed directly in front of the device.

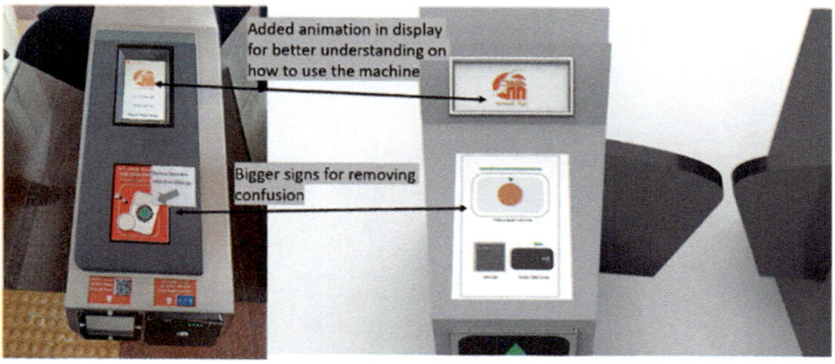

Fig. 5 Difference between the existing entry AFC gate and redesigned AFC gate

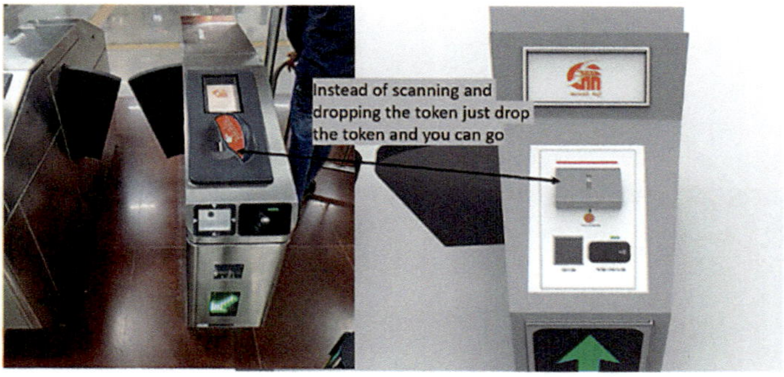

Fig. 6 Difference between the existing exit AFC gate and redesigned AFC gate

The coin dropping area at AFC gates has been improved with larger coin slots for convenience, sensor technology for precise coin recognition, instantaneous feedback through visual and auditory cues, improved user education with explicit instructions, and a public awareness campaign. These measures aim to streamline fare collection and enhance the traveller's experience (see Figs. 5 and 6).

The passengers are finding it difficult to find platforms due to the absence of floor path markings in this metro station, coupled with the challenge of small signs. This highlights the need to introduce a whole navigation system to enhance station accessibility and the user experience. In response to these challenges, a system of user-friendly navigation was introduced at the station so that passengers could enjoy a pleasant and seamless travel experience. To facilitate effective and smooth navigation, this system provides easily recognizable icons. To make the system efficient, it is necessary to carefully select icons that are universally understood (see Figs. 7, 8 and 9).

Fig. 7 Placement of signage on floor for navigation system for metro

Fig. 8 Placement of signage on floor for navigation system towards ticket counter

Fig. 9 Signage board for metro

7 Conclusion

The objective of this research project is to improve the Ahmedabad Metro system by addressing key problems relating to automated fare collection and AFC implementation to provide a user-centred navigation system. The proposal acknowledges that the metro system is plagued with problems, among others congestion, language barriers, and budget inefficiency, which have an impact on the user experience and overall efficiency. Two major solutions to these problems have been proposed by the research team. Two key solutions to these problems are suggested by the research team. To make it more accessible for all passengers, they propose first that the AFC's gates be upgraded with modern hardware and allow smart card payments. This change will seek to reduce traffic and improve the overall user experience. A user-focused navigational system is also proposed by the group to be installed in metro stations. This system would have clear signs, maps, and information displays that are strategically installed to ensure that passengers can be easily guided. To make navigation easy and user-friendly, the strategy also places a particular emphasis on concepts relating to brain ergonomics.

According to the impact analysis, these improvements would help passengers by making things more convenient and cutting down on wait times. Local companies and vendors may see growth potential, while metro authorities can anticipate increased revenue and better passenger flow management, although there may be brief delays during implementation. It is also necessary to engage actively with the public and stakeholders to address these challenges, as well as technological difficulties and opposition to change. Finally, a comprehensive solution to improve the operation of the Ahmedabad Metro is provided by this study project, which will make city transport more efficient, accessible, and user-friendly.

References

1. https://www.gujaratmetrorail.com/passenger-information/route-fares/
2. Yang X et al (2019) Recognizing the critical stations in urban rail networks: an analysis method based on the smart-card data. IEEE Intell Transp Syst Mag 11(1):29–35. Available at: https://doi.org/10.1109/MITS.2018.2884492
3. van Maarseveen M, Martinez J, Flacke J (eds) (2019) GIS in sustainable urban planning and management: a global perspective. Taylor & Francis. Available at: https://library.oapen.org/handle/20.500.12657/27516. Accessed: 2 Sept 2023
4. Sun Y, Jungang S, Schonfeld P (2016) Identifying passenger flow characteristics and evaluating travel time reliability by visualizing AFC data: a case study of Shanghai Metro. Public Transp 8. Available athttps://doi.org/10.1007/s12469-016-0137-8

Promoting Primary Healthcare Center Through Instructional Animation: A Case Study of Chengdi

Kayem Ali and Subash Rai

1 Introduction

The distribution of basic healthcare through Primary Health Centers (PHCs) is a cornerstone of India's healthcare system, ensuring that necessary healthcare is both accessible and affordable. It is impossible to overestimate the importance of PHCs in preserving the health and welfare of communities. However, a major obstacle—a pervasive lack of awareness—continues to hamper the effectiveness of these important healthcare facilities in reaching the rural residents of Chengdi and its surrounding villages in the Barpeta region of Assam.

The urgent issue of the rural population's poor awareness of PHC services in the Barpeta region of Assam are taken into account, especially in Chengdi and its environs. The primary goal of promoting healthcare access to underprivileged communities is seriously hampered by the lack of knowledge. Therefore, this study paper promotes the use of an original strategy—instructional animation—to raise rural inhabitants' understanding and knowledge of the variety of services provided by PHCs [1]. The primary goal of this working on is to close the knowledge gap that currently exists by using animation as a medium for spreading important information regarding healthcare benefits offered by the government. This research project aims to provide rural populations with information through careful implementation of animated content so they may decide on their healthcare needs with understanding.

A thorough study plan was implemented to support the viability and efficacy of this unique technique. This method included two major phases: a preliminary survey conducted by medical specialists with in-depth knowledge of Assam's healthcare system and outreach to the rural residents of Chengdi and the surrounding areas. The

K. Ali (✉) · S. Rai
Central Institute of Technology, Kokrajhar, Assam, India
e-mail: kayemali703@gmail.com

S. Rai
e-mail: s.rai@cit.ac.in

© The Author(s), under exclusive license to Springer Nature Singapore Pte Ltd. 2025

S. Das et al. (eds.), *Human-Centered Digitalization, Volume 3*, Design Science and Innovation, https://doi.org/10.1007/978-981-96-4383-7_10

purpose of these surveys was to identify the essential data required for the production of educational animation content. Subsequently, the research sought feedback from the target audience following the presentation of the instructional animation Chengdi has a total population of 1314 peoples [2], out of all the population researcher did 50 people survey in percentage 26.28% people.

The results of these surveys, which are discussed in more detail in the next sections of this research study, provide persuasive proof of the significant potential involved in raising healthcare awareness among rural communities. This research aims to dramatically improve the current healthcare environment in Assam and, by extension, the larger healthcare framework in India by leveraging the instructional animation's communicative ability. The ultimate goal is to promote more informed communities and more equitable healthcare access for all, which will enhance the general health and wellbeing of the population.

2 Literature Review

The objective of an instructional design model is to "...provide both an appropriate destination and the right road to get you there," or to "...assist educators in ensuring that they are teaching the appropriate material in an optimal manner." The ADDIE model is one example of an instructional design model. It's been used to create curriculum in across a range of fields, including as library training and online continuing education. This methodology's phases include analysis, design, development, implementation, and assessment [3].

The ADDIE model provides a valuable and straightforward basis for instructional design. Its systematic and adaptable nature allows for its application in various settings. This framework offers a structured approach for identifying the specific requirements of the intended audience and facilitating the incorporation of this knowledge into program design and development. Designers utilize the ADDIE model to gather essential data, including formative and summative assessments, to continually assess the program's performance during both its implementation phase and upon completion. Subsequently, developers can make necessary adjustments or take corrective actions to ensure the program's successful implementation. The evaluation phase plays a pivotal role in program execution, as it ensures that the program effectively addresses the needs of the target audience [4].

3 Background

The following data is collected to highlight the information's in order to gain a better perspective on the services offered by the government and the opinions of the general public:

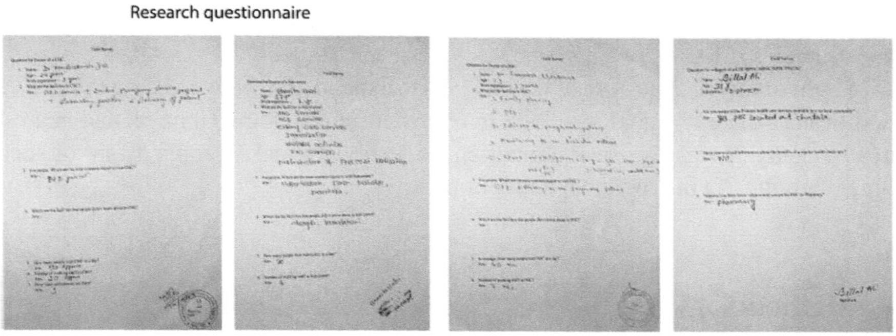

Fig. 1 (Research questionnaire)

A. **Primary Data**

50 persons were surveyed out of 1314 people, or 26.28% of the population. Out of 50 people, 39 decided to go to a pharmacy, in percentage 78% of them. 9 decide to go to PHC, in percentage 18% of them. 1 decide to go to CHC or 2% of them. And 1 decided to go to a sub-center. See Fig. 1 to view the field survey questionnaires.

B. **Secondary Data**

With a total of 252 households living there, Chengdi is a medium-sized hamlet in the Sarthebari Circle of the Barpeta district in Assam. According to the Population Census of 2011, there are 1314 people living in the Chengdi village, 684 of them are men and 630 women.

3.1 Instructional Animation to Increase Human Awareness

People frequently prefer to remember information from videos. The engagement in the videos will cause viewers to watch and watch and breakdown it. Understanding how to separate primary ideas or concepts from instructions so they may later be used as the foundation for additional processing is essential to building a coherent representation from them. How well learners focus their attention will determine how successful they are at this task. In overall, learners frequently lack the expertise necessary to differentiate between significant and irrelevant information. As a result, they run the danger of focusing on unimportant details or making erroneous assumptions. For example, learners who are inexperienced with the subject area may find it challenging to pick out the key ideas in a text or the right parts in an image. By emphasizing content that needs deliberate processing, instructional designers can direct the process of attention allocation. Thus, the primary purpose of instruction is to draw attention to specific information that is meant to demonstrate the significance of the triggered item. Regarding the information that is cued, this type of instruction is very precise as well as straightforward (Lorch 1989). To highlight the value of instructed information, for instance, color shifts in images or underlining in text are two strategies [5].

In light of this, the purpose of this study is to expand knowledge among the PHC's facilities. This study's goal is to look into the data surrounding Chengdi and nearest area by using 2D instructional animation, as well as to make people aware of such things through 2D instructional animation.

4 Methodology

4.1 Research Work-Flow

This research work-flow was created to show the steps takes when designing this 2D animation for the animated film "Instructional Animation for Increasing Primary Healthcare center Awareness" (Fig. 2).

Pre-production

At this phase, conducted interviews and a literature review to gather information about the subject of this study. And then perform an analysis by looking over and understanding the data that has been collected. Then, the process begin developing the idea for the plot and combining the concepts from the animation into a script, which will then be transformed into a storyboard. Character designs will also be created in accordance with the overall plot concept and additional assets.

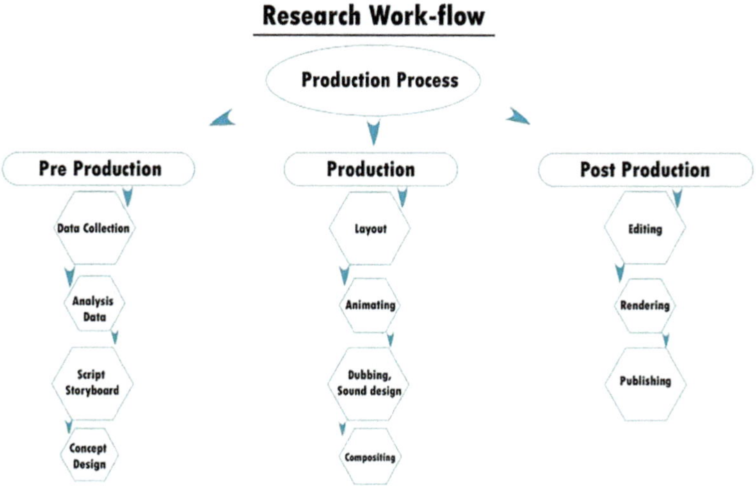

Fig. 2 Research work-flow

Production

At this point, starts by deciding how the animation will be laid out. After that, starts creating the animation using the develop program and a frame-by-frame method. After all the images have been integrated into one clip and have been animated, and adds sound effects with a calm and soft tone.

Post-production

A key phase that follows the actual animation development in 2D animation is post-production. Before the animation is prepared for distribution or viewing by the target audience, this phase includes a number of crucial tasks to polish and finished it.

At this point, begins altering the animated movie by changing the light and contrast as well as adding filters and effects. All the animation components will be rendered together once the editing process is finished. A 2D animation with a ratio of ~~00:00~~ will be the study's final output and be posted on the YouTube platform [6].

The ADDIE model is one of the models that is most frequently utilized in the field of instructional design as a guide for actual design, according to Thim-Mabrey (2006). It is a method that supports designers in producing useful teaching and design. In any environment, ADDIE model phases are applied. Every stage is interconnected and affects the others [7] (Fig. 3).

Figure 3 displays the cycle of the ADDIE model. It is still one of the most widely used Instructional System Design (ISD) models today and is employed by numerous significant organizations. Since it is common to refer to a methodical approach to teaching growth by this name, ADDIE does not strictly follow a linear progression through each level. The model's stages are divided into various sections of the process (Idris et al. 2018). Figure 4 displays the project development activities and output using the ADDIE paradigm (Fig. 4).

A. **Phase 1: Analysis**

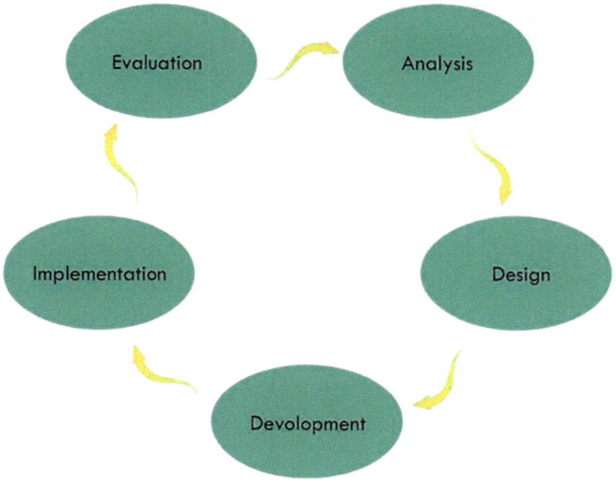

Fig. 3 ADDIE model cycle

Fig. 4 Framework of ADDIE model

This study identifies issues with awareness providing benefits of Primary Healthcare Center. As it will reveal the project's issue, the analysis phase is the project's starting and primary stage. The problem is identified during this phase, along with its cause and a possible solution (Bhushan 2006).

The present investigation also examines the value of instructional animated videos for consumers while in this analysis phase.

B. **Phase 2: Design**

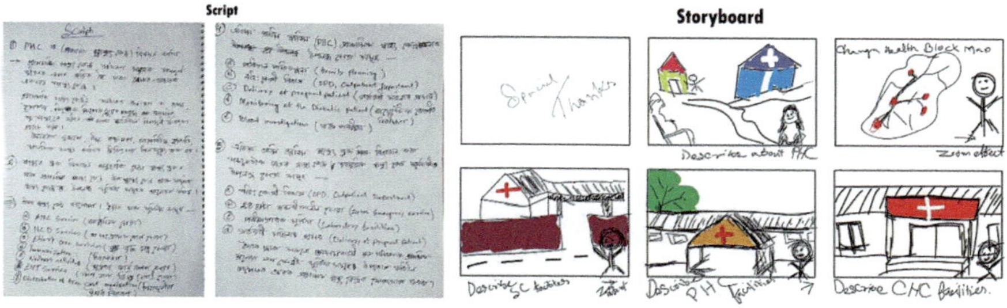

Fig. 5 (Script and storyboard)

The outputs from the analysis phase are used in the design phase to establish project development strategies. Then layout the plan for achieving the objective during the analysis phase during this phase of development. Flowcharts and storyboards for designing projects before the development phase are also covered in the current study. By using storyboards to describe how the project will look, it was designed to demonstrate the steps in the ongoing process. script and storyboards and scripts will help the project become more organized.

The script and storyboard drawing for the project development is shown in Fig. 5. The theme and concept for the project have been established by the study after constructing the screenplay and storyboard. In this project, the user interface design is concentrated on the visual design components of color, typography, and layout. The user experience design, however, is concerned with how the user feels when using the interface.

C. **Phase 3: Development**

In order to ensure a successful implementation, all project processes will be based on scripts and storyboards during the development phase. Based on what was discovered during the design process, designed criteria are created. Figure 6 depicts the project layout that was created using a 2D design.

D. **Phase 4: Implementation**

The implementation stage of the method is where everything is put together. To improve the finished animation, music, color grading, and even cutting or trimming certain video were used. The instructional animation video is examined several times at this stage to see if any modifications need to be made. Then, the project to the intended audience during the implementation phase and take feedback from the villagers.

E. **Phase 5: Evaluation**

A crucial stage toward the process's conclusion is the evaluation phase. In this project, I conduct a summative evaluation of the study utilizing a survey form and numerical analysis. The analysis phase of quantitative research methodologies has been completed to acquire findings by distributing questionnaires, and they are related to

Fig. 6 Project layout

quantification variables. By using a questionnaire survey to gather information, the assessment phase has been concentrating on audience awareness in this phase.

F. Phase 6: Conclusion

The government has been providing aid to the general public for a long time, but a lack of advertising has kept the majority clueless and in the dark, which has been remedied by this multi-lingual instructional animation video. As promised, this instructive animation video is capable of connecting the bridge between the government and the populace. Through this, the concept of connecting with the public who are less informed through audio and video media has been successfully established. In summary, the majority of respondents agree that the 2D animation, instructional animation video, and it has been successful in improving awareness of PHC (Benefits), especially in Chengdi, Assam, where it used to below.

References

1. Zakaria RS, Jaafar R (2022) Public service announcement (2D)-Insomnia (sleep disorder). J Comput Technol Creative Content (JTec) 7(1):48–54
2. https://villageinfo.in/assam/barpeta/sarthebari/chengdi.html
3. Cheung L (2016) Using the ADDIE model of instructional design to teach chest radiograph interpretation J Biomed Educ 2016:1–6
4. Peterson C (2003) Bringing ADDIE to life: instructional design at its best J Educ Multimedia Hypermedia 12(3):227–241
5. De Koning BB et al (2009) Towards a framework for attention cueing in instructional animations: guidelines for research and design. Educ Psychol Rev 21 (2009):113–140
6. Deli D, Winna W (2022) Short 2d animation for increasing depression awareness. In: CoMBInES-conference on management, business, innovation, education and social sciences, vol 2, no 1
7. Kamil MHFM, Abidin ISZ, Osman FF (2020) Design process: 2D interactive video animation development for cosmetic awareness among women. J Crit Rev 7(8):891–894
8. Idris S, Malim T, Ab W, Wan A, Development G, Jeli K (2018) Adaptation of ADDIE instructional model in developing an educational website for language learning. Global J Al-Thaqafah 8(2):7–16
9. Bhushan P, Bhushan P (2006) Module 4. Dermatology in a Week," pp 68–68
10. Thim-Mabrey C (2006) ADDIE model. 5(6) 2015. ADDIE model 107(3):361–373

Exploring the Efficacy of Cultural Documentaries in Informing and Preserving Heritage: A Case Study of Secret Shrines in Kamrup, Assam

Namrata Sarmah and Subash Rai

1 Introduction

Sacred shrines embody cultural, spiritual, and historical significance, maintaining traditions and beliefs. They bring together communities through worship and reflection, fostering a shared identity. With their architecture and artifacts, they retell ancient stories, preserving collective memory. Devotees find solace and connection, developing a deep link to their spiritual heritage. These shrines offer a tangible link to ancestral wisdom, anchoring communities in a profound continuity. With deep cultural significance, the sacred shrines in India attract tourists seeking spiritual solace and cultural immersion. Notable examples include Varanasi's ancient ghats along the Ganges, the Golden Temple in Amritsar, and the intricate carvings of Khajuraho temples. These sites offer a profound glimpse into India's rich religious heritage. This form of tourism also exerts a positive influence on local economies by driving up visitor numbers, bolstering ancillary sectors such as hospitality, transportation, and indigenous enterprises. During the research it was observed that the northeast state of Assam has notable numbers of holy shrines some of which are very popular among the visitors and many remained unknown due to lack of proper information and awareness. This paper endeavors to tackle the challenge of disseminating information effectively by conducting a documentary experiment focused on the Navagraha Temple situated in the Kamrup District of Assam, a location that has traditionally received less attention from visitors.

N. Sarmah (✉) · S. Rai
Department of Multimedia Communication and Design, Central Institute of Technology Kokrajhar, Balajan, Kokrajhar, BTR, Assam, India
e-mail: namratasarmah426@gmail.com

© The Author(s), under exclusive license to Springer Nature Singapore Pte Ltd. 2025

S. Das et al. (eds.), *Human-Centered Digitalization, Volume 3*, Design Science and Innovation, https://doi.org/10.1007/978-981-96-4383-7_11

2 Literature Review

Interestingly, despite having sufficient historic, archeological and artistic value, many famous holy shrines are still there without scientific protection either by central agency or by state authority. This is because of their traditional conservation and management system run by local management bodies such as board, trust, committee, etc., e.g. main temple of Goddess Kamakhya in Guwahati city [3]. Information about protected sites and monuments is disseminated through various publications by government and private agencies, including books, leaflets, and pamphlets. However, these resources are not widely accessible to the public, tourists, or devotees. Both central and state tourism departments promote these sites for their tourism potential, regardless of their protection status. Unfortunately, there is a scarcity of comprehensive documentaries on these sacred places, with many lacking the depth needed to provide a thorough understanding of all aspects, focusing solely on either the mythological and historical or the archeological and architectural elements. Some documentaries are focused on tourism potential of a monument only. In social media platform also (YouTube video, Facebook, Instagram reel, blog etc.), we found many such uploads, which are not crafted in a proper way and sometime with misleading information [2].

On the basis of their historical background, the following well-known holy shrines of undivided Kamrup District of Assam (presently Kamrup Metropolitan District and Kamrup Rural District) have been identified [3–5].

Sl. no.	Name of shrines	Era/year of inception	Location	Descriptions
01	Kamakhya Temple	1565	Nilachal Hills, Guwahati	The Kamakhya Temple at Nilachal hills in Guwahati, Assam is one of the oldest and most revered centers of Tantric practices having all India reputation, dedicated to the goddess Kamakhya
02	Ugratara Temple	1725	Uzanbazar, Guwahati	Ugratara Devalaya is a temple dedicated to Goddess Ugratara located in the western side of Jorpukhuri tanks in the heart of Guwahati city. Legend has said that the navel of Sati, first consort of Shiva, is related to this temple

(continued)

(continued)

Sl. no.	Name of shrines	Era/year of inception	Location	Descriptions
03	Umananda Temple	1694	Peacock Island, Guwahati	The Umananda temple, located on the Peacock Island, is one of the important temples of Guwahati. Situated in the middle of River Brahmaputra, this temple is dedicated to the worship of Lord Shiva
04	Navagraha Temple	1752	Chitrachal Hills, Guwahati	The Navagraha temple, located on Chitrachal hills, is devoted to the nine planets of the solar system. Each of the nine planets in this temple is represented by a Shivalinga inside the temple
05	Basistha Temple	1751	Sandhyachal Hills, Guwahati	Bashistha Temple, located in the Sandhychal hills at the southern corner of Guwahati city is an important pilgrimage center. Constructed in the second half of the eighteenth century, the temple was built by the Ahom king Rajeswar Singha
06	Sukreswar Temple	1744	Panbazar, Guwahati	The Sukreswar temple, located at Panbazar area of Guwahati city, is a temple dedicated to Lord Shiva. The temple is famous for housing the sixth Jyotirlinga of India
07	Chatraakar Devalaya	1785	Uzanbazar, Guwahati	It is built on top of a hillock at south bank of Brahmaputra River in Uzanbazar area of Guwahati city originally by Ahom King Kamaleswar Singha. As the dome of the hillock had similarities with an umbrella, the temple is named as Chatraakar Devalaya

(continued)

(continued)

Sl. no.	Name of shrines	Era/year of inception	Location	Descriptions
08	Ulubari Maajar	1960	Ulubari, Guwahati	In the heart of the Guwahati city Pir Hazarat Zahir Auliya Khuwajagan or popularly known as Ulubari Maajar is an embodiment of unwavering faith and religious harmony. This dargah is a reflection of strong religious harmony and Assam is a perfect example where different communities have lived together creating a strong cultural identity
09	Burha Jame Mosque	1850	Ambari, Guwahati	Situated at Ambari area of Guwahati, this is one of the oldest and largest mosques of Assam. Nobody exactly knows how the name Burha Masjid originated
10	Christ Church	1845	Panbazar, Guwahati	The Christ Church, which sits in the middle of Nehru Park opposite Cotton University, Panbazar, is a heritage site and an icon of Guwahati and northeastern India. People of all walks of life and faiths visit the church
11	Guwahati Baptist Church	1836	Panbazar, Guwahati	Located at the Panbazar Mission Compound, near the Mahendra Mohan Choudhury Hospital, Panbazar. It is the second oldest church of undivided Assam and first of Baptist sect under American Baptist Mission
12	Guwahati Sikh Gurudwara	1902	Fancybazar, Guwahati	Situated in the heart of Guwahati city Fancy Bazar, this gurudwara is also known as Sri Guru Singh Sabha and popularly as Sikh Temple. This is the oldest and one of the prominent gurudwara of Kamrup District

(continued)

(continued)

Sl. no.	Name of shrines	Era/year of inception	Location	Descriptions
13	Aswaklanta Temple	1720	North Guwahati Town	Located at Majgaon, North Guwahati at a distance of 24 km from city, AswaklantaTemple is an ancient Hindu temple on the banks of River Brahmaputra. It is one of the popular places of pilgrimage in Assam and among the best places to visit in and around Guwahati
14	Doul Govinda Temple	British Era (post 1826)	North Guwahati Town	Located at the north bank of the mighty river Brahmaputra. It is a place of historic importance. The temple and shrines, copper plates and rock inscription are scattered to suggest a glorious heritage in and around North Guwahati
15	Dirgheswari Devalaya	1737	North Guwahati Town	Another attraction of North Guwahati town, Dirgheswari Devalaya is located on the north bank of the Brahmaputra and linked by a motorable road. This shrine is considered as one of the supreme Devi Tirthas of the state
16	Manikarneswar Devalaya	1755	North Guwahati Town	Situated at the top of Manikhalaya hill in North Guwahati, Manikarneswar Devalaya is a temple dedicated to God Shiva. The present brick temple was built by Ahom King Rajeswar Singha in 1755 CE upon a star-shaped ground plan of a stone temple of the 10–eleventh century and made arrangements for regular worship at this temple

(continued)

(continued)

Sl. no.	Name of shrines	Era/year of inception	Location	Descriptions
17	Rudreswar Devalaya	1748	North Guwahati Town	The Rudreswar Temple or Devalaya is a temple dedicated to Lord Shiva in the village of Rudreswar, under North Guwahati town on northern bank of the river Brahmaputra. Built in 1748 CE by Ahom king Pramatta Singha, in memory of his father Rudra Singha, the temple is a fine example of a mixed style of Ahom-Mughal architecture
18	Madan-Kamdev Devalaya	Pala Dynasty (750–1161)	Baihata Chariali	Perhaps this is the oldest shrine of Northeast India excavated till date. Barely 40 kms away from Guwahati, and 3 kms from Baihata Chariali on National Highway 15, Madan Kamdev is an enigma, a mystery, a marvel which speaks about the ancient Kamrup
19	Hoyagrib-Madhava Temple	1583	Hajo	Hoyagrib-Madhava Temple is situated on the Monikut hill near Hajo of Kamrup Rural District. It is known that the temple was constructed by the King Raghu Deva Narayan in 1583. According to some other historians, king of Pala Dynasty constructed it in sixth century
20	Kedar Temple	1581	Hajo	Kedar Temple (also known as Kedareswara Temple) is a medieval Hindu temple dedicated to Lord Shiva located at top of Madanachala Hill in Hajo, Kamrup Rural, Assam. The central chamber has a shivlinga
21	Powa-Mecca Shrine	1897	Hajo	Powa Mecca is a place of pilgrimage for the Muslims and is located at Hajo town near Guwahati. It is the tomb of seer Giasuddin Auliya. It is believed that by offering prayers here the faithful gain one fourth of the spiritual enlightenment of what could be gained at Mecca

(continued)

(continued)

Sl. no.	Name of shrines	Era/year of inception	Location	Descriptions
22	Chandika Devalaya	1647	Chaygaon	Chandika Devalaya at Chaygaon, about 37 km to the west of Guwahati in Assam's Kamrup Rural district is one of the most important and popular Shaktipeeths in Assam

Following the creation of Nanook of the North, the initial documentary crafted by American filmmaker Robert Flaherty in 1922, the prominence of documentaries in illuminating and safeguarding vibrant cultural legacies becomes evident. Documentary films have been used as a historical source to construct and preserve visual chronicles of society, culture, and national identity also documentary films continue to be a valuable resource for understanding and interpreting historical events and cultures [6].

3 Background Study

The sacred Navagraha Temple, nestled on Chitrachal Hills of Guwahati, idolizes the nine celestial bodies of our solar system. Within its hallowed halls, each planet finds representation in a Shivalinga, cloaked in distinct hues. This unique embodiment serves as a tribute to their cosmic influence. The temple, selected for case study, holds intrigue due to its relative obscurity in public knowledge. Its spiritual significance and celestial symbolism make it a compelling subject for exploration, offering a glimpse into the intersection of faith and astronomy in this hidden gem atop the Chitrachal Hills.

4 Primary Research

During field visits, discussions were held with senior government officials, office bearers of shrine management committee, priests, bishops and imams, devotees, owners of business establishment in shrine premises and local citizens. Most of this discussion and interview were confined to the availability of cultural documentaries and allied audiovisual documents on the concerned shrines and its impact on the overall preservation and development of the same. The conversations brought to light the following points:

a. Insufficient availability of audiovisual material
b. Limited access to information regarding the shrines
c. Contemporary content creators propagating inaccurate information.

5 Methodology

For this case study, the following approach have been adopted.

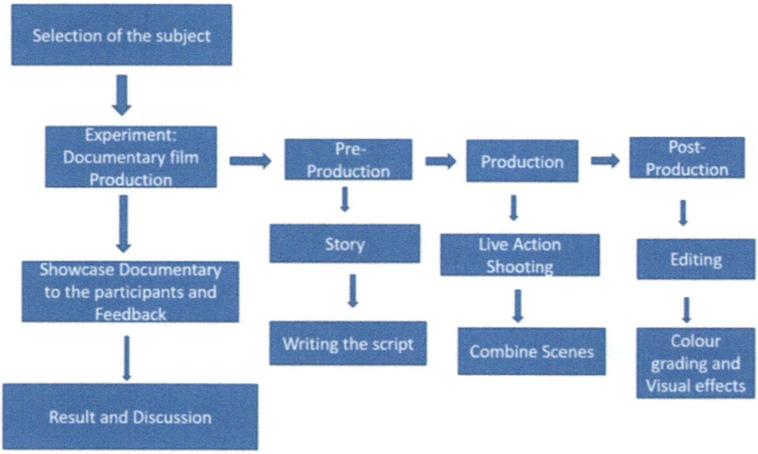

5.1 Selection of Subject for Experiment

A brief field visit was conducted before selecting the subject of experiment. For collection of authentic data and to take stock of current situation, the following shrines and places were visited and hold interview with various stakeholders from the view of information, preservation, and audiovisual documentation.

- Kamakhya Temple
- Ugratara Temple
- Navagraha Temple
- Basistha Temple
- Sukreswar Temple
- Chatraakar Devalaya
- Ulubari Maajar
- Burha Jame Mosque
- Christ Church
- Guwahati Sikh Gurudwara
- Aswaklanta Temple
- Doul Govinda Temple
- Madan-Kamdev Devalaya
- Hoyagrib-Madhava Temple
- Kedar Temple
- Powa-Mecca Shrine

Fig. 1 From left to right–Kamakhya Temple, ASI Guwahati Circle Office, Directorate of Archeology, Assam

- Directorate of Archeology, Assam
- Archeological Survey of India, Guwahati Circle
- Department of Archeology, Cotton University
- Archive of Assam State Museum.

Picture of a few sites from the visit is given below (Fig. 1).

5.2 *Experiment: Documentary Production on Navagraha Temple in Three Distinct Stages I.E. Pre-production, Production, and Post-production*

Documentary centered on the revered Navagraha Temple in Guwahati city titled **'The Echo of Pragjyotishpur'** was crafted with the following stages:

5.2.1 Pre-production Stage

In the initial stages of documentary filmmaking, comprehensive research, meticulous planning, and thorough organization were vital. This encompassed the conceptualization of ideas, crafting the script, financial planning, scouting locations, and obtaining essential permits. Furthermore, this phase encompassed the coordination of interviews, procurement of equipment, and the establishment of a production schedule to guarantee a seamless filming procedure (Fig. 2).

5.2.2 Production Stage

During this phase, we focused on meticulously capturing footage, conducting interviews, and assembling vital visual components. Maintaining seamless operations was of utmost importance, as the quality of shots and the richness of content are pivotal factors for a documentary's success (Fig. 3).

Fig. 2 Navagraha Temple

Fig. 3 Establishing shot of nine celestial bodies inside the temple

5.2.3 Post-production Stage

During this phase, the raw footages was meticulously edited, refined the sound quality, and carefully arranged visual elements. This crucial step breathed life into the story through precise editing, sound design, color grading, and visual effects. It marked a pivotal moment in refining and perfecting the documentary.

5.2.4 Feedback and Discussion

After presentation of the documentary to a randomly select 40 participants who were visitors to different shrines situated in Guwahati, the following questions were asked to know the efficacy of documentary film crafted for informing and preserving heritage.

Q1. Did you watch informative videos and documentary?
Q2. Were you able to understand the significance and history of the temple after watching the documentary?
Q3. Did the documentary effectively showcase the architectural and artistic features of the temple?
Q4. Were the visual elements (footage, images, and graphics) well-presented and relevant to the content?
Q5. Was the narration or voiceover clear and easy to follow?
Q6. Did the documentary cover all the essential aspects of the Shrine Navagraha Temple?
Q7. Were there any parts of the documentary that you found confusing or unclear?
Q8. Do you believe the documentary effectively disseminated proper information about the Shrine Navagraha Temple?

On review of the above questionnaire, it was found that 90% people watch informative videos and documentary (Q.1). 0.85% people understand the significance and history of the temple (Q.2).0.80% feel the documentary effectively showcases the architectural and artistic features of the temple (Q.3). 0.85% viewers agree visual elements were well-presented and relevant to the content (Q.4). 0.82.5% think that voiceover was clear and easy to follow(Q.5). 75% observes that documentary cover all the essential aspects of the Shrine (Q.6). 0.90%feel that documentary was clear to understand (Q.7). Finally, 85% feel that documentary effectively disseminated proper information about the Shrine Navagraha Temple (Q.8).

Graphical representation of the feedback received from the participants:

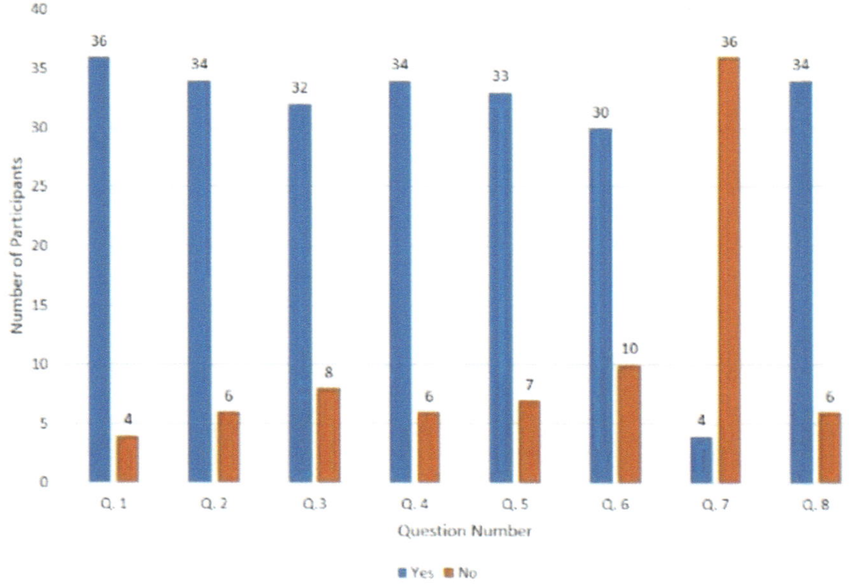

6 Conclusion

The study underscores the profound cultural and historical significance of sacred shrines, particularly in the Kamrup District of Assam. It revealed a noticeable lack of awareness among tourists regarding these revered sites. To address this gap, a carefully designed documentary focused on the esteemed Navagraha Temple in Guwahati was created. The positive feedback from participants affirms the documentary's success in not only educating audiences but also in preserving our rich cultural heritage. This research emphasizes the pivotal role cultural documentaries play in disseminating knowledge and safeguarding our cultural legacy.

References

1. Vivekananda Kendra Institute of Culture, Heritage of Kamakhya on the Nilachala Hills
2. Directorate of Archeology, Assam, 'Pragjyotishor Jilingoni' (a series of documentaries on Archeological Sites in and around Guwahati. https://assam.mygov.in/en/talk/ashwaklanta-temple-where-divinity-meets-history/)
3. Adhikary G (2001) A history of the temples of Kamrup and their management, Chandra Prakash, Guwahati
4. Sarma PC (n. d.) Holy Shrines of Assam
5. Barua SK (2020), Axomor Kirtisihna (Monuments of Assam), Axom Sahitya Sabha
6. Syska A (2021) Using documentary film as a historical source

Work Rest Cycle for Manual Material Handling in Rice Mills

B. Surya Kumar Chhetry and K. N. Dewangan

1 Introduction

The manual material handling (MMH) tasks in rice mills involves lifting and carrying of heavy load for long duration of time [1]. The mass of paddy bags varies from 30.5 to 56.5 kg and the mass of milled rice bags is either 25–50 kg. Three to eight workers perform MMH tasks in rice mills. Tasks in rice mills are categorized as moderate to heavy [2]. Owing to heavy workload, the MMH tasks in rice mills are performed by male workers only. Carrying heavy load results in early fatigue, however, long duration of heavy load carrying may cause low-back pain (LBP) [3, 4].

Lifting and material handling tasks are important risk factor for the development of work-related musculoskeletal diseases (WMSDs) [1, 3, 5, 6]. About 46.5% of rice mills workers of India experience WMSDs due to load carrying activities and WMSDs was correlated with duration of work [7]. In India, lifting-related occupational activities account for 63% of non-fatal injuries and 35% of fatal injuries in the workplace [8]. The rice mill worker experiences WMSDs on their neck, back, waist and buttocks [1].

The rice mill managers assign the tasks to the workers. The workers perform tasks one after the other. The rice mill operates from 8:30 am to 6 pm with a lunch break of 1 h from 12 noon to 1 pm. The rice mills are operated for 56 to 74 h a week. Due to long working hours, the worker may experience overexertion in performing the task. The workers get less resting time which can impair physical and mental capacities and limit job effectiveness [9]. Overexertion is the possible causes for injury to the workers [10–12]. About two thirds of all overexertion causes occupational injuries to the worker during MMH [12]. Studies on MMH have been

B. S. K. Chhetry · K. N. Dewangan (✉)
Department of Agricultural Engineering, North Eastern Regional Institute of Science and Technology, Nirjuli, Arunachal Pradesh, India
e-mail: kndewangan2001@yahoo.co.in

© The Author(s), under exclusive license to Springer Nature Singapore Pte Ltd. 2025

S. Das et al. (eds.), *Human-Centered Digitalization, Volume 3*, Design Science and Innovation, https://doi.org/10.1007/978-981-96-4383-7_12

done on the WMSDs, however, limited studies are available for interventions to reduce WMSDs and development of work rest cycle.

Work rest cycle of the rice mill workers in MMH tasks was determined in the present study. The heart rate (HR) responses of the workers in different MMH tasks were measured. The oxygen uptake (VO_2) of the workers was estimated and work rest cycle was developed using Murrell's equation.

2 Material and Methods

A total of eight commercial rice mills located in Lakhimpur and Dhemaji was selected for the study. The selected rice mills have an average milling capacity of 0.7 tons/hour. The seven tasks, namely unloading of paddy bags from the vehicle, assistive in unloading of paddy bags from the vehicle to the godown, assistive in unloading of paddy bags, handling of paddy bags from the godown to feeding section, assistive in handling of paddy bags, filling of rice bags and loading of rice bags in the vehicle and assistive in loading of rice bags were selected for the study. Three to four workers carry load and one or two assistive workers perform unloading of paddy bags from the vehicle to the godown, handling of paddy bags from godown to the feeding pit and loading of rice bags from rice mill to the vehicle. One worker holds the bag and other workers fill the bag with milled rice using container for filling of rice bags.

2.1 Measurements

Selection of Subjects

A total of three workers were selected for each task in every rice mill. The most important criteria for selecting workers were the absence of musculoskeletal disorder and at least one year of experience in load carrying. Workers of 18 years and older were considered for the study. The workers willing to participate in the study and fulfilling all the selection criteria were selected as subjects. Written consent was obtained from the workers willing to participate in the study. Approval from the Ethics Committee of the institute for the study was obtained. The mean age, stature, body mass and working experience of the selected workers were 36.7 years, 1644.5 mm, 53.6 kg and 9.58 years, respectively.

Heart Rate

The HR of the workers was measured during various MMH tasks using polar heart rate monitor (Model: H10, Make: Polar Electro Oy, Finland). Before experiment, 5–10 min rest was given to the subjects to obtain resting HR. The working HR of the subjects were measured for 30 min during load carrying. The maximum HR (HR_{max}) was calculated using Robergs and Landwehr equation [13]

$$HR_{max} = 220 - age \quad (1)$$

Oxygen Uptake

The measurement of VO_2 under actual field conditions requires sophisticated instrumentation as well as skilled technical support [14]. The prediction equation to estimate VO_2 consumed for the male Indian agricultural workers has been developed by Nag (1981) using working HR and given in Eq. 2 [15, 16]. This equation was used to calculate VO_2

$$y = 0.0183\text{HR} - 1.28 \quad (2)$$

where
y = oxygen uptake, l/min.

The VO_2 was estimated for all the selected subjects and for all the tasks. The VO_2 corresponding to HR_{max} was termed as VO_{2max}.

Energy Expenditure Rate

The energy expenditure rate (EER) of all the subjects and all the tasks was calculated using Eq. 3 [15]. The EER was calculated based on the volume of oxygen consumed.

$$EER = VO_2 \, \text{X} \, 20.88 \quad (3)$$

where
EER = energy expenditure rate, kJ/min.
VO_2 = volume of oxygen consumed, l/min.

Total Work Time

Total work time is the total period the worker performs the MMH task in the rice mill. The time-motion study was conducted on three workers for each task from every rice mill. The time-motion study was conducted by continuous observation through visual inspection using a digital stopwatch.

Work Rest Calculation

The total rest allowance during MMH task was calculated using Murrell's equation (Eq. 4) [17, 18].

$$R = \frac{T(E_{work} - E_{rec})}{(E_{work} - E_{rest})} \quad (4)$$

where R = resting time, min; T = total work time to complete the task, min; E_{work} = energy expenditure during the task; E_{rec} = recommended energy expenditure level and E_{rest} = energy expenditure during rest.

Bink recommended that an acceptable energy expenditure rate during a task should not exceed the energy expenditure corresponding to 33% of the VO_{2max} [19]. Saha et al. suggested that energy expenditure corresponding to 35% VO_{2max}, as the acceptable work rate [20]. In the present study, the E_{rec} corresponding to 35% VO_{2max} was calculated. The mean resting and mean working EER of each subject during different tasks were calculated to determine work rest schedule. VO_2 calculated from linear regression equation between HR and VO_2 was used to determine resting time.

3 Results and Discussion

3.1 Heart Rate of the Workers During MMH Tasks

The mean working HR during unloading of paddy bags from the vehicle to godown, assistive in unloading of paddy bags, handling of paddy bags from the godown to feeding section, assistive in handling of paddy bags, filling of rice bags, loading of 50 kg rice bags and assistive in loading of rice bags from rice mills to the vehicle varies from 109.9 bpm to 128.5 bpm. The minimum HR was observed during filling of rice bags and the maximum HR was observed during loading of 50 kg rice bags in the vehicle. The mean mass of paddy bags was 44.4 kg. Since the rice bags is heavier than the paddy bag therefore the heavy load carried by the subjects from the rice mills to the vehicle might be the reason for higher HR. During filling of rice bags, the subject lifts about 11 kg milled rice in a container from the floor thus the lower load may be the reason for relatively lower HR. According to Varghese et al. [21], the workload during MMH tasks in the rice mills can be classified as moderate to heavy. The value of VO_2 and EER during MMH tasks in rice mills varied from 0.95 to 1.17 l/min and 19.75 to 28.85 kJ/min, respectively.

3.2 Total Work Time

The total work time during different MMH tasks is presented in Fig. 1. It was observed that total work time varies from 11.2 to 152 min during unloading of the paddy bags from the vehicle to the godown. The total work time during unloading of paddy bags from the vehicle to the godown depends on the number of subjects for unloading paddy bags, the capacity of the vehicle and thus the number of paddy bags in the vehicle, the distance between the tailgate of the vehicle to the godown and the location of paddy bags stacking in the godown. Total work time during handling paddy bags from the godown to feeding pit varies from 91.8 to 163.5 min. The total work time during handling paddy bags from the godown to feeding pit depends on the number of bags to be milled, operation time of the rice mill and distance between the godown and feeding pit. Total work time during filling of milled rice in bags varies from 67.6 to 179.7 min. The number of bags to be filled depends on the quantity of rice milled and number of subjects employed. The total work time during loading of the rice

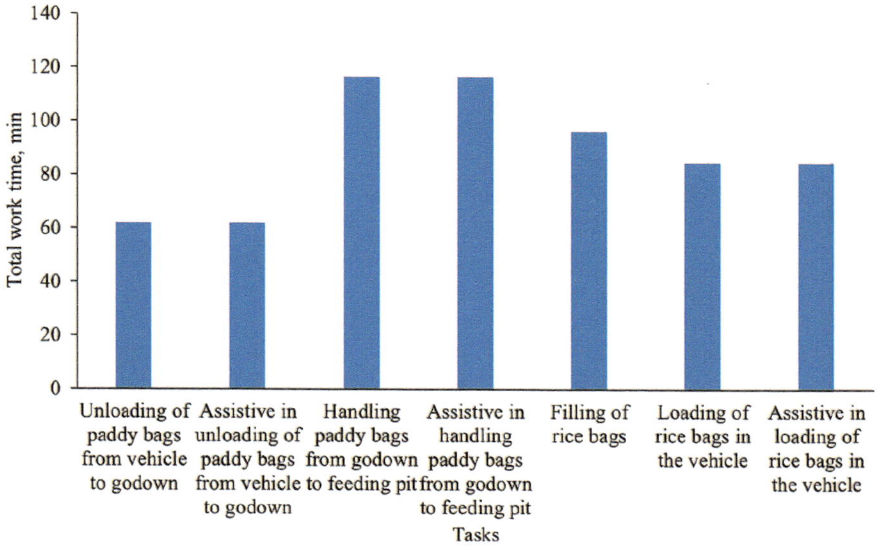

Fig. 1 Total working time for different tasks in rice mills

bags from the rice mills to the vehicle in varies from 30.4 to 164.9 min and it depends on the capacity of vehicle, number of subjects employed and distance between the godown to the vehicle. The total work time was relatively lower during unloading of the paddy bags from the vehicle to the godown, however, total work time was more during handling of paddy bags from the godown to feeding pit.

3.3 *Rest Allowance*

The rest allowance required by the subjects carrying the load and the assistive worker for seven different tasks is presented in Fig. 2. The rest allowance required by the subjects during unloading of paddy bags from the vehicle to godown varied from 3.06 to 69.86 min while for the assistive subjects for the unloading of the paddy bags varied from 4.79 to 44.36 min. The rest allowance required by the subjects during carrying paddy bag from the vehicle to godown and the assistive subjects during handling of paddy bags from the godown to feeding pit varied from 23.92 to 47.05 min and 23.35 to 47.05 min, respectively. The rest allowance required for filling of 50 kg rice bag varied from 8.74 to 44.78 min. The rest allowance required during carrying 50 kg rice bags from the rice mill to vehicle and their assistive subjects varied from 6.6 to 97.93 min and 12.98 to 60.49 min, respectively. The subjects carrying the load required higher amount of rest than their assistive subjects to the level of 9.11%, 2.54% and 7.01% during unloading of paddy bags from the vehicle

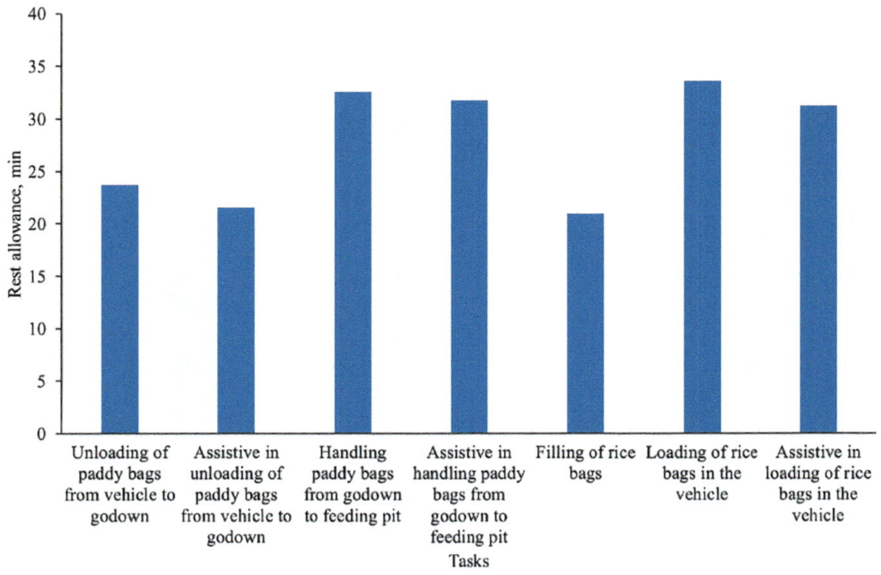

Fig. 2 Rest allowance for different MMT tasks in rice mill

to godown, handling of paddy bags from the godown to feeding pit and loading of 50 kg rice bags from the rice mills to vehicle, respectively.

The mean rest allowance required by the subjects carrying paddy bags from the vehicle to godown and their assistive subjects was 33% and 36.6% of the total work time, respectively. The mean rest allowance for the subjects carrying paddy bags from the godown to feeding pit and their assistive subjects was 29.2% and 26.8% of the total work time, respectively. The rest allowance required for filling of 50 kg rice bag was 21.3% of the total work time. Similarly, carrying 50 kg rice bags from the rice mill to vehicle and their assistive was 45.5% and 35.2% of total work time, respectively. The rest allowance for different tasks in rice mill varied from 10.46% to 61.1% of total work time.

4 Conclusion

The study was conducted to estimate the work rest cycle during MMH tasks in rice mills. Maximum working HR was observed during loading of rice bags to the vehicle. Work intensity during the MMH tasks falls under moderate to heavy workload. The mean working time was relatively higher during handling of the paddy bags from the godown to feeding pit. The rest allowance during different MMH tasks varies from 10.4 to 63.1% of total work time.

References

1. Ruliati LP, Adiputra N, Sutjana IDP, Sutajaya IM (2015) The ergonomics improvement on work conditions reducing fatigue and musculoskeletal disorders of rice milling workers in J village. In: Proceedings of the joint international conference on electric vehicular technology and industrial, mechanical, electrical and chemical engineering, ICEVT 2015 and IMECE. pp 124–127. https://doi.org/10.1109/ICEVTIMECE.2015.7496675.
2. Pradhan CK, Thakur S, Chowdhury AR (2007) Physiological and subjective assessment of food grain handling workers in West Godavari District. India Ind Health 45:165–169. https://doi.org/10.2486/indhealth.45.165
3. Waters TR, Putz-Anderson V, Garg A, Fine LJ (1993) Revised NIOSH equation for the design and evaluation of manual lifting tasks. Ergonomics 36:749–776. https://doi.org/10.1080/00140139308967940
4. Waters T, Yeung S, Genaidy A, Callaghan J, Barriera-Viruet H, Deddens J (2006) Cumulative spinal loading exposure methods for manual material handling tasks. Part 1: is cumulative spinal loading associated with lower back disorders? Theor Issues Ergon Sci 7:113–130. https://doi.org/10.1080/14639220500111392
5. Kim MH, Yoo WG (2015) Effect of a worktable position on head and shoulder posture and shoulder muscles in manual material handling. Work 51:289–292. https://doi.org/10.3233/WOR-141866
6. Njue RM (2019) Assessment of the level of awareness in occupational safety and health among rice mill workers in central Kenya (Mwea). Int J Innov Res Adv Stud 6:24–28
7. Darbastwar MA, Kumar B, Ravinder A (2016) A study of prevalence of musculoskeletal disorder among the rice mill workers in Karimnagar. J Evol Med Dent Sci 5:1106–1111
8. Girish N, Iqbal R, Khanzode V, De A (2015) Manual material handling and occupational low back disorders: a narrative literature review emphasising maximum acceptable weight of load. Int J Hum Factors Ergon 3:376–392. https://doi.org/10.1504/ijhfe.2015.073010
9. Anggraini YD, Mulyati GT, Ainuri M (2019) Analysis of workload and long rest periods on mobile rice grain milling operator at Sidolelono Pleret Community Bantul. J Phys Conf Ser 1367:012082
10. Nag P, Nag A, Vyas H, Shah P (2014) Occupation, participation, inclusion, work role Postural control in different load handling tasks. Work 49:577–584. https://doi.org/10.3233/WOR-131735
11. Mital A, Pennathur A, Kansal A (1999) Nonfatal occupational injuries in the United States part I–overall trends and data summaries. Int J Ind Ergon 25:109–129. https://doi.org/10.1016/S0169-8141(98)00102-4
12. Wu SP (2006) Psychophysically determined 1-h load carrying capacity of Chinese females. Int J Ind Ergon 36:891–899. https://doi.org/10.1016/j.ergon.2006.06.009
13. Robergs RA, Landwehr R (2002) The surprising history of the "HRmax=220-age" equation. J Exerc Physiol Online 5:1–10
14. Tiwari PS, Gite LP (2002) Physiological responses during operation of a rotary power tiller. Biosyst Eng 82:161–168
15. Nag PK (1981) Predicting maximal oxygen uptake of workers engaged in agricultural tasks. J Hum Ergol (Tokyo) 10:25–33
16. Nag PK, Gite LP (2020) Energy cost of human labour in farming. In: Human-centered agriculture : ergonomics and human factors applied. pp 61–89
17. MÜller EA (1962) Occupational work capacity. Ergonomics 5:445–452. https://doi.org/10.1080/00140136208930615
18. Maiti R (2008) Workload assessment in building construction related activities in India. Appl Ergon 39:754–765
19. Bink B (1962) The physical working capacity in relation to working time and age. Ergonomics 5:25–28. https://doi.org/10.1080/00140136208930548
20. Saha PN, Datta SR, Banerjee PK, Narayane GG (1979) An acceptable workload for Indian workers. Ergonomics 22:1059–1071. https://doi.org/10.1080/00140137908924680

21. Varghese MA, Saha PN, Atreya N (1994) A rapid appraisal of occupational workload from a modified scale of perceived exertion. Ergonomics 37:485–491. https://doi.org/10.1080/00140139408963665

Discomfort in Poultry Rearing Among Workers in Northeast India

Sarju Thokchom and K. N. Dewangan

1 Introduction

Globally, the poultry sector is growing in terms of production and entrepreneurs because of the growing human population and the growing demand for animal protein due to its affordable price, affordable label, increased spending power of consumers, variety of products, and urbanisation [1]. In the last four decades, Indian poultry industry has begun to witness remarkable growth from backyard to the poultry industry, and this industry has been growing at annual rates of 7.3% [2]. India ranks third largest in terms of eggs production and ranks seventh-largest producer of poultry broiler meat in the world [3]. Despite its significant contribution to the nation's gross domestic product and its prevalence in rural areas, the chicken business is labour-intensive, involving a lot of repetitive chores that demand force.

The back injury from physical material handling and excessive exertion of body segments cause the accident on a chicken farm [4]. Lee and Lee employed OWAS and rapid entire body assessment (REBA) to avoid musculoskeletal disorders (MSDs) in a Korean chicken farm [5] and the study found that the tasks produced tension or stress in the arm, knee, shoulder, and trunk as a result of mismatch between humans and work and heavy workload.

Feeding the birds is an essential task and is performed manually. To perform this task, the workers carry a feed bag and fill it in feeders by pouring the feed. Carrying a heavy load for a long duration of time demands higher physical effort, and it causes discomfort to the workers in the back, neck, and arm [6]. Bending and catching poultry birds are performed in repetitive motions, and therefore, it is cause musculoskeletal problems in the long run [7–9].

S. Thokchom · K. N. Dewangan (✉)
Department of Agricultural Engineering, North Eastern Regional Institute of Science and Technology, Nirjuli, India
e-mail: kndewangan2001@yahoo.co.in

© The Author(s), under exclusive license to Springer Nature Singapore Pte Ltd. 2025

S. Das et al. (eds.), *Human-Centered Digitalization, Volume 3*, Design Science and Innovation, https://doi.org/10.1007/978-981-96-4383-7_13

The workers in the poultry industries have reported work-related musculoskeletal disorders [10, 11] and about 86% of the women workers reported pain and discomfort on their hands [12].

Broiler farms, or farms that raise hens for meat production, are typically large-scale operations using intensive agricultural methods. Given that egg-laying hens and broilers are reared in separate facilities, it is expected that farmers will require different positions when working, so different discomfort issues may arise. The ergonomic issues in boiler poultry farms in India have not been studied. Therefore, a study was performed to analyse the perception of physical pain on poultry farm workers in three tasks: Feeding poultry, cleaning poultry houses, and catching poultry birds, as well as its association with gender, age, and duration of work.

2 Material and Methods

2.1 Selection of Site

A total of 15 poultry farms located in the Lakhimpur district of Assam were selected for the study. The farms were selected randomly.

2.2 Selection of Task

Various tasks are performed in the poultry farms, from purchasing chicks to selling poultry. Three tasks, namely cleaning the poultry housing, feeding the poultry, and catching poultry for sale, were selected in poultry rearing farms. Figure 1 shows the tasks performed by poultry farm workers: (a) Cleaning of farms, (b) feeding birds, and (c) catching birds for sale. The above-mentioned task is labour-intensive and repetitive.

Cleaning of Poultry House

The first cleaning of the poultry housing is done 21 days after the placement of the chicks, and subsequent cleaning is done every week till the end of the sale of the poultry. The litter of bedding material is shovelled manually and packed. Spades and shovels are used to remove bedding material. Complete cleaning of the poultry house is done after selling all the poultry from the house. Two to three days are required to completely clean the poultry houses. For complete cleaning of the farms, after removing the bedding material, the floor is sprayed with lime, bleaching powder, and disinfectants to kill microorganisms so that the house can be used for the next batch of poultry.

Fig. 1 Task performed by poultry farm workers: **a** Cleaning of farms, **b** feeding birds, and **c** catching birds for sale

Feeding of Poultry

Feeding of the poultry includes distributing feeds and drinking water. Distribution of feeds is performed manually, while water is supplied automatically or manually to the feeders. Supply of drinking water manually is strenuous, and it is being performed three times a day in the summer season and two times a day in the winter season.

Catching of Poultry Birds

Poultry birds are sold after desired growth and for selling the birds are caught manually. The workers adopt an awkward posture for catching the birds. There is a need to bend the torso to reach the bird on the ground while the birds run in the house. Generally, two to three birds are caught together and then transferred to the basket. This process is repeated for about three to four hours a day.

2.3 *Selection of Subjects*

A total of 120 subjects, 50 women and 70 men, were selected from 15 poultry farms. Initially, subjects were screened, and the presence of disability, major surgeries, and sickness were excluded from the study. The criteria for selection of the subjects were at least 1 year of work experience in the selected tasks in the poultry farm and being employed in the farm. Written consent was obtained from all the selected subjects. Prior approval from the Ethical Committee of the Institute was obtained for the study.

2.4 Instruments

A human body diagram was used as a measuring instrument for discomfort assessment [13]. In this method, the subjects reported and/or showed perceived pain/discomfort in the body region. Discomfort levels on nine body parts, such as upper back, neck, shoulder, arms, wrists, legs, lower back, feet, and hips were considered in the survey.

2.5 Data Analysis

The collected data were analysed to study the effect of gender, experience, and rearing methods on discomfort using odd ratio and Chi-square test. Significant at 0.05 was the level.

3 Results

The mean age of women and men poultry farm workers in the present study was 33.4 years and 34.6 years, respectively. The workers were employed in the farm for a duration of one year to thirty-five years with mean experience of 8.4 years. The physical characteristics of the selected subjects are presented in Table 1.

Table 2 shows the perception of body discomfort. The results show that 73.3%, 85.8%, and 65.8% of workers reported discomfort in at least one body region in feeding poultry, cleaning poultry houses, and catching poultry birds, respectively. The body discomfort was more on the shoulder (73.3%) followed by lower back (40.8), upper back (37.5%), and wrist (19.2%) in the feeding of poultry. Cleaning poultry houses was more strenuous, and the discomfort in the body region was more than 77.5%. Most workers reported discomfort on the shoulder (65.8%), followed by lower back (35.0%) in catching poultry birds.

Table 1 Physical characteristics of the selected subject

Subject	Male			Female		
	Min	Mean	Max	Min	Mean	Max
Age (years)	18.0	33.6	62.0	20.0	35.2	52.0
Body mass (kg)	41.0	58.0	80.0	39.0	49.3	67.0
Stature (mm)	1490	1622	1760	1490	1561	1650
BMI	16.0	21.9	31.3	15.6	20.2	29.0
Experience (years)	1.0	7.6	19.0	3.0	11.0	25.0

Table 2 Perception of body discomfort among the poultry farm workers

Body parts	Feeding		Cleaning		Catching	
	Yes, n (%)	No, n (%)	Yes, n (%)	No, n (%)	Yes, n (%)	No, n (%)
Neck	22 (18.3)	98 (81.7)	103 (85.8)	17 (14.2)	20 (16.7)	100 (83.3)
Shoulder	88 (73.3)	32 (26.7)	104 (86.7)	16 (13.3)	79 (65.8)	41 (34.2)
Upper back	45 (37.5)	75 (62.5)	103 (85.8)	17 (14.2)	38 (31.7)	82 (68.3)
Arms	21 (17.5)	99 (82.5)	100 (83.3)	20 (16.7)	20 (16.7)	100 (83.3)
Wrists	23 (19.2)	97 (80.8)	99 (82.5)	21 (17.5)	21 (17.5)	99 (82.5)
Lower back	49 (40.8)	71 (59.2)	101 (84.2)	19 (15.8)	42 (35.0)	78 (65.0)
Hips	15 (12.5)	105 (87.5)	93 (77.5)	27 (22.5)	17 (14.7)	103 (85.8)
Legs	19 (15.8)	101 (84.2)	102 (85.0)	18 (15.0)	20 (16.7)	100 (83.3)
Feet	19 (15.8)	101 (84.2)	103 (85.5)	17 (14.2)	17 (14.2)	103 (85.8)

Table 3 Effect of gender, experience, and rearing methods on bodily discomfort

Parameter	Feeding of poultry		Cleaning of poultry house		Catching of poultry birds	
	OR (95% CI)	*p*	OR (95% CI)	*p*	OR (95% CI)	*p*
Gender	2.41 (0.9–6.7)	0.073	1.23 (0.44–3.4)	0.443	4.10 (1.6–10.4)	< 0.05
Experience	3.77 (1.2–11.3)	0.013	3.40 (1.1–10.3)	0.024	1.26 (0.53–3.0)	0.38
Rearing methods	15.04 (5.4–41.8)	< 0.05	3.93 (1.5–10.2)	< 0.05	7.332 (2.8–18.9)	< 0.05

Table 3 shows the effect of gender, experience, and rearing methods on body discomfort. Body discomfort during the feeding of poultry and cleaning of the poultry house did not have any association with gender. The relationship between gender and physical discomfort was significant ($p < 0.05$), in catching poultry birds (OR = 4.10; 95% CI: 1.61 to 10.42). Body discomfort during catching of poultry birds did not have any association ($p = 0.38$) with the experience (OR = 1.26; 95% CI 0.53 to 3.00). A significant ($p < 0.05$) correlation was observed between the experience of physical discomfort and rearing methods in feeding poultry, cleaning poultry houses, and catching poultry birds.

4 Discussion

Gumasing reported that poultry workers experience work-related musculoskeletal disorders (WMSDs). Lower back pain is the most prevalent among the workers, followed by wrist/hands, upper back, shoulders, neck, and elbow pain [11]. The results of the present study are in line with the results of [11]. In the present study, 65.8% of the workers experience discomfort in one of the body segments. Tirloni [14] reported that 67% of poultry abattoir processing workers complain about discomfort. About 63% of workers reported pain in the shoulders, and 43% reported pain in the neck [14]. In the present study, it was found that about 88% of workers were engaged in repetitive tasks, and 61% of workers used hand tools. The workers use shovels and spades to remove the bedding material in cleaning poultry houses in the region. Overtime workers had more pain due to their fast-paced work schedules and repetitive tasks [15]. Women are more prone to have back and arm pain and discomfort when working with poultry. [1].

5 Conclusions

The main outcomes of the study are as follows

1. About 65.8% to 86.7% of workers reported discomfort in the shoulder in all three selected tasks.
2. Cleaning of poultry house is considered as the most discomfort tasks and majority of the workers reported discomfort in all the body parts in performing this task.
3. Gender, experience, and rearing methods have significant effects on the body part discomfort in performing poultry feeding.

References

1. Harmse JL, Engelbrecht JC, Bekker JL (2016) The impact of physical and ergonomic hazards on poultry abattoir processing workers: a review. Int J Environ Res Public Health 13
2. Puram PN, Shinde AT, Padghan PV (2022) Feeding and management practices followed on broiler farm in Latur tahsil of Latur district 11:532–535
3. Chatterjee RN, Rajkumar U (2015) An overview of poultry production in India. Anim Hlth 54:89–108
4. Wang H, Kong YK, Jung MC (2012) Postural evaluation in a poultry farm for broiler chickens. Int J Occup Saf Ergon 18:67–75
5. Lee KS, Lee SH (2007) Musculoskeletal stress disorder problems at a chicken farm. J Korean Soc Saf 22:90–95
6. Caieiro TTM, De Assis DB, Mininel VA, Rocha FLR, Hortense P (2019) Musculoskeletal pain: comparison between administrative and production employees of a poultry farming company. Rev Bras Med do Trab 17:30–38
7. Mora DC, Arcury TA, Quandt SA (2016) Good job, bad job: occupational perceptions among Latino poultry workers. Am J Ind Med 59:877–886
8. Quandt SA, Grzywacz JG, Marín A, Carrillo L, Coates ML, Burke B, Arcury TA (2006) Illnesses and injuries reported by Latino poultry workers in western North Carolina. Am J Ind Med 49:343–351
9. Quandt SA, Arcury-Quandt AE, Lawlor EJ, Carrillo L, Marín AJ, Grzywacz JG, Arcury TA (2013) 3-D jobs and health disparities: The health implications of latino chicken catchers' working conditions. Am J Ind Med 56:206–215

10. Mohammadi G (2012) Risk factors for the prevalence of the upper limb and neck work-related musculoskeletal disorders among poultry slaughter workers. J Musculoskelet Res 15:1–8
11. Gumasing MJJ, Robielos RAC (2018) Ergonomie intervention addressing musculoskeletal disorders among poultry layer workers. In: 2018 5th International conference on industrial engineering and applications ICIEA, pp 429–434
12. Dokuztug F, Acik F, Aydemir A, Issever H, Yilmaz I, Erer M (2006) Early symptoms of the work-related musculoskeletal disorders in hand and upper extremity in the poultry industry. J Med Sci 6:305–313
13. Corlett EN, Bishop RP (1976) A technique for assessing postural discomfort. Ergonomics 19:175–182
14. Tirloni AS, Dos Reis DC, Dos Santos JB, Reis PF, Barbosa A, Moro ARP (2012) Body discomfort in poultry slaughterhouse workers. Work 41:2420–2425
15. Schulz MR, Grzywacz JG, Chen H, Mora DC, Arcury TA, Marín AJ, Mirabelli MC, Quandt SA (2013) Upper body musculoskeletal symptoms of Latino poultry processing workers and a comparison group of Latino manual workers. Am J Ind Med 56:197–205

Health Hazards in Fish Harvesting System of Apatani Paddy Cum Fish Cultivation

Tapi Tada and K. N. Dewangan

1 Introduction

The paddy cum fish cultivation system is a unique practice done by Apatani tribes of Ziro Valley in Arunachal Pradesh. In Apatani dialect is it known as Ajie Ngyie Asonii.

Various tasks in the Apatani paddy cum fish culture are being performed manually by female workers. This culture includes proper land and water management system [1]. Water from the hilly area of the forest is channelized into a central canal and from the central canal, the water is diverted to the field using bamboo, plastic and pine wood pipes. Trenches are constructed in the middle of the field to protect fish from wild animal, birds and also for settlement of fish during hot summer, weeding and harvesting as the level of water is low during this period [2]. Large dykes or bunds are built to maintain water level/preserve water and prevent soil erosion, supported by bamboos, concrete and wooden clips. In this culture, often fish harvesting coincide with the field weeding, depending upon growth rate of fish. The first batch of fish is stocked during late March to early April before transplantation of paddy saplings [3, 4] and harvested in mid-June to early July, followed by second batch in the month of July and harvest in mid-September to early October, before harvesting of paddy [5, 6].

Fish harvesting comprehend five steps, i.e., placing of traps in the outlet pipe, releasing of water, cleaning of trap, pushing and harvesting of fish near the outlet pipe. It requires two or more persons to perform the tasks. For harvesting fish, water from the field needs to be drained through one or two trench outlets by placing fish trap at the exit point. Farmers use different types of fish trap made of different materials as shown in Fig. 1a. Foreign matter deposits in the fish trap and thus block the water

T. Tada · K. N. Dewangan (✉)
Department of Agricultural Engineering, North Eastern Regional Institute of Science and Technology, Nirjuli, India
e-mail: kndewangan2001@yahoo.co.in

© The Author(s), under exclusive license to Springer Nature Singapore Pte Ltd. 2025
S. Das et al. (eds.), *Human-Centered Digitalization, Volume 3*, Design Science and Innovation, https://doi.org/10.1007/978-981-96-4383-7_14

(a) (b) (c)

Fig. 1 Various tasks in paddy cum fish cultivation: **a** various types of fish traps; **b** cleaning of fish traps; **c** pushing and harvesting of fish

outlet. Therefore, cleaning of fish trap is done frequently with hands which causes cut or pierce type injury in the fingers as shown in Fig. 1b. Injury to the farmers are also caused by aquatic insect bites which comprises of water beetle (*Dytiscus marginalis*) and water leech (*Hirudo medicinalis*). Prolong standing in the water causes trench foot leading to skin and tissue breakdown and therefore infections to the workers are common. Stagnant water in the paddy field contains various parasitic worm which can cause serious infections and allergies. Harvesting of fish and pushing of fish toward the outlet pipe are carried out in protracted bending and awkward posture [7, 8] (Fig. 1c), which may cause musculoskeletal disorders (MSDs) [9], dizziness and shortness of breath.

The various injuries among the workers of paddy cum fish cultivation have a negative effect on farmers' health. Ngajilo and Jeebhay reported that the occupational exposures, hazards and their associated adverse health effects among fish farming workers are merely recognized [9]. Deterioration in farmers' health has negative effect on paddy cum fish cultivation and the farmers are renouncing this method of cultivation. Furthermore, there is no study on health risks among the farmers of paddy cum fish cultivation in the Ziro Valley. Therefore, a study is performed data on occupational health risks among the rice cum fish cultivating farmers by collecting workers perceptions.

2 Material and Methods

2.1 Selection of Site

Apatani paddy cum fish cultivation is done in the Ziro Valley, therefore the study was conducted in Ziro of Lower Subansiri District in Arunachal Pradesh. Five major villages, namely Hong, Hari, Bulla, Hija and Dutta were selected for the study.

2.2 Selection of Subject

The workers with minimum one year experience in paddy cum fish cultivation system were selected as subjects. The workers with disability, major surgeries and sickness were excluded from the study. The workers were informed about the purpose of study and questionnaire for data collection. The workers who agreed to be part of the study were selected as subjects. Written consent was obtained from the selected subjects. Prior approval from the Ethical Committee of the Institute was obtained for the study. A total of 58 female subjects from five selected villages in the age between 28 and 76 years participated in this study.

2.3 Questionnaire Development

A questionnaire was developed to collect data on health hazards in fish harvesting system of paddy cum fish cultivation system. The questionnaire was designed such that it is comprehensive. The questionnaire was typed so that it is easily legible. The developed questionnaire consists of demographic information, health problems and requirements of medical assistance in injuries. The health problems and requirements of medical assistance in injuries were semi closed-ended questions. Demographic information included name, age, gender, stature, body mass, dominant hand, experience and education. Questions on health problems included dizziness; shortness of breath; allergies to field water; and injuries due to paddy plant, fish fins, tail splashing, slip, trench foot, parasite or worm, aquatic insect bite and cut in finger.

2.4 Collection of Data

The information on the health hazard on fish harvesting system was collected through personal interviews using developed questionnaire. Random sampling was followed to collect data. The workers were approached and their willingness to participate in the study was asked. The demographic information was collected initially from the subjected willing to participate in the study and submitted written consent. Subsequently, data on health problems related to fish harvesting in paddy cum fish cultivation system were collected. In the last, data related to requirement of medical assistance was collected. The questions were explain in the native language to the subjects and responses were obtained native language and were recorded. The responses obtained from the subjects were later translated in to English language. In the data collection, the participants were allowed to discuss the topic freely. Data was collected in one-to-one personal interview.

Table 1 Demographic characteristics of the subject. ($n = 58$)

Variables	Minimum	Mean (SD)	Maximum
Age (years)	28	47.55 (13.85)	76
Stature (ft.)	4.1	4.84 (0.43)	5.9
Body mass (kg)	46	54.72 (6.39)	70
Experience (year)	3	30.55 (15.35)	60

2.5 Statistical Data Analysis

The information obtained from the villagers were statistically analyzed using IBM® Statistical Package for the Social Sciences (SPSS) version 26 and Microsoft Excel software. All answers to the questionnaire were entered into the excel database and encoded with numerical values such as 1 for "Yes" and 2 for "No." The demographic characteristics of the subjects were divided into two groups viz. less than or equal to 50 years age and more than 50 years age. Similarly, stature, body mass and experience were also divided in two groups for the values of 152.4 cm, 55 kg and 30 years, respectively. The subject's dominant hand and education were categorized into right and left, literate and illiterate, respectively. For all statistical tests, a p-value less than 0.05 was taken to indicate a significant difference. To describe the socio-demographic characteristics, minimum, maximum, mean, standard deviation (SD) and percentage were used. Chi-square test was performed. Odd ratios (OR) with a 95% confidence interval (CI) along with p-value were analyzed to study the effect of age, stature, body mass, dominant hand, experience and education on health problems of the workers.

3 Results

3.1 Demographic Characteristics

Demographic characteristics of the subjects are shown in Table 1. Among the subjects, majority were right handed (82.76%) and rest were left handed (17.24%). About 55.17% subjects were literate and 44.83% were illiterate.

3.2 Health Problems

The percentage of various health problems related to fish harvesting in paddy cum fish cultivation is shown in Fig. 2. The study revealed that a substantial proportion, i.e., approximately 91.4% subjects experienced injuries in the face during catching fish, followed by injuries due to paddy leaves among 79.3% subjects. Overall 45% subjects have sought medical attention for the various injuries.

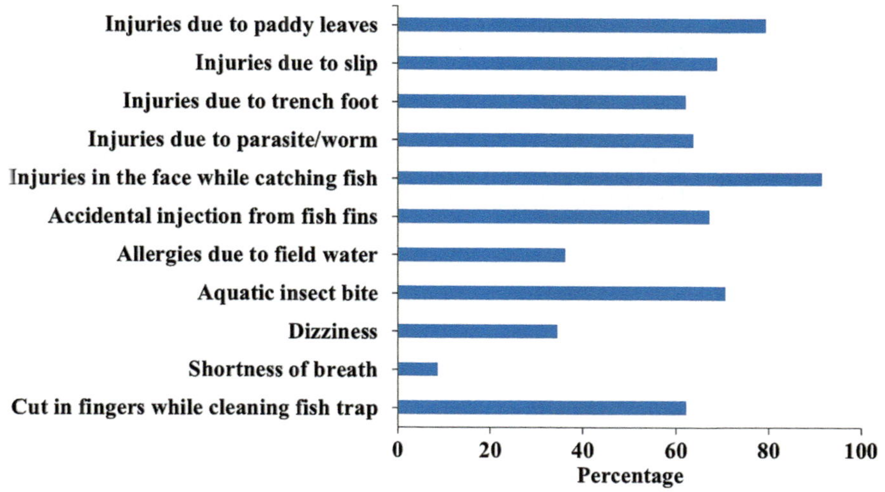

Fig. 2 Percentage of various health problems

3.3 *Effect of Demographic Characteristics on Health Problems*

Table 2 shows the statistical analysis showing OR, CI at 95% and *p*-value on the effect of age, stature, body mass, dominant hand, experience and education on various health problems. The statistical analysis shows that there was a significant association between age of the female workers and the risk of injuries due to trench foot and cut in fingers during cleaning a fish trap ($p < 0.05$). The effect of stature on health problems was significantly ($p < 0.05$) higher on two injuries, i.e., trench foot (OR: 4.971 and 95% CI: 1.532–16.13) and dizziness (OR: 4.121 and 95% CI: 1.031–16.48). The body mass of the respondent was significantly ($p < 0.05$) higher on trench foot with OR 2.600 (95% CI: 0.858–7.883). The dominant hand played a significant ($p < 0$ 05) association with injuries due to paddy leaves showing an OR 2.308 (95% CI: 0.076–1.243). Furthermore, experience exhibited a considerably lower injuries caused by trench foot, parasites/worms, aquatic insect bites and finger cuts during fish trap cleaning ($p < 0.05$). The effect of education was notably significant ($p < 0.05$) with injuries due to trench foot displaying an OR 11.205 (95% CI: 2.778–45.199), while dizziness and finger cut during cleaning fish traps also exhibited significant associations with OR 3.571 (95% CI: 1.145–11.137) and 3.333 (95% CI:1.060–10.482), respectively.

Table 2 Effect of demographic characteristics on health problems and statistical analysis showing OR, CI at 95% and p-value

Health problems	OR (95% CI)	p-value
Effect of age		
Injuries due to trench foot	0.038 (0.005–0.315)	< 0.001
Cut in fingers while cleaning fish trap	0.158 (0.040–0.629)	0.005
Effect of stature		
Injuries due to trench foot	4.971 (1.532–16.132)	0.007
Dizziness	4.121 (1.031–16.481)	0.033
Effect of body mass		
Injuries due to trench foot	2.600 (0.858–7.883)	0.037
Effect of dominant hand		
Injuries due to paddy leaves	2.308(0.076–1.243)	0.041
Effect of experience		
Injuries due to trench foot	0.069 (0.017–0.284)	< 0.001
Injuries due to parasite/worm	0.273 (0.086–0.863)	0.023
Aquatic insect bite	0.218 (0.061–0.785)	0.015
Cut in fingers while cleaning fish trap	0.166 (0.050–0.556)	0.002
Effect of education		
Injuries due to trench foot	11.205 (2.778–45.199)	< 0.001
Dizziness	3.571 (1.145–11.137)	0.025
Cut in fingers while cleaning fish trap	3.333 (1.060–10.482)	0.033

4 Discussion

The Apatani method of fish harvesting in Ziro Valley has many sustainable and economic advantages, at the same time it has various health challenges for farmers involved in paddy cum fish cultivation system. Though the absence due to work-related illness and health concerns are everywhere, the subjects of paddy cum fish cultivation reported that various health problems decreases their working efficiency and due to which the next generation is not willing to practice this traditional method of cultivation. Furthermore, the increase in expenditure for medical causes economic worries. Moreover, the subjects are frequently doing the potentially hazardous work including lifting heavy basket containing fish, standing in the muddy field, awkward posture and monotonous labors [8]. To mitigate this challenges Das [10], suggested the use of appropriate safety equipment such as gloves, full sleeve uniform, thick water boots with socks and take regular breaks to avoid physical strain and dizziness. The use of modern technology could ease the health problems related to paddy cum fish cultivation [2].

5 Conclusions

The following conclusion can be drawn from the study:

1. Injuries in the face during catching fish from the trench in paddy cum fish cultivation is more prevalent followed by injuries due to paddy leaves.
2. There is a significant association between injuries due to trench foot with various demographic characteristics such as age, stature, experience, education and body mass.
3. About 45% of the farmers have sought medical assistance for the injuries during cultivation.

References

1. Chaudhuri KN, Chaudhuri S (2022) The art of 'organic' living: lessons from Apatani women in Ziro valley of Arunachal Pradesh in Northeast India. In: Back to basics: a multidisciplinary approach to organic living and waste management. pp 75–90
2. Rai SC (2005) Apatani paddy-cum-fish cultivation: an indigenous hill farming system of North East India. Indian J Tradit Knowl 4:65–71
3. Dabral PP (2002) Indigenous techniques of soil and water conservation in north eastern region of India. In: 12th ISCO conference, Beijing, China. pp 26–31
4. Saikia S, Das D (1970) Rice-fish culture and its potential in rural development: a lesson from Apatani farmers, Arunachal Pradesh. India J Agric Rural Dev 6:125–131. https://doi.org/10.3329/jard.v6i1.1667
5. Nimachow G, Rawat Dai JS, Loder O (2010) A sustainable mountain paddy-fish farming of the Apatani tribes of Arunachal Pradesh, India. Aquac Asia Mag 15
6. Taka T, Tangjang S (2015) Sustainable approach: paddy-cum-fish and bamboo-cum-pine agroforestry in Ziro valley, Arunachal Pradesh. India J Bioresour 2:50–57
7. Thorvaldsen T, Kongsvik T, Holmen IM, Størkersen K, Salomonsen C, Sandsund M, Bjelland HV (2020) Occupational health, safety and work environments in Norwegian fish farming–employee perspective. Aquaculture 524:735238. https://doi.org/10.1016/j.aquaculture.2020.735238
8. Moreau DTR, Neis B (2009) Occupational health and safety hazards in Atlantic Canadian aquaculture: laying the groundwork for prevention. Mar Policy 33:401–411. https://doi.org/10.1016/j.marpol.2008.09.001
9. Ngajilo D, Jeebhay MF (2019) Occupational injuries and diseases in aquaculture—a review of literature. Aquaculture 507:40–55. https://doi.org/10.1016/j.aquaculture.2019.03.053
10. Das DN (2018) Farming of fishes in rice fields of Northeast India: a review. J Coldwater Fish 1:27–41

Driving Manufacturing Excellence: Unleashing People Potential Through Process Refinement and Workforce Optimization

Shilpa Narayanswamy and Nikhil Ghantial

1 Introduction

Indian GDP is expected to grow at 8.2% in FY2023 while it is expected to be in a range of 6.5–7% between 2023 and 2027. The key drivers for growth are rising domestic consumption, India's push for domestic manufacturing, and favorable policy support. The pharmaceutical industry is characterized by stringent quality requirements and continuous process improvement efforts to ensure efficient production. Healthcare expenditure in India is currently low and is expected to increase owing to demographic reasons like an increase in population, policy factors increase in health insurance penetration in India and policy like Ayushman Bharat, a national health policy that aims to provide healthcare services to low-income class have opened primary health center and community health centers. Aggregating the above factors on the demand side for the pharma industry primary drivers are the aging population, increase in generic medicines, government support for providing health care services, growing prevalence of chronic diseases, and increase in coverage of health insurance.

On the supply side, pharma industry in India has a high entry barrier due to regulatory pressure, licenses required, and highly intensive capital inclusion. Covid-19 has caused disruption in the supply chain. It has helped to highlight the need to make the supply chain resilient with the development of vendors locally and having a backup resource identified in case of any discrepancy. There are many players in the pharma industry and a lot of market share is taken by players who have better quality products, putting pricing pressure on the competitors. Pricing pressure is mostly derived from manufacturing processes used by the company without compromising

S. Narayanswamy · N. Ghantial (✉)
Operations Management Professional, Mumbai, India
e-mail: ghantialnikhil@gmail.com

S. Narayanswamy
e-mail: shilpa.n18@gmail.com

© The Author(s), under exclusive license to Springer Nature Singapore Pte Ltd. 2025
S. Das et al. (eds.), *Human-Centered Digitalization, Volume 3*, Design Science and Innovation, https://doi.org/10.1007/978-981-96-4383-7_15

on quality, since in a pharma company quality is given utmost importance otherwise this may affect the health of people consuming it.

Our study focuses on one such pharma company which has top-quality products, which are developed with the latest cutting-edge technology/machines and with the use of good quality raw materials. The company has good presence locally as well as in the global pharmaceutical market. There was a need to look at process optimization as well as manpower utilization to ensure manufacturing excellence for the company. The product considered for this case involved various critical processes done to achieve the final product. The primary focus will be on how the manpower in the company was optimally utilized, the approach followed for it, and the impact it had both tangible and non-tangible.

2 Literature Review

This literature review focuses on the strategies adopted by the company to address the problem by refining the four key factors in production: Method, Machine, Material, and Manpower, with a special emphasis on the human resource aspect. The study aims to understand the significance of aligning human resources with key processes, employing data-driven approaches, and fostering employee engagement to drive productivity and return on investment. Excellence is the term that stands for perfection and quality [4]. Organizations across the globe thrive for achieving excellence, and operation excellence is one of the tools for it. Operation excellence involves major functions of an organization like manufacturing, finance, HR, SCM, etc., and aligning them in a way to achieve better outcomes by reducing non-value-adding activity. The need for achieving operation excellence can be related to increasing global competition among peers to better serve better the customers and gain market share. The way for achieving operation excellence is by following a step-by-step methodology which includes various methods like Kaizen, Overall Equipment Effectiveness (OEE), TQM, and lean manufacturing [6]. One of the ways for driving manufacturing excellence is by visiting Gemba (The place where the actual value addition is done) and observing the 3 M in Lean which are MUDA, MURA, and MURI and identify the ways to remove them helping in increasing the productivity and thus contributing to the overall output. Simultaneously, calculating actual cycle time, transportation time, built inventory, and waiting time of raw material to get processed helps to do the time study of a particular manufacturing line [5].

In quality management systems, one of the topics related is process improvement, before thinking of improvement first thing to understand is a process, a process means a set of actions carried out on an input to change its shape, size, and form to output for which customer is willing to pay. Thus, the output is the value-adding product in which the customer is interested. Process improvement is a systematic and periodic approach by which tangible improvement is made in the process in the form of increased output, better quality, or reduction in cycle time. One of the pioneers of process improvement was the PDCA cycle introduced by W. Edwards Deming. The

Deming PDCA cycle shows how to achieve continuous improvement in a process [7].

- Plan—Identify the problem and potential sources of system weakness or error. Ask the right question. And accordingly, develop a plan for improvement.
- Do—Implement whatever plans have been developed in short put the action plan.
- Check—A kind of monitoring process is used to access the effectiveness of the action taken.
- Act—Take corrective action if required and check whether the results are as expected.

Work smarter captures the essence of workforce optimization, as the current era of manufacturing has changed from resource optimization to customer delight the way by which the workforce reacts to this change is also important. Hybrid work culture is getting in and organizations are facing challenges in benchmarking appropriate levels of productivity for the workforce considering employee retention as well. Workforce optimization can be attributed to lower operating costs, higher service levels to customers, increase efficiency and productivity, and learning development. Managers are one of the crucial linkages with the employee as they are responsible for creating good working culture, a feedback mechanism, and implementation of the feedback from the employees, managers are also involved in developing training programs. Effective training reaps benefits in the long run where employees can add more value and have a better understanding of the technological development, gaining full command of skills required to perform their task [8].

Effective human resource management is crucial for manufacturing success. Research shows that aligning human resources with production processes can positively impact manufacturing performance and overall productivity [3]. By focusing on the Manpower aspect, the significance of investing in employee skills and development is understood. Strategic human resource practices, such as targeted training programs, identifying high-potential employees, and providing them with higher responsibilities, can lead to improved job performance and increased productivity [2]. The case study demonstrates that the development and utilization of human capital can play a pivotal role in overcoming manufacturing challenges.

Value chain analysis (VCA) is a powerful tool used by businesses to identify critical areas in their value chain where improvements can be made to enhance overall efficiency. The study showcases how the company employed VCA to identify the areas requiring high skills and ability in their manufacturing process. It analyzed the factors of skill, responsibility/accountability, and effort, identified key jobs and tasks crucial for the processes.

Performance-potential matrix (PPM) for talent management is a talent management tool that assesses employee performance and potential, allowing organizations to identify high performers and develop their potential further [1]. PPM is used to segregate employees based on their performance and potential. Placing high performers with high potential in critical areas, such as on a particular process, and providing on-the-job training and guidance, helps to strategically utilize manpower for achieving maximum productivity.

3 Live Case Background

The company under study had a manufacturing process for pharmaceutical tablets and capsules that involves various critical stages, and one such critical stage was the Compression process. The Compression process involves compressing powdered pharmaceutical blends into tablets or capsules. Achieving consistent and high-quality compression is vital for meeting production targets and maintaining profitability. However, in the case under investigation, the company faced a continuous decline in yield, impacting its value chain and overall productivity.

The specific issues that emerged were twofold: the Compression output did not align with the planned budget, and the yield continuously declined. This situation became increasingly concerning, especially considering the negative yield of approximately Rs. -44 lacs observed for continuous months of production. Such setbacks are particularly distressing for low-margin products, as they directly impact profitability and overall performance.

Despite having top-of-the-line machines, cutting-edge technologies, and well-established processes in place, the root cause of this problem proved to be multi-faceted and not immediately apparent. Initial attempts at finding immediate solutions proved futile, and the company realized that this challenge required a more in-depth analysis and a thoughtful, long-term approach. The company decided to take a step back and focus on understanding the intricacies of the Compression process to identify the right path to move forward.

The broad and explicit problems were that the Compression output was not getting generated as per the plan/budget and the yield was continuously going down. This was not at all a good sign for any manufacturing process especially when it comes to low-margin products. With the best of the best machines, technologies, processes, etc. in place, the reason was slightly complex to understand. So, as a team of researchers, we stopped looking for immediate short-term solutions to this problem and rather we started looking for the right path to move on.

3.1 Methodology

Quantitative data was systematically collected using surveys, performance metrics, and operational records. Complementing this, qualitative insights were meticulously extracted from in-depth interviews and thought-provoking focus group discussions involving employees, supervisors, and management personnel. The team of researchers spent a couple of months on the shop floor to thoroughly understand the process and interact with the team involved.

3.2 *Root Cause Analysis*

The identified root causes were further analyzed which helped in understanding the impact of each process stage on the overall manufacturing efficiency and financial performance which was resulting in many tangible and intangible losses, e.g., Machine (Compression Area) stoppages for more than 40 times a month resulting into damage of 120 man hours, Compression output getting hampered and much below potential, Yield was running in negative in almost all months, etc. The analysis highlighted the critical role of the human element in the Compression process and its alignment with key production factors.

3.3 *Addressing the Problem*

The company decided to focus on refining the four key factors in production: Method, Machine, Material, and Manpower, to address the identified issues. The Methods being followed were best in class and was also in line with the processes in the industry. The Machine being utilized was also the best available in the market and the most suitable one for the process. As the Method and Machine were already optimized, the primary emphasis shifted to Manpower (Table 1).

During the observation phase at first instance, the method was ruled out as the plant was already USFDA certified, and the method/SOP are well defined and cannot be deviated at any point in time. Considering machines, they were also of superior quality and so changing them was not advisable as a lot of cost was already incurred in the process. The material is of high quality and is approved post-internal inspection hence that cannot be doubted as well, so the only M left was the manpower.

So, the thinking went on how to observe, ask, and answer why and what to change in manpower as it is important to provide valid suggestions as any wrong steps may tamper the sentiment of the workforce, and there is an understanding that most serious mistakes are not being made because of wrong answers, the truly dangerous thing is asking the wrong questions. While the first question of identifying the right 'M'

Table 1 3M of production

	Key factors	Analysis
	Machine	Working with one of the best machines available in the market
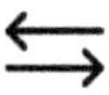	Method	No major changes can be done due to regulatory impact
	Man	What to do with them? Some improvement?

which needs to be focused on was answered as manpower, now the second hurdle was what do we do with it now?

Some of the answers to the second question came as.

a. Hire more specialized/expert people from the external market in the Compression area who can help with the process/execution.
b. Work on improving the quality of skill sets of existing people and develop them to become experts.

But for following option B. extensive technical skills needed to be invested on the manpower. But there are chances of train and retraining for solutions which will again form a viscous cycle. This was not recommended as we came to an understanding that training was not the panacea for everything because already the company had been investing a lot both on and off the job in terms of both cost and time on training the employees.

So, the idea was to think beyond the training which can be considered an all-season solution to people-related problems. The same thinking will provide the same set of solutions, so the need to think out of the box was required. To proceed in this process, two things were started. One had been to analyze the value chain and find critical areas by value chain analysis (VCA) using a factor analysis model.

The main objective of VCA was to:

a. Identify the areas which require high skills and ability.
b. Identify the highest Value-adding point and Value addition in each process.
c. List down the jobs/tasks done in that area.
d. List down the skills required for each of the jobs.

The second was to plot the manufacturing process against the 5 factors (Table 2). The factors employed were as follows:

1. Mental requirement
2. Physical requirement
3. Skill requirement
4. Responsibility
5. Working conditions

Table 2 Current processes plotting against factors employed

Area	Mental requirement	Physical requirement	Skill requirement	Responsibility	Working conditions
Dispensing	4	4	2	2	1
Sifting	3	4	2	1	4
Granulation	3	3	3	2	3
Compression (Target Area)	3	3	3	4	3
Coating	4	2	4	4	1
Inspection	3	4.5	3	3	3

The weighted average was taken from the final comparison based on the criticality of the job. This was done in discussion with the production shift in charge, floor supervisors, and the area leads for each area under study. The final scores were also cross verified with the factory head, production head and quality head.

1. Skill (Key factor): Education and training required, breadth/depth of experience required, social skills required, problem-solving skills, degree of discretion/use of judgment, creative thinking.
2. Responsibility/accountability: Breadth of responsibility, Specialized responsibility, Complexity of the work, Degree of freedom to act, Number and nature of the subordinate staff, Extent of accountability for equipment/plant, Extent of accountability for product/materials; and.
3. Effort: mental demands of a job, physical demands of a job, and degree of potential stress.

In line with VCA, product stratification was also done, and we divided the products into 3 broad categories:

I. High-value products: These were 25% of total products and were targeted for profit. These products were found to have the potential of generating 80% of the target profit.
II. Low-value high volume products: These were 25% products and were targeted to maintain zero loss.
III. Others—neither high value nor high volume not much stressed—Comprised of 50% products.

The idea was to prioritize and focus on critical ones. The weighted average method (WAM) was used to find the criticality and importance of every job.

Once VCA was done, the second phase was to identify the potential of people and plot them on a 3*3 performance-potential matrix (PPM). The main purpose for doing this was to identify the people who were high performers and have great potential and identify where they were currently working in the system. The objective of doing this activity was to find answers to the following questions:

- Do we have good followers?
- Do we have good leaders?
- Do we have a good mix of doers and thinkers?

And the process followed for performing the potential matrix was as follows:

1. Identification and segregation of employees based on performance.
2. Analysis of the potential based on knowledge level, skills, attitude, and behavior.
3. Development of the performance-potential matrix (PPM)

For doing this activity, we also first defined the terms performance as the accomplishment of a given task measured against preset known standards of accuracy, completeness, cost, and speed and potential as having or showing the capacity to develop into something in the future. This was a crucial stage in the development of the entire PPM.

These terms and definitions were then thoroughly defined and explained to the first-line supervisors with whom employees were interacting daily. Sessions were carried out for various shifts in the factory to ensure everyone spoke the same language and understood the same meaning for both the terms: Performance and potential. Following this, the first-line supervisors were then asked to rate each of their team members on a scale of 1 to 5 on the factors of performance and potential.

Needless to mention, these ratings were discussed on an individual level with the first-line supervisors and then ratified at three levels above them in the hierarchy. This was to ensure that the errors were minimized, and chances of subjectivity removed as well (Table 3).

The population (N) of 210 people was then plotted on the performance-potential matrix and the distribution of the employees was as follows:

While 49% ($n = 105$) of the population was placed under the high performance and high-potential quadrant, 32.70% (n = 69) of the population was under moderate performance and moderate potential quadrant followed by 14.69% (31) in high potential and moderate performance. The overall coordinates of the performance potential were at 7 and 6, respectively, (Per, Pot) = (7, 6) (Table 4).

The correlation (r) between the performance and potential was found to be 0.89. This was a significant score for the team. This helped in segregating the employees on a potential and performance level using structured thinking and scientific process.

Once both the processes of VCA and performance-potential matrix were completed, the final phase of putting/placing the right people in the right place was started. Slowly and gradually, the team of researchers in alignment with the first-line supervisors and their hierarchies, started moving the Quadrant 1 (high performance and potential, i.e.) people to the Compression area which was the main target (Fig. 1).

Table 3 Performance-potential matrix

Potential (Knowledge, skill, and attitude)				
		High	Medium	Low
Performance	High			
	Medium			
	Low			

Table 4 Performance-potential matrix as per distribution of employees

Potential (Knowledge, skill, and attitude)				
		High	Medium	Low
Performance	High	105 (49%)		
	Medium	31 (14.69%)	69 (32.7%)	3 (1.42%)
	Low		1 (0.47%)	1 (0.47%)

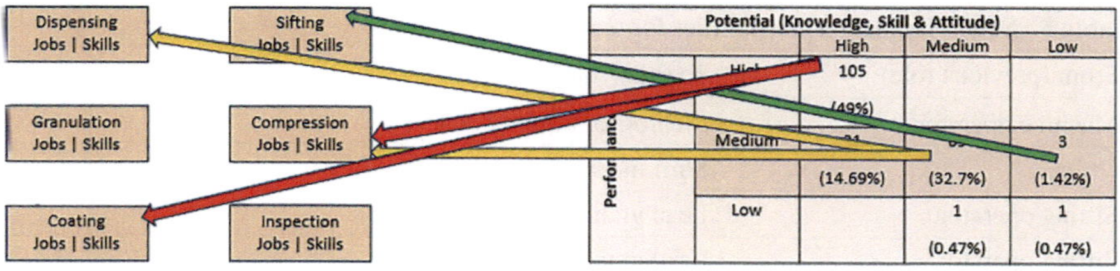

Fig. 1 Potential performance matrix plotted against the manufacturing processes

Along with this, the team also started expanding the roles of the Quadrant 1 people and started giving them higher responsibilities. The Quadrant 5 (moderate performance and potential, i.e.) employees were attached to the Quadrant 1 employees so that they continuously get on-the-job training and guidance.

The results obviously cannot be seen overnight especially when people are involved. While there were lot of changes being carried out, it was also necessary to make sure that employee's motivation and attitude at work remains positive while changing their roles and profiles. All changes were done after detailed discussions with the employees and the reasons and data discussions were highlighted, and only after the final agreement, the final movements were initiated.

A sample table of the role changes and movements: From (previous role) and to (new role) is shared below. This was the format used to finally make the announcements of the movements to all involved at the shop floor.

The names of the employees are not displayed due to confidentiality requirements (Table 5).

This whole process gave us the much-needed results and it had a direct impact on the yield. While the current state was at a negative figure of 44 lacs. Within the next 6 months, the number moved up to a positive figure of Rs.160 lacs resulting in a total net savings of Rs. 200 lacs.

This was an amazing improvement of 300%. This project also helped us in putting a lot of systems in place for products as well. A Yield Monitoring System was placed at every stage for each product.

Where on the one hand the yield improved by threefold, and the Compression output too was increased on a month-on-month basis. From 635.06 to 702.00 million. In six months, without any change of process or acquisition of a new machine which was remarkable. People were playing the critical role here and that was showing now. This was a straight increase/improvement of 10.55%. This was also crucial for the people on the shop floor to understand how placing the right person at the right job could be important to get the results.

The method followed also proved right when during the People Engagement survey, the company received a very positive score of 3.99 on a 5-point scale on Overall Satisfaction, a score of 4.05 on Leadership Confidence, and a score of 4.11 on the Overall Engagement parameter.

Table 5 New roles and responsibilities for operator

From (previous role)	To (new role)
Machine operator	Mentor for other operators
Officer	Shift in Charge
Sifting operator	Lead granulation operator
Sifting operator	Lead granulation operator
Operator	Mentor for all other operators
Machine operator	Mentor for other operators
Machine operator	Lead operator in Shift + Bin movement control
Machine operator	Lead operator in shift + Bin movement control
Officer	Additional given training responsibility
Machine operator	Lead operator in shift + Bin movement control
Machine operator	Lead operator in shift + Bin movement control
Officer	Shift in Charge
Machine operator	Lead operator in shift + Bin movement control
DGS section head	DGS + Coating in charge
BMR review	Inspection in charge

Along with all these direct improvements, the company was also able to bring the Machine stoppage from 40 times a month to 2 times which was an improvement of over 95%, and hence was able to save 114 man hours per month.

Just by doing VCA in combination with PPA and aligning people to the right process backed by data, the company was not only able to engage them but also impact overall ROI and productivity.

4 Key Findings

- **Skill Alignment and Training**: Manufacturing facilities that emphasized aligning right employee skills with right and relevant job roles and provided targeted training programs demonstrated higher productivity rates. Skill gaps were identified as a significant hindrance to achieving manufacturing excellence.
- **Flexible Workforce Deployment**: Facilities that adopted flexible workforce deployment strategies, such as cross-training and multi-skilling, reported increased operational agility. This flexibility enabled quicker response to production demands and reduced downtime during skill shortages.
- **Employee Engagement**: Positive correlations were observed between employee engagement and manufacturing excellence. Facilities that encouraged employee involvement in process improvement initiatives and decision-making showcased improved overall performance.

- **Performance Measurement**: Implementation of Key Performance Indicators (KPIs) and real-time performance tracking systems contributed to heightened accountability and efficiency. Regular performance reviews helped identify bottlenecks and areas for improvement.
- **Continuous Improvement Culture**: Facilities that fostered a culture of continuous improvement through regular feedback loops and recognition programs exhibited sustained manufacturing excellence. This culture encouraged innovation and process optimization.

5 Conclusion

The case study shows how having a well-organized and data-focused plan can help solve problems in manufacturing. By working on improving the four main factors of production (Manpower, Machine, Material, and Method), with a focus on the manpower part, the company made impressive progress in terms of output, Compression, and machine stoppages. Smart management of people, holistically looking at the steps in the process, and tools to manage talent, like the performance-potential matrix, all played a big role in getting employees involved and, making things more efficient. This study really underlines how important it is to analyze and involve the people side of the business to impact productivity.

What the research found is important to know for making things better in manufacturing:

- Getting the people part right—matching skills, training, and where people are needed—really helps things run smoother and faster.
- When employees feel included and are part of making decisions, it helps everything work better and makes the whole manufacturing process excellent.
- If there are clear ways to see how well things are going and find problems early, it helps stop things from getting stuck and makes things get better.
- Always trying to improve and having a culture that keeps looking for ways to do things better really helps make things excellent and keeps new ideas coming for making medicines.

6 Recommendation

Based on the conclusions drawn from this study, the following recommendations are put forth for manufacturing facilities seeking to drive excellence through workforce optimization:

- **Invest in Skill Development**: Prioritize employee training and skill development programs to ensure a competent and adaptable workforce.

- **Promote Engagement**: Encourage employee participation and input in decision-making processes to foster a sense of ownership and commitment.
- **Implement Performance Metrics**: Establish relevant KPIs and real-time monitoring systems to track progress and identify areas needing improvement.
- **Cultivate Continuous Improvement**: Create an environment that promotes continuous learning, innovation, and a culture of iterative enhancements.

In conclusion, this study underscores the pivotal role of workforce optimization in achieving manufacturing excellence. The insights gathered offer a roadmap for manufacturing facilities to unlock their people potential and elevate their operational performance to new heights.

References

1. Church AH, Rotolo CT, Ginther NM, Levine R, Golladay MJ (2015) The practical and legal realities of performance-implicit matrix operation. Pers Psychol 68(3):575–613
2. Guest DE (2011) Mortal resource operation and performance Still searching for some answers. Mortal Res Manag J 21(1):3–13
3. Jackson SE, Schuler RS (2003) Managing mortal coffers through strategic hookups. In Dunnette MD and Hough LM (Eds.), text of artificial and organizational psychology. Wiley, vol 4, pp 271–326
4. Porter ME (1985) Competitive advantage creating and sustaining superior performance. Free Press
5. Dysko D (2012) Gemba kaizen-application of mortal eventuality to achieving nonstop enhancement of company. J Trans Log, Issue 24/2012
6. Adnan AN, Arbaai NA, Ismail A (2016) Enhancement of overall effectiveness of product line by using line balancing 11:7752–7758
7. Deming WE (1986) Out of the crisis, ISBN 978- 0262541152
8. Brinkerhoff RO (2006) Adding impacts of training investments an evaluation strategy for building organizational learning capability, industrial and commercial banking. Emerald Group Pub 38(6):302–307

A Study on Prioritizing the QoE Features Based on Dynamic User Requirements

Karur Krishna Sahithi, In Seok Heo, Alivia Kirana Hartono Putri, Beom Su Kim, and Sang Ho Kim

1 Introduction

Emerging technologies like extended reality (XR) and collaborative robots are rapidly becoming part of daily life, elevating user expectations. Designers face the responsibility of incorporating top-tier system attributes in line with Industry 5.0 principles for emphasizing a user-centered design. User requirements are shifting from a system-centric to a user-centric focus, paralleling Aarron Walter's hierarchy of human needs for technology, which moves from functional to pleasurable experiences [1]. Quality of Experience (QoE) has great potential to help understand what factors influence the total satisfaction of the users while interacting with novel technologies.

The International Telecommunication Union-Telecommunication Standardization Sector (ITU-T) delves into the study of quality [2]. QoE represents a human-centered perspective, contrasting with Quality of Service (QoS) [3]. Users' needs have evolved in sync with the QoE concept's development. Mean Opinion Score (MOS) and other measurements are used by ITU-T to evaluate QoS [4], whereas QoE incorporates contextual, system, and human factors [5].

There are three views to understand QoE. These definitions all share a common theme: they view QoE from a user-centered perspective. However, they approach it from three angles—one emphasizing acceptability [6], another taking a holistic view [5], and the third being grounded in performance-based aspects [7]. QoE is usually understood through three key influence factors: system, context, and human factors. System factors involve technical parameters like bandwidth, resolution, and noise. Context factors include location, time, and cultural context, while human

K. K. Sahithi · I. S. Heo · A. K. H. Putri · B. S. Kim · S. H. Kim (✉)
Department of Industrial Engineering, Kumoh National Institute of Technology, Gumi, South Korea
e-mail: kimsh@kumoh.ac.kr

B. S. Kim
e-mail: kimsh@kumoh.ac.kr

© The Author(s), under exclusive license to Springer Nature Singapore Pte Ltd. 2025

S. Das et al. (eds.), *Human-Centered Digitalization, Volume 3*, Design Science and Innovation, https://doi.org/10.1007/978-981-96-4383-7_16

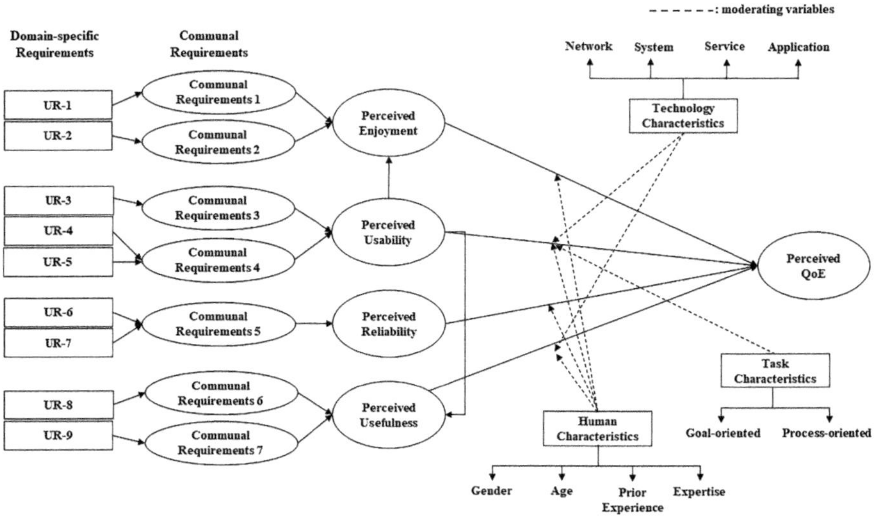

Fig. 1 Conceptual model for QoE measurement in XR environments [9]

factors are factors such as age, gender, and experience [8]. To address these perspectives and construct a comprehensive QoE model for extended reality (XR) environment, a model was developed [9] (see Fig. 1). This user-centric approach abstracts user requirements into four perceptual dimensions: Perceived Usefulness (PUF), Perceived Reliability (PR), and Perceived Usability (PU), Perceived Enjoyment (PE) aligned with Aarron Walter's Hierarchy of Needs. Influence factors are considered moderating variables within the framework rather than mediating ones. It, hence, forms a multi-layered network of factors contributing to QoE formation.

To emphasize the importance of deriving a quantitative and comprehensive QoE for practical use, there are several theoretical models, including one in [10]. Various mediating and moderating factors, such as technological, business, and contextual characteristics, as well as age, gender, human roles, and context [11], are explored in these models. The study of QoE, hence, reveals it as a multidimensional, abstract concept capable of explaining users' quality perception when quantified. However, there is no universally accepted method that fully encompasses the diverse QoE theories and definitions. This paper conducts a non-exhaustive review of QoE measurement literature to identify commonalities and differences among the methodologies employed. This research proposes a user requirement-based QoE calculation approach. The paper is further divided into two sections: Sect. 2 presents literature review findings and XAI modeling proposals, and Sect. 3 discusses research limitations and concludes the paper.

2 Methodology and Results

Using keywords such as "Quality of Experience" or "QoE" in combination with "modeling" or "measurement," we conducted a search and selected papers that most effectively encapsulated the diverse range of methodologies employed. Detailed information on these selected papers can be found in Table 1.

We also investigated QoE evaluations in extended reality (XR) environments as part of our investigation into QoE computation in novel technologies [20–24]. Two main viewpoints were drawn from the study: Section 2.1 data collection and nature of data and Sect. 2.2 data analysis methods. The advantages and disadvantages of the methods are described below.

2.1 Data Collection

The most prominent method for collecting user data often involves using questionnaires. To closely align with users' perspectives, questionnaires offer effective ways of gathering data, ensuring standardized data collection from a large number of participants. Using questionnaires is also cost-effective and generates data to quantify QoE. To support questionnaire development, multi-disciplinary theories and

Table 1 Summary of methods for measurement of QoE

References	Input variables/dimension	Method
[12]	Perceptual usability, aesthetics, setup duration	Multiple linear regression with beta coefficients as weights. MOS prediction
[13]	Perceived quality, system factors, context factors, human factors	Linear regression with beta coefficients as weights. MOS prediction
[14]	Network-related factors (delay, jitter, packet loss)	Machine learning models for MOS prediction
[15]	Legal-regulatory, socio-contextual, technological-process, stock-company aspects	Regression and boosted decision trees for MOS prediction
[16]	Acceptability, perceptual quality, presence dimensions	Linear and non-linear models with beta coefficients
[17]	QoS and UX parameters	Even weight distribution, linear regression analysis, principal component analysis
[18]	Predicted quality scores, rebuffering metrics, memory-related data	Various recurrent neural networks
[19]	Technical quality (TQ), content quality (CQ)	Surveys for TQ, CQ. Regression

rigorous exploratory and confirmatory factor analysis are used. However, because different theories use different jargon to define the dimensions of QoE, it can get confusing.

Typically, QoE in extended reality (XR) encompasses three dimensions: Presence [20], Immersion [21], and Cyber-sickness [22]. However, some studies [21] delve deeper into immersion, exploring sub-dimensions like engagement, engrossment, total immersion, flow, presence, emotional involvement, focus of attention, usability, and interest. This can be confusing when other studies [22] consider presence and cyber-sickness as separate dimensions within QoE and do not overlap each other. Additionally, variations in the number of measurement items used (e.g., 32 items [23] vs. 14 items [24]) in measuring the same dimension make it uncertain about the dimensional weights and their impact on QoE. This also makes user requirements to be diluted, which poses the risk of not considering certain requirements.

2.1.1 Nature of Data Collected

The independent variables collected are predominantly focused on technical factors and reflect a system-centric approach to calculating QoE [12–14, 17–19]. Although it is agreed upon that parameters like bandwidth, jitter, delay, and device type significantly affect users' QoE, this perspective fails to explain a human-centered QoE design fully. Nowadays, users generally assume a certain level of fulfillment with QoS parameters.

2.2 Data Analysis Methods

We have also observed patterns in data analysis categorized into two broad categories: traditional approaches and machine learning approaches.

2.2.1 Traditional Approaches

Multiple linear regression is a commonly used method for assessing QoE and determining weights. Principal component analysis has also been explored [17]. Multiple linear regression provides easy reporting, and the β coefficients offer insights into the importance of factors in QoE determination [12, 13]. It is easy to visualize what factors are more important in determining QoE by looking at the weight coefficients.

The unique contributions of each variable to an intrinsically multidimensional QoE, however, might be challenging to ascertain when independent variables are significantly correlated with one another. This can result in inaccurate coefficient estimates and make the model more challenging to interpret. It is challenging to calculate a multi-factored QoE as outlined in Sect. 1 when utilizing techniques like regression since they presuppose a linear, non-interdependent procedure.

2.2.2 Modern Approaches

These approaches involve machine learning-based methods [14, 15, 18] for predicting QoE values. Some use decision trees [15], while some utilize neural networks [18].

While machine learning efficiently handles extensive datasets with numerous input features, because of the "black-box" nature, interpreting them is a challenge. This black-box nature also limits our comprehension of interactions between the different features. If this problem has a possible solution, it can enhance our understanding of the weights of the QoE dimensions.

Toward a New Interpretable Model for Determining QoE

To address aforementioned problems, we suggest the use of a standardized questionnaire that includes both user-related factors and QoS parameters and build a QoE measurement model that considers the dynamics of user needs and their intricate connections within a network structure. This approach can reduce differences across various domains and alleviate confusion. Given the challenges associated with data analysis and to solve the limitations of machine learning models, we propose that employing explainable AI (XAI) presents a feasible solution.

The basic philosophy behind the use of XAI is to make artificial intelligence systems more transparent, interpretable, and understandable to humans. The different techniques and features that are supported by XAI are organized in Table 2.

We are keenly focused on using the "*interpretability*" feature among the various features supported by XAI, which enables us to understand the reasons why a particular AI decision was taken. The "post-hoc" ability of XAI hence makes sure to provide us with explanations and insights after a particular decision or prediction

Table 2 Summary of XAI techniques and advantages

XAI technique	Description and advantages
Decision trees	Hierarchical decision-making with interpretable structure, easy to understand
Rule-based systems	Transparent, rule-based decisions provide clear explanations
Linear models	Simplicity for linear relationships, easy to interpret
Feature importance	Identifies key driving factors and highlights influential inputs
Counterfactual explanations	Demonstrates input changes' impact and aids understanding of decision boundaries
Local vs. Global explanations	Provides insights into specific predictions and overall model behavior
Local interpretable model-agnostic explanations (LIME)	Creates interpretable surrogate models for specific predictions
SHapley Additive exPlanations (SHAP)	Offers fair and consistent feature attributions using cooperative game theory

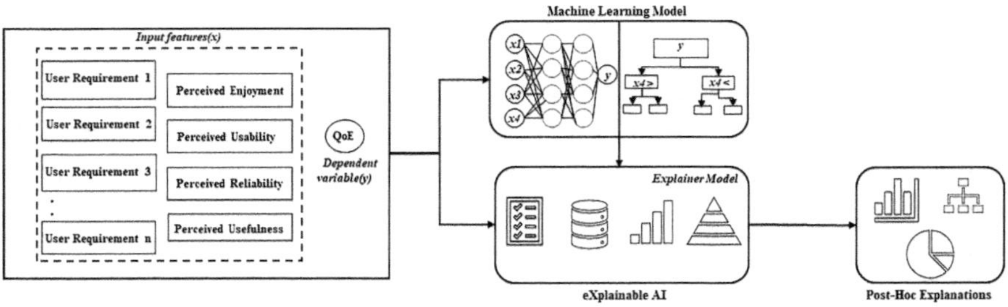

Fig. 2 Simple representation of the measurement model

has been made by the AI. This will help us know how the user requirements' priorities manifest onto the perceptual dimensions and, in turn, understand the effect on QoE. Establishing that the conceptual model is a hierarchical network model, quantifying QoE as a representation of its multidimensions in a successive hierarchical manner would establish the representation of QoE as:

$$\text{QoE} = f(PDi \ldots PDn)$$

The perceived dimensions are Perceived Usefulness (PUF), Perceived Reliability (PR), Perceived Usability (PU), and Perceived Enjoyment (PE). These perceptual dimensions are further represented as:

$$\text{PD}_i = f(UR_1 \ldots UR_m)$$

UR represents the different user requirements. A feasible method could be to use the input features (x) as the scores for the user requirements (UR) along the perceptual dimensions (PD), as mentioned in Fig. 2, to measure the output feature (y) QoE while using the moderating variables as control variables. An explainer model like SHAP or LIME could be used and features like SHAP values permutation feature importance can then be applied to assess the priorities of user requirements and perceptual dimensions and, hence, produce a holistic QoE measurement model.

The weights obtained in both hierarchical stages will prevent us from overlooking the significance of QoE at a detailed user requirement level, thereby helping us achieve our goal. Utilizing this hierarchical mapping of user requirements' impact on QoE, the "feature importance" aspect of XAI can be employed.

3 Conclusion and Future Work

It is crucial to identify the complex nature of QoE and the evolving user requirements with emerging technologies to enhance user-centric QoE modeling. We have identified the recurrent patterns in the reviewed QoE quantification methods that must be addressed to achieve a comprehensive QoE model. We emphasize the importance of quantifying QoE from a user-needs perspective, going beyond Quality of Service (QoS) metrics. We argue that in order to improve systems to align with users' expectations, calculating weights for user requirements and perceptual dimensions within the QoE framework is critical for a user-centric understanding. We are currently looking into whether the functions are linear or non-linear in order to represent the weights optimally. Incorporating interpretable machine learning into QoE modeling clears up the user requirements' dynamics and hence has the capability to provide solutions and insights into the user experience in the evolving technology landscape. Integrating the Kano model enriches our understanding of QoE, which would have the ability to capture both basic needs and attractive features, thus creating a holistic QoE framework. However, this study has certain limitations, including a non-exhaustive literature review. The nature of the functions being linear or non-linear and the optimal weight determination is currently being studied. Future work will explore additional QoE calculation methods, questionnaire development, and subsequent validation of the measurement model.

References

1. Walter A (2011) Designing for emotion. A book apart, New York
2. International Telecommunication Union Telecommunication Sector (ITU-T). Recommendation E.800. https://handle.itu.int/11.1002/1000/14832. Last accessed 2023 Sept 19
3. Rodrigues D, Silva D, Cerqueira E, Monteiro E (2008) Quality of service and quality of experience in video streaming. In: Proceedings of the international workshop on traffic management and traffic engineering for the future internet (FITraMEn2008). EuroNF NoE, Porto, Portugal, pp 11–12
4. International Telecommunication Union Telecommunication Sector (ITU-T). Recommendation P.800.1 http://handle.itu.int/11.1002/1000/12972. Last accessed 19 Sept 2023
5. Brunnström K, Beker SA, De Moor K, Dooms A, Egger S, Garcia MN, Zgank A (2013) Qualinet white paper on definitions of quality of experience. In: Output from the fifth Qualinet meeting. Novi Sad
6. International Telecommunication Union Telecommunication Sector (ITU-T). Recommendation. https://handle.itu.int/11.1002/1000/13408. Last accessed 19 Sept 2023
7. Reiter U, Brunnström K, Moor KD, Larabi M, Pereira M, Pinheiro AM, You J, Zgank A (2014) Factors influencing quality of experience, quality of experience. 55–72
8. Putri AKH, Sahithi KK, Kim BS, Heo IS, Woo JH, Kwon MS, Shin JG, Kim SH (2023) A conceptual model for assessing QoE in XR environment. In: The 25th Japan–Korea joint symposium and international conference on human factors and ergonomics. Japan Ergonomics Society, Japan, pp 9–10
9. Brooks P, Hestnes B (2010) User measures of quality of experience: why being objective and quantitative is important. IEEE Netw 24(2):8–13
10. Laghari KU, Connelly K (2012) Toward total quality of experience: A QoE model in a communication ecosystem. IEEE Commun Magazine 50:58–65
11. Barakovic J, Baraković S, Krejcar O, Marešová P (2021) Modeling of quality of experience for web-based unified communications with perceptual dimensions. SIViP 15:977–985
12. Barakovic J, Baraković S (2021) Multidimensional modelling of quality of experience for video streaming. Comput Hum Behav 129:107155

13. Begluk T, Husic JB, Baraković S (2018) Machine learning-based QoE prediction for video streaming over LTE network. In: 2018 17th international symposium INFOTEH-JAHORINA (INFOTEH), pp 1–5
14. Banjanin MK, Maričić G, Stojčić M (2023) Multifactor influences on the quality of experience service users of telecommunication providers in the Republic of Srpska, Bosnia and Herzegovina. Int J Qual Res 17(2):369–386
15. Tran HT, Ngoc NP, Thang TC (2021) Towards an overall QoE model for 360-degree video. In: 2020 IEEE eighth international conference on communications and electronics (ICCE), pp 327–331
16. Hamam A, El Saddik A (2013) Toward a Mathematical model for quality of experience evaluation of haptic applications. IEEE Trans Instrum Meas 62:3315–3322
17. Wu W, Arefin A, Rivas R, Nahrstedt K, Sheppard RM, Yang Z (2009) Quality of experience in distributed interactive multimedia environments: toward a theoretical framework. In: Proceedings of the 17th ACM international conference on multimedia, pp 481–490
18. Spachos P, Li W, Chignell MH, Leon-Garcia A, Zucherman L, Jiang J Acceptability and quality of experience in over the top video. In: 2015 IEEE international conference on communication workshop (ICCW), pp 1693–1698
19. Shahid Anwar M, Wang J, Ahmad S, Ullah A, Khan W, Fei Z (2020) Evaluating the factors affecting QoE of 360-degree videos and cybersickness levels predictions in virtual reality. Electronics 9(9):1530
20. Georgiou Y, Kyza EA (2017) The development and validation of the ARI questionnaire: an instrument for measuring immersion in location-based augmented reality settings. Int J Hum Comput Stud 98:24–37
21. Brunnström K, Sjöström M, Imran M, Pettersson M, Johanson M (2018) Quality of experience for a virtual reality simulator. In: Human vision and electronic imaging (HVEI). Burlingame, California USA
22. Witmer BG, Singer MJ (1998) Measuring presence in virtual environments: a presence questionnaire. Presence 7:225–240
23. Schubert T, Friedmann F, Regenbrecht H (2001) The experience of presence: factor analytic insights. presence: teleoperators and virtual environments 10(3):266–281
24. Molnar C (2019) Interpretable machine learning. A guide for making black box models explainable. https://christophm.github.io/interpretable-ml-book/

Can Pareidolic and Abstract Shape Affect the Product Preference of Consumers?

Subhankar Banerjee, Rajarshi Paul, Saunak Banerjee, Sanjit Dey, and Somnath Gangopadhyay

1 Introduction

In this digital era of marketing, every brand is competing with each other to sell a product [1]. Manufacturing industries are prioritizing on consumer-product interaction [2]. To enhance the consumer-product interaction, designing industries are developing interactive products. Visual attributions can contribute significantly in consumer-product interaction and can affect perception of the usability [3, 4]. One of the well-established approach is to develop anthropomorphic designs especially for user-driven products [5]. Anthropomorphism is the tendency of human-like characteristics to inanimate objects [6]. It has been effectively incorporated into the domains of product design and manufacturing [7]. Previous studies suggest the occurrence of consumers' affective behaviour and preference towards anthropomorphic products, brands, foods, marketing, and emotions [8]. Applications of anthropomorphism have become popular because it increases consumers' emotional satisfaction [9, 10]. However, the effect of another similar kind of phenomenon named 'Pareidolia' is not much explored in the field of affective design and manufacturing. Pareidolia has been derived from two Greek words para (wrong or false) and eidolon (image) [11, 12]. Thus, pareidolia can be defined as the false imagination or perception of a pattern or meaning in an object.

The 'Kansei Engineering' (KE) was coined by Mitsuo Nagamachi in 1995 [13]. KE is a reliable user-oriented technology based on the principles of ergonomics for

S. Banerjee · S. Dey · S. Gangopadhyay (✉)
Department of Physiology, University of Calcutta, Kolkata, West Bengal, India
e-mail: ganguly1961@gmail.com; sgphys@caluniv.ac.in

R. Paul
Department of Physiology, Serampore College (UG & PG), Affiliated to University of Calcutta, Serampore, West Bengal, India

S. Banerjee
Vice-President Brand Marketing, HiFuture Group, Shenzhen, China

© The Author(s), under exclusive license to Springer Nature Singapore Pte Ltd. 2025

S. Das et al. (eds.), *Human-Centered Digitalization, Volume 3*, Design Science and Innovation, https://doi.org/10.1007/978-981-96-4383-7_17

product development. Kansei Engineering is one of the popular methodologies that construes customers' affective feelings and emotional demands on new or existing products or concepts into design solutions [14]. In recent years, different products like digital camera, cars, web pages, and many more have been designed and evaluated using KE approach [15–17].

Like other user-driven product, pen stand can be a research object in the domain of Kansei affective science and Kansei engineering because it is a user-centric product. Hence, the purchase preference of a pen stand is directly related to its affective meaning.

1.1 Aim

The purpose of this study is to compare the effect of pareidolic and abstract shape on the pen-stand preferences of consumers using Kansei engineering approach. The secondary goal of this study is to predict the association between Kansei affective scale (KAS) score and Product Purchase Preference scale (PPS) score in pareidolic and abstract-shaped pen stands.

2 Materials

2.1 Sample Size Calculation

A-priori sample size has been calculated using G*power software where effect size, probability of α error and power (1-β error probability) were 0.5 (medium), 0.05 and 0.80, respectively, [18]. The calculated total sample size was 34 (df = 33).

2.2 Participant Selection

Seventy-five participants between the ages of 18 to 60 years (25.4 $\pm$ 4.30 years) were chosen at random from several cities in India. Each participant had a normal or corrected vision, and none of them were colour-blind. All participants were aware of using of the pen stand. No handedness of the products was involved in the study.

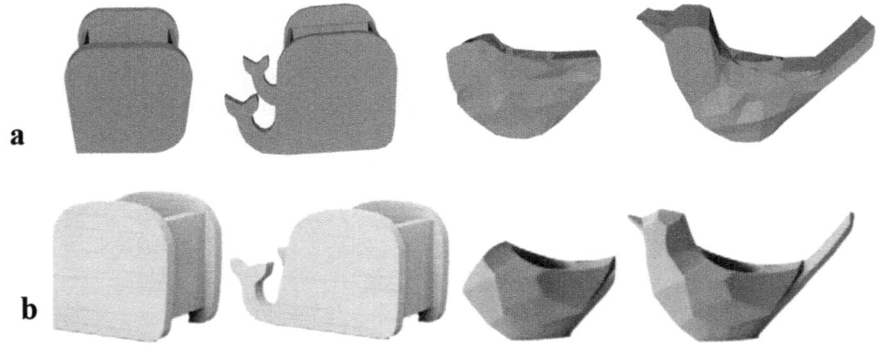

Fig. 1 Pareidolic and abstract-shaped black and white pen stands. A. Primary design concepts and B. Final design concept

2.3 *Stimuli*

Two pairs of black and white pen-stands (2 pareidolia and abstract designed each) images were selected from online market to use as the stimuli. Images were finalized and developed in the blender and Cinema 4D software and edited in Adobe Photoshop 17. The resolutions of the images were 1800 × 1200 pixels. Preference for the colour was eliminated by developing black and white images (Fig. 1).

2.4 *Kansei Affective Scale*

After collection of sixteen Kansei words from different journals, a seven-point Kansei affective scale (KAS) was applied to assess the quantitative affective preference of the pen stands [15, 19, 20]. Reliability was measured using Cronbach's alpha coefficient. Cronbach's alpha coefficient of KAS for pareidolic and abstract-shape products were 0.986 and 0.989, respectively. Composite reliability (CR) of KAS for pareidolic and abstract-shape products were 0.986 and 0.989, respectively, (>0.7). Average variance explained (AVE) of KAS for pareidolic and abstract-shape products were 0.781 and 0.854, respectively, (>0.5). Reliability and Validity Coefficients were above the acceptable limit [21, 22].

2.5 *Product Preference Assessment*

Seven-point purchase preference scale (PPS) was performed to assess the product preferences of the consumers. The scale contained four items on 'Prefer to Buy', 'Prefer to Spend Money', 'Prefer to Recommend', 'Prefer to See' a product in the online or offline market [23, 24]. In this present study, the above-mentioned scale was used to assess the purchase preference of the pareidolic and abstract-shaped pocket sanitizer containers. Cronbach's alpha coefficient of PPS for pareidolic and abstract-shape products were 0.943 and 0.966, respectively. Composite reliability (CR) of PPS for pareidolic and abstract-shape products were 0.943 and 0.966, respectively, (>0.7). Average variance explained (AVE) of PPS for pareidolic and abstract-shape products were 0.805 and 0.876, respectively, (>0.5). Reliability and Validity Coefficients were above the acceptable limit [21].

2.6 Statistical Analysis

The items of the Kansei affective scale (KAS) and purchase preference scale (PPS) have been summated and represented as the total KAS score and total PPS score. Normality tests of the total KAS score and PPS Score were assessed using Shapiro-Wilk method. Paired sample Wilcoxon signed rank test was performed to compare the distribution of the total KAS and PPS scores. Kendall's tau (τ) was used to find out the association between KAS and PSS [25]. The partial least square (PLS) algorithm of structural equation model (SEM)-based because the data did not meet normality [26].

Data collection and statistical analysis have been performed using Microsoft excel and SPSS 26.0 implementing standard procedure.

3 Results

Figure 2 represents individual KAS items of pareidolic and abstract-designed pen stands. Mean ratings of pareidolic pen stands were higher than the abstract pen stands.

Results of the Normality test using the Shapiro–Wilk test suggest that the total KAS and PPS scores were a distribution-free dataset ($p < 0.001$). Thus, non-parametric tests will be applied for hypothesis testing.

Paired sample Wilcoxon test suggests that KAS scores were significantly higher in pareidolic pen stands (Median = 67.0) than in abstract-shaped pen stands (Median = 59.0), z-score = −4.954, $p < 0.001$, effect size = 0.493. Similarly, purchase preferences scores were significantly higher in Pareidolic pen stands (Median = 14.0) in Aastract-shaped pen stands (Median = 11.5), z-score = −6.187, $p < 0.001$, effect size = 0.674 (Fig. 3).

Product Purchase Preference scale (PPS) was significantly associated with Kansei affective scale (KAS) in pareidolic pen stands, Kendall's tau (τ) = + 0.505, 95% confidence interval (95% CI) = +0.412 and +0.593, $p < 0.001$.

Product Purchase Preference scale (PPS) was significantly associated with Kansei affective scale (KAS) in abstract-shaped pen stands, Kendall's tau (τ) = + 0.529, 95% confidence interval (95% CI) =+ 0.445 and +0.604, $p < 0.001$ (Fig. 4).

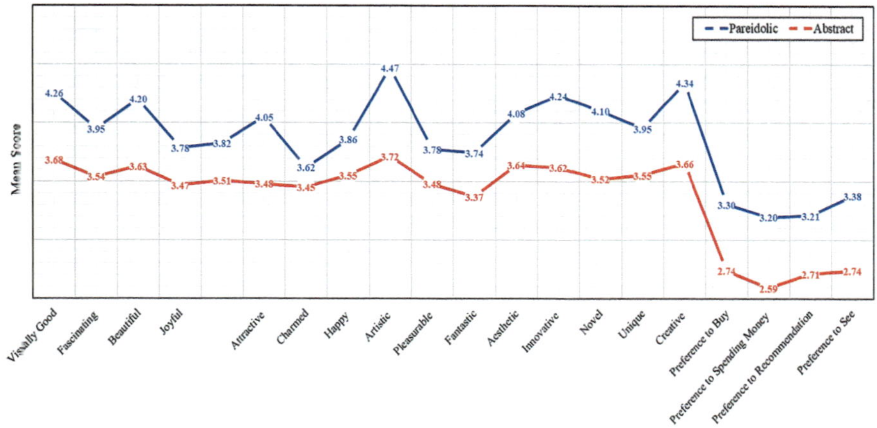

Fig. 2 Mean KAS and PPS rating pareidolic (blue dash) and abstract-shaped (red dash) black and white pen stands

Fig. 3 Dot plot with standard error bar of Total KAS and PPS score of pareidolic and abstract-shaped pen stand

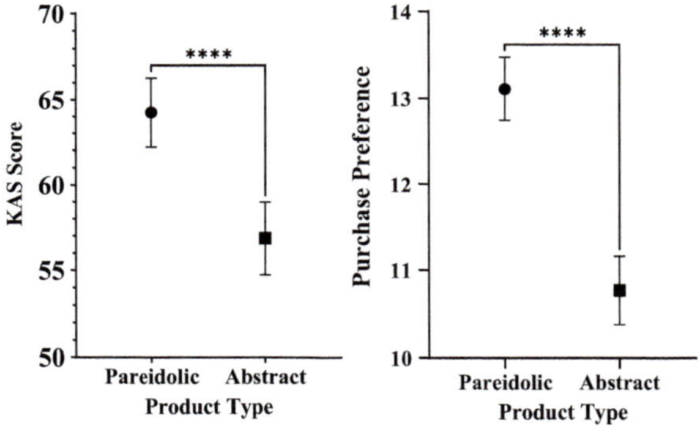

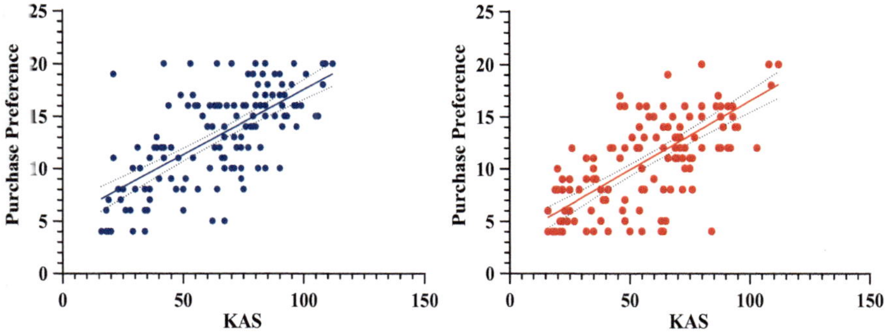

Fig. 4 Scatter plot with standard error lines of Total KAS and PPS score of pareidolic (blue, in left) & Abstract-shaped (red, in right) pen stands

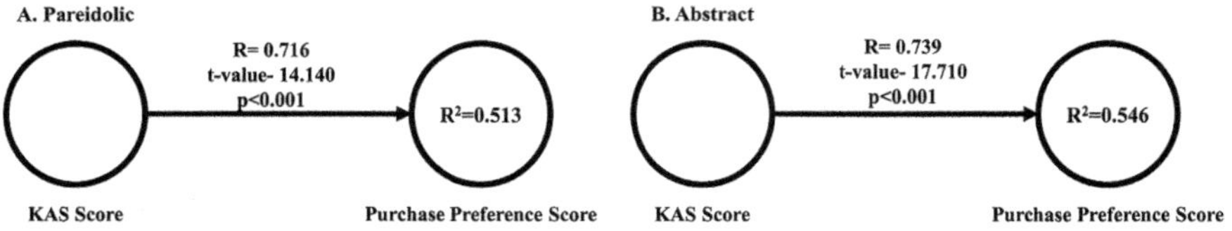

Fig. 5 PLS-SEM model of the PPS score (dependent variable) and Total KAS (Independent variable) in Pareidolic & Abstract-shaped pen stands

There is a presence of statistically significant Structural Equation Model (SEM) of pareidolic ($R^2 = 0.513$, $t = 14.140$, $p < 0.001$) and abstract-shaped ($R^2 = 0.546$, $t = 17.710$, $p < 0.001$) pen-stand implementing KAS and PPS. Hence, PLS-SEM has successfully predicted the relationships between PPS and KAS (Fig. 5).

4 Discussion

KE has been proposed to associate the emotional (Kansei) needs of the customers and responses to the unique characteristics attribution of a product or service [27]. Thus, products should be designed to put forward the intended consumers feelings and emotions. This study has found a significant effect of the pareidolic and abstract shape on customer emotions by using Kansei affective science. KAS has revealed pareidolic pen stands were significantly preferred over abstract-designed pen stands. Purchase Preference for the pareidolic pen stand was significantly higher than abstract-designed product. There is a significant correlation between Kansei affective scale (KAS) score and the Product Purchase Preference scale (PPS) score in pareidolic and abstract- shaped pen stands. Higher product purchase preference is associated with higher perceived Kansei affective meaning of the products. Further, PLS-SEM has been predicted the total PPS score from the total KAS score. Both were statistically significant for pareidolic and abstract-shaped pen stand.

Pareidolic pen stands have animal-like impressions. This pareidolic forms emotionally triggers the consumers and gets attentions of the potential buyers [16] [28]. This emotional connection between product and consumer results a higher consumer-product interaction. Hence, this interactive design causes a preference towards pareidolic products. Like Anthropomorphic products, pareidolic products have also established a positive relationship with consumers [29, 30]. This cognitive process leads to the preference for pareidolic products over abstract products. Kansei design emphasizes users' affective needs and the affective considerations must be assessed and predicted [31]. Interactive pareidolic products engage customers psychologically and emotionally more than the abstract-shaped products [32]. Thus, product designers need to consider pareidolic features while developing a product [11]. Moreover, pareidolic products can evoke raised human emotions, perceptions, and sensory experiences than the abstract-shape products [33]. Further, the present study has revealed that affective hedonic meaning will take the upper hand over pragmatic quality because pen stands are more user-centric and do not need precise functionality. The results are crucial not only for the pen-stands design that affects the customers' emotions but also for other user-centric products [34]. Therefore, designers and product innovation management professionals need to consider pareidolic products more specifically pen stands for better acceptance in the market.

5 Conclusion

Findings suggest that pareidolic pen stands were evaluated as more emotionally affective than abstract pen stands. The consumers preferred pareidolic pen stands over abstract-shaped pen stands. Further study is proposed to investigate product preferences based on a larger population. Different methods (apart from Kansei engineering, like event related potential, Galvanic skin responses, eye tracking, etc.) should be applied to get quantification of affective design [35, 36]. It is important to combine two or more (both subjective and objective) methods to improve the analysis of the design elements based on users' psycho-physiological preferences.

Acknowledgements Authors would like to acknowledge the participants and the University BI grant for the fulfilment of this research. Authors would also like to thank Mr. Romit Majumder, Ms. Sutithi Dey and Ms. Madhuri Dutta (Research fellow, Department of Physiology, University of Calcutta) for his contribution to proof reading the manuscript. Authors would also like to thank Mr. Nitish Kumar Singh (Project Trainee, Centre of Excellence, Footwear Design and Development Institute, Kolkata) for his valuable insights in image editing.

Authors Contribution SB[1]–data collection, writing and editing of the manuscript, statistical analysis, conceptualization of the work. RP[2]–SB[1]–data collection, writing. SB[3]–graphics designing, SD[1]–conceptualization of the work. SG[1]–correction of the manuscript, conceptualization and standardization of the work.

Disclosure of Interests Nil.

References

1. Luo S, Lin H, Hu Y, Fang C (2022) Preliminary study on the aesthetic preference for taillight shape design. Int J Ind Ergon 87:103240. https://doi.org/10.1016/j.ergon.2021.103240
2. Cheung ML, Pires GD, Rosenberger PJ, Leung WKS, Salehhuddin Sharipudin MN (2021) The role of consumer-consumer interaction and consumer-brand interaction in driving consumer-brand engagement and behavioural intentions. J Retail Consum Serv 61:102574. https://doi.org/10.1016/j.jretconser.2021.102574
3. Shi A, Huo F, Hou G (2021) Effects of design aesthetics on the perceived value of a product. Front Psychol 12:1–11. https://doi.org/10.3389/fpsyg.2021.670800
4. Banerjee S, Dey S, Gangopadhyay S (2022) Evaluation of system usability scale as a marker of non-human computer interface's usability: a sanitizer container-based study. J Ergon Adv Res 2:1–20

5. Ulrich KT, Eppinger SD (2012) Product design and development. McGraw-Hill, New York
6. Epley N, Waytz A, Cacioppo JT (2007) On seeing human: a three-factor theory of anthropomorphism. Psychol Rev 114:864–886. https://doi.org/10.1037/0033-295X.114.4.864
7. Aggarwal P, Mcgill AL (2007) Is that car smiling at me? Schema congruity as a basis for evaluating anthropomorphized products. J Consum Res 34:468–479. https://doi.org/10.1086/518544
8. Schroll R (2022) "Ouch!" When and why food anthropomorphism negatively affects consumption. J Consum Psychol, pp 561–574. https://doi.org/10.1002/jcpy.1316
9. Chowdhury A, Karmakar S, Ghosh S, Chakrabarti D (2014) Purchase intention of anthropomorphic chair is 4:133–140
10. Niu D, Terken J, Eggen B (2018) Anthropomorphizing information to enhance trust in autonomous vehicles. Hum Factors Ergon Manuf 28:352–359. https://doi.org/10.1002/hfm.20745
11. Wodehouse A, Brisco R, Broussard E, Duffy A (2018) Pareidolia: characterising facial anthropomorphism and its implications for product design. J Des Res 16:83–98. https://doi.org/10.1504/JDR.2018.092792
12. Noble E, Wodehouse A, Robertson DJ (2023) Face pareidolia in products: the effect of emotional content on attentional capture, eagerness to explore, and likelihood to purchase. Appl Cogn Psychol, pp 1–14. https://doi.org/10.1002/acp.4105
13. Nagamachi M (1995) Kansei engineering: a new ergonomic consumer-oriented technology for product development. Int J Ind Ergon 15:3–11. https://doi.org/10.1016/0169-8141(94)00052-5
14. Nagamachi M (2016) Kansei/affective engineering and history of kansei/affective engineering in the world
15. Nagamachi M, Lokman AM (2015) Kansei innovation. CRC Press, Taylor & Francis Group
16. Guo F, Qu QX, Nagamachi M, Duffy VG (2020) A proposal of the event-related potential method to effectively identify kansei words for assessing product design features in kansei engineering research. Int J Ind Ergon 76:102940. https://doi.org/10.1016/j.ergon.2020.102940
17. Guo F, Liu WL, Cao Y, Liu FT, Li ML (2016) Optimization design of a webpage based on kansei engineering. Hum Factors Ergon Manuf 26:110–126. https://doi.org/10.1002/hfm.20617
18. Erdfelder E, FAul F, Buchner A, Lang AG (2009) Statistical power analyses using G*Power 3.1: tests for correlation and regression analyses. Behav Res Methods 41:1149–1160. https://doi.org/10.3758/BRM.41.4.1149
19. Hadiana A, Permana B, Tjahjadi D (2019) Kansei approach in development of application interface design based on user's emotional feeling. Eur J Eng Res Sci 4:121–126. https://doi.org/10.24018/ejers.2019.4.10.1566
20. Schütte STW, Eklund J, Axelsson JRC, Nagamachi M (2004) Concepts, methods and tools in kansei engineering. Theor Issues Ergon Sci 5:214–231. https://doi.org/10.1080/1463922021000049980
21. Cronbach LJ (1951) Coefficient alpha and the internal structure of tests. Psychometrika 16:297–334. https://doi.org/10.1007/BF02310555
22. Hayes AF, Coutts JJ (2020) Use omega rather than cronbach's alpha for estimating reliability. But Commun Methods Meas 14:1–24. https://doi.org/10.1080/19312458.2020.1718629
23. Peña-García N, Gil-Saura I, Rodríguez-Orejuela A, Siqueira-Junior JR (2020) Purchase intention and purchase behavior online: a cross-cultural approach. Heliyon 6. https://doi.org/10.1016/j.heliyon.2020.e04284
24. Zhou R, Tong L (2022) A study on the influencing factors of consumers' purchase intention during livestreaming e-commerce: the mediating effect of emotion. Front Psychol 13:1–15. https://doi.org/10.3389/fpsyg.2022.903023
25. Field A (2017) Discovering statistics using IBM SPSS statistics. SAGE Publications Ltd., London
26. Hair JFJ, Black WC, Babin BJ, Anderson RE (2018) Maltivariate data analysis. Cengage Learning India Pvt. Ltd
27. Hartono M (2020) The modified Kansei engineering-based application for sustainable service design. Int J Ind Ergon 79:102985. https://doi.org/10.1016/j.ergon.2020.102985

28. Mamaghani NK, Rahimian E, Pishro AJ, Mortezaei SR, Mortezaei S-R (2014) Kansei engineering approach for consumer's perception of the ketchup sauce bottle. Int Conf KANSEI Eng Emot Res Kansei, pp 1487–1494
29. Airenti G (2018) The development of anthropomorphism in interaction: inter subjectivity, imagination, and theory of mind. Front Psychol 9:1–13. https://doi.org/10.3389/fpsyg.2018.02136
30. Cao Y, Zhang Y, Ding Y, Duffy VG, Zhang X (2021) Is an anthropomorphic app icon more attractive? Evid Neuroergonomomics Appl Ergon 97:1–13. https://doi.org/10.1016/j.apergo.2021.103545
31. El Marghani VGR, Da Silva FC, Knapik L, Verri MA (2013) Kansei engineering: methodology to the project oriented for the customers
32. Demirbilek O, Sener B (2003) Product design, semantics and emotional response. Ergonomics 46:1346–1360. https://doi.org/10.1080/00140130310001610874
33. Naeini H, Heidarpour M (2011) Kansei engineering and ergonomic design of products. Int. J. Occup. Hyg. 3:81–84
34. Kittidecha C, Yamada K (2018) Application of Kansei engineering and data mining in the Thai ceramic manufacturing. J Ind Eng Int 14:757–766. https://doi.org/10.1007/s40092-018-0253-y
35. Olofsson JK, Nordin S, Sequeira H, Polich J (2008) Affective picture processing: an integrative review of ERP findings. Biol Psychol 23:1–7
35. Laparra-Hernández J, Belda-Lois JM, Medina E, Campos N, Poveda R (2009) EMG and GSR signals for evaluating user's perception of different types of ceramic flooring. Int J Ind Ergon 39:326–332. https://doi.org/10.1016/j.ergon.2008.02.011

Raagrang: A Designer's Perspective on Enhancing the Indian Classical Music Listening Experience Using Emotion Musicality-Based Hue Mapping

Kedar Dicholkar, Sanket Dutta, and Ritwij Bhowmik

1 Introduction

Colour-visual and music-audition interweaving give humans aesthetic and psychological pleasure. Synaesthesia describes this psychological condition. Traditional synaesthesia is the ability to 'see' music and 'hear' colours. Only 1–4% of the population experiences this [1], but what about those without synaesthesia? Do most individuals view sounds and hues similarly? [2]. How do people form these relationships? When the sound source can be identified and visually recalled, sounds are often associated with colours (e.g. ocean waves with blue). Non-synesthetes tend to have high music-colour correlations across ethnicities and societies. When asked to choose a hue that 'imbibes' with a classical musical sample, U.S. and Mexican participants chose lighter, more saturated, yellower colours for fast-paced major mode works [3].

If applied to music, can these colour mappings improve the listening experience as a correlation? Some music players fulfil this demand. However, most give illumination unrelated to the music, which might disrupt the experience. Matching a hue to real-time music is tough. To solve the mismatch and automatically match colours to any music played arbitrarily, a primary set of correlations must be established, including the mapping between musical parameters and colours, recognition and mapping of music being played onto the colour wheel, and RGB-LED mixed light colours. This study addresses the same subject but focuses on Hindustani Classical Music. We proceed to explore the history of visualizing music and prior attempts to map and codify sound to visual translation.

K. Dicholkar (✉)
Symbiosis Institute of Design, Symbiosis International (Deemed University), Pune, India
e-mail: kedarnehadicholkar@gmail.com

S. Dutta · R. Bhowmik
Indian Institute of Technology Kanpur, Kanpur, India

© The Author(s), under exclusive license to Springer Nature Singapore Pte Ltd. 2025
S. Das et al. (eds.), *Human-Centered Digitalization, Volume 3*, Design Science and Innovation, https://doi.org/10.1007/978-981-96-4383-7_18

2 History and Related Work

Cutietta and Haggerty [4] presented an assertion from the point of view of age demographics, stating that specific colours are associated with particular music kinds by a high number of people from different ages. Due to the plethora of affective responses that music stimulates, they draw the conclusion that colour, and music associations are very common among individuals. Furthermore, this study attempts to explain how people from diverse backgrounds associate specific musical genres with specific colours, based on their emotions, visual memory, and imagination.

In their investigation, Ward [5] identified distinct differences between sound-colour associations as experienced by synaesthetes and non-synaesthetes. Consistency of colour associations, specificity of colour selection, and automaticity of colour associations were the primary distinctions. Nonetheless, both groups utilize the same cognitive mechanism for mapping the auditory to the visual, which involves mapping between intonation and brightness.

Pridmore [6] proposed that hue corresponds to tone via wavelength and that loudness corresponds to brightness via amplitude. Lipscomb and Kim [7] discovered that pitch corresponded to vertical position, timbre to shape, and volume to size. In contrast, the emotional mediation hypothesis proposes that colour and music are indirectly related through shared, higher-level emotional associations [3, 8–12]. This view associates colours and music based on their shared emotional content. For instance, jolly sounding music would be associated with vibrant and jovial looking colours and sad-sounding music would be associated with depressive and pensive looking colours. The majority of the evidence for emotional mediation comes from studies in which non-synaesthetes were asked to select colours based on how well they went with various pieces of western classical music [3, 13]. Participants then rated each musical selection and each colour on several emotion-related scales1 (e.g. happy/sad, angry/calm, strong/weak) in order to assess the role of emotion in their colour–music associations. Strong correlations were found between the emotion-related evaluations of the musical excerpts and the emotion-related ratings of the colours that participants deemed to go best/worst with the music (e.g. 0.97 for happy/sad and.96 for strong/weak; [3]).

The above discussions use western classical or other western genres as their audio signals to discuss the correlation. Our study revolves around Indian classical music as we strongly feel that Ragas have a rich context and character. HCM has stricter rules and regulations in terms of exclusivity of the notes in the scale and the purity of the Raga. HCM prohibits any usage of notes beyond the predetermined scale. Hence, we feel that Raga samples would give better results in terms of stronger correlations. Kelkar [14] discusses visualization of the tonal spaces in which Ragas exist by constructing a graph with a force-directed configuration. She evaluates a radial layout for these tonal spaces in comparison with a colour harmony profile and by applying the methodologies used to derive colour schemes, describe some cross-Raga relationships.

3 Ragas and Thaats

Ragas can be considered HCM equivalents of modes but there are quite a few additional rules and regulations in terms of phrasing and presentation. These include prominence of certain notes, refrain of others, signature phrasing, octave emphasis, and peculiar inflections on certain notes to give character to the Raga. Other common features like harmonization, transposition, chromatic scales which are commonly used in western music are absent in HCM.

3.1 Swaras

Similar to the western system of solfège, there are 12 notes in HCM; 'Shadja' (S), Rishabh (R), Gandhar (G), Madhyam (M), Pancham (P), Dhaiwat (D), and Nishad (N). Notes have a flat (R, G, D, N; known as Komal Swaras) or sharp (Augmented) form (M, known as Teevra Swara) except the first and fifth (S and P) which stand unaltered.

3.2 Ragas

We will now discuss some descriptors of the Raga that determine the behaviour to some extent and would be relevant for our future discussions.

Vadi-Samvadi. The Vadi note is defined as the most significant note in the Raga, with the Samvadi note ranking second. It is usually a norm that if the Vadi is present in the first half of the octave (S, R, G, M, and P) the Samvadi is usually found in the second half of the scale (P, D, N, S). This provides balance in terms of presentation of the Ragas.

Anga Pradhanta (Octave Prominence). There are certain Ragas which are primarily articulated in the lower half of the octave, these Ragas are known as Poorvanga Pradhan Ragas. These Ragas mostly have the Vadi in the first half of the octave. Poorvang Pradhan Ragas usually tend to strike a pensive, sorrowful, or distressed chord. Whereas, Ragas with Vadi in the upper half of the octave are usually known to be Uttarang Pradhan Ragas. These Ragas tend to emanate a vibrant aura.

Thaat. The Thaat system, developed by Pt Bhatkhande, is a succinct way to classify Ragas using their notes and Vadi and Samvadi notes, preserving structure and practicality [15]. The Circle of Thaats (see Fig. 1), akin to the Circle of Fifths in the west, it shows the relationship between Thaats. Since all thaats are heptatonic, the number of minor notes increases clockwise around the circle, starting with Bilawal (all major notes) and ending with a melodic arrangement that doesn't belong to Thaat system but has the most minor intervals. Bilawal and the unidentified thaat are diametrically

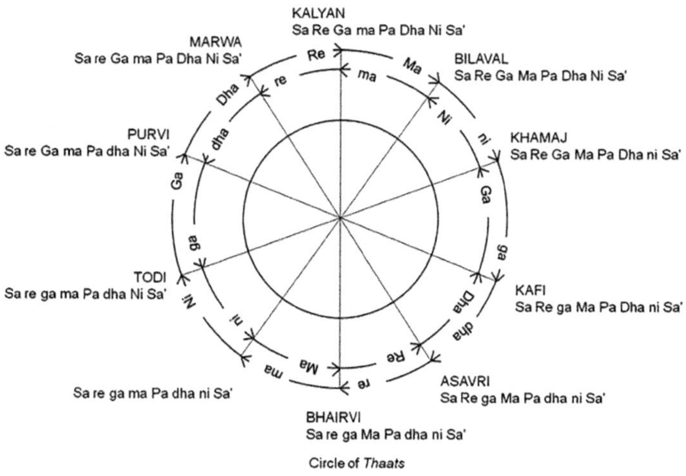

Fig. 1 Circle of Thaats [16]

opposite. As one progresses away from Todi, the number of major notes increases until one substitutes the Augmented 4th of Kalyan Thaat with the Natural 4th to reach Bilawal again. Interestingly, a fundamental separation with M as a parameter is also visible with the natural and Augmented 4th, Thaats are situated on the left and the right side of the circle, respectively.

4 Methodology

For this experiment, a computer-based subjective test was designed to assess the choice of colours that the listeners felt that 'visualizes' the Raga the best. 10 Samples of Ragas were played for the test subjects to listen to aid with the visualization. After listening to an audio sample, they were told to choose the colour that visualized the Raga to their accordance. The test sheet design (see Fig. 2) can be seen in the figure. Rather than providing the listeners with a selective palette of basic hues [17, 18], a subtractive colour wheel was used for the listeners to choose a hue of their choice. An eye dropper tool was used by the listener to choose the colour of their choice, giving them a greater control over their choices.

4.1 Music Set Sampling

A total of 30 non-synaesthetic subjects (having an aural exposure and experience of listening and appreciating HCM) gave their inputs for this test. The subjects, listeners

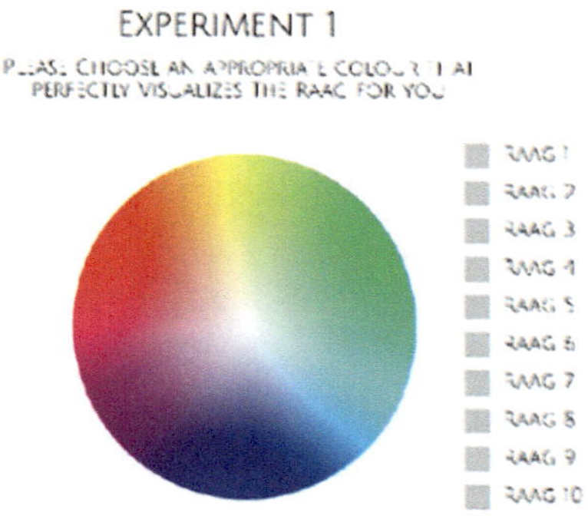

Fig. 2 Test designed to conducted user data on music-colour association

in this case, were divided into 2 groups A and B, who tested with the 2 sets of Ragas with 10 Ragas each, with the results being documented separately (Table 1).

For the context to remain unaltered and avoid dissonance, all the samples collected were set at the same pitch, D#. Aiding to the coherence, all the Ragas were sequenced according to their positions in the Thaat circle, starting from Bilawal and ending at Kalyan.

4.2 Colour Wheel Model

As shown in Fig. 3a, we use the HSV model to depict colours in this paper. This layout includes Hue, Saturation, and Value parameters, with hue distributed around the angle of the circle, saturation varying radially outwards, and value on a vertical axis depicting the depth. The colours are distributed along the light spectrum, with the principal hues occupying triadic positions 120 degrees apart. This creates the cylindrical colour wheel depicted in Fig. 3a.

Since we aim to present Heat Map primarily based on 2D colour wheel best represented by Hue and Saturation in a plane with maximum value. Points outside this plane need to be projected to have maximum value. Simply removing the value

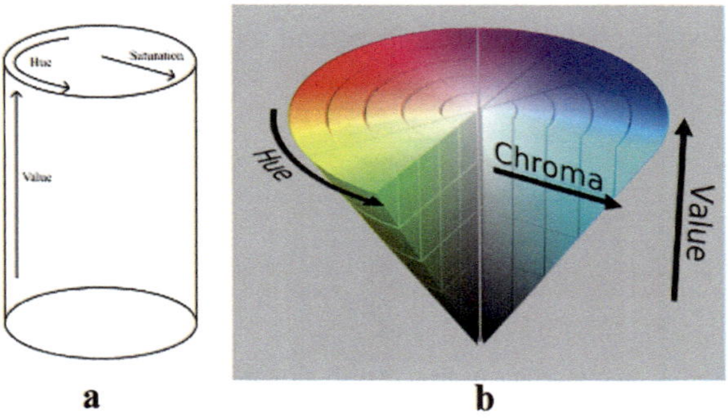

Fig. 3 **a** HSV colour model and, **b** HSV value versus chroma (range of RGB values) rather than saturation (chroma over maximum chroma for that segment) yields a bicone or cone as the resulting solid [19]

component and plotting Hue and Saturation may cause misinterpretation of data due to incorrect projection onto the 2D plane. To counter this misinterpretation, we translate the HSV model into a conic form, Hexone Model with Chroma (Brightness), which is a multiple of Saturation and Value (S*V), as its Radial Dimension (see Fig. 3b).

5 Analysis

We will now look at trends seen in the choices of the subjects and discuss whether a possible correlation can be drawn from the visualizations of the data. We are using bar gradients (in sequence derived from the circle of Thaats) and Heat Maps (Cumulative heat areas of individual Gaussian distribution of each Raga plot) for our analysis.

5.1 Theory 1-Circular. If not, then Complementary

Figure 4 depicts mean values of the heat distribution for the combined Set, Set A and Set B. For our analysis, let us consider Bhairav as a different case of non-conformity. On keen inspection, a clear progression of colours can be noticed in the combined set. The mean colour coordinates of each Thaat tend to travel clockwise from the bright shade hue of a bright Crayola (#C3E870) to return to a similar zone with Kalyan having a hue of Dark Khaki (#BAC86D), both of which are in the Yellow-Green region.

Referring to the colour wheel in Fig. 3b, you will notice that the colour pairings of Set A and B in Thaat Khamaj, Kafi and Asawari appear to be roughly complementary to each other. Hence, the hues of Grey appear in the combined set mean value for the following Thaats.

Based on qualitative analysis of the collected data, it can be concluded that, when matched with the sequences of Ragas as prescribed in the Thaat circle, colour preferences tend to move roughly in a circular direction, with Bilawal Thaat falling within the Yellow-Green to green region. Secondly, if the subject's preferred hue does not belong to the subsequent sequential region of the colour wheel, it will likely

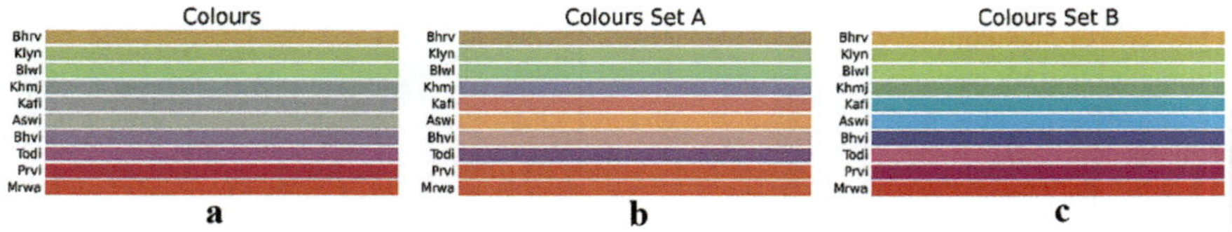

Fig. 4 Mean values of heat distribution for, **a** combined distribution, **b**. distribution for Set A and **c**. Distribution for Set B

fall into the complementary region. Consequently, it can be summed up in just one phrase: Circular. If not, then complementary. The cause of complementarity shifts in certain cases of hue preferences will be discussed in the next section.

5.2 *Theory 2—Anga Pradhanta, Shuddha M Versus Teevra M*

As described in Sect. 3.2, the Vadi-Samvadi of a Raga can determine whether the Raga is intended to be in the lower or upper half of the Octave. This emphasis on specific regions of the spectrum, in our opinion, plays a significant role in explaining the complementary shift. Ragas Sahana, Adana, Bilaskhani Todi, Sohni, and Shankara are Uttarang Pradhan Ragas, either owing to traditional progressions of presenting the Raga or the presence of the Vadi Swara in the upper half of the octave.

For Shuddha M Thaats, we have Kafi and Asawari with their Uttarang Pradhan Raga, i.e. Raag Sahana and Adana, respectively, both in Set A. If we look at the Figs. 5d and 6d, gradient bar chart clearly indicates a clear dominance of hues of pink and Magenta for Sahana. We also find majority of subject choices lying in the region of vibrant oranges and yellow. Both these regions are complementary to their sequential circular counterparts. This conforms to the notions of previous well documented studies which stated that higher pitched tones are associated with lighter, brighter colours [5, 20, 21]. On the other hand, Marwa, which is a Teevra M Raga, shows no complementary jump in either of it sets, despite Sohni (Set B) being an Uttarang Raga (Fig. 7a, b).

The team of author strongly feel that the complementary shift in Shuddha M Thaats in cases of Uttarang Ragas is due to the vibrant and bright sounding features of the Ragas. Since the Thaats to which these Raag belong are tentatively located in the cool hemisphere when superimposed colour wheel (as per theory 1), the Raga shows a tendency of moving towards warmer region of the colour wheel and hence the evident complementary jump. On the other hand, the Ragas which belong to thaats with Teevra M sound very pensive in nature, irrespective of octave affinities, if any. Since, the Thaat is already tentatively located in the hottest region of the

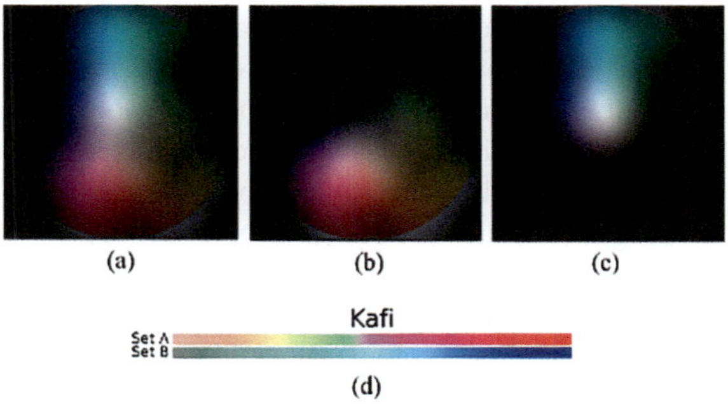

Fig. 5 **a** Heat Map for combined distribution for Kafi Thaat, **b**. map for Sahana (Set A) and c. map for Megh Malhar (Set B). d. Gradient bar charts for Sahana and Megh Malhar

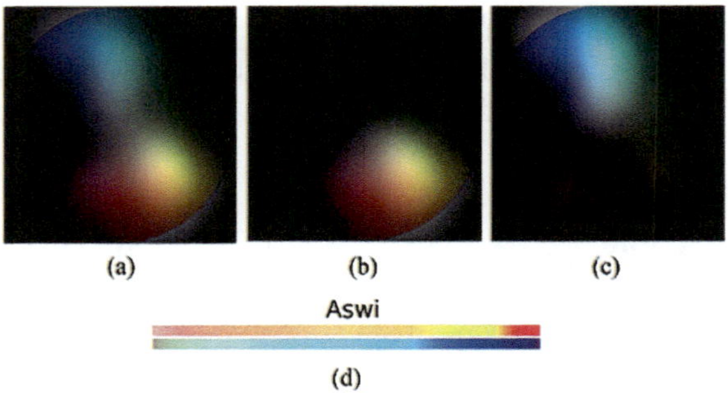

Fig. 6 **a** Heat Map for combined distribution for Asawari Thaat, **b** map for Adana (Set A) and c. map for Jaunpuri (Set B). d. Gradient bar charts for Adana and Jaunpuri

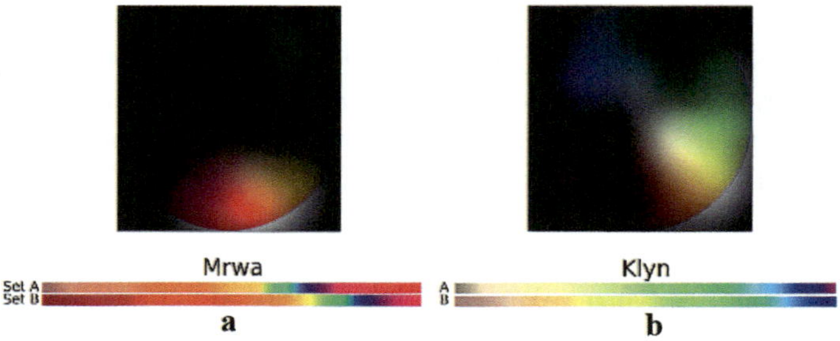

Fig. 7 **a** Heat Map and gradient bar charts for combined distribution for Asawari Thaat, **b** Heat Map and gradient bar charts for combined distribution for Kalyan Thaat

colour wheel in accordance with the circular sequence of Thaat Alignment, there is no noticeable jump seen towards complementary regions as seen in Shuddha M Thaats. Hence, we conclude this section by stating that the first half of the thaat circle is heavily dependent on Octave affinities (Ang Pradhanta) in terms of determination of the hue for a particular Raga belonging to the Thaat in question. Teevra M Thaats stand relatively unaffected.

6 Discussions

Though we have qualitatively explained the phenomena being observed using principles of HCM and previous results obtained for octave-based scale variations being reflected across colour hues [22], more consolidated data was required to quantitatively assert the significance of the correlations seen as per Theory 1 and Theory 2. We strongly feel that an expansive form of the test with greater number of individuals and lengthier, context rich Raga samples could provide us with data to quantitatively assert the statements proposed in our theories. A few Research studies demonstrate that music-to-colour associations are better characterized as emotionally mediated (music-to-emotion to colour) rather than direct (music-to-colour) using more conclusive methods than previous studies [3, 10, 13]. It seems to be a challenge to identify

Table 1 Raga lists for set A and set B

	Thaat	Set A	Set B
1	Bilawal	Hansadhwani	Shankara
2	Khamaj	Jaijaiwanti	Rageshree
3	Kafi	Sahana	Megh malhar
4	Asawari	Adana	Jaunpuri
5	Bhairavi	Bilaskhani Todi	Bhairavi
6	Todi	Multani	Todi
7	Poorvi	Poorvi	Puriya dhanashree
8	Marwa	Marwa	Sohani
9	Kalyan	Yaman	Kedar
10	Bhairav	Bhairav	Ramkali

any low-level perceptual characteristics shared by music, colours, and emotional faces that could reasonably explain the common associative patterns. However, it is simple to explain these outcomes if they are all mediated by emotion [23]. Hence in our estimation too, correlation results between a colour music and emotion module would be similar to our study. But it could give a richer context as to how and why Ragas get mapped according to such an association.

7 Conclusion

This investigation in the field of colour-music association from an HCM purview disclosed that every Raga generates a colour atmosphere in the perception of listeners, whether consciously or unconsciously. We have proposed two theories of Circularity-Complementarity and Octave emphasis shifts with substantial proof in data as well visualization. The conscious association between colours and music could open doors to imagination and inspiration and expand people's perception of the interpretations of these Ragas leading to a richer and much more holistic understanding of Ragas and their nuances. The outcomes of this research could be applied to an audio-visual performance, an animation or film, or a graphical interface in order to obtain an aesthetically pleasing and well-balanced combination of music, colours, and light.

References

1. Simner J, Mulvenna C, Sagiv N (2006) Synaesthesia: The prevalence of atypical cross-modal experiences. J Percept London 35(8):1024–1033
2. Hubbard TL (1996) Synaesthesia-like mappings of lightness pitch, and melodic interval. Am J Psychol 109(2):219–238

3. Palmer SE, Schloss KB, Xu Z, Prado-León LR (2013) Music–color associations are mediated by emotion. Proc Natl Acad Sci
4. Cutietta RA, Haggerty KJ (1987) A comparative study of color association with music at various age levels. J Res Music Educ 35(2):78–91
5. Ward J, Huckstep B, Tsakanikos E (2006) Sound-colour synaesthesia: to what extend does it use cross-modal mechanisms common to us all? Cortex 42(2):264–280
6. Pridmore RW (1992) Music and color: relations in the psychophysical perspective. Color Res Appl 17:57–61
7. Lipscomb SD, Kim EM (2004) Perceived match between visual parameters and auditory correlates: an experimental multimedia investigation. In: Proceedings of the 8th international conference on music perception and cognition, pp 72–75. Casual Productions, Adelaide, Australia
8. Barbiere JM, Vidal A, Zellner DA (2007) The color of music: correspondence through emotion. Empir Stud Arts 25:193–208
9. Bresin R (2005) What is the color of that music performance? In: Proceedings of the international computer music conference (pp 367–370). International Computer Music Association, Barcelona, Spain
10. Lindborg P, Friberg A (2015) Colour association with music is mediated by emotion: evidence from an experiment using a CIE Lab interface and interviews. PLoS One 10
11. Palmer SE, Langlois TA, Schloss KB (2016) Music-to-color cross-modal associations of single-line piano melodies in non-synesthetes. Multisensory Research 29:157–193
12. Sebba R (1991) Structural correspondence between music and color. Color Res Appl 16:81–88
13. Isbilen ES, Krumhansl CL (2016) The color of music: emotion-mediated associations to Bach's Well-Tempered Clavier. Psychomusicology: Music, Mind, and Brain 26:149–161
14. Kelkar T (2015) Colour wheel for music. https://doi.org/10.13140/RG.2.1.1327.8880
15. Chordia P, Rae R (2007) Modeling and visualizing tonality in North Indian classical music. Neural Inf Process Syst, Music Brain Work
16. Jairazbhoy NA (1995) The raags of North Indian music: their structure and evolution, 2nd edn. Popular Press Pvt. Ltd., Mumbai
17. Patera M, Tsiounta M, Staniland M (2013) Why is classical music yellow: a colour and sound association study
18. Collier WG, Hubbard TL (2004) Musical scales and brightness evaluations: Effects of pitch, direction, and scale mode. Music Sci 8:151–173
19. HSL and HSV. Wikipedia. https://en.wikipedia.org/wiki/HSL_and_HSV. Accessed 21 Aug 2023
20. Marks LE (1987) Cross-modal similarity: auditory–visual interactions in speeded discrimination. J Exp Psychol Hum Percept Perform 13:384–394
21. Ward J, Huckstep B (2006) Tsakanikos, E: sound-colour synaesthesia: to what extent does it use cross-modal mechanisms common to us all? Cortex 42:264–280
22. Spence C (2011) Cross-modal correspondences: a tutorial review. Atten Percept Psychophys 73:971–995
23. Whiteford K, Schloss K, Helwig N, Palmer S (2018) Color, music, and emotion: bach to the blues. i-perception 9:204166951880853. https://doi.org/10.1177/2041669518808535

User Survey of Music Streaming Apps

Shubhankar Joshi and Bhutkar Ganesh

1 Introduction

Music occupies a significant and cherished position in the lives of individuals worldwide. The diverse and vibrant culture of India has given rise to a rich tapestry of musical traditions, each resonating with its own unique charm. From classical ragas to Bollywood hits, and folk melodies to devotional songs, the Indian populace showcases a remarkable affinity for various types of music.

Bollywood music, known for its catchy tunes, lively rhythms, and cinematic appeal, has an immense fanbase in the country. The Digital Music Study report of 2022 [1] included a user survey involving 4000 respondents. According to their results, Bollywood music was heard by the most of respondents in the user survey. To get a clear view, they divided Bollywood music into five subsections. Bollywood (Modern) with the latest songs released in the past two years, Bollywood (Classic) with a list of classic songs sung by legendary singers like Kishore Kumar and Lata Mangeshkar, Bollywood (New Age 2000–2020) with songs from the past twenty years, Bollywood (80–90 s) with songs from the 90 s era and Bollywood (Retro) with songs between the late 50 s and 60 s. Around 49% preferred Bollywood (Modern), 40% chose Bollywood (Classic), followed by 33% Bollywood (New Age 2000–2020), 28% Bollywood (80-90 s), and 20% Bollywood (Retro). This survey highlights the classification of popular Bollywood music in India.

India is a place of multiple states where all regional cultures live in harmony. Folk music is a symbol of this harmony. Folk music represents the cultural fabric of different regions/states and adds a distinct flavor to India's musical landscape. The folk traditions from various states like Punjab, Rajasthan, Maharashtra and

S. Joshi (✉) · G. Bhutkar
Vishwakarma Institute of Technology, Pune, India
e-mail: joshishu16@gmail.com

G. Bhutkar
e-mail: ganesh.bhutkar@vit.edu

© The Author(s), under exclusive license to Springer Nature Singapore Pte Ltd. 2025

S. Das et al. (eds.), *Human-Centered Digitalization, Volume 3*, Design Science and Innovation, https://doi.org/10.1007/978-981-96-4383-7_19

Fig. 1 Popular music streaming apps in India

others, find immense popularity among people. According to [1], around 27% of their users listen to Punjabi music. Apart from these dominant genres, there is also a growing interest in Western music, including pop, rock, hip-hop, and electronic music. These genres collectively make up around 35% of the music preference in India, representing the influence of international music on the younger generation.

The major growth in the music market was seen after the effective use of Music streaming services like Spotify, YouTube Music, and Apple Music on their smartphones. Users were able to access music anytime and anywhere on these apps. Some apps offer free services with advertising, but all offer 'premium' services for a month or year. Spotify, YouTube Music, and Gaana are among the companies that provide free services with ads, as well as paid services without ads. It is also called Hybrid Service when the company uses free and paid models [2]. However, there are also a few music companies that offer only paid services, such as Apple Music and Tidal.

The trend of music apps came into the limelight because of the free model, also known as the freemium model [3]. According to [4], Music streaming apps are offering a 'free' version to attract many users. Another reason may be that companies continue to push the free version to control or conquer the Music streaming market.

To gain a better understanding of the landscape of Music streaming services, we conducted a user survey to evaluate the popularity, features, and user rating of different Music streaming apps. In Fig. 1, logos of Spotify, Gaana, Jio Saavn, YouTube Music, Amazon Music, Resso, and Apple Music are presented. These apps have common features like 'Creating Playlist,' 'Finding Favorites,' 'Song Recommendation,' and 'Artist Album.' This research paper provides valuable factors that users take into consideration before choosing the Music streaming app and their premium versions.

2 Literature Review

This research work identifies 2 research papers, 3 marketing reports, and 2 online questionnaires. All of them belong to a time span of 2012–2022. The keywords used for the literature review are 'Music streaming apps,' 'digital music,' 'music survey,' and 'music survey questionnaire' with Google and Google Scholar.

The Jackdaw Research team [5] conducted two surveys on Music streaming apps to understand the patterns and habits of users. The surveys were in the groups of 200 and 500 users. The patterns found in both groups were consistent. The survey stated the popular features of music apps. According to their findings, users loved to explore the feature—'favorites' in the apps. The survey was focused on how people were using Apple Music and its features. It also discussed the number of premium and free users and their retention rates.

Christian van Dijk [6] conducted a user survey of 216 users using different Music streaming apps. The aim of the survey was to study a type (free/ paid) of service preferred by the users. It also answers the factors that influence the user for the premium version. It compared the usage time of paid users versus free users. It was found that users who chose paid service tended to listen for hours compared to free users. User Income was one of the factors which influenced users for premium services.

The Indian Music Industry (IMI) [7] conducted a user survey of 4000 respondents in India. The survey was focused majorly on users aged between 16 and 44 years. The survey stated that around 54% of the respondents enjoyed music in live streams. The survey also highlighted that on other platforms such as video streaming, short-form video services, and social media, around 70% of the respondents listen to Indian artists. The survey also discusses the relation of music to mental health.

Mohesen et al. [8] conducted a user survey of 222 participants to understand the listening and management behaviors of users. The key findings from the survey indicate that factors like song familiarity, level of distraction, and alignment with the listener's mood, are influenced by the specific activity during which music is being enjoyed. The survey also concluded that users preferred passive listening and compared different recommendation systems.

Ruma et al. [9] compare different Music streaming services in India. The comparison is based on the premium fee, features, and plans provided by different Music streaming apps. The [10, 11] are different questionnaires for Spotify and Apple Music, in which the users are asked about how many hours they spend on these apps. The questionnaire also interrogates the most attractive app features to the users.

3 App Review

Before creating the questionnaire, 7 popular Music streaming apps were compared to find the common trends among the apps. Spotify, Gaana, Jio Saavn, Amazon Music, YouTube Music, Resso, and Apple Music were compared on the basis of their features, cost of premium versions, device limit, and many more. The comparison is shown in Table 1. In Table 1, 'M' stands for Millions, 'K' stands for Thousands, and 'NA' stands for Not Available.

From this app review, it is seen that all the apps provide some basic features like creating playlists, recommendations, and discovery of new music. Except for Resso, all the other apps are available in more than 100 countries. Indian Currency is mainly accepted for payment options. All the apps have more than 50 million users. Spotify has the highest, 205 million premium users, out of its 500 million users. Spotify has around 20 million reviews on the Google Play Store. These apps have

Table 1 Comparison of popular music streaming apps

Items	Spotify	Gaana	Jio Saavn	Amazon music	YouTube music	Resso	Apple music
Vendor	Spotify	Times Internet	Reliance Industries	Amazon	Google	Byte Dance	Apple Inc
Start year	2008	2010	2007	2007	2015	2020	2015
Globally available	184 Countries	Global	Global	Global	101 Countries	India, Brazil, Indonesia	175 countries
Free access	Yes	Yes	Yes	Partial	Yes	Yes	Yes (Radio)
Premium charges in INR/year	1189	0339	0749	0999	1290	1150	0999
Currencies	INR, USD, and AUD	INR and USD	INR	INR, USD, and EURO	USD and INR	INR	USD, CAD, and INR
App size	25 MB	32 MB	27 MB	74 MB	20 MB	7.7 MB	45 MB
No. of tracks	100 M +	45 M	55 M +	75 M	100 M +	NA	100 M +
No. of users	500 M	201 M	100 M	82.2 M	2500 M	50 M	88 M
Premium users	205 M	NA	NA	NA	80 M	NA	88 M
Offline mode	Yes	Yes	Yes	Yes	No	Yes	Yes
Track Downloads	Yes	Yes	Yes	Yes	Yes	Yes	Yes
Podcast	Yes	Yes	Yes	Yes	Yes	No	Yes
Login required	Yes	Limited	Limited	Yes	No	Yes	Yes
App rating	4.3	4.3	4.3	3.7	4.2	4.3	3.7
Total reviews	20 M	5 M	5 M	2.7 M	4.4 M	3.4 M	0.5 M

an average rating of more than 4.1 stars on the Google Play Store. All the apps have different login systems. For example, apps like Gaana and Jio Saavn have a Guest login option, where users do not have to mention their personal details like email or mobile number. While other apps like Spotify or Amazon Music need some kind of personal verification from the user. It is observed that most of the apps followed the freemium version except Apple Music. In Apple Music, the only free tool available is the Radio.

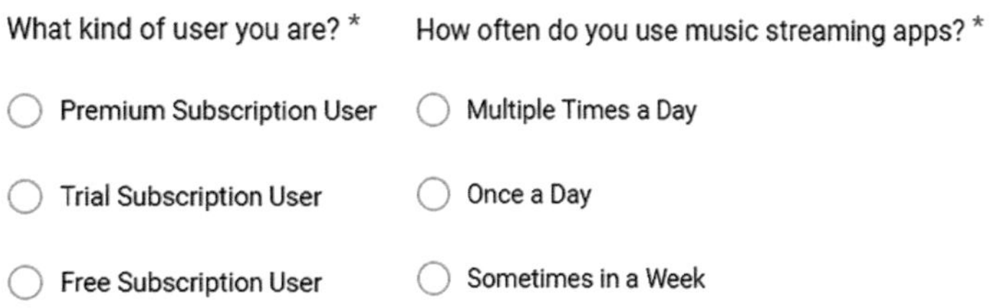

Fig. 2 Sample questions in the user survey

4 Research Methodology

This section discusses the research methodology followed, including the design of the questionnaire, data collection process, and demographic representation of participants.

4.1 Design of Questionnaire

After the App review, a structured questionnaire of 26 questions was created and divided into three sections. The respondents were directed to the second or third section depending on their responses in the first section. All the questions are objective multi-optional questions with 2 to 4 options. The first section included questions about their demographic information such as gender, age, occupation, and others. The second section consisted of questions related to Music streaming apps. This section was focused on obtaining what features of music apps are essential to the users. Figure 2 provides a couple of sample questions in this second section. The third section consisted of questions for users not using Music streaming apps. This section is focused on knowing the reasons for not using the Music streaming apps and also, the possibility of using them in the future.

4.2 Collection of Data

The user survey is mainly targeted at Indian users, but it also has respondents from foreign countries. The survey has been targeted at different age groups, which are (16–24), (25–44), (45–60), and 60 + years. The survey has a total 210 respondents, out of which 180 respondents are considered for the analysis, as seen in Table 2. The other 30 respondents are not considered in the rest of the results and in further analysis. Out of these 30 users, 10 users don't listen to music on these apps and the other 20 are invalid users. In order to validate our responses, some questions are repeated in the questionnaire. If the user gives the same answer for those questions, then only it's a valid response. There were 20 such users in our survey and those invalid users are not considered in the analysis.

The online tool used to design this survey is Google Forms. It is used since it is easy to understand, easy to design questionnaires and also, supports data analysis effectively. The questionnaire has been circulated on different social media platforms such as Instagram, Twitter, Whatsapp, and Reddit.

Tables 2, 3, 4, 5, and 6 depict the demographic information of users such as the number of app users, age groups, gender distribution, and occupations-wise distribution.

In Table 3, the age-wise distribution of participants is given. The maximum respondents to the user survey belonged to the age group of 16–24 years. The users in this category are students from schools and colleges. The second major category is of 25–44 years have 60 users, which are mostly working professionals.

In Table 4, it is observed that there are 68% male users and 32% female users participated in this user survey.

Table 2 Respondents to user survey

Respondents using music streaming apps	Respondents not using music streaming apps	Invalid users	Total users
180	10	20	210
85.5%	5%	9.5%	100%

Table 3 Age group of respondents of user survey (in years)

16–24	25–44	45–60	Above 60	Total
105	60	08	07	180
58%	34%	4%	4%	100%

Table 4 Gender of respondents to user survey

Male respondents	Female respondents	Total users
123	57	180
68%	32%	100%

Table 5 Occupation of the respondents of the user survey

Occupation	No. of respondents	Percentage (%)
Business	28	16
Working professional	64	36
Student	82	45
Retired employee	06	03
Total	180	100

Table 6 Role of respondents to the user survey

User type	No. of respondents	Percentage (%)
Music enthusiast	106	59
Part-time singer/instrumentalist	24	13
Professional artist/performer	15	08
Nothing specific	35	20
Total	180	100

Table 5 displays the occupations of the users who participated in the survey. Around 45% of users are students and 36% are working professionals. Only 3% of users were retired and belonged to the 60 + years age group.

In order to get better results, the users are asked to describe themselves. In Table 6, this distribution is stated. Around 59% of the respondents classify themselves as music enthusiasts. In this survey, 15 professional artists participated and shared their experiences.

5 Results and Discussion

In this section, data analysis is performed according to the responses given by the respondents, during the user survey. The section discusses parameters like popularity, usage, and charges of Music streaming apps. Popularity is discussed according to age groups, location, and gender. The weekly usage of premium and free users is presented. This section also highlights important features of Music streaming apps. Finally, the app ratings by users are discussed.

5.1 Popularity of Streaming Apps with Different Factors

In this section, popularity of Music streaming apps is discussed with factors like age, location, and device. The key reasons for the results obtained from the user survey are also presented.

5.1.1 Popular Music Streaming App

As seen in Fig. 3, a user can use multiple Music streaming apps. Spotify is the most popular app in the user survey closely followed by YouTube Music. Spotify is used by 97 out of 180 users and 94 users use YouTube Music.

The main reason for their popularity is that Spotify and YouTube Music each provides more than a hundred million tracks free to their users [12]. Also, this survey

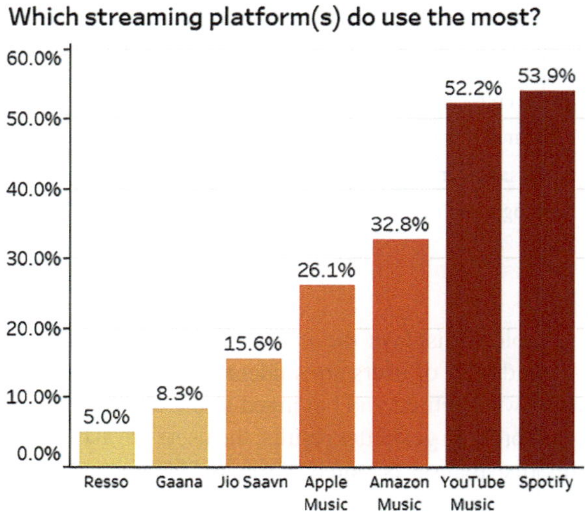

Fig. 3 Bar graph of music streaming apps popularity

has a maximum number of respondents from the age group of 16–24 years. The users from this age group are generally students who can't afford paid apps like Apple Music.

5.1.2 Location-Wise Popular Music Streaming App

In Fig. 4, the location-wise distribution of respondents in the user survey is provided. In the survey, the respondents are asked a question—Are they in India or Abroad? As the survey is mainly focused on Indian users, the option for the location of users outside of India is given the title as 'Abroad.' There are 130 Indian users and 60 Abroad users. According to Indian respondents, Spotify and YouTube Music are in the top 2 positions. Spotify is preferred by 66% of Indian respondents which is depicted in green, whereas YouTube Music is preferred by 49%, which is depicted in red. The respondents from Abroad preferred Amazon Music and YouTube Music over Spotify. YouTube Music is preferred by 60% of respondents from Abroad, whereas Apple Music is also used by 44%. However, Apple Music is not preferred by many Indian respondents. The main reason for that is—Apple Music doesn't provide any free membership to its users [12]. While apps like Spotify and YouTube Music provide millions of songs free to their users.

5.1.3 Age-Wise Popular Music Streaming App

In Fig. 5, the age-wise distribution of the respondents is presented. According to the above graph, in the age range of 16–24 years, Spotify has 70% of users which is depicted in green, whereas in the age range of 25–44 years, YouTube Music has 54%

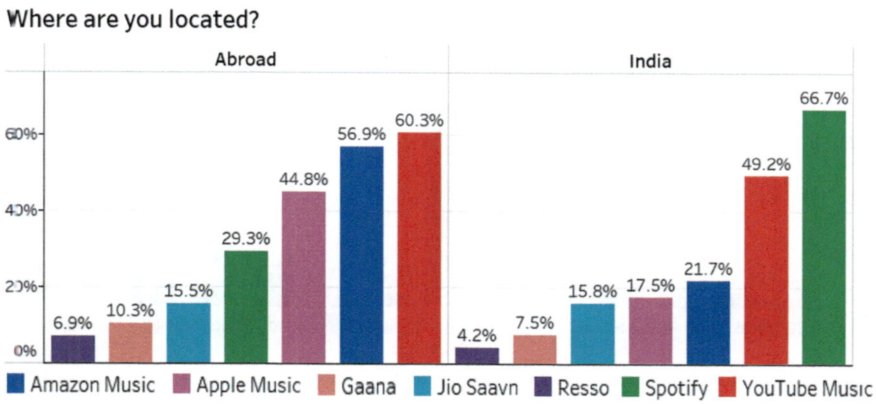

Fig. 4 Graph of location-wise popularity of music apps

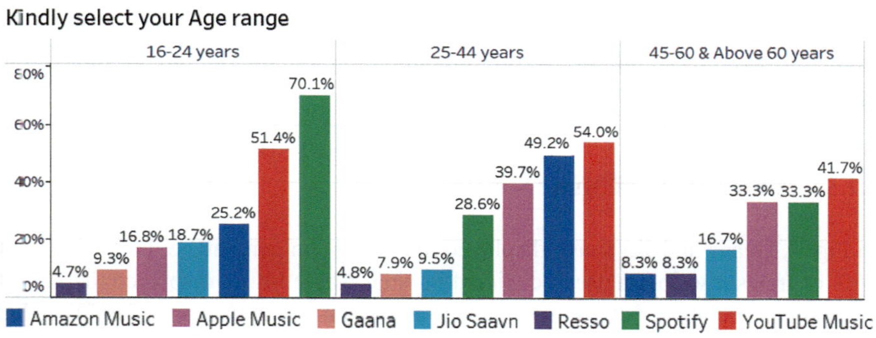

Fig. 5 Graph of age-wise popularity in music streaming apps

of users which is depicted in red. It is observed that YouTube Music was preferred over Spotify by the older age group. The main reason for that is—YouTube is also a video streaming platform which relatively older than Spotify. Many people from the older generation are already familiar with YouTube so that's why they have preferred YouTube Music as their music platform [13]. Additionally, YouTube has a large database of old Bollywood classic songs which are favorites of older generations. From the above two graphs, it is very clear that YouTube Music and Spotify have dominated the arena of Music streaming apps.

5.1.4 Cell Phone Wise Popular Music Streaming App

In Fig. 6, the graph of Android vs iPhone users is provided. It is observed that about 73% of Android users prefer Spotify which is depicted in green and around 48% prefer YouTube Music which is depicted in red. Whereas about 56% of iPhone users

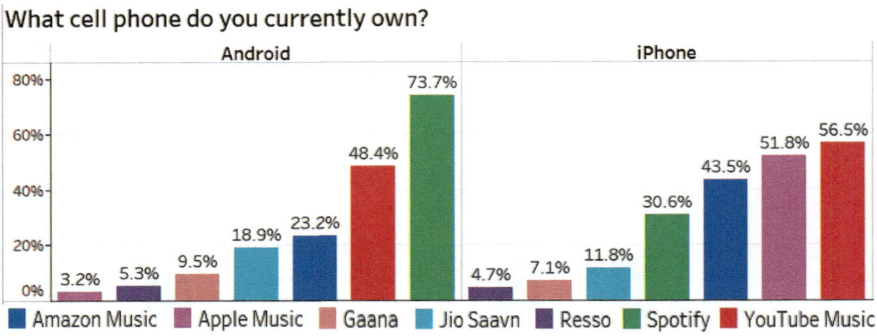

Fig. 6 Graph of android versus iPhone in music streaming apps

prefer YouTube Music and around 52% use Apple Music which is depicted in pink. The main reason for Spotify having a large number of Android users is that Spotify comes pre-installed on many Android devices which makes it more accessible to its users [14, 15]. Also, Android devices often provide users with more customization options compared to iOS devices.

5.1.5 Subscriptions of Music Streaming Apps

In Fig. 7, Gaana and Jio Saavn displayed the maximum percentage, i.e., over about 66% of free subscription users, depicted in blue. Apple Music displayed the lowest percentage, i.e., about 32% of free subscription users. The main reason for this is that Gaana and Jio Saavn have Guest Login systems that enable users to stream music without logging in or giving their personal information [16]. Also, is observed that Apple Music has the maximum percentage, i.e., about 59% of premium subscription users, depicted in orange. From the chart of India vs Abroad, it is known that the majority of these premium users are from Abroad.

5.2 Usage of Music Streaming Apps

In this section, usage of Music streaming apps is discussed. The graphs discuss the duration of usage of Music streaming apps. The possible reasons for the obtained results are also mentioned.

5.2.1 Duration of Usage of Music Streaming Apps

Music streaming apps have totally changed the trend of listening to music. Almost 50% of users in this survey have been using these apps for more than 3 years, depicted

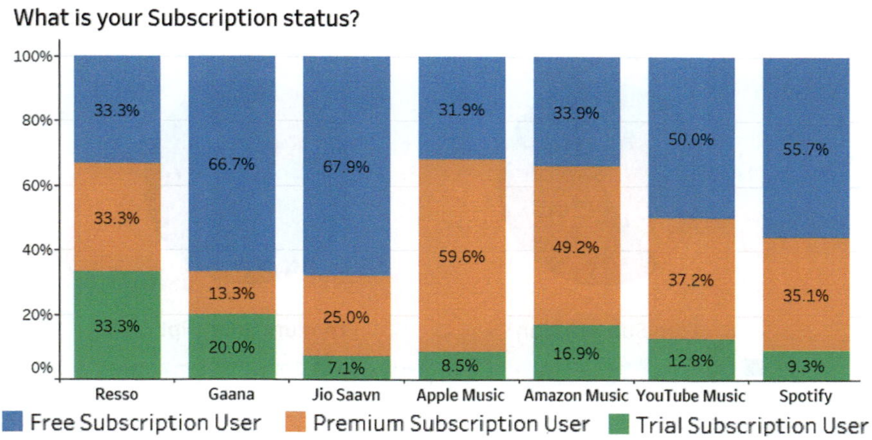

Fig. 7 Graph of subscription-wise distribution in music streaming apps

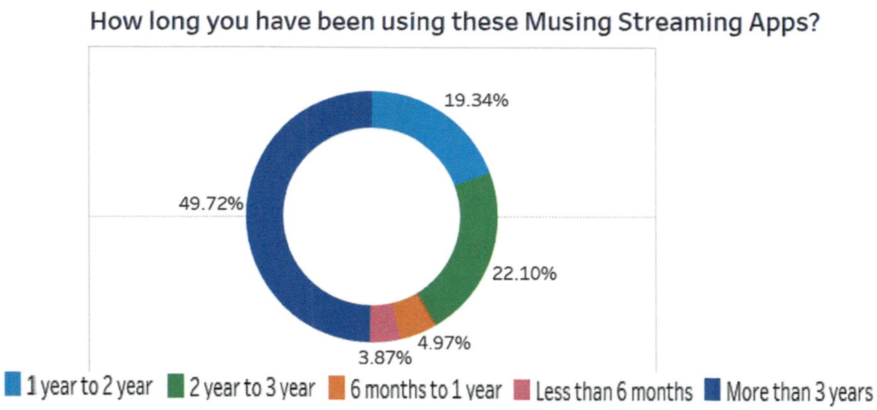

Fig. 8 Chart of duration of usage of music streaming apps

in dark blue. Some reasons for this could be the easy availability of the Internet over the last 7–8 years and the other reason is the lockdown of the COVID-19 pandemic. During the lockdown of 2020, people had enough time to listen to music, watch movies, and other recreational activities [17] (Fig. 8).

5.2.2 Comparison of Free vs Premium Subscription in Music Streaming Apps

In Fig. 9, the usage of premium and free subscription users of the Music streaming apps is presented. It is observed that users use the Music streaming app multiple times a day. Around 60% of the free subscription users use these apps multiple times

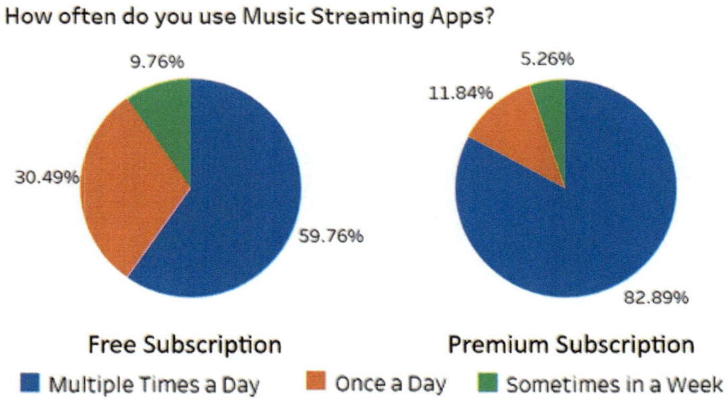

Fig. 9 usage of free and premium users

a day whereas around 83% of the premium subscription users use these apps various times a day. The major reason for this is the ads played in the free subscription. Most of the users get annoyed by the ads and close the app. However, most people buy the premium version to avoid annoying ads [18]. Also, the premium version has different features like recordings of live shows and downloading track files locally on the device [18, 19].

5.3 *Features of the Music Streaming Apps*

Music apps offer various features to enhance the user's music listening experience. They have a vast music library with songs from various genres and artists. Users can search for songs, albums, artists, and playlists using the search function. Personalized playlists and curated radio stations can help users discover new music [19]. In this section, the popular features preferred by users are discussed.

5.3.1 Features Enjoyed by Premium Subscribers

In Fig. 10, features preferred by premium users are presented. Around 78% of premium users like ad-free music. This can also be the reason for purchasing the premium version of these Music streaming apps. Also, about 43% of users enjoy access to exclusive content like live sessions, concerts, or podcasts. This graph also complements the graph of the comparison of free vs premium subscriptions, depicted in Fig. 9.

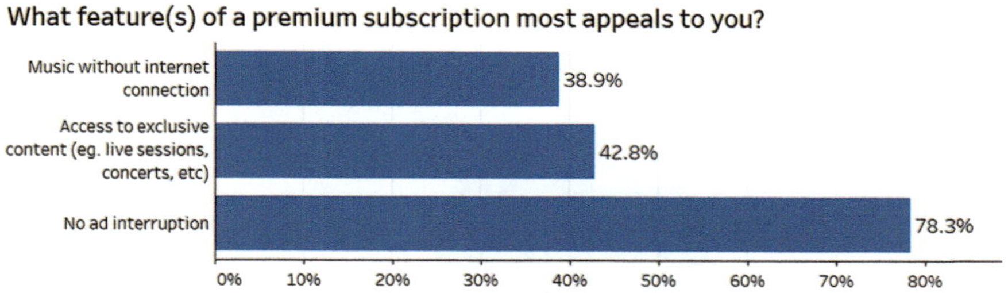

Fig. 10 Graph of features of premium subscriptions

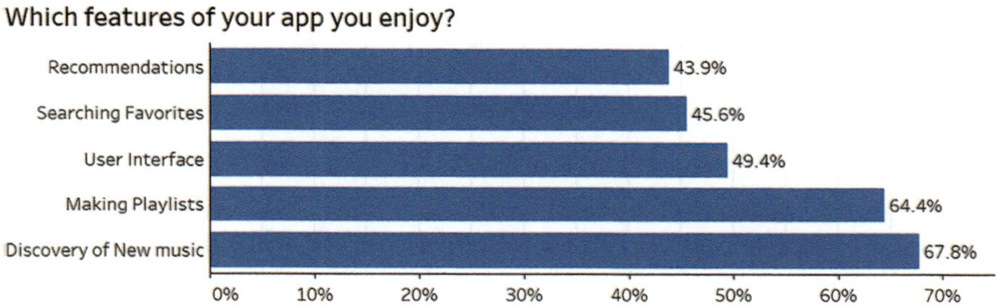

Fig. 11 Graph of features enjoyed by user

5.3.2 Popular Features in Music Streaming Apps

In Fig. 11, the popular features used by users are discussed. Around 68% of users use the 'Discovery of New Music' in the Music streaming app, whereas about 64% of users use the 'Playlist' feature. These two features are the most common features in all Music streaming apps. Users use these Music streaming apps to find their favorite songs easily on these platforms. These features can be considered trends for launching a new music app in the market [19]. People love to create and share their playlists.

5.4 Charges for Premium Subscriptions in Music Streaming Apps

Premium charges for Music streaming apps vary depending on the platform and the features offered. The monthly subscription fee ranges from 100 to 250 Rs. depending on the app and selected package. Many Music streaming apps offer free trials to new users to test out the premium features before committing to a subscription. Premium charges are a way for Music streaming apps to generate revenue and provide users with a better listening experience.

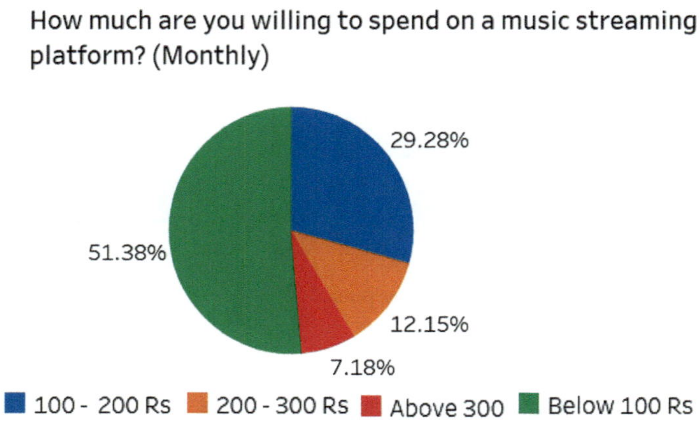

Fig. 12 Chart of users willing to spend on the apps

Spotify has a monthly subscription of INR 119/month and YouTube Music has a premium subscription of INR 129/month [9]. Amazon Music has a premium of INR 299/month, which is quite high for students. In this user survey, around 51% of users have preferred the charges to be below INR 100/month, whereas about 29% of users prefer between INR 100–200/month. The reason for Youtube Music and Spotify have a larger number of premium subscribers is that both these music streaming apps have their subscription rates close to INR 100/month (Fig. 12).

6 Conclusion

In conclusion, the survey results suggest that music streaming apps are becoming an increasingly popular way for people to listen to music and discover new music. The majority of respondents reported using Music streaming apps regularly, with Spotify and YouTube Music being the most popular platforms. The key factors that influence users' choice of Music streaming apps are the availability of a vast music library, personalized playlists, and a user-friendly interface. Respondents who do choose a paid streaming service tend to listen to more hours of music per week. The main reason to choose the premium service was to escape from the annoying ads. Respondents from India favored Spotify and respondents from Abroad favored Apple Music along with YouTube Music. In this survey, the people who said no to Music streaming apps were from the age of 25–44 and 45–60. There were around 10 respondents and most of them liked their music on the local device. Overall, Music streaming apps have become an integral part of people's music listening habits and are likely to continue growing in popularity in the coming years.

References

1. Report of IMI and IFPI—'Digital music study—2022'
2. Kim J, Nam C, Ryu MH (2017) What do consumers prefer for music streaming services? A comparative study between Korea and US. Telecommun Policy 41(4):263–272. https://doi.org/10.1016/j.telpol.2017.01.008
3. Wagner TM, Benlian A, Hess T (2013) The advertising effect of free - do free basic versions promote premium versions within the freemium business. In: Paper presented at 46th annual hawaii international conference on system sciences, Wailea, Maui, HI, United States. https://doi.org/10.1109/HICSS.2013.21
4. Wlömert N, Papies D (2016) On-demand streaming services and music industry revenues—Insights from Spotify's market entry. Int J Res Mark 33(2):314–327. https://doi.org/10.1016/j.ijresmar.2015.11.002
5. Dawson J (2015) Jackdaw research music survey report
6. Dijk CV. BSc thesis management and consumer studies, Music streaming services 2021 456143 (wur.nl)
7. Report of IMI and IFPI—'Digital Music Study—2021'
8. Kamalzadeh M, Baur D, Möller T (2012) A survey on music listening and management behaviours, pp 373–378
9. Chakraborty R (2022) Digital downloads & streaming in Indian music 9:498–503
10. Survey on Spotify. https://www.smartsurvey.co.uk/s/MXQSH8
11. Apple music versus Spotify survey. https://www.surveymonkey.com/r/HX6Z368
12. A (2023) Spotify versus apple music. https://www.makeuseof.com/spotify-vs-apple-music-key-differences/
13. Beuscart J-S, Coavoux S, Garrocq J-B (2022) Listening to music videos on YouTube. Digital consumption practices and the environmental impact of streaming. J Consum Cult 0(0). https://doi.org/10.1177/14695405221133266
14. Pre-installed apps on Xiaomi, Oppo, Samsung and other phones: What you can install and what you can't. (n.d.). Gadgets Now. https://www.gadgetsnow.com/featured/pre-installed-apps-on-xiaomi-oppo-samsung-and-other-phones-what-you-can-install-and-what-you-cant/articleshow/93106987.cms
15. Rangongo T (2019) Finding a spot in the local music streaming market. finweek 2019(6):12–12
16. Lal A, Hesmondhalgh D, Umney C (2023) The changing shape of the Indian recorded music industry in the age of plat formisation. Contemp South Asia 31(2):291–310
17. Gazmer SP, Bandopadhyay S, Mohan KR (2020) COVID-19 pandemic lockdown and music listening: a case study in Sikkim Himalayas. J Soc Sci 48(3):1751–1765
18. Fernandes T, Guerra J (2019) Drivers and deterrents of music streaming services purchase intention. Int J Electron Bus 15(1):21–42
19. Mäntymäki M, Islam AKMN, Benbasat I (2020) What drives subscribing to premium in freemium services? A consumer value-based view of differences between upgrading to and staying with premium. Inf Syst J 30(2):295–333

Oral-Pictorial Tradition a Communication Tool: Its Cultural Dimensions, Revival, and Shifting Paradigm

Ragini Ranjana and Mohammad Shahid

1 Introduction

Communication, an integral part of human progression, is commonly defined as an exchange of message between two or more people. The message could serve various purposes, from informing to educating, entertaining, interacting, motivating, etc., and use various mediums/channels to transmit it. Categories in three types: Verbal, non-verbal, and visual communication. Where verbal communication uses oral dialects, non-verbal communication depends on body language and gestures. Visuals, being dominantly used in today's world, communicate through images.

Unlike standard communication systems described above, communication within a cultural and social context depends on mutual beliefs and values. The beliefs within a community shared by people in the form of a story (oral/pictorial), followed by a reception from the audience in the form of a bow or offering, lead to a set of practices. Repetition of such practices within society engages communities in forming rituals. Rituals combining and coding these beliefs develop identity. With these identities, we find our belonging and flourish.

Oral-pictorial traditions refer to its dual communication system of storytelling; one is in visual form, and the other in its oral narrative. Both co-existed and supported the communication system to provide a vivid imagination. Various such traditional storytelling is found in almost all parts of India. However, their forms change from vertical to horizontal scrolls to wooden shrines. In the form of a wooden shrine, it is popularly known as Kaavad, and storytelling is called "Kaavad Baanchana," which belongs to the west of India, Rajasthan [4]. In another, it is elaborately worshipped

R. Ranjana (✉) · M. Shahid
Indian Institute of Technology Hyderabad, Hyderabad, Telangana, India
e-mail: md21resch11001@iith.ac.in

R. Ranjana
National Institute of Fashion Technology Patna, Patna, Bihar, India

© The Author(s), under exclusive license to Springer Nature Singapore Pte Ltd. 2025
S. Das et al. (eds.), *Human-Centered Digitalization, Volume 3*, Design Science and Innovation, https://doi.org/10.1007/978-981-96-4383-7_20

by its patrons in the form of horizontal Patuas depicting stories of Indian mythology of Ramayana and Mahabharata as Patuas in the east of India, West Bengal [15].

Folk and tribal practices representing oral-pictorial storytelling can also be found in all parts of India, deeply embedded in daily life, rituals, and practices. Some depict the usual routine of life; some serve the purpose of suing away evil spirits and animals, and some are devotional practices on sacred occasions. The forms are often sung in songs while representing it on platforms (mud wall/floor/cloth). These embodied forms are divine and sacred, sometimes representing the community deities. Beyond the reach of outsiders, these traditions flourished for generations till they got recognized and introduced to the world.

The three traditions mentioned above, Kaavad, Patuas (Oral-Pictorial Tradition), and Maithili/Mahbubani painting, are among the most celebrated forms of India [4, 7, 14, 15]. Recognized by the world, these traditions have seen transitions while adapting their reception to the world. Further evolving with the pace of time, it may adapt to various other forms.

The study aims to understand the community's traditional oral-pictorial communication system and trace the events and their impact across a timeline. It briefly describes the traditional structure of practices and eventually talks about the initiative of researchers and practitioners across the time frame divided into four major timelines (Table 1).

Tracing the changing synergy of oral-pictorial tradition in the community, its story/subject deviation due to modification in the communication system. This research paper attempts to record the accords of the shifting paradigm of oral tradition and folk art through the selected model of Kaavad and Bengal Patachitra for oral-pictorial tradition and Madhubani as a model form for folk art.

2 Communication System in Oral-Pictorial Tradition

Oral-pictorial communication system in its storytelling tradition of Kaavad and Patuas involves multiple communities: The maker/painter, the storyteller, and the patrons. The painter/maker is of the community who makes the pictorial representation. They are asked to develop the images by the community of storytellers. The storyteller is the person who carries these pictorial narrative scrolls/shrines to the patrons and narrates the stories. Patrons are the community who welcome the storyteller with the sacred scroll/shrine to their houses to bring auspiciousness and prosperity. In return, they offer/pay the storyteller with goods and grains. They also commission the new panel to be added in their name. The storyteller takes back the sacred scroll/shrine with an untold promise to return the auspiciousness with them next time. In these traditional communication systems, where the message is told through story, using the oral form of communication, the pictorial/visual representation carried by the storyteller (Kaavadiya Bhat) vividly widens the imagination of the patrons. In the process of communicating the message, the storyteller performs acts through gesture and tonal variation, performing non-verbal communication [4].

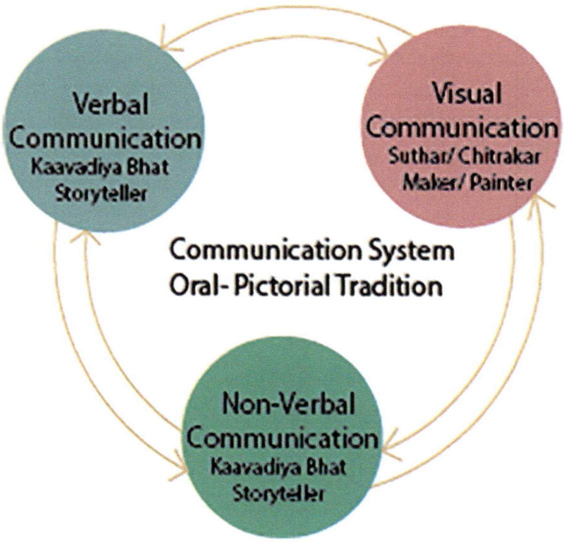

Fig. 1 Communication system of oral-pictorial tradition and practices

Thus, as shown in Fig. 1, the use of different types of communication, verbal, non-verbal, and visual, is evident. The symbiotic approach of co-existence of all three ways of communication generates the curiosity of understanding the synergies between them.

2.1 Synergies Between the Community in the Oral-Pictorial Tradition (Before 1960)

Nina Sabnani, in her book The Kaavad Tradition of Rajasthan, stated that,

> "The maker acquires their uniqueness as they are the only ones who make the Kaavads, depending on the storyteller to some degree to continue making them. The storytellers achieve their professional identity from the very name Kaavadiya Bhat and depend on their patrons to sustain the tradition. The patrons claim their identity through the recitation of the genealogies. The invocation of identity is achieved through the narration of stories that are populated by individuals cherished by the community, or who represent the community, consolidated through the narration of genealogies of patron's family [4]."

The above description reveals the pride taken in performing the given role in the communication system, elaborating the synergies between the communities. The tradition of Kaavad Baachanam (Kaavad Recitation) validates their generations' existence and memories and establishes their purpose of living on this earth. The practiced act of oral-pictorial tradition goes beyond the purpose of and provokes the recognition of their own identity within a community. Kaavad for patrons is the most trustworthy source of their ancestral memories, which get invoked, revived, and relived every time the storyteller visits their house. Whereas the storyteller "Kaavadiya Bhat" recognizes themselves by connecting the mortal life to the divine and revisiting the ancestors' memories. The act of recognition by patrons was not limited to the depicted images. It is the storyteller's words that connect them to the figures. They invest meaning in images through rhymes, chants, and memory of their stories. The community synergy may provide them with economic sustenance for each other. However, more significant than that, it is part of their identity without which they will not be able to return to their rememberable past, behold the identity of today and bestow the integrity of the community to the future. Patua having the similar format get painted by Chitrakar and narrated at the door of patrons on different auspicious occasions. In Patuas unlike Kaavad the Chitrakar and Narrator could be same or belongs to same community or family.

Unlike Oral-pictorial Tradition, folk arts follow a linear structure of transmission of knowledge from generation to generation. Madhubani painting of Mithila region, Bihar, is a ritual for life stage celebration, where women of the community paint the walls on auspicious occasions and celebrate Solah Samskara (sixteen stages of human life), Dussehra and others [14]. Among the Solah Samskara, marriage is the most celebrated occasion. The interior walls of the wedding chamber, "Khobbar Ghar," were depicted with deities and symbolic representations of auspiciousness as a blessing for the bride and groom at their marriage ceremony as a betrothal gift, protecting the newly wedded couple from all bad omens and spirits and showering blessings. In the Hindu religion, marriages are full of rituals, songs, and embedded meaning, Mithila being part of such rituals rich in symbolism, aesthetic, and cultural beliefs [8]. The message encoded by the elderly women of the community in the form of visuals (signs and symbols) decoded while performing the ceremonial act of marriage through prescribed gestures and reciprocating to the songs supporting the ceremonies.

2.2 Revival, Recognition, and Commercialization (1960–1980)

After world war-I, the need to conserve the world's cultural heritage become a concern for humanity. In 1946 United Nations Educational, Scientific and Cultural Organization came into force for international collaboration in education, science, and culture. The promotional effort drew the attention of countries worldwide, and India was not left untouched either.

Pika Ghose, in her article "The Story of a Storyteller's Scroll," talks about the Patua scroll exhibited in 1964 in the Philadelphia Museum of Art and expresses the reach of Patuas on the world platform. The forms of practices expanded its limited sacred access from doors and verandas of Indian patrons and reached international space creating a presence among new patrons [15].

The shift in the recognition platform led to the rise of demands among new patrons (art admirer and curators) of the forms as art pieces demanding further interactions and involvement. The urge was sensed, and efforts were made to accommodate the new patrons. The traditional wall got replaced by a new medium (paper) which was convenient to transport, easy to work with, sharable and saleable. The adaption of new media became a great success and provided the platform to earn a livelihood, bridging the access gap between the forms and the new patrons.

Soon, more commissioned works started knocking on the door of the Mithila household. A few famous commissioned works were the Ashoka Hotel, a mural at the prime minister's home, and paintings for Indian Railways [7].

In all the efforts made, the message imbibed in the communication system shifted from a symbiotic co-existence of verbal, non-verbal, and visual communication to visual-based communication. Despite all the efforts to retain the quality of visuals, the absence of non-verbal and verbal communication was left unattended. The decoding of the imbibed message now depends on the interpretation of the new patrons.

Kaavad and Bengali Patuas have not remained untouched either. The Kaavad form, a wooden box with enchanting images (sacred shrine), unfolds the new world with every door purely a work of art. The new patrons were introduced to the realm of creation, which was vast enough to discover for ages. And without the guidance of the "Kavaadiya Bhaat" (the storyteller) with imbibed meanings in "Kaavad Baachanam" (Kaavad Recitation) oral communication and non-verbal communication remained undiscovered or less known to the new patron. The realm of creation has little voice and magic in the absence of its sacred space, ritual, and memories which is imbibed in the invested meaning by the storyteller.

2.3 *Change in Paradigm from Oral-Pictorial Traditional Art/ Folk Art to Craft/Fine Art (1980–2000)*

By the 1980s, the riches of Indian folk art reached the globe—the demand for Indian craft and folk-art triumphs in the commercial markets. The new patrons were growing and expanding. Patrons and market rising needs are not limited to the quantity of craft and folk art; the patrons ask for something new every moment. The craftsperson adapted to the scenario by reacting to the growing market and buyer's demand. Themes like "Nari Shakti" depicting goddesses like Kali, and Radha start appearing with individual stories and panels [1, 1, 1]. The artist also started recorded their experience and expressed their memorable encounter in their art form, giving an individual expression.

The development in the field expanded their scope further. The artist starts telling classical stories and epic and tries their hands at contemporary subjects like feminism, environmental issues, and journey. Folk art slowly and steadily transitioned to fine art [6].

Drawing parallels, a similar pattern of expansion could be seen in oral-pictorial tradition of Kaavad where the size and panel of Kaavad is adapted to make it convenient for transportation and affordable to tourist. The deviation in the story could also be traced were Suthar the maker of the Kaavad started depicting global context, stories of other religion and content of secular theme (life of Christ—Mahatma Gandhi—Girl Child and Microfinance) [4]. The believes in conserving craft by making it relevant to society and market become the norms. The government of India, seeing the scope, introduces training programs to teach the craft of Madhubani painting. Many benefited, and the art crossed the gender line; man became the new practitioner of the art [6, 6].

The new patrons have created a direct connection with the painter/makers. And the storyteller and the old patron were not actively involved in the developing synergies. They are still part of traditional practices and continue practicing community beliefs in their space.

The visual communication practiced by the new patron was more common and had larger global audiences than the previous localized communication system between communities. Though the "Khobar Ghar" is still painted by Madhubani practitioners, its context and purpose are disconnected from the global audience.

2.4 Extended Forms and Shifting Paradigm (2000–2018)

The scope of expansion was immense; every day, new possibilities arose, and catering to them led to further diversified demands. Governments and individuals created the new fair and market segment catering to many craft/tradition-oriented products. Delhi Haat (1994) and Dastakar (1981) rooted their ground in providing authentic experiences and products to tourists contributing to flourishing cultural tourism. The different stake of society also recognizes the immense possibilities that lie with folk art and traditional mediums. Designers from all disciplines (fashion, product, interior, etc.) explored the field to develop the commercialized market. The practitioner drew new possibilities in developing new products. Madhubani (or any other art/craft form) is no longer limited to walls or paper. It could be seen in fashion merchandise, tabletop accessories, decorative furniture, and many more. Product Design researchers and others worked on various possibilities of mass production [3, 13]. The commercial market keeps the expansion flexible to accommodate all. Furthermore, the high cultural relevance of these forms and its importance in communicating the belief system lead to successfully spreading the government's, policymaker's, and educator's words to the masses [2].

The advancement in communication mediums to print and digital has opened the scope of exploring storytelling. Tara's publication initiative brought books like Sita Ramayan and Tsunami, beautifully painted and narrated stories by traditional artists [11]. The publication's experimentation with forms and pictorial graphics while maintaining the aesthetic with mass production allures the reader's attention. Communication scholars took the initiative to develop communication tools for conserving and preserving storytelling with its visual language. Animation application is explored and documented [5, 9, 12]. The animated film either represents or is influenced by folk art and becomes part of the watch list with mainstream content.

The transition in practice is evident. The researcher, social bodies, and practitioners sense the shifting paradigm. Art admirers, art critics, and researchers were also aware of changes. Reviews of this gradual change are reported in India and from different parts of the world [1, 6, 7, 9]. The recorded studies drew attention, brewing concerns, and further propagation scopes. The reviews and critics also illuminated the need for more attention to storytelling, an integral part of the earlier oral tradition and folk art (songs) practice.

Later in the timeline, the question of proper practices and ethos also arises.

3 Conclusion: Emerging Further Direction

Figure 2 shows the expanding number of new patrons and the varied engagements they are involved in with craft. Several new fields also come up to inspire, work, and collaborate to serve the raised demand and potential market. The new market expanded the circumference of new patrons larger than the existing traditional/localized community. The practice is now known in the visual and form representation rather than traditional belief and cultural practice. The threat could be felt. UNESCO added the 5th C in its strategic objective; besides credibility, conservation, capacity building, and communication, the community got added. The objective was to enhance the role of the community in implementing the World Heritage Convention. It is in addition to the 2003 Convention for Safeguarding Intangible Cultural Heritage.

The shifting paradigm of folk art and oral-pictorial tradition came across different timeline stages. The expansion happened to the needs and demands of society and the economy itself. The progression was organic and adapted following opportunities. The iterated form does give us ground to critics, but at the same time, in the absence of such iteration and adaptation, it might have gotten lost long ago.

The design is a progressive and iterative process. The effort to rejuvenate Indian art and craft has supported the survival and flourishment of form and community. With community, the tradition is still alive and does exist in its traditional form rooted deep in their daily practices. The shift did increase its circumference of new patrons and will keep expanding in the future. With every expansion, we will discover new opportunities and find gaps. An attempt is required to fill the gaps before they become void. Varied gaps of representation in oral-pictorial communication system and synergies need to be introduced and spread among the new patrons. Making new patrons experience the synergies may reveal the importance of the painted icon, the ideology, the beliefs, and its relevance in the community.

Further transformation in our courses of adapting leaves us with our next step. The question arises: Will the gaps in the shifting paradigm and its impact and implication get detected and filled? Can the next material/technology/human effort reconnect and develop the preexisting synergies?

The whispers of Intangible culture we believed for years needs to reach us again in the noised market. The next communication system needs to be designed to sync

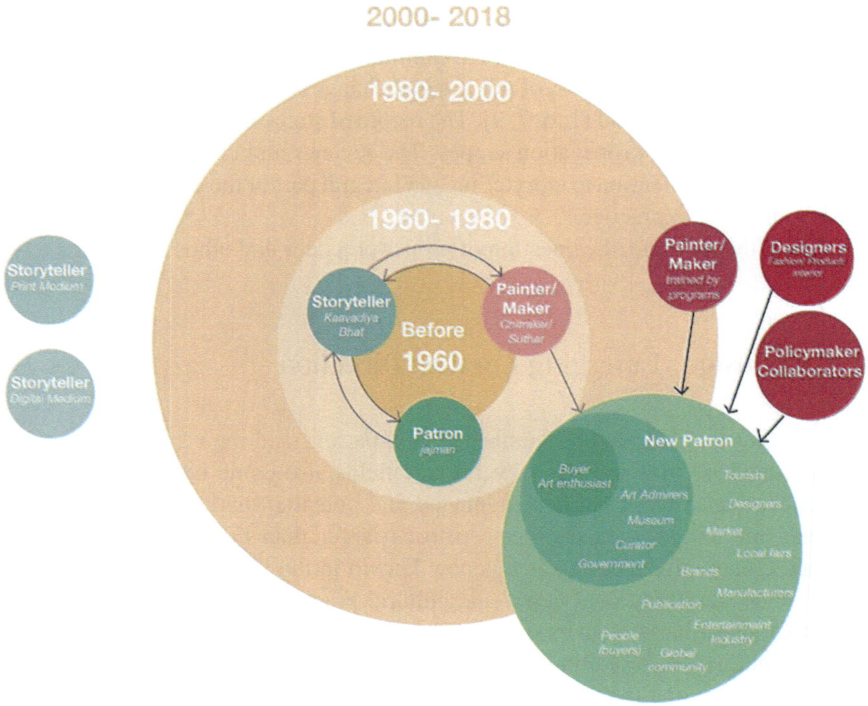

Fig. 2 Expanded market and involvement of new patrons and trained creators

Table1 Timeline for studying the shifting paradigm of oral tradition and folk art

Before 1960	Initial communication system
1960–1980	Revival, recognition, and commercialization
1980–2000	Change in paradigm from oral traditional art/folk art to craft/fine art
2000–2018	Different forms, the shifting paradigm

the synergies and reestablish the verbal, non-verbal, and visual co-existence where cultural heritage flourishes in its both tangible and intangible forms.

References

1. Beatrix H (2022) From oral tradition to folk art: reevaluating Bengali scroll painting, Asian Folklore Studies
2. Michael B, leeson L, Rathor MS, Roy A, Sabnani N (2018) Interdisciplinary research in Rajasthan, India: exploring the role of culture and art to support and rural development and water management develop. Water Alternative
3. Kashyap KK, Shashibala ML (2019) New Delhi, India: national book trust

4. Sabnani N (2015) Kaavad tradition of Rajasthan. Niyogi Books, Ahmedabad
5. Douglas T (2015) The tales of tribe: animation as a tool for indigenous representation
6. Rekha N (2010) The form folk art to fine art: changing paradigms in the historiography of maithil painting
7. Devis RH (2008) From the wedding chamber to the Museum. What's the use of Art?, pp 77–95
8. Jain J (1997) Ganga Devi: tradition and expression in mithila painting. Maplin Publication Pvt. Ltd., Ahmedabad
9. Sabnani N (2005) The challanges of the sleeping gaint. Design Issues
10. Sabnani N (2009) Kaavad storytelling tradition of Rajasthan
11. Chitrakar M, Arni S, Ramayan S (2018) Chennai, India: Tara books Pvt. Ltd.
12. Kolay S, Roy ST (2015) Designing alterative paradigm for traditional visual storytelling. Springer, India
13. Prakash K (1994) Madhubani, Bombay. The Design Point, India
14. N. M. o. Nature, Aditi: a celebration of Life, Washington D. C: Smithsonian Institution Press, 1985
15. Ghose P (2013) The story of a storyteller's scroll. Anthropology and aesthetic, pp 166–185

Music Genre Preference of Non-musicians

Ankita Verma, R. D. Sambath, and Manish Kumar Asthana

1 Introduction

Music is a universal language that transcends borders and culture [44]. It can evoke a wide range of emotions and set the tone for various situations [39, 69]. In daily life, music plays a significant role in enhancing our experiences and shaping our moods [74]. Whether we are consciously aware of it or not, music significantly shapes our emotions, behaviors, and overall well-being. This emotional resonance is a fundamental reason why music is so prevalent in our daily lives [74].

As a result, people turn to music as a tool for emotional regulation [12]. In times of stress or sadness, listening to soothing music can provide comfort and solace [7]. Another way in which music is integrated into daily life is as a productivity enhancer [3]. Many individuals listen to music while working, studying, or performing tasks that require concentration [40]. Research by Lesiuk [43] has shown that music can improve productivity by reducing stress and anxiety, which can be detrimental to focus. Additionally, music can create a pleasant auditory environment that helps individuals stay engaged in their work. Its power lies in its ability to connect with people deeply and personally.

We can take another interesting life situation where people of diverse backgrounds come together, often with the shared purpose of traveling. Railway stations and bus stands are bustling hubs of activity where people listen to music or sounds in the form of announcements, buskers, or ambient background music. Upbeat tunes may

A. Verma · M. K. Asthana (✉)
Department of Humanities and Social Sciences, Indian Institute of Technology Roorkee, Roorkee, India
e-mail: m.asthana@hs.iitr.ac.in; asthanakm@gmail.com

M. K. Asthana
Department of Design, Indian Institute of Technology Roorkee, Roorkee, India

R. D. Sambath
TKM School of Architecture Design and Planning, Kollam, Kerala, India

© The Author(s), under exclusive license to Springer Nature Singapore Pte Ltd. 2025

S. Das et al. (eds.), *Human-Centered Digitalization, Volume 3*, Design Science and Innovation, https://doi.org/10.1007/978-981-96-4383-7_21

be played over the loudspeakers to create a sense of excitement and anticipation among passengers [78]. Similarly, there are also circumstances where people often find themselves in awkward silence. For instance, when a person is in an elevator, the music in elevators is intended to soothe passengers, reduce anxiety, and make the ride more pleasant. In daily life, sometimes people have to visit hospitals and therapeutic centers where they face pain, anxiety, and stress. Researchers suggested that playing soothing background music in such an environment can enhance therapeutic qualities [11]. In addition, another compelling aspect of music is its ability to motivate and help people focus on physical activities. For instance, upbeat and highly energetic music can inspire and energize people to focus on their fitness goals [35, 36].

Music also significantly creates soothing atmospheres in hotels and fine dining places [56]. Music is a powerful tool that enhances daily life experiences in various settings [4]. Whether it's setting the mood at railway stations [78] and airports [2], providing solace in hospitals, [24], motivating at the gym [76], or influencing consumer behavior in restaurants [54, 64] and shopping malls [51, 75, 85]. Studies have shown that music has personal, cultural, and social value for listeners. Hence, music helps to change their mood and reduce the stress and anxiety of a person and society [13, 27]. Its ability to evoke emotions, shape environments, and connect people is a testament to its universal appeal. In an increasingly hectic world, the strategic use of music in these everyday situations can profoundly impact our well-being and the quality of our experiences [62]. Hence, the choice of music varies from person to person according to the context. Some may prefer instrumental music or classical compositions, which are often recommended for tasks that require deep concentration.

Literature suggests that music's most widely researched emotion variable is valence and arousal [66]. This dimensional model of emotion is a two-fold circumplex model, which proposes that all affective states arise from two independent neurophysiological systems. Emotional valence is a continuum on which pleasantness is located at one end while unpleasantness is on the other. The arousal dimension denotes a continuum with high physiological arousal at one end (i.e., excited) and low physiological arousal at the other end (i.e., calm). Researchers like Bai et al. [5] and [46] used these two dimensions to understand the relationship between music and emotion.

According to the researchers' findings, there are three primary music categories that various researchers have used to understand the effect of music on listeners. The categories are Western music [23, 49], Indian classical music [10, 68], and Natural sound music [9, 60]. There are numerous studies found in western context, but there are limited studies on the perceptions of Indians toward music and sounds. The current study also examines the kind of (genre) music that can be played in Indian contexts according to user perception. Hence, the present paper aims to understand the perception of Indians among various genres of music that can be played in their daily lives. This pilot study focused on the perceptions of Indians of three different genres of music (Indian, Western, and Natural sound), which were mainly studied by previous researchers. The study also aims to understand the difference in perception of natural sounds over Indian ragas and Western music in the Indian male and female population.

2 Methods

2.1 Participants

In total one hundred and eight participants (males = 61, females = 47; mean age = 24.95 years, SD = 4.62), were recruited through a purposive snowball sampling technique method. They submitted their response through the Google form, which was shared beforehand. Informed consent form was signed from all individual participants included in the study. Then the data were further considered for the statistical measure (47 females, age $M = 24.71$ years = 3.11 and 61 males, age $M = 24$ years, SD = 3.9). Participants were instructed to sit in a distraction-free environment and to use headphones, earphones, and a laptop as a device to complete the study successfully while they were performing the task. The experimenter was monitoring them through zoom meeting.

3 Materials and Apparatus

3.1 Stimuli

In the experiment, participants rated the valence and arousal of each music stimulus on an online questionnaire. The music stimuli are selected based on three famous genres of music and are instrumental in nature. The first category of music stimuli consist of Western classical music, which include the famous works of Beethoven music and Mozart, like Symphony No. 40 [1, 21, 23, 49, 55, 58, 80], Serenade No. 13 [19, 45, 71, 72], Fur Elise [20, 28, 41, 50, 83, 84].

The second category of stimulus used in this study consists of Indian Classical Music, like India Raga Anand Bharavi [10, 22, 33, 63, 65] (Merrill and Amin n.d.), Raga Yaman, [6, 15, 16, 33, 79], Raga Mohanam [17, 61, 63, 68] (Iyengar et al., n.d.). Finally, the researchers emphasize natural sound as a third category of music stimuli. The natural sound of Blossom (sound of a chirping bird) [42, 52, 73], Healing Flute [9, 48] River Sound [38, 37, 47, 60], and Tibetan healing sound [14, 30, 59, 70] (Ryka, n.d.) were used. The music stimuli used in this study have already been paid for. The details of the music stimuli are explained in Table 1.

Table 1 List of music stimuli used in the study; the table indicates the mean and standard deviation of ratings done by male and female participants

Valence scale				
Stimuli	Male (61) M (S.D.)	Female (47) M (S.D.)	*p*	*Cohen's d′*
Symphony No. 40	6.76 (1.75)	6.57 (1.81)	0.39	0.06
Serenade No. 13	6.25 (1.91)	6.15 (2.19)	0.67	0.05
Fur Elise	6.88 (1.74)	6.63 (1.89)	0.23	0.12
Raga Anand Bharavi	6.21 (2.09)	6.10 (2.26)	0.68	0.05
Raaga Yaman	5.92 (2.37)	5.97 (2.67)	0.86	0.01
Raaga Mohanam	6.71 (1.77)	6.44 (2.13)	0.21	0.14
Blossom	6.91 (1.60)	6.55 (1.89)	0.07	0.20
Healing flute	7.03 (1.78)	7 (2.07)	0.88	0.01
River sound	7.24 (1.72)	6.97 (2.04)	0.20	0.14
Tibetan healing sound	6.64 (1.67)	6.55 (1.62)	0.90	0.05
Arousal scale				
Stimuli	Male (61) M (S.D.)	Female (57) M (S.D.)	*p*	*d′*
Symphony No. 40	4.67 (2.67)	4.37 (2.03)	0.57	0.13
Serenade No. 13	6.06 (1.75)	6.15 (1.80)	0.80	0.05
Fur Elise	7.00 (1.59)	6.73 (2.10)	0.50	0.14
Raga Anand Bharavi	4.43 (2.54)	3.60 (2.44)	0.13	0.33
Raaga Yaman	3.67 (2.52)	3.44 (2.44)	0.68	0.09
Raaga Mohanam	5.08 (1.98)	5.31 (2.32)	0.60	0.10
Blossom	4.61 (2.43)	3.58 (2.46)	0.05	0.42
Healing flute	3.55 (2.86)	2.95 (2.29)	0.29	0.23
River sound	3.84 (2.88)	4.03 (2.83)	0.77	0.07
Tibetan healing sound	6.64 (1.67)	6.55(1.62)	0.90	0.05

It also shows the statistical significance and Cohen's d′ for the appropriate effect size for comparing means

p = statistical significance, M = mean, S.D = standard deviation

4 Research Design

The Institute of Human Ethics (IHEC) committee of the Indian Institute of Technology, Roorkee, has ethically approved this experiment. The protocol number of approval is (IHEC No. BT/IHEC-IITR/2020/7004/14). The Google form provided to the participants consisted of three sections: the demographic section, a questionnaire consisting of music stimuli, a rating scale, and a user feedback section. Participants listened to the music by clicking on the link either on their mobile phone, laptop, or notepad and rated their experience based on valence and arousal on a 9-point Likert scale. To enhance the user experience, along with the rating scale, a Self-Assessment Manikin (SAM) approach was also used [8].

5 Result

To analyze the significant mean difference in valence and arousal dimension in three specific genres of music (Indian, Western, and Natural sound), we conducted an Analysis of variance (One-way ANOVA) using SPSS 20.0 (Statistical Package for the Social Sciences) software. For the Indian genre (raga yaman, raga mohanam, and raga anand bharavi), the Western music genre (symphony no. 40, serenade no.13 and furelise), and the nature sound category (healing flute, river sound, and sound of chirping bird), stimuli were used, significant difference was found on the valence and arousal dimensions (Fig. 1).

The result of One-way ANOVA revealed that the mean valence dimensions ($M = 6.332$, SD = 1.407) of three different genres of music show a significant difference (F (2, 107) = 25.4282, $p = 0.01$). Also, the data suggests that participants in the study felt pleasant and calm after listening to the natural sound ($M = 7.006$, SD = 1.215). Similarly, the mean arousal dimensions ($M = 4.742$, SD = 1.8987) of three different genres of music also showed a significant difference (F (2, 108) = 67.48104, $p = 0.01$). They experienced high arousal ($M = 6.63$, SD = 1.14) and moderate valence ($M = 5.94$, SD = 1.26) when listening to western music. For the Indian music category, the experience of arousal was low ($M = 4.86$, SD = 1.65) and the ratings for valence was comparatively high ($M = 6.25$, SD = 1.42). For the nature sound category, the arousal rating was lowest ($M = 3.42$, SD = 1.81). Therefore, it is evident from the dataset that different genres of music have different levels of emotional perception. However, when the data was analyzed to investigate the gender difference between the rating of male and females on valence and arousal dimension, respectively, no gender difference was observed.

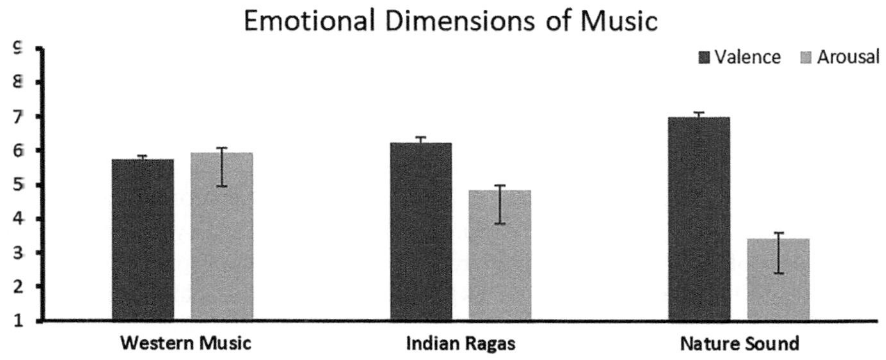

Fig. 1 Preferred emotional dimensions (valence and arousal) level (1–9) on selected genres of music between participants, where the preferred valence level (pleasant to unpleasant) is shown on the X-axis and music genre on the Y-axis

6 Discussion

This study analyzes Indians' perceptions of numerous kinds of music that are often experienced in their daily lives across the valence and arousal dimensions. We focus specifically on Indian perceptions regarding the three primary genres of music: Indian, Western, and Natural sound. Having an in-depth understanding of this situation is essential, as it can enhance researchers' ability to choose suitable background music in accordance with contextual circumstances, an area that currently lacks comprehensive understanding. In accordance with the afore-mentioned conceptual framework, a total of 108 individuals were selected to participate in the study. These participants were tasked with evaluating music samples that fell into three distinct categories: Western, Indian, and Natural sounds. The evaluation was based on two aspects, namely valence and arousal. On a nine-point Likert scale, the valence category ranges from one, indicating a state of sadness, to nine, representing a state of intense happiness. Similarly, the arousal dimension ranges from one, indicating a state of calmness, to nine, representing a state of high arousal.

The result of the current study also aligns with the biophilia theory put forth by Grinde and Patil [25] and Wilson [81]. According to biophilia theory, individuals inherently possess a proclivity to establish associations with nature and various living forms. This connection can be extended to the realm of music, as music often draws inspiration from the sounds and rhythms found in nature. In the field of music therapy, the use of biophilic principles is becoming increasingly important. For instance, [18] investigated how music therapy can delve into biophilia by incorporating natural rhythms and sounds into therapeutic sessions. Patients who were exposed to music created with natural instruments, such as wooden flutes or drums, reported reduced stress levels, and a stronger sense of connection to the environment [18].

In this conversation, the researcher seeks to extend the discussion by proposing the concept that the selection of musical stimuli should be relevant to the environmental situation. For example, listening to music with high valence and high arousal can help people to boost their mood and thus can be more motivated and attentive to their surroundings. In his book named "The science and Psychology of music" Thompson [76] wrote how music is processed by our brains, affects us emotionally, shapes our personal and cultural identities, and can be used in different environmental context. Building upon the existing body of research, it has been observed by researchers that the presence of high-valence and high-arousal music inside retail contexts can create a vibrant and energetic atmosphere. This genre of music has the potential to augment the shopping experience by rendering it more pleasurable and energizing. It may lead to increased exploration, longer time spent in stores, and a greater likelihood of making unplanned purchases [82]. In other situations, like while working out in gym the presence of high-valence and high-arousal music has been seen to enhance individuals' mood and motivation levels. Such music tends to be fast-paced, rhythmic, and exciting, which can help individuals maintain an elevated level of energy and enthusiasm throughout their workout. It may lead to improved performance and endurance, making the gym experience more enjoyable and effective [35, 36].

However, music with high valence and low arousal, often characterized by its positive and soothing qualities, is typically used in various environmental settings to create a calm and pleasant atmosphere. This genre of music is frequently employed at high-end restaurants to cultivate a refined and cultured atmosphere. According to [57], including this particular genre of music can enhance the dining experience by promoting a relaxed and pleasurable atmosphere [57]. Researchers also found that high-valence and low-arousal music creates a serene and contemplative environment in cultural settings like museums and art galleries. It can encourage visitors to engage with art and exhibits calmly and focused [31].

In conclusion, it can be asserted that the factors of valence and arousal exert a notable influence on the choice of musical stimuli across many contextual settings. Hence, it is imperative to conduct replications of this Western research on valence and arousal dimensions in the above-mentioned Indian settings. This could enhance the efficacy of music-based interventions and produce a more emotionally potent musical environment, thereby improving and facilitating the user's experience [34, 74, 77].

7 Limitations

Even though all precautions were taken to minimize the extraneous factors in the study, it is essential to list out the possible limitations of this study. Since it utilizes online surveys as a method for data collection, it has inherent drawbacks like lesser control over participants, extreme ratings, fakability, and low clarity of confusion for respondents. Another restriction is the survey's length and duration, and respondents' user feedback (which indicated that 62% of people found the current online study to be lengthy) has confirmed this. Regarding the sample population, there is a bias toward male participants since they responded in more significant numbers than female participants (56.48% male and 43.51% female participation). Another limitation is that the current study was a more generic one, the study revealed that our subjects preferred western music very less while their preference to natural sounds are higher. However, it is still unknown what might be perfect background music for various daily life contexts to reduce the user stress and improve well-being. In our next research, we intend to employ multimodal analysis in a virtual or real-world environment to assess the appropriateness of music selection for various contexts, with the aim of enhancing and optimizing the overall user experience.

References

1. Abdelhakim AM, Samy A, Abbas AM (2019) Effect of music in reducing patient anxiety during colposcopy: a systematic review and meta-analysis of randomized controlled trials. J Gynaecol Obstet Hum Reprod 48(10):855–861
2. Adkins M, Cummings S (2019) Music beyond Airports—appraising ambient music 254

3. Allen R (Rich), Wood WW (2012) Rock 'n' roll classroom: using music to manage mood, energy, and learning, 313
4. Anon (2012) Music and everyday life. Cult Study Music, 149–158
5. Bai J, Peng J, Shi J, Tang D, Wu Y, Li J, Luo K (2016) Dimensional music emotion recognition by valence-arousal regression. In: 2016 IEEE 15th International conference on cognitive informatics & cognitive computing (ICCI*CC). IEEE, pp 42–49
6. Bardekar AA, Gurjar AA (2016) Study of Indian classical ragas structure and its influence on human body for music therapy. In: 2016 2nd International conference on applied and theoretical computing and communication technology (iCATccT). IEEE, pp 119–23
7. Bogt TFM, Vieno A, Doornwaard SM, Pastore M, Van Den Eijnden RJJM (2017) You're not alone: music as a source of consolation among adolescents and young adults. Psychol Music 45(2):155–171
8. Bradley MM, Lang PJ (1994) Measuring emotion: the self-assessment manikin and the semantic differential. J Behav Ther Exp Psychiatry 25(1):49–59
9. Burns CE (2019) Therapeutic uses of the flute within music therapy practice. Music Ther Perspect 37(2):169–175
10. Carter J, Pyati S (2018) 14 Nonpharmacologic management of postsurgical pain. Perioperative Pain Manag Gen Plast Surg 128
11. Clements-Cortés A, Yip J (2020) Social prescribing for an aging population. Act Adapt Aging 44(4):327–340
12. Cook T, Roy ARK, Welker KM (2017) Music as an emotion regulation strategy: an examination of genres of music and their roles in emotion regulation 47(1):144–154. https://doi.org/10.1177/0305735617734627
13. Corbijn van Willenswaard K, Lynn F, McNeill J, McQueen K, Dennis CL, Lobel M, Alderdice F (2017) Music interventions to reduce stress and anxiety in pregnancy: a systematic review and meta-analysis. BMC Psychiatry 17(1):1–9
14. Crowe BJ, Scovel M (1996) An overview of sound healing practices: implications for the profession of music therapy. Music Ther Perspect 14(1):21–29
15. Desai BU, Tandel KI, Patel RM (2020) Essential features extraction from Aaroh and Avroh of Indian Classical Raag–YAMAN. Int Res J Eng Technol (IRJET), 7. https://www.Irjet.Net/Archives
16. Desai NM, Sanjay HS, Desai DM, Kunikullaya K, Sinha AK (2021) An overview of the Indian system of Raagas and their positive effects on health. J Drugs Addict Ther. SRC/JDAT-112 Page 3:3–6.
17. Dev A, Smitha KV, Pillai RR (2015) Music therapy for institutionalised elderly persons with depression. Open J Psychiatry Allied Sci 6(1):15–19
18. Fancourt D, Finn S (2019) What is the evidence on the role of the arts in improving health and well-being? A scoping review
19. Fedotchev A, Parin S, Polevaya S, Zemlianaia A (2021) Human body rhythms in the development of non-invasive methods of closed-loop adaptive neurostimulation. J Personalized Med 11(5)
20. Frederick N (2018) Using music to recognize and regulate emotions with at-risk adolescent girls
21. Fredrickson WE (2000) Perception of tension in music: musicians versus nonmusicians. J Music Ther 37(1):40–50
22. Fu VX, Oomens P, Klimek M, Verhofstad MHJ, Jeekel J (2020) The effect of perioperative music on medication requirement and hospital length of stay: a meta-analysis. Ann Surg 272(6):961
23. Galińska E (2015) Music therapy in neurological rehabilitation settings. Psychiatr Pol 49(4):835–846
24. Gallagher LM, Huston MJ, Nelson KA, Walsh D, Steele AL (2014) Music therapy in palliative medicine. Support Care Cancer 2000 9(3):156–161
25. Grinde B, Patil GG (2009) Biophilia: does visual contact with nature impact on health and well-being? Int J Environ Res Public Health 6(9):2332–2343

26. Hagen M, Pruyn A, Galetzka M, Sauren J (2010) Music at railway stations: The influence of music on waiting experience
27. Hanser SB (1985) Music therapy and stress reduction research. J Music Ther 22(4):193–206
28. Hishida H, Kasahara R, Hishida K, Komatsu Y, Etoh H, Hishida Y, Hishida M (2021) Experiment of Music Therapy Conducted at a Classical Music Recital-Measurement of Saliva Amylase, Hand Sweat and Muscle Hardness. Complex, Inf Cybern: IMCIC 2021:59
29. Hoover SA (n.d.) Music as care: Artistry in the hospital environment: CMS emerging fields in ... - Sarah Adams Hoover - Google Books. Retrieved 11 Oct, 2023
30. Humphries K (2010) Healing sound: contemporary methods for Tibetan singing bowls
31. Husain G, Thompson WF, Schellenberg EG (2002) Effect of musical tempo and mode of arousal, mood and spatial abilities. Correspondencias & Análisis 20(15018):1–23
32. Iyengar KM, Prasad MV, Henkin R (n.d.) A study of Asian Indian and Asian Indian American Carnatic music students on emotive responses to six Carnatic Ragas: qualitative analysis of student responses
33. Jayamala AK, Preethi Bangalore Lakshmanagowda, G. C. M. Pradeep, and Jaisri Goturu (2015) Impact of music therapy on breast milk secretion in mothers of premature newborns. J Clin Diagn Res: JCDR 9(4):CC04
34. Juslin PN, Sloboda JA (2010) Handbook of music and emotion: theory, research, and applications
35. Karageorghis CI, Priest D-L (2012a) International review of sport and exercise psychology music in the exercise domain: a review and synthesis (part i), international review of sport and exercise. Psychology 5(1):44–66
36. Karageorghis CI, Priest DL (2012b) Music in the exercise domain: a review and synthesis (part I). Int Rev Sport Exerc Psychol 5(1):44–66
37. Kenny C (2015) Performing theory: playing in the music therapy discourse. J Music Ther 52(4):457–486
38. Kenny C (2014) The field of play: an ecology of being in music therapy. In: Voices: a world forum for music therapy, vol 14
39. Koelsch S (2014) Brain correlates of music-evoked emotions. Nat Rev Neurosci 15(3):170–180
40. Kumar N, Wajidi MA, Chian YT, Vishroothi S, Swamy Ravindra S, Ashwini Aithal P (2016) The effect of listening to music on concentration and academic performance of the student: cross-sectional study on medical undergraduate students. Res J Pharm, Biol Chem Sci 7(6):1190–1195
41. Kupski G (2020) Musiktherapie Mit Elise: Kann Klavierunterricht Therapie Sein? Reproduzieren Statt Improvisieren? Music Therapy With Elise. Can Piano Lessons Be Therapy? Reproducing Instead of Improvising? Musiktherapeutische Umschau 41:107–119
42. Landis-Shack N, Heinz AJ, Bonn-Miller MO (2017) Music therapy for posttraumatic stress in adults: a theoretical review. Psychomusicology: Music, Mind, and Brain 27(4):334
43. Lesiuk T (2005) The effect of music listening on work performance. Psychol Music 33(2):173–191
44. Letts R (1997) Music: universal language between all nations? 29(1):22–31.https://doi.org/10.1177/025576149702900104
45. Lin O, Yliopisto J (2020) Exploring therapeutic process through music therapy-the experiences of an adult 'third culture kid'
46. Liu Y, Liu G, Wei D, Li Q, Yuan G, Wu S, Wang G, Zhao X (2018) Effects of musical tempo on musicians' and non-musicians' emotional experience when listening to music. Front Psychol 9:2118
47. Loewy J (2004) Integrating music, language and the voice in music therapy. In: Voices: a world forum for music therapy, vol 4
48. Loombe D, Tomlinson J, Oldfield A (2015) Flute, accordion or clarinet? Using the characteristics of our instruments in music therapy. Jessica Kingsley Publishers.
49. Magnaye BP, Abalos NR, Kurt PD, Driz MK, Limbo CP, Tabuga MN (2020) Music therapy and labour pain among primigravida women. Eur J Mol Clin Med 7(2):4141–4153

50. Matney B (2017) The effect of specific music instrumentation on anxiety reduction in university music students: a feasibility study. Arts Psychother 54:47–55
51. Mattila AS, Wirtz J (2001) Congruency of scent and music as a driver of in-store evaluations and behavior
52. McFerran K (2011) Music therapy with bereaved youth: expressing grief and feeling better. Prev Res 18(3):17–21
53. Merrill R, Amin MT (n.d.) Rhythmically-enhanced music as analgesic for chronic pain: a pilot non-controlled observational study
54. Milliman RE (1986) The influence of background music on the behavior of restaurant patrons. J Consum Res 13(2):286–289
55. Mirenkov N, Kanev K, Takezawa H (2009) Mobile music therapy with multimedia quality of life supporters for elderly and disabled. J Mob Multim, 29–44
56. North AC, Hargreaves DJ, Hargreaves JJ (2004) Uses of music in everyday life. Music Percept 22(1):41–77
57. North AC, Hargreaves DJ, McKendrick J (1999) NorthEtAl99. J Appl Psychol 84(2):271–276
58. Pavlicevic M, Wood V (2005) Music therapy in children's hospices: Jessie's Fund in action. Jessica Kingsley Publishers
59. Perret D (2005) Roots of musicality: music therapy and personal development. Jessica Kingsley Publishers
60. Pfeifer E (2017) Music-nature-therapy: outdoor music therapy and other nature-related approaches in music therapy. In: Environmental expressive therapies. Routledge, pp 177–203
61. Priyadarsini AC, Rohini NS (2015) Effect of camatic music and pranayama on depression and resilience in mild depressives. Indian J Health Wellbeing 6(9)
62. Ransom PF (2015) ScholarlyCommons message in the music: do lyrics influence well-being?
63. Reddy BU, Phanisree P, Priyanka M, Kavitha D, Indira S, Bhandarkar P, Samudrala VD, Agrawal A (2017) Effect of music therapy in patients with moderate-to-severe traumatic brain injury. J Datta Meghe Inst Med Sci Univ 12(1):51
64. Rehman NA, Shakir K, Noorani I (2021) Effect of music on consumer emotions—an analysis of Pakistani restaurant industry. JISR Manag Soc Sci Econ 19(2):89–105
65. Rohith MN, Lokesh BS (2017) Non-conspicuous strong reaction for therapeutic Disarray using music. I-Manager's J Pattern Recogn 4(2):22
66. Russell JA (1989) Measures of emotion. In: The measurement of emotions. Elsevier, pp 83–111
67. Ryka LC (n.d.) Healing sounds of Tibetan singing bowls
68. Sangeetha KS (2019) Status and problems of music education in Kerala at school level
69. Schaefer HE (2017) Music-evoked emotions-current studies. Front Neurosci 11:245482
70. Shrestha S (2009) How to heal with singing bowls: traditional Tibetan healing methods. Sentient Publications
71. Shunqin Y (2018) Comparative study on the effect of different music therapy on auxiliary treatment of college students' anxiety. Biomed Res (0970-938X)
72. Skaggs R, Lense M, Clayton J (2017) Building an evidence base for using music as a treatment for children with autism spectrum disorder
73. Sufie SNM, Sidik R (2017) What is medical music therapy in Islamic civilization? Int J Bus Soc Sci 8(3):195–199
74. Särkämö T, Tervaniemi M, Laitinen S, Forsblom A, Soinila S, Mikkonen M, Autti T, Silvennoinen HM, Erkkilä J, Laine M, Peretz I, Hietanen M (2008) Music listening enhances cognitive recovery and mood after middle cerebral artery stroke. Brain 131(3):866–876
75. Tendai M, Crispen C (2009) In-store shopping environment and impulsive buying. Afr J Market Manag 1(4):102–108
76. Thompson WF, Olsen KN (eds) (2021) The science and psychology of music: from Beethoven at the office to Beyoncé at the gym. Bloomsbury Publishing, USA
77. Van Den Tol A, Edwards J (2013) Listening to sad music in adverse situations: how music selection strategies relate to self-regulatory goals, listening effects, and mood enhancement final draft

78. van Hagen M, Pruyn A, Galetzka M, Sauren J (2010) Music at railway stations: the influence of music on waiting experience. In: European transport conference 2010. Association for European Transport (AET)
79. Vishwakarma SD, Kulkarni VD, Shinde AD, Virkar OM, Barkade VM (2017) IOT-BEAT: an intelligent nurse for the cardiac patient with music therapy. Int Res J Eng Technol (IRJET) 4(12):1679–1685
80. Whittemore J (2019) A study of music: music psychology, music therapy, and worship music
81. Wilson EO (1984) Biophilia. Harvard University Press, Massachusetts
82. Yalch R, Spangenberg E (1990) Effects of store music on shopping behavior. J Serv Mark 4(1):31–39
83. Yang X (2020) Effects of music on chronic pain and associated mood disorders in rats. Rev Cient Fac Cien Veterinarias 30(2):1111–1119
84. Yew S, Ng AW (2020) Music as medicine: a literature review on the effectiveness of Mozart's Music on Epilepsy. Intersect: Stanford J Sci, Technol, Soc 14(1)
85. Yi F, Kang J (2019) Effect of background and foreground music on satisfaction, behavior, and emotional responses in public spaces of shopping malls. Appl Acoust 145:408–419

Comfort and Stress: Assessing the Role of Furniture in the Well-Being of Geriatric Inpatients

Mohamad Yasser Arafat and Sonal Atreya

1 Introduction

Hospitalization poses various challenges for elderly patients, such as the loss of their home environment, the adjustment to new routines, and the anxiety of illness. These challenges can negatively affect their well-being and functional status, as they may experience psychological distress, sensory deprivation, and immobility. Prolonged bed rest can lead to irreversible decline in functional ability and quality of life. The built environment, which includes the physical features and design of healthcare facilities, can influence the health and well-being of individuals, especially older adults.

The way our surroundings are built can have impacts, on our mental and emotional well-being [8]. Previous studies indicate that the layout of our environment can influence our mood thinking abilities and overall health. For instance, research has found that healthcare facilities incorporating elements like plants, daylight and water features can enhance patients recovery time pain perception and medication usage [14]. Other research has shown that the design of hospitals can affect patients stress levels and how they perceive the quality of care they receive [6]. Furthermore, the design of a facility can impact factors such as safety measures, satisfaction levels, and staff productivity [12].

The furniture, in hospitals plays a role in shaping the experience of patients. It has an impact on their comfort, ability to move around, safety, privacy, and overall satisfaction. However, there is a lack of research on how furniture design can improve the health outcomes and quality of life of geriatric patients. Most studies examining the relationship between the built environment and health tend to focus on elements like lighting, noise levels, ventilation systems and access to surrounding nature.

M. Y. Arafat · S. Atreya (✉)
Department of Architecture and Planning, Indian Institute of Technology, Roorkee, India
e-mail: sonal.atreya@ar.iitr.ac.in

© The Author(s), under exclusive license to Springer Nature Singapore Pte Ltd. 2025

S. Das et al. (eds.), *Human-Centered Digitalization, Volume 3*, Design Science and Innovation, https://doi.org/10.1007/978-981-96-4383-7_22

Moreover, most of the studies are conducted in general hospital settings, rather than in geriatric wards or units.

The aim of this research is to improve the design of healthcare facilities by exploring how the furniture within these spaces impacts patients. The ultimate goal is to enhance patient care by reducing anxiety. The findings from this study can inform the creation of evidence-based design guidelines, helping healthcare professionals develop environments that promote health, healing, and overall well-being.

1.1 Furniture

The comfort provided furniture in hospital ward indoor environments can significantly impact the stress levels of geriatric patients. Furniture design, material, and arrangement can influence the psychological and physiological well-being of patients in healthcare facilities [7, 10]. Ergonomically design furniture for hospitalized patients can improve their level of comfort and satisfaction. Adequate support for the body, easy-to-use features, and the ability to adjust seating positions can contribute to a patient's comfort, leading to improved well-being [1–4, 9, 11]. Having a comfortable and supportive bed is crucial for patients a satisfactory sleep at night which greatly impacts their overall well-being and recovery. On the hand, if the bed is uncomfortable or inadequate it can cause restlessness, discomfort and even pain which can have an effect, on both the patients recovery process and their satisfaction, with their hospital stay.

Properly designed furniture can ensure patients' safety, reduce the risk of falls, support patients' mobility and help maintain independence. Additionally, furniture with rounded edges can reduce the risk of injury and provide a sense of security [4]. Moreover, furniture that is aesthetically pleasing can contribute to a cozy atmosphere thereby decreasing feelings of anxiety and encouraging a sense of calm. In addition utilizing materials that're safe and hypoallergenic can effectively minimize the risk of reactions and potential health concerns [4, 5].

Objective: The objective of this research is to enhance the design of healthcare facilities by examining the relationship between the furniture within the built environment of hospital wards and its implications for geriatric patients, with the ultimate goal of improving patient care.

2 Method of Research

2.1 Study Design

The study adopted a cross-sectional design.

2.2 Context Overview

This study investigates how ward furniture affects the anxiety levels of geriatric inpatients in different types of healthcare facilities in Delhi, such as government, private, and NGO-based ones.

2.3 Sample Size Estimation

Delhi's population was 16.78 million in 2011, with 6.8% over age 60. The determination of the sample size (*n*) was carried out by applying the following formula

$$n = \frac{Z^2 * p * (1 - p)}{C^2} \tag{1}$$

where

'*n*' represents the sample size;

'*z*' denotes the *z* value corresponding to a 95% confidence interval, which is a standard value of 1.96;

'*p*' signifies the proportion of the target population expected to exhibit a specific characteristic, set at 6.8%;

'*c*' stands for the margin of error, which is 5%, and is represented by a standard value of 0.05;

'NRR' refers to the non-response rate, which is accounted for at 4%.

The above calculation yielded a total of 97 samples; however, to account for potential non-availability, non-response, or dropouts during the interview process, a 4% non-response rate was factored in, resulting in the inclusion of 100 geriatric inpatients admitted to the general wards of government, private, and NGO-based healthcare facilities in Delhi for the study.

2.4 Data Collection

The survey of geriatric acute care inpatients was conducted across hospitals that included a public hospital, a non-profit medical college hospital, and a private hospital. The study obtained the necessary permissions from each of the hospitals. The institution granted the approval from the Institutional Human Ethics Committee (IHEC) to collect data from the intended respondents. The study provided information to the directors, chief medical officers, and other relevant authorities of the selected hospitals regarding its purpose and methodology, and informed consent was obtained from the respondents prior to conducting the interviews.

2.5 Research Instrument

An observational checklist was employed to document the architectural layout and features of the ward. The degree of comfort, control, and safety while utilizing the furniture available in the ward by the geriatric inpatients was recorded using the Likert Scale instrument (Table 1). The level of anxiety among geriatric inpatients was measured using the State Anxiety Checklist, which was an adapted version of the Spielberger State-Trait Anxiety Inventory [13]. The degree of the patients' likelihood of agreeing with the anxiety statement was recorded using a 4-point rating scale ranging from not at all (1) to very much so (4). A total of 20 items were used to assess the level of anxiety of geriatric inpatients in the sampled hospital [13]. The degree of comfort, control, and safety while utilizing the furniture available in the ward by the geriatric inpatients was recorded using a self-developed Likert Scale instrument (Table 1). The instruments were validated through the Delphi method, incorporating feedback from 150 experts, including practicing psychologists and healthcare architects.

Table 1 Perception of geriatric inpatients

Indicators	Likert scale instrument	Category
Patient bed	Very uncomfortable (1) to very comfortable (5)	Furniture comfort
Bed's adjustment features	Very difficult (1) to very easy (5)	Furniture ease
Getting over bed	Very unsafe (1) to very safe (5)	Furniture safety
Utilizing patient bed	Very unsafe (1) to very safe (5)	Furniture safety
Utilizing food table	Very unsafe (1) to very safe (5)	Furniture safety
Adequacy of the furniture allotted on each bed	Very uncomfortable (1) to very comfortable (5)	Furniture quality
Quality of the furniture	Very uncomfortable (1) to very comfortable (5)	Furniture quality
Ward's storage space	Very uncomfortable (1) to very comfortable (5)	Furniture quality
Furniture for family and visitors	Very uncomfortable (1) to very comfortable (5)	Furniture quality

2.6 Analysis

The data collected was analyzed using Chi Square statistical tests to determine the correlation between furniture comfort and anxiety levels among geriatric inpatients.

2.7 Architectural Overview of Hospital Wards

The hospitals exhibited variations in the design and capacity of their general wards. A notable example is a 300-bed private super specialty hospital located in Delhi NCR. The infrastructure contained 10 general medicine wards, with each ward having the capacity to accommodate up to six patients. Additionally, it provides four communal restroom facilities, shared between male and female patients. Another establishment is Delhi's public hospital with a capacity of 200 beds. Within this hospital, there exists a 17-bed general medicine division that offers two communal toilets, catering to both genders. Finally, a non-profit medical college hospital with a total of 300 beds, situated in the Delhi NCR region, was also included in the review. This hospital has two general medicine wards, and each of these wards is designed to accommodate up to 10 patients. These wards are serviced by two shared restroom facilities, accessible to both male and female patients. An architectural checklist was utilized to observe and evaluate the wards in all aforementioned institutions for the purpose of data collection.

3 Result and Analysis

3.1 Demographic Characteristics of the Respondents

Survey respondents included 100 Geriatric inpatients from super specialty private hospital (52), public hospital (25), and non-profit medical college hospital (23). In the entire interviewed sample, there were 36 (36%) women and 64 (64%) men. There were 19 women (36.54%) and 33 men (63.46%) in hospital super specialty private hospital with a mean of 66.8 years and standard deviation of 4.0 years. There were 11 women (44%) and 14 men (56%), with a mean age of 67.0 years and standard deviation of 4.5 years in the public hospital. There were 6 women (26.09%) and 17 males (73.91%) in the non-profit medical college hospital with a mean age of 66.8 years and standard deviation of 3.8 years. The demographic characteristics of the respondents is shown in tabular form in Table 2.

Table 2 Demographic characteristics of the respondents

Hospital type	Number of respondents	Percentage of women (%)	Percentage of men (%)	Mean age (years)	Standard deviation (years)
Super specialty private hospital	52	36.54	63.45	66.8	4.0
Public hospital	25	44.00	56.00	67.0	4.5
Non-profit medical college hospital	23	26.09	73.91	66.8	3.8
Total	100	36.00	64.00	66.9	4.1

3.2 Stress Levels of Geriatric Patients in Relation to Their Perceptions Regarding Furniture and Its Features Provided in the Ward

The anxiety levels among the patients appeared to be influenced by their comfort with the adequacy and quantity of furniture allocated for each patient bed in the ward. The study found that patients who reported comfort with the available furniture had the highest positive anxiety levels (86.2%), while those who reported very discomfort had the lowest positive anxiety levels (64.3%). The Chi Square test results (χ^2 = 9.100, df = 4, p = 0.059) indicated a marginally significant association between patients' comfort regarding the number of usable furniture items and their anxiety levels. This observation implies that providing an appropriate amount of functional furniture for each patient in the ward may affect on patients' anxiety levels, although the relationship was not definitively statistically significant. But there was still a noticeable trend suggesting its potential influence. Ensuring an adequate amount of functional furniture in the ward may contribute to improving patients' well-being and reducing their anxiety.

The level of comfort that patients experienced in their beds seemed to have an impact, on how stressed they felt. Patients who found the provided beds to be very comfortable had highest positive anxiety levels at 94.4% whereas those who were uncomfortable reported the lowest levels at 63.2%. The results of the Chi Square test (χ^2 = 11.600, df = 4, p = 0.021) indicated a correlation between patients comfort in their bed and their anxiety levels. This finding highlights the importance of having comfortable beds available, in the ward to help alleviate patients anxiety levels.

Positive anxiety levels experienced by patients varied, between 70.0% for those who felt uncomfortable with the furniture quality in the ward and 78.9% for patients who remained undecided, no differences observed across different comfort levels. The Chi Square test results (χ^2 = 1.100, df = 4, p = 0.894) did not find a relationship between patients comfort with the furniture quality in the ward and their anxiety levels. This suggests that other factors may play a role, in determining patients anxiety levels.

Patients who felt highly comfortable, with the storage facilities experienced peaked positive anxiety levels (100.0%) whereas those who were uncomfortable had the lowest levels (66.7%). The results of the Chi Square test ($\chi^2 = 34.900$, df $= 4$, $p < 0.001$) indicated a correlation between patients comfort with storage facilities and their anxiety levels. These findings emphasize the importance of providing comfortable storage facilities within the ward to alleviate patients anxiety levels.

The data indicates that there is no substantial difference in positive anxiety levels across different levels of safety while using the food table in the ward. The study found no significant difference in the positive anxiety levels among patients who reported different levels of safety while using the food table ($\chi^2 = 6.300$, df $= 4$, $p = 0.178$). The percentage of positive anxiety ranged from 70.8% for those reporting high safety to 78.6% for those reporting low safety with only minor variations across different safety levels. This suggests that other factors might have an impact, on anxiety levels related to food table usage among elderly patients.

The positive anxiety levels among patients, in the ward remained fairly consistent regardless of the safety levels provided by the footstep stool. The levels of anxiety experienced by patients, in the ward were relatively consistent regardless of the safety levels provided by the footstep stool. The data showed that the positive anxiety levels ranged from 63.6% for undecided patients to 78.6% for both safe and unsafe patients. This indicates that the degree of safety provided by the footstep stool does not have a significant impact on the anxiety levels of geriatric patients in the ward. The statistical analysis using a Chi Square test ($\chi^2 = 19.300$, df $= 4$, $p = 0.001$) confirmed a relationship between feeling safe while using the footstep stool and anxiety levels. This finding underlines the importance of ensuring stability and safety measures for footstep stools in order to alleviate anxiety among patients.

The level of comfort and safety experienced by patients, in the hospital bed has an impact on their anxiety levels. When patients reported comfort and safety 65.6% experienced positive anxiety levels, which increased to 90.9% for those who reported very high comfort and safety. There is a trend across levels of comfort and safety. This difference in anxiety levels is $\chi^2 = 11.100$, df $= 4$, $p = 0.025$) indicating that how patients perceive the safety and comfort of using the bed greatly affects their anxiety levels. These findings highlight the importance of prioritizing patient safety and comfort when designing and selecting beds for hospital wards. Beds should be designed to offer support minimize the risk of falls and allow patients to easily get in and, out of bed.

Among patients those who expressed comfort, with the furniture provided for their family or visitors reported the levels of anxiety (85.0%). On the hand, undecided patients had the levels of positive anxiety (50.0%). The Chi Square test results ($\chi^2 = 26.000$, df $= 4$, $p < 0.001$) indicated an association between comfort with furniture for family or visitors and anxiety levels. These findings imply that ensuring furniture for patients families and visitors could play a role, in reducing patients anxiety levels.

The findings suggest that patients who perceived the furniture in the patient area to be sturdy had higher positive anxiety levels (78.5%) than those who did not (68.6%). The Chi Square test outcomes ($\chi^2 = 9.000$, df $= 1, p = 0.003$) indicated an association between the sturdiness of furniture and anxiety levels emphasizing the importance of providing furniture in the patient area to reduce patients anxiety levels. These findings underscore the necessity of ensuring that the furniture in the area is both sturdy and well maintained creating an secure environment for elderly patients, in the ward.

It was discovered that there is a difference, in the levels of anxiety experienced by patients depending on whether their beds had adjustable features or not. Patients whose beds had features showed higher levels of positive anxiety (76.0%) compared to those without such features (72.0%). However, the results of the Chi Square test ($\chi^2 = 25.000$, df $= 1, p < 0.001$) indicated a correlation between adjustment features on patients' beds and their anxiety levels. This suggests that providing beds with adjustment features could potentially help reduce patients anxiety levels. Therefore, it is recommended to consider offering beds to patients, in the ward as a means to alleviate their anxiety.

The study found that patients who had furniture with round edges, in the patient's area had higher levels of positive anxiety (76.0%) compared to those who didn't (72.0%). The results of the Chi Square test ($\chi^2 = 25.000$, df $= 1, p < 0.001$) showed a connection between the presence of edges on furniture and anxiety levels. This emphasizes the importance of providing furniture with round edges, in the patient's zone as it can potentially reduce patients anxiety levels and enhance their overall well-being. Hospitals and healthcare facilities should prioritize comfort and safety by ensuring that furniture has rounded edges in its design and construction.

4 Conclusion

The study revealed that the comfort and safety of furniture have an impact, on the anxiety levels of patients staying in different healthcare facilities in Delhi. The study suggests that by offering comfortable furniture in areas we can enhance their overall well-being and lower their anxiety levels. The results of the Chi Square test demonstrated connections between various furniture factors and patients anxiety levels. Providing beds ample storage options, sturdy furniture, adjustable bed features and furniture with round corners may all contribute to reducing patients anxiety levels. Additionally, ensuring the safety of beds and step stools while also providing furniture for patients families and visitors can help alleviate anxiety within the ward. These findings have implications, for hospital administrators and healthcare professionals who should consider the significance of offering comfortable furniture in patient areas to promote patient well-being while reducing anxiety.

References

1. Architectural Services Department (2018) Elderly-friendly design guidelines. Thesis
2. Biancheri R, Landi S (2016) Gender and healthcare environments: a proposal of gender-sensitive methodology for improving the environmental quality in the existing heritage. Tria-Territorio Della Ricerca Su Insediamenti E Ambiente 9(2):69–81
3. Blackler A, Brophy C, O'Reilly M, Chamorro-Koc M (2018) Seating in aged care: physical fit, independence and comfort. SAGE Open Med 6. https://doi.org/10.1177/2050312117744925
4. Calkins MP (2009) Evidence-based long term care design. NeuroRehabilitation 25(3):144–154. https://doi.org/10.3233/NRE-2009-0512
5. Chaudhury H, Cooke HA, Cowie H, Razaghi L (2018) The influence of the physical environment on residents with dementia in long-term care settings: a review of the empirical literature. Gerontologist 58(5):e325–e337. https://doi.org/10.1093/geront/gnw259
6. Devlin AS, Arneill AB (2003) Health care environments and patient outcomes: a review of the literature. Environ Behav 35(5):665–694. https://doi.org/10.1177/0013916503255102
7. Evans GW, Lepore SJ, Schroeder A (1996) The role of interior design elements in human responses to crowding. J Pers Soc Psychol 70(1):41–46. https://doi.org/10.1037/0022-3514.70.1.41
8. Hartig T, Mang M (1991) Restorative effects of natural environment experiences. Environ Behav 23(1):3–26. https://doi.org/10.1177/0013916591231001
9. Huisman ERCM, Morales E, van Hoof J, Kort HSM (2012) Healing environment: a review of the impact of physical environmental factors on users. Build Environ 58:70–80. https://doi.org/10.1016/j.buildenv.2012.06.016
10. Lawton CA (1996) Environmental Journal of. J Environ Psychol, 137–145
11. Obeidat IM, Ph D (2000) Designing for the elderly: comfort, health and well-being research problem research question research methodology research objectives, 59–70
12. Shepley MMC, Watson A, Pitts F, Garrity A, Spelman E, Kelkar J, Fronsman A (2016) Mental and behavioral health environments: critical considerations for facility design. Gen Hosp Psychiatry 42:15–21. https://doi.org/10.1016/j.genhosppsych.2016.06.003
13. Spielberger CD (2015) State-trait anxiety inventory instrument and scoring key
14. Ulrich R, Quan X, Systems H, Architecture C, Texas A (2004) The role of the physical environment in the hospital of the 21st century: a once-in-a-lifetime opportunity. Environment, 439:69. http://www.saintalphonsus.org/pdf/cah_role_physical_env.pdf

An Analysis of Driving Hours on the Driver Preferential Use of the Dashboard: A Pilot Study

Vrashal Verma and Manish Kumar Asthana

1 Background

1.1 *User Experience of the Driver Dashboard*

Dials and displays give information about the system status of a machine. The dashboard serves as an interface for transactions between humans and machines [1]. These interactions in driving involve complex multitasking activities, which require successful integration of the driver's cognitive, physical, sensory, and psychomotor skills [1–4]. Driving involves the use of an infotainment system usage [5]. Many drivers perceive driving as a medium for enjoyable experiences [6]. Adverse use and misunderstanding may cause errors in the system [5]. The general problems with the driving dashboard are often linked to the dashboard. The layout of the dashboard affects driving performance and can cause errors and performance [1]. The automobile industry witnessed the dashboard design changes frequently. These changes have a phenomenal effect on driving hours. The driving hours vary according to the demography and work allotted to the user group selected.

V. Verma · M. K. Asthana (✉)
Department of Design Indian Institute of Technology Roorkee, Roorkee, India
e-mail: m.asthana@hs.iitr.ac.in; asthanakm@gmail.com

V. Verma
e-mail: vrashal_v@design.iitr.ac.in

M. K. Asthana
Department of Humanities and Social Sciences Indian Institute of Technology Roorkee, Roorkee, India

© The Author(s), under exclusive license to Springer Nature Singapore Pte Ltd. 2025

S. Das et al. (eds.), *Human-Centered Digitalization, Volume 3*, Design Science and Innovation, https://doi.org/10.1007/978-981-96-4383-7_23

1.2 The Demography of the Drivers

The drivers selected were from two different states in India. The drivers of Madhya Pradesh had driving taxis as their primary occupation. Secondly, drivers from Uttarakhand state were institute drivers of the Indian Institute of Technology Roorkee. The Madhya Pradesh drivers also had a group of people who were serving the police department in Indore city. The drivers in Indore city generally use the Maruti Suzuki Swift Dzire. The in-vehicle technologies used in the automobile are aimed at increasing safety, but driving distractions also increase with the in-vehicle functions [4].

Long and short driving hours can be broken into two sets: 1–7 h as short and 7–15 h as short driving hours [2]. The two driver groups (long and short working hours) also have different experiences when considering the control panel. There are differences between the selected demographic locations, as Indore is inclined toward a logistic location, and Roorkee is more academia-centric. Various internal and external factors can influence interaction quality [3].

The usage of dashboard features varies. Also, mobile phone usage in the car dashboard and usage depends on the (1) Long Driving Hours, (2) Short Driving Hours, (3) More Experienced, and (4) Less Experienced.

1.3 The Importance of Driver-Centered Design and Development

A driver-centered perspective is required to communicate information to the drivers. In adverse conditions, information overload might occur; driver training and support are required to counter this. Human factors are essential to the design and introduction of technology to drivers [3].

Various other studies show that automobile user interfaces are becoming complex with the introduction of new technologies [7]. Many researchers have done ergonomic error evaluation tasks, such as OPTIVe, driving errors, and visual standardized methods in the car display. ISO 9241 part 11 is related to the ergonomic requirements of the display from the user satisfaction point of view [8].

1.4 Type of Work

The police department in Indore City has a working culture analogous to the government job. The working shift starts from 9 AM to 4 PM. The vehicles commonly used are gypsy and bolero, which are meant to serve the logistics purpose for the Police force. Some sedan-type cars, which are used for higher officials, are also used.

Table 1 Demography of the drivers

Parameters	N	N%
Age (years)	< 25	2, 8.3
	26–35	3, 12.5
	36–42	11, 45.8
	> 43	8, 33.4
Experience (years)	< 9	3, 12.5
	10–19	6, 25
	20–29	12, 50
	> 30	3, 12.5
Driving (hours)	< 5	54.2
	6–11	29.2
	> 12	16.6
Gender	Male	24, 100
	Female	0, 0

N = Number of users
N% = Percentage of users.

The vehicles used in Roorkee City, Uttarakhand State, are Similar to these Sedan cars. Regular taxi drivers use these taxis for intercity travel. The driver groups had different tasks according to their placement in the demographic area. These tasks reflect their engagements, whether with the academia, taxi, or police department.

Types of dashboards: Most of the dashboards come up with various hybrid and electronic components. The dashboard upgraded to digital interaction and displays. Police drivers have preferences aligned to analog displays and fewer features on digital displays (Table 1).

1.5 The Features Which Are Significant Over the Others

See Table 2.

Table 2 Features on the dashboard and significance according to result (Maruti Suzuki Dzire)

Feature on the car	Importance in the long/short driving hours	Significance of the result
Hours	Important in the long driving hours	Yes
Air conditioner	Important in the long diving hours	Yes
Time on the dashboard	Important in the long driving hour	Yes
Outside temperature	Important in the long driving hour	Yes

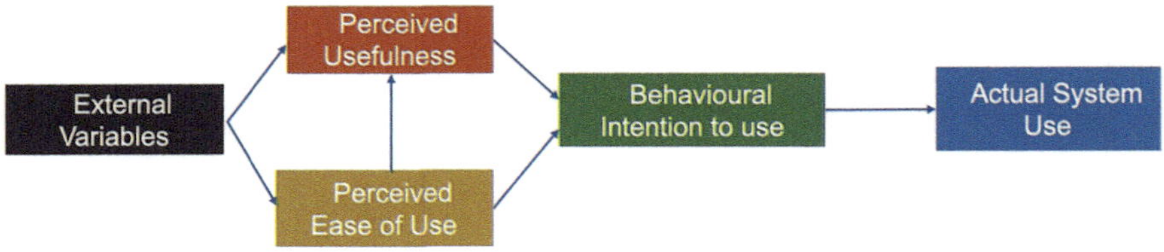

Fig. 1 Technology acceptance model. Adapted with permission from Davis et al. [9]

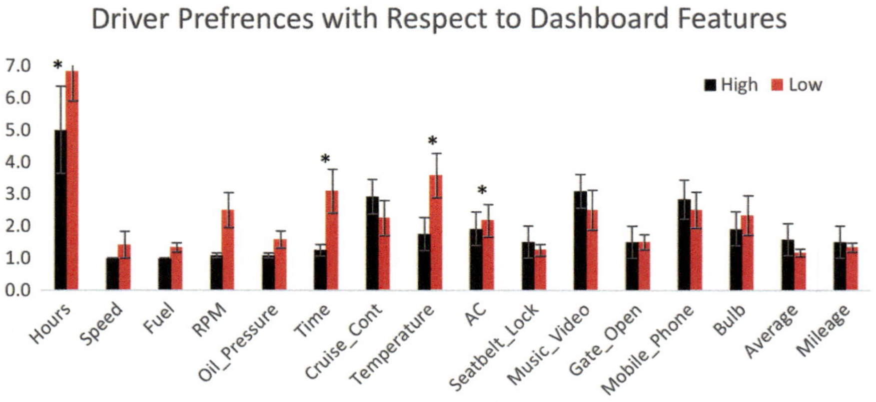

Fig. 2 Graphical representation of the comparison of the mean by one-way ANOVA for the groups having short and long driving hours. Short driving hours had 9 participants, and long driving hours had 15 participants. Total subjects $N = 24$

Model of Driver Acceptance. Technology acceptance model (TAM): This model is applied in studies assessing the acceptance of driving assistance systems. The model includes acceptance based on information attributes. Trust in travel information, socio-demographics, and cognition of alert routes. TAM model is found reliable for the measurement of instrument design [9]. Such theories help in assessing the aptness of the driving dashboard systems (Figs. 1 and 2).

One-way ANOVA was conducted to compare the effect of driving hours on the dashboard's features.

2 Methodology

2.1 OPTIVe Survey Outcomes

The study was done on $N = 24$ participants using the Optimized System Integration for Vehicle Safety Interaction (OPTIVe) [10]. The survey outcomes also suggested that most of the drivers do not know any language (this is English or Hindi). The survey is referred to by Lulea University [10]. This OPTIVe survey was chosen because it dictates the dashboard's driver information questions, instrument cluster questions, and warning symbols [10]. The amendments made regarding this study include the number of hours of the daily commute, driver education, and driver experience for a better understanding of the divide between the two groups. This survey aimed to evaluate the user's preference for the use of the dashboard. Questions were asked general in nature to understand the current problems in reading the dashboard features and usage concerning the dashboard features [8].

2.2 *Qualitative Methods*

For this study, drivers were chosen based on their experiences with driving hours. The user acceptance was used to test the user experience with the driving hours. In the process of data collected by the survey, the first step was the OPTIVe questionnaire. The second step includes thematic analysis to interpret the questionnaire. Thematic analysis helps create the themes and patterns in qualitative data [11]. We hypothesize that driver driving for long driving hours may not use the dashboard effectively. On the contrary, drivers driving for short-driving hours use the dashboard effectively. By exploring the relationship between the driving dashboard feature usage and driving hours, the selective action of drivers is studied. Based on the driving hours, themes are created. These themes are long driving hours and short driving hours. Theme creation is based on reflecting the choice of scenario and opinion for the driver dashboard [3]. The scenario includes driving in distraction, fatigue, day conditions, and night condition.

Once the data was collected, it was analyzed using the Statistical Package for Social Science (SPSS) program to identify the significant features of driver preferential use (Fig. 3).

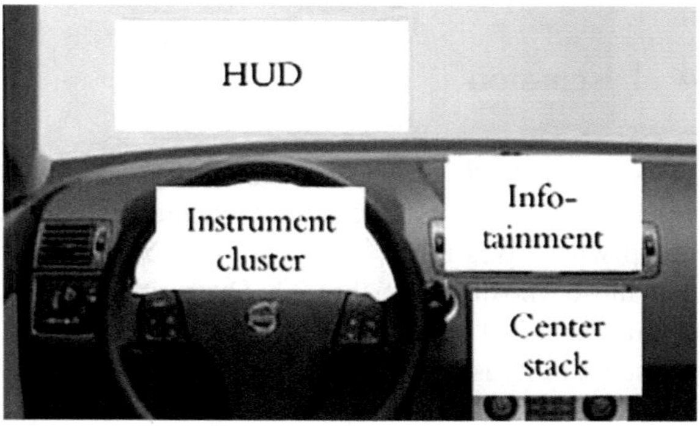

Fig. 3 Visual head-up display, instrument cluster, infotainment, and center stack locations used for the OPTIVe survey [12]

Table 3 One-way ANOVA results

Variable (features)	df	Mean	F	P-value	Partial Eta Squared (η)
How many hours in a day they drive	1	24.00	63.75	**< 0.001**	**0.139**
	22	17.00			
Temperature outside the car	1	4.66	21.21	**< 0.001**	**0.491**
	22	1.46			
Time on the car	1	3.11	3.90	**0.061**	**0.151**
	22	1.60			
Air conditioner and humidity in the car	1	3.00	5.22	**0.032**	**0.192**
	22	1.46			

The partial eta squared and p value are bold to ensure visual clarity for the reader

3 Results

The results indicated the importance of dashboard features and frequency of use. The outside temperature feature in the control panel is essential to understand the user preferences and mode of driving. Work and occupation also play a vital role in the dashboard preferential use. The long working hours drivers need music and an infotainment system. There was a significant effect of driving hours on how many hours they drive in a day $[F(1, 22) = 63.75, p = <0.001, \eta p^2 = 0.139]$, temperature on dashboard usage $[F(1, 22) = 21.21, p = <0.001, \eta p^2 = 0.491]$, time on the car $[F(1, 22) = 3.898, p = 0.061, \eta p^2 = 0.151]$ and air conditioner and humidity in the car $[F(1, 22) = 5.220, p = 0.032, \eta p^2 = 0.192]$ (Table 3).

According to the one-way ANOVA, it was realized that there is a significant difference between the responses of the participants during long working hours and short working hours, $p < 0.5$ these features are the amount of fuel left in the car, engine RPM, time on car, temperature outside, and how many hours they drive.

4 Discussion

In the present study, we show that the user groups with more and fewer working hours behave differently with the automobile human machine interaction (HMI). The user group who drives more uses RPM, dashboard clock, and speedometer frequently than those who drive less than 7 h. Our results concur with the previous studies done in this regard of human–machine interaction (HMI) and driver preferences [12]. These usability issues can impact driving performance and usage of the dashboard. These features sometimes negatively affect driving performance. The uncluttered and refined HMI determines the driver's ability to perform the primary driving task using in-vehicle devices [12]. The dashboard preferential use is reconfirmed with previous eye-tracking studies [13]. These results align with the validation from the eye-tracking results obtained on the dashboard [5, 13, 14]. When drivers use the car dashboard in various conditions, the preferential use of the dashboard varies [15]. The preferential use dashboard raises many questions regarding driving behavior and use concerning day and night conditions [16]. The ergonomic aspects of the passenger car are influenced by the type of work and driving hours of the car. This study further raises several questions regarding investigating the optimized dashboard layout design according to the frequently raised question of driving hours. Interplay of the placement of the preferential features on the dashboard might increase the driver's performance [17]. Further work can be done from the interaction design perspective, where the interaction designers can better comprehend the existing use and context of the use of cars [18]. The study is limited to males only; we don't know how females would have responded. Future studies must highlight the significance of gender in dashboard preference and driving hours. The time for this behavioral study is not covered. Hence, the day and night condition and selection of features are not covered.

5 Conclusion

The research paper aimed to investigate user preferences and their impact on the dashboard preferential use. It is essential to remind the researchers about human–machine interaction. This paper also clarifies the thought process of accepting features on the car control panel. The two groups of drivers use the control panel very differently. The significance of the features relies on the long and short hours of operation, demography, and education. This qualitative study shows that the user experience role is important in dashboard design [17]. Hence, optimizing the driver information and measuring the preferential use are necessary. The actual use of the system may vary depending on the driving scenarios and aesthetic arrangement of the control panel and its features. The study analyzes the preferential use of dashboards from drivers' perspectives of the existing in-vehicle functions. The recent trends in the twenty-first century indicated the changes in the driving experience of autonomous cars. Thus, a new set of design methods is required to understand the newly features added to the car dashboard.

References

1. De Carvalho RC, Soares MM (2012) Ergonomic and usability analysis on a sample of automobile dashboards. Work: A J Prev Assess Rehabil 41:1507–1514. https://doi.org/10.3233/wor-2012-0345-1507
2. Regan MA, Horberry T, Stevens A (2016) Driver acceptance of new technology, 2nd edn. CRC Press, New York
3. Meng F, Li S, Cao L, Li M, Peng Q, Wang C, Zhang W (2015) Driving fatigue in professional drivers: a survey of truck and taxi drivers. Traffic Inj Prev 5(16):474–483

4. Young K, Regan M, Hammer M (2003) Driver distraction: a review of the literature. Monash University Accident Research Centre, Victoria, Australia
5. Gellatly AW, Kleiss JA (2000) Visual attention demand evaluation of conventional and multi-function in-vehicle information systems. In: Proceedings of the human factors and ergonomics society annual meeting/proceedings of the human factors and ergonomics society. Annual meeting, vol 44(20), 3–285. https://doi.org/10.1177/154193120004402015
6. Harvey C, Stanton NA, Pickering CA, McDonald M, Zheng P (2011) In-vehicle information systems to meet the needs of drivers. Int J Hum-Comput Interact 27(6):505–522. https://doi.org/10.1080/10447318.2011.555296
7. Rundqvist M, Persson L (2007) MSc. Thesis: design of instrument cluster for automobiles
8. Prabhakar G, Biswas P (2021) A brief survey on interactive automotive UI. Transp Eng 6
9. Davis FD, Venkatesh V (1996) A critical assessment of potential measurement biases in the technology acceptance model: three experiments. Int J Hum Comput Stud 45(1):19–45. https://doi.org/10.1006/ijhc.1996.0040
10. Francois M, Osiurak F, Fort A, Crave P, Navarro J (2017) Automotive HMI design and participatory user involvement: review and perspectives. Ergonomics 60:541–552
11. Braun V, Clarke V (2006) Using thematic analysis in psychology. Qual Res Psychol 3(2):77–101. https://doi.org/10.1191/1478088706qp063oa
12. Tretten P (2008) Ph.D. Thesis: the driver and the instrumental panel. Lulea University of Technology
13. Kim S, Dey AK, Lee J, Forlizzi J (2011) Usability of car dashboard displays for elder drivers. In: Proceedings of the SIGCHI conference on human factors in computing systems. https://doi.org/10.1145/1978942.1979011
14. Lee Y, Lee JD, Boyle LN (2007) Visual attention in driving: the effects of cognitive load and visual disruption. Hum Factors 49(4):721–733. https://doi.org/10.1518/001872007x215791
15. Marcus A (2002) Information visualization for advanced vehicle displays. Inf Vis 1(2):95–102. https://doi.org/10.1057/palgrave.ivs.9500016
16. Fort E, Chiron M, Davezies P, Bergeret A, Charbotel B (2013) Driving behaviors and on-duty road accidents: a French case-control study. Traffic Inj Prev (Online) 14(4):353–359. https://doi.org/10.1080/15389588.2012.719091
17. Designing dashboard of national electric cars SmarT EV.2 regarding to the ergonomic aspects (2015). IEEE Conference Publication | IEEE Xplore. https://ieeexplore.ieee.org/abstract/document/7496651/figures#figures
18. Pettersson I, Ju W (2017) Design techniques for exploring automotive interaction in the drive towards automation. In: Proceedings of the 2017 conference on designing interactive systems. https://doi.org/10.1145/3064663.3064666

An AI-Driven Multifaceted Approach to Mental Health: Leveraging Technology for Mental Well-Being

Sabyasachi Biswas, Bhakti Patil, and Sandipan Bhattacharjee

1 Introduction

The COVID-19 pandemic has profoundly impacted global mental health, creating a crisis necessitating urgent attention [1]. India, with its massive population, faces unique challenges, including disrupted routines, loss of loved ones, fear, uncertainty, and isolation, leading to deteriorating mental well-being [2]. This study delves into these post-pandemic mental health challenges and proposes multifaceted strategies for mitigation.

The pandemic-induced stressors, such as fear of infection, social isolation, financial strain, and disrupted routines, have intensified mental health conditions like anxiety, depression, PTSD, and substance abuse disorders [3]. Vulnerable communities, including LGBTQ+ individuals and college students, have been disproportionately affected [4]. Cultural beliefs and stigmas related to mental health in India significantly influence help-seeking behaviors and coping mechanisms during the pandemic [5].

Despite challenges, the pandemic spurred innovations in mental health interventions, leading to the rapid expansion of telehealth services and digital mental health platforms [5]. Understanding the effectiveness and limitations of these emerging approaches is crucial for inclusive and accessible mental health services post-pandemic.

S. Biswas
School of Architecture and Planning, DIT University, Dehradun, India
e-mail: sabyasachi.biswas@ditiuniversity.edu.in

B. Patil
Department of Industrial Design, Karnavati University, Gandhinagar, India

S. Bhattacharjee (✉)
Department of Communication Design, Karnavati University, Gandhinagar, India
e-mail: sandipan@karnavatiuniversity.edu.in

© The Author(s), under exclusive license to Springer Nature Singapore Pte Ltd. 2025

S. Das et al. (eds.), *Human-Centered Digitalization, Volume 3*, Design Science and Innovation, https://doi.org/10.1007/978-981-96-4383-7_24

This research advocates a comprehensive approach focusing on psychotherapy, complementary medicine, technology-driven interventions, and governmental support to enhance mental well-being [1]. The paper emphasizes the significance of awareness campaigns, mental health prioritization, and acknowledging cultural nuances in India for a holistic approach [2]. By critically evaluating the effectiveness of these interventions, this study aims to inform future mental health policies and practices, addressing the diverse mental health challenges exacerbated by the pandemic.

2 Literature Review

The onset of the COVID-19 pandemic inflicted a detrimental toll on mental health, marked by an array of stressors encompassing financial strain, health complications, loss of loved ones, and the pervasive fear of infection. Coping mechanisms such as emotional eating and excessive social media usage escalated, compounding the challenges faced by individuals. Within this landscape, vulnerable communities like the LGBTQ+ population experienced a disproportionate impact, grappling with anxiety, depression, and an alarming rise in suicidal tendencies, alongside issues of substance abuse [6]. Furthermore, a spectrum of mental health conditions ranging from Adverse Childhood Events (ACEs), Autism, PTSD, and ADHD to complex disorders like various forms of dementia demanded nuanced and tailored interventions for fostering mental well-being.

In response to these challenges, psychotherapy has emerged as a pivotal approach, providing targeted support for diverse groups affected by mental health disorders. A significant adjunct to psychotherapeutic methods is music therapy, classified under Complementary and Alternative Medicine (CAM). Rooted in the rich heritage of Indian Classical Music, this therapy transcends conventional boundaries, facilitating communication, nurturing relationships, enhancing learning, and bolstering cognitive functions such as memory, language, and decision-making [7]. The vast repertoire of Ragas and Taal forms a foundation for innovative services, including teleconsulting, online support groups, and mental health surveys, thus utilizing technology and tradition to bridge gaps in mental healthcare.

Additionally, Virtual Reality (VR) technology has emerged as a ground-breaking therapeutic tool, offering interactive interventions tailored for cognitive impairment treatment and symptom assessment. The immersive nature of VR enables comprehensive evaluations of cognitive performance, paving the way for nuanced and effective interventions [8].

The Government of India has recognized the urgent need to address mental health concerns and has implemented strategic initiatives like Opening Minds, Nalanda Way Foundation, and SVARAM Sound Experience. These initiatives, coupled with targeted campaigns, play a crucial role in raising awareness about mental health. By emphasizing that mental health needs are a priority, these campaigns not only reduce stigma but also foster a supportive environment where individuals are encouraged to seek help.

In summary, the multifaceted impact of the COVID-19 pandemic on mental health necessitates a comprehensive and diverse approach. Integrating psychotherapy, music therapy rooted in Indian Classical Music, cutting-edge VR interventions, and governmental initiatives fosters a resilient mental health ecosystem. By embracing innovation, tradition, and policy support, society can not only address the immediate challenges posed by the pandemic but also create a foundation for sustainable mental well-being.

2.1 *Mental Health Consequences of the COVID-19 Pandemic Financial Struggles and Stressors*

The COVID-19 pandemic has induced severe economic and social disruptions, pushing millions into extreme poverty and exacerbating hunger, potentially affecting 132 million more people by year-end [9, 10]. This crisis has triggered secondary mental health effects, raising suicide and alcohol-related death rates.

Financial woes, stemming from job loss, reduced income, or debt, intensify anxiety and chronic stress, amplifying pre-existing mental health conditions. Such challenges often lead to depression, hopelessness, and diminished self-esteem. The entertainment sector witnesses elevated mental health issues due to constant rejection and unpredictable career opportunities [11]. Financial stressors heighten the risk of mental health disorders, including depression, anxiety, and substance abuse. These pressures can also contribute to suicidal thoughts. Strained finances strain relationships, fostering conflicts. Vulnerable groups, such as those with chronic mental illnesses or low-income families, suffer disproportionately due to downsizing, layoffs, and limited access to healthcare [12]. The pandemic's impact underscores the urgent need for comprehensive support and mental health services.

Loss and Grief, Fear and Anxiety

The COVID-19 pandemic has induced widespread fear, amplifying psychological and social challenges. Studies reveal elevated depression, health anxiety, financial worries, and loneliness due to prolonged home confinement [13]. The fear of infection has led to irrational thinking, stigmatization, and exclusion, particularly among professionals in close-contact professions [14]. Healthcare workers, teachers, and others in customer-facing roles experience heightened fear, potentially escalating their depression, anxiety, and stress levels [15]. Individuals with chronic illnesses or infected family members face heightened pandemic-related fear, impacting their ability to respond rationally. Prolonged exposure to negative and inconsistent COVID-19 news, often consumed by people at home via television, internet, and social media, intensifies worries and fosters COVID-19-related phobia [16]. Physical and psychological health play a significant role, individuals already struggling with health issues are more susceptible to heightened anxiety, despair, and stress due to pandemic-induced fear. Witnessing loved ones contracting the virus intensifies these negative effects, exacerbating psychiatric symptoms and overall damage caused by the virus.

Psychological Impact on Specific Populations

Different populations are uniquely impacted by psychological challenges influenced by factors like age, gender, cultural background, and life circumstances. Young individuals often struggle with emotional regulation, stress, and identity development, facing issues like anxiety, depression, bullying, and academic pressure. College students, disrupted by pandemic-related closures and remote learning, have experienced heightened mental health concerns [5]. Older adults, contending with aging-related changes and losses, commonly grapple with depression, anxiety, loneliness, and cognitive disorders [17]. The LGBTQ community faces additional stressors due to societal discrimination, leading to elevated rates of depression, anxiety, substance abuse, and suicidal thoughts [18]. People with chronic illnesses encounter psychological distress due to physical limitations and social stigma, leading to conditions like depression, anxiety, and reduced quality of life [19]. Creating inclusive, supportive environments is vital for the well-being of these diverse populations.

2.2 Role of Psychotherapy in Mental Health Intervention

A variety of treatments that aim to help people identify and change troubling emotions, thoughts and behavior are known as psychotherapy, also sometimes called talk therapy. Psychotherapy mostly takes place when a licensed mental health professional and a patient meet individually or in group setting with other patients.

Importance of Targeted Psychotherapeutic Approaches

Targeted psychotherapeutic approaches play a vital role in mental healthcare, addressing specific needs and symptoms of individuals alongside or as an alternative to medication [20]. Even when medication alleviates symptoms, psychotherapy aids in tackling issues such as irrational fears, self-defeating thoughts, and difficulties in social interactions or coping with various life situations. Cognitive Behavioral Therapy (CBT), a key approach, helps individuals challenge harmful thoughts, understand their impact on emotions and behavior, and change self-defeating patterns [21]. Exposure therapy, a type of CBT, assists those with anxiety disorders by gradually desensitizing them to distressing stimuli, diminishing associated fears [22]. People seek psychotherapy for diverse reasons: Managing chronic stress from work, family, or bereavement, addressing unexplained symptoms like changes in sleep or appetite; managing persistent irritability, excessive worry, or hopelessness; or supporting family members diagnosed with mental health conditions. Psychotherapy provides crucial tools for enhancing mental well-being and managing complex life challenges.

Music Therapy as a Complementary and Alternative Medicine

Music's role in human expression spans millennia, with ancient instruments dating back 40,000 years, revealing its deep-rooted significance [23]. Music therapy's origins trace back to Ancient Greece and gained traction post-World War II. Modern applications are diverse, with therapists operating in private care, education, and social sectors. Music therapy, a component of Complementary and Alternative Medicine (CAM), utilizes music's power to enhance well-being and aids in the treatment, rehabilitation, and education of individuals with physical, mental, or emotional disorders. Conditions like Autism Spectrum Disorder (ASD) and Attention Deficit Hyperactivity Disorder (ADHD) involve challenges in social interaction and communication, which music therapy addresses through tailored interventions [7]. Indian Classical Music, encompassing North and South Indian traditions, boasts numerous Ragas and Taals, evoking emotions and providing a foundation for therapeutic compositions. This unique blend of Indian classical music and traditional healing practices distinguishes music therapy in India [24].

2.3 *Leveraging Technology for Mental Health Support*

Many services were no longer available to people after the epidemic. Because COVID-19 has a significant impact on mental health in several ways, leveraging technology as a means to make these services available to people was critical. This section uses several online and technological approaches to make mental health services more accessible and interactive.

Natural Language Processing (NLP) Methods for Mental Health Detection and Support

Textual Emotion Detection (TED) in Natural Language Processing (NLP) has emerged as a vital field focusing on recognizing emotions expressed in text, distinct from general sentiment analysis [25]. In healthcare, TED is pivotal for detecting emotions in areas like depression, suicidal ideation, and mental states of patients with various conditions such as asthma, Alzheimer's, cancer, and diabetes. Data sources encompass social media, healthcare services, and counseling centers [26]. Deep learning-based NLP techniques, particularly in the context of social media, have shown superior performance. Embracing this, recent advancements in AI and chatbot technologies have transformed healthcare interactions [9, 10]. A diagnostic chatbot has been developed utilizing machine learning algorithms and NLP to aid users in diagnosing their conditions based on symptoms [27]. This chatbot, informed with disease-related data, interprets user queries through NLP, offering tailored responses akin to virtual assistants like Siri or Alexa but specialized in disease diagnosis.

Moreover, computer vision plays a crucial role in understanding emotions. Facial expressions, gestures, voice, and speech patterns serve as indicators. Computer vision analyzes facial features and interprets them for healthcare goals [28]. It tracks body language and gestures, identifying signs of emotional distress like slouched posture or fidgeting. These affective interfaces enhance user experience in Human–Computer Interaction, Text-to-Speech synthesis systems, and Computer-Mediated Communication [28]. Such technological integration offers multifaceted insights into both mental and physical health, revolutionizing the healthcare landscape.

3 Methodology

This research paper delves into an innovative AI-driven multifaceted approach designed to address the complex landscape of post-COVID mental health. Leveraging new age technologies, this approach seeks to provide a comprehensive solution that not only aids in early detection and monitoring of mental health concerns but also offers personalized interventions to promote mental well-being. In this section, we elucidate the research methodology that underpins this study, detailing the strategies employed to achieve our research objectives. The method of the double diamond process is used in this design project to effectively pinpoint areas of concern and focus on critical elements for improvement.

Phase 1 (Discover and Define): The study aimed to gather information on the problem area related to mental health, COVID-19 pandemic, machine learning, artificial intelligence, NLP, and music therapy.

Secondary sources were used, and primary research was conducted through interviews (Table 1) and online surveys (Fig. 1). The data was analyzed to identify patterns, pain points, and opportunities for improvement. Two persona groups were identified (Fig. 1), capturing diverse user characteristics, behaviors, needs, and goals. Empathy mapping was used to understand user emotions (Fig. 2). A hypothetical scenario was developed to identify possibilities for improving the user's journey, which was then prioritized in the customer journey mapping (Fig. 3).

Phase 2 (Develop and Deliver): After prioritizing the points of improvement in the user experience, the goal is to explore different approaches to address the defined problem areas. Ideation was carried out using the study conducted in Phase 1 on current trends and technologies. Hub and Spoke method has been used to represent the idea (Fig. 4).

Approach 1: Awareness—Incorporating mental health education into school curricula to help children and young adults develop an understanding of mental health from an early age. Organizing workshops and seminars in schools, colleges, workplaces, and communities to provide information about mental health, its importance, and strategies for maintaining well-being. In this day and age, when people get their news and updates online, we can launch public awareness campaigns through TV, social media, and other platforms to disseminate accurate information about mental health and reduce stigma.

Table 1 Qualitative data collected by interviewing students of age group 15 to 20

Sr. No.	Gender	Living status	Remarks
1	Female	Neutral	"I live in a very strict parenting environment, that's why I don't feel comfortable taking about my mental health to my parents"
2	Male	Joint	"Personally, I don't think I have suffered from major anxiety issues, people talk about it but I can't relate"
3	Female	Neutral	"I need counseling to tackle negative thoughts, overthinking and build confidence in my work"
4	Female	Neutral	"I am currently taking counseling but I want to finish it quickly because the per session cost for each counseling session is 5000 rupees"
5	Female	Joint	"I prefer my own company when I feel low. I don't think I need counseling"
6	Male	Neutral	"I like to distract myself from my overthinking thoughts by occupying myself with work"
7	Male	Neutral	"I believe in being tough to handle my own problems"
8	Female	Neutral	"I talk to some of my close friends when I feel anxious, I think they are ones who understand me and I feel safe around them"
9	Male	Joint	"I am very open about my mental health to my family and I have tried counseling which helped a lot"
10	Trans	Neutral	"My family is very excepting about my sexuality but the people around me are not, which makes me not want to seek help whereas I actually want to"
11	Female	Joint	"I have friends that suffer a lot mentally, in those situations I want to be a help in hand but I don't know what to do or say to comfort them"
12	Female	Joint	"Moving to hostel made it very uncomfortable for me emotionally, I am very scared to trust people here and I also don't want to trouble my family members"
13	Male	Neutral	"I am not a very expressive person in general, that's way I use my own coping mechanism like listening to music or playing video games to tackle with anxiety"

Approach 2: Regular mental health check-ups can be facilitated through a kiosk database incorporating technologies like Natural Language Processing (NLP) for sentiment analysis, machine learning for personalized diagnosis, and computer vision for facial expression recognition. Deployed in schools, universities, businesses, and hospitals (Fig. 7), individuals can undergo check-ups at the kiosk booth, with results sent to their mobile app for easy access (Figs. 5, 6, 8, and 9). The app also offers features like note writing, affirmations, online mental health dialogues, counseling appointments, and a connecting chatbot.

Approach 3: Emergency—In the event of an emergency where the user is unable to comfort themselves, the phone's vibration feature will activate and send impulses

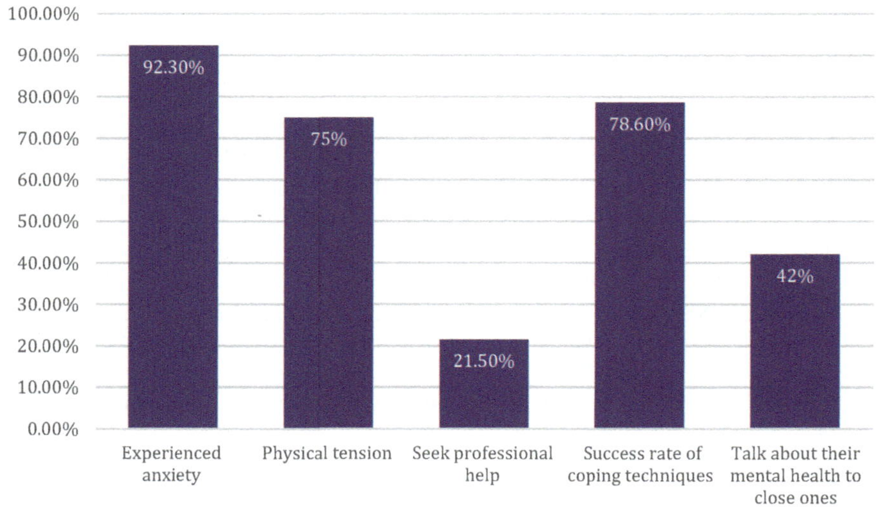

Fig. 1 Quantitative data collected through survey results

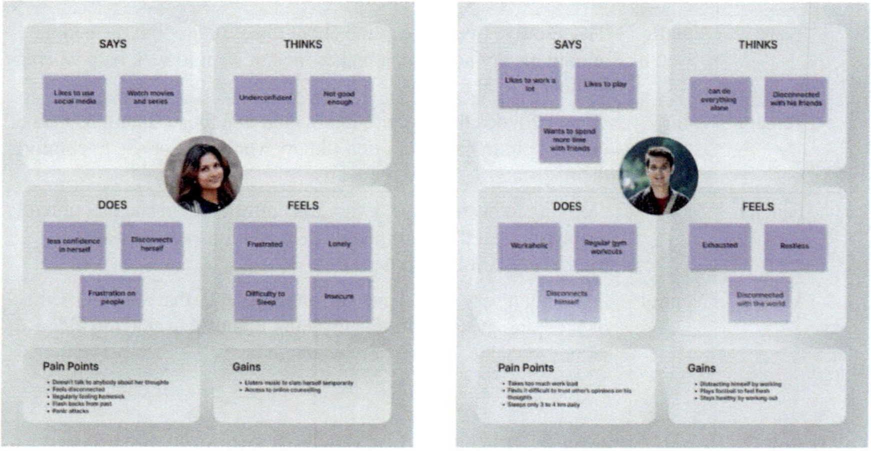

Fig. 2 Empathy mapping of the users groups to identify the pinpoints of the user

through the phone to calm the user down. These vibration patterns can be tailored to the user's preferences.

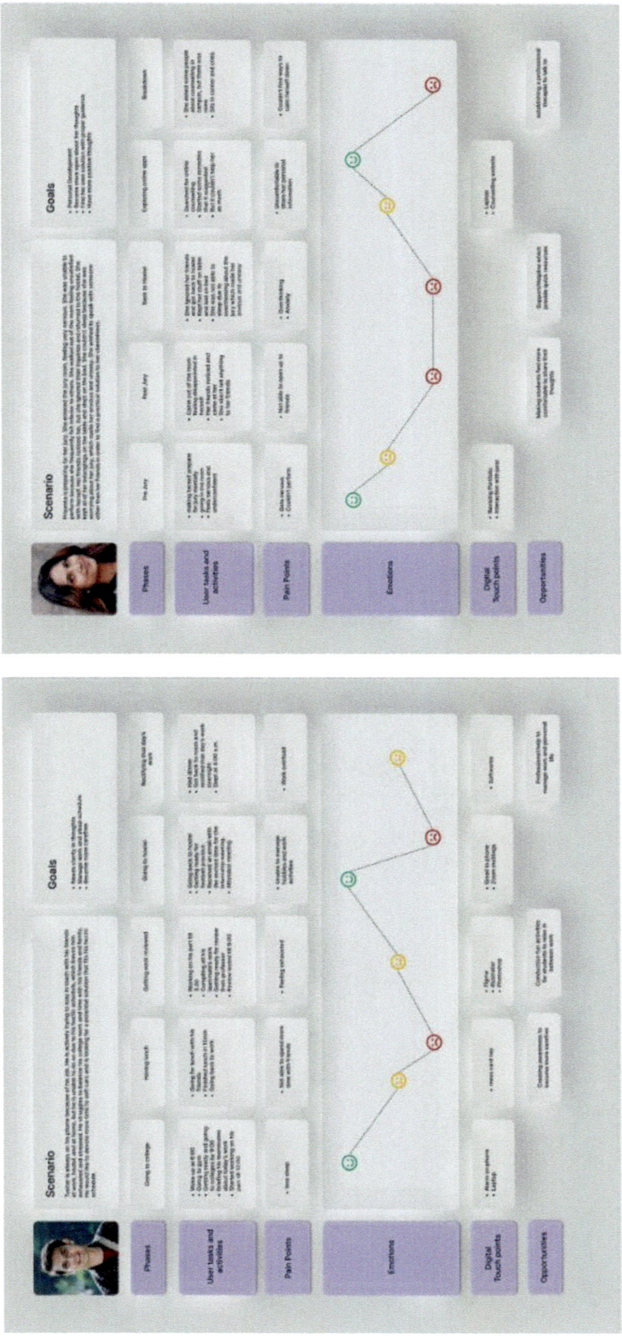

Fig. 3 Customer journey mapping: To identify the opportunities for enhancing user experience

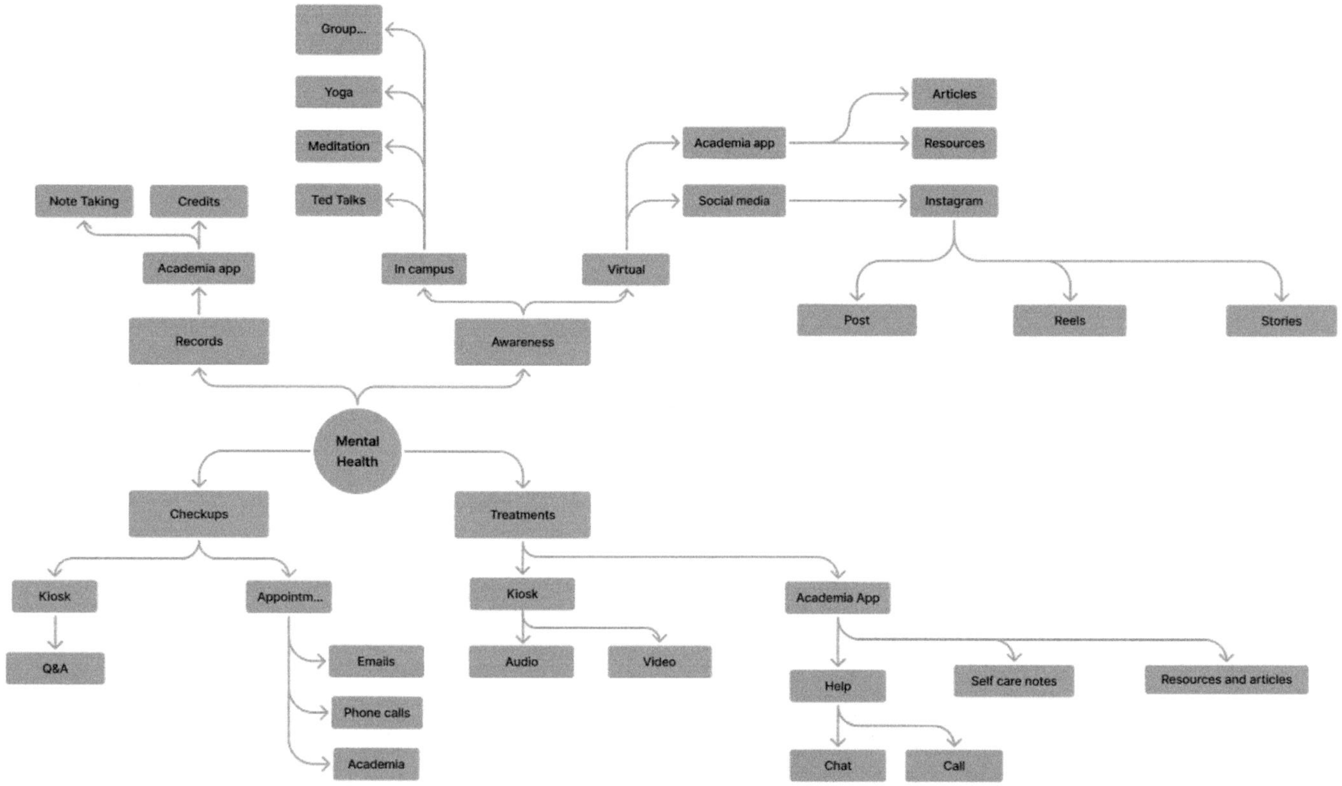

Fig. 4 Hub and Spoke method to plot the idea

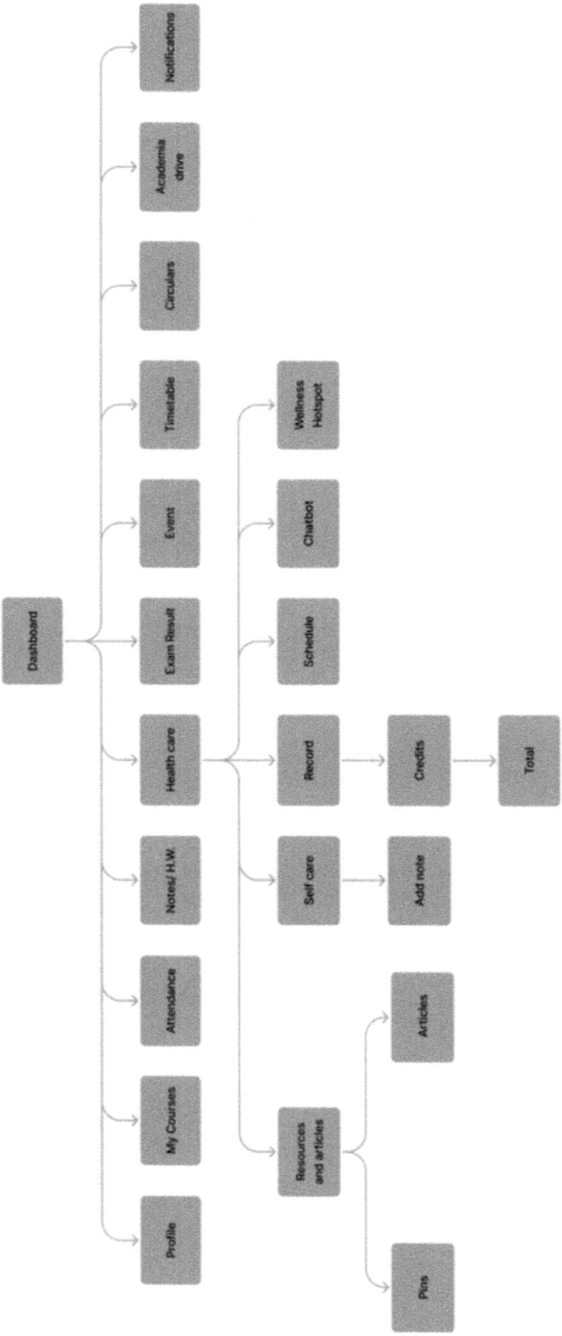

Fig. 5 Information architecture of the mobile app

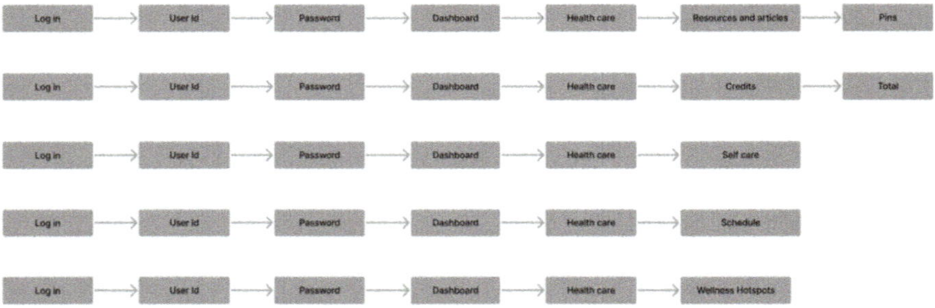

Fig. 6 Task flow of mobile app

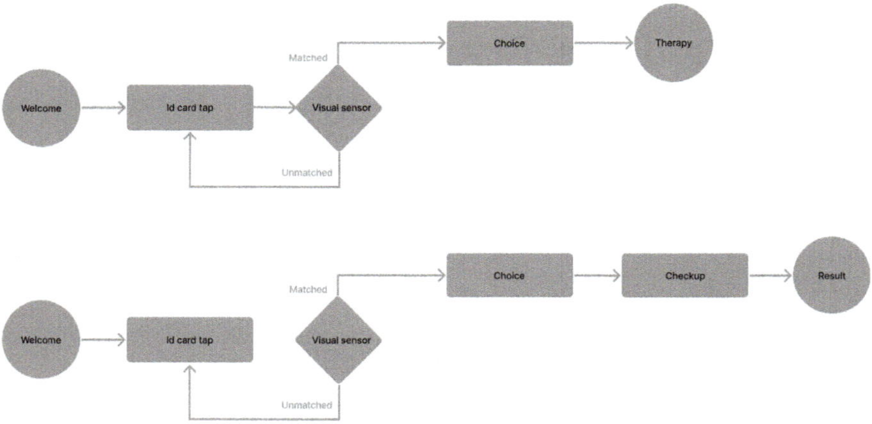

Fig. 7 Task flow of kiosk booth

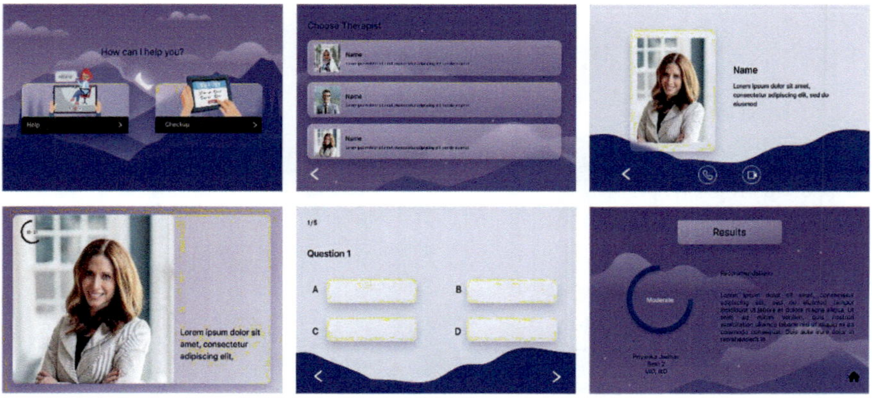

Fig. 8 High fidelity of kiosk booth

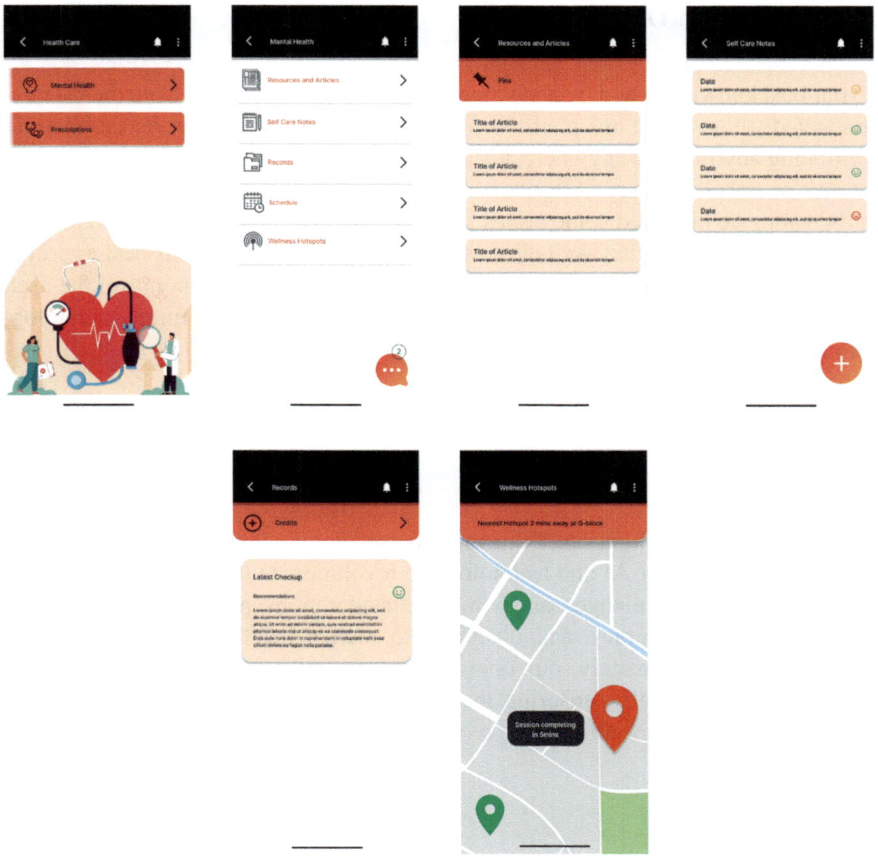

Fig. 9 High fidelity of mobile application

4 Usability Testing

In a user testing involving 20 diverse participants, our mental health app demonstrated a strong usability score of 87%. Users found the app's navigation intuitive, successfully completing tasks like accessing self-care notes and relaxation exercises. Although 15% experienced minor technical glitches, these did not significantly impact the overall positive user experience. This testing, conducted on mobile, provided valuable insights for app refinement, ensuring its effectiveness in enhancing users' mental well-being.

5 Results and Discussions

To our best knowledge and by analyzing the key insights collected from the research conducted we designed this platform for people who can seek help for their mental well-being anytime and anywhere, keeping the data collected from the users miscellaneous to ensure privacy. After conducting usability testing of our mental health mobile app, we observed some issues faced by the users that can be solved to make the user experience better. We believe there is a future scope for this design project with constant room for improvement. By understanding the touchpoints of the users further, we aim to keep updating and add new features as per future technology commences.

6 Conclusion

In conclusion, our exploration of this multifaceted approach to mental health underscores the potential for AI and technology to revolutionize the landscape of mental healthcare. While challenges and ethical considerations persist, the promise of improved access, accuracy, and outcomes in mental health treatment remains a beacon of hope. Ultimately, our commitment should be to leverage technology responsibly and ethically, ensuring that it becomes a valuable ally in the pursuit of mental well-being for all.

References

1. Kumar A, Nayar KR (2021) COVID 19 and its mental health consequences. J Mental Health
2. Grover S, Sahoo S (2020) Psychological impact of COVID-19 lockdown: an online survey from India
3. Nadkarni A (2021) Alcohol use among adolescents in India
4. Das SC, Shikha D (2023) A narrative review on priorities of mental health issues among transgenders: "so near, yet so far", 9–14
5. Apurvakumar Pandya K (2020) Innovative mental health initiatives in India: a scope for strengthening primary healthcare services, 502–507
6. Gaur PS, Saha S (2023) Mental healthcare for young and adolescent LGBTQ+ individuals in the Indian subcontinent. Front Psychol, 14
7. Sehgal S (2022) An overview of the practice and potential of music therapy in India: biographical approaches and art-based interventions in mental disorders. Int J Res Eng, Sci Manag, 2581-5792
8. Skurla MD (2021) Virtual reality and mental health in older adults. Int Psychogeriatr 34(2):143–155
9. Babita Pandey DK (2022) A comprehensive survey of deep learning in the field of medical imaging and medical natural language processing: challenges and research directions. J King Saud Univ—Comput Inf Sci, 5083–5099
10. Ghanshyam Pandey SM (2022) COVID-19 and its effects on the mental health of people living in urban slums in India. J Inf Knowl Manag

11. Gayatri Bhatia SP (2023) Movies and matters of the mind-portrayal of psychiatric illnesses in Bollywood over last two decades: a content analysis. Asian J Psychiatry
12. Ravikumar T, Sriram M (2022) Financial stress, financial literacy, and financial insecurity in India's informal. Investment Manag Financ Innov
13. Mortazavi SS, Assari S (2022) Fear, loss, social isolation, and incomplete grief due to COVID-19: a recipe for a psychiatric pandemic. Basic Clin Neurosci, 225–232
14. Kokou-Kpolou CK-A (2020) Prolonged grief related to COVID-19 deaths: do we have to fear a steep rise in traumatic and disenfranchised griefs? Psychol Trauma: Theory, Res, Pract Policy, S94–S95
15. Kiran Shafiq Khan MA (2022) The mental health impact of the COVID-19 pandemic. Int J Ment Health Addict, 380–386
16. Naslund JA, Bondre A (2022) Social media and mental health: benefits, risks, and opportunities for research and practice. J Technol Behav Sci, 245–257
17. da Cruz WM, D'Oliveira A (2022) Mental health of older people in social isolation: the role of physical activity at home during the COVID-19 pandemic. Sport Sci Health, 597–602
18. Chakrapani V, Scheim AI (2022) Affirming and negotiating gender in family and social spaces: stigma, mental health and resilience among trans masculine people in India. An Int J Res, Intervention Care, 951–967
19 Jonathan Ee BS (2021) A systematic review of the knowledge, attitudes and perceptions of health and social care professionals towards people with learning disabilities and mental health problems. Br J Learn Disabil
20. Ajit Avasthi SG (2022) Ethical and legal issues in psychotherapy. Indian J Psychiatry S47–S61
21. Cuijpers P, Miguel C (2023) Cognitive behavior therapy vs. control conditions, other psychotherapies, pharmacotherapies and combined treatment for depression: a comprehensive meta-analysis including 409 trials with 52,702 patients. Official J World Psychiatr Assoc, 105–115
22. Margarita Surmai ED (2022) Cognitive behavioural therapy: a strategy to address pandemic-induced anxiety. J Nurse Pract, 36–39
23. Junrui Huang XL (2022) Effects and applications of music therapy on psychological health: a review. Atlantis Press SARL, 2352-5398
24. Nag S, Basu M (2022) On the application of deep learning and multifractal techniques to classify emotions and instruments using Indian Classical Music. Phys A: Stat Mech Its Appl, 0378-4371
25. Zhang T, Schoene AM (2022) Natural language processing applied to mental illness detection: a narrative review. npj Digit Med
26. Nijhawan T, Attigeri G (2022) Stress detection using natural language processing and machine learning over social interactions. J Big Data
27. Jana Sedlakova MT (2022) Conversational artificial intelligence in psychotherapy: a new therapeutic tool or agent? Am J Bioeth, 4–13
28. Ketan Sarvakar RS (2023) Facial emotion recognition using convolutional neural networks. Mater Today: Proc, 3560–3564

Nature-Based Learning for Aesthetic Development in Young Children: A Case Study

Prakriti Joshi, Mudit Aswal, Jyoti Kumar, and Jatin Bhatt

1 Introduction

This study highlights the critical role that early childhood education plays, stressing the value of play-based learning and outdoor activities. Children's development can be hindered by urbanization and limited access to nature, which can have an adverse effect on their physical and mental health. The research investigates the advantages of nature-based learning experiences in promoting aesthetic development in order to close this gap. Teachers can foster creativity and a closer bond with nature by utilizing natural materials and stimulating situation [1]. This strategy encourages sustainability and environmental responsibility [2]. Children's aesthetic development also heavily depends on the use of colour and aesthetic principles like proportion, harmony, and balance [3]. The goal of the project is to pinpoint practical methods for introducing outdoor learning experiences into early childhood education, laying the groundwork for a sustainable future.

P. Joshi (✉) · M. Aswal · J. Bhatt
Department of Design and Innovation, Jamia Millia Islamia, New Delhi, India
e-mail: joshi.prachi881@gmail.com

M. Aswal
e-mail: muditaswal@gmail.com

J. Bhatt
e-mail: thinkjatin@gmail.com

P. Joshi · J. Kumar
Department of Design, Indian Institute of Technology, New Delhi, Delhi, India
e-mail: meetjyoti@gmail.com

© The Author(s), under exclusive license to Springer Nature Singapore Pte Ltd. 2025
S. Das et al. (eds.), *Human-Centered Digitalization, Volume 3*, Design Science and Innovation, https://doi.org/10.1007/978-981-96-4383-7_25

2 Methodology

Parent's perspectives on nature-based learning and the integration of aesthetics into early childhood education were gathered through quantitative research. Given the emphasis on curriculum development, the research primarily targeted parents of children between the ages of three and six. Next a set of activities were designed for the children of this age group. This was followed with conduct of the activity with children in school. The observations from the play-based learning have been reported in this paper.

2.1 *Quantitative Research Findings*

Based on the survey conducted through Google Forms, which garnered responses from 128 parents, the participants primarily belonged to the 28 to 45-year age group. These parents had children in the same age range or had recently transitioned beyond it. The demographic representation was predominantly from Delhi. The observations from the survey have been tabulated in Table 1.

Table 1 Parental perspectives on education

Interest in aesthetic development	93.7% of parents out of 128 parents, comprising a substantial majority, recognize the significance of aesthetic and cultural development in fostering their children's overall growth and well-being
Nature-based learning	61.3% of parents out of 128 parents, frequently take their children to natural environments and emphasize the importance of nurturing an appreciation for nature and the environment in early childhood
Mixed satisfaction with education system	Satisfaction with the current education system for children aged 3–6 is divided among parents, suggesting potential room for improvement to better meet the needs of young children with 29.1% satisfied, 34.6% dissatisfied, and 36.2% being neutral towards it
Engagement in art and creative activities	61.8% of parents out of 128 parents, actively engage in art and creative activities with their children, indicating a strong interest in fostering creativity
Interest in specialized schools	A substantial percentage of 86.6% of parents out of 128 parents, express interest in enrolling their children in schools that emphasize aesthetic and cultural development alongside nature-based learning experiences
Advocating for integration	The research highlights a significant interest of 89% out of 128 parents, and potential for incorporating nature and art into early childhood education, warranting further exploration and advocacy for such opportunities

2.2 *Findings from Observational Research*

Primary data was gathered for this qualitative observational study in a school context using naturalistic observations, note-taking, and recorded observations. Children between the ages of three and six were the study subjects. They were observed in a school environment with an emphasis on playtime, social interactions, and classroom activities using recorded observations, note-taking, and naturalistic observations to better understand students' behaviours and attitudes towards learning. Key conclusions from the analysis include students' short attention spans and susceptibility to distractions, a noticeable curiosity and enthusiasm for exploration, limited exposure to natural elements and outdoor environments, a lack of interest in nature, a preference for indoor and technological activities, social engagement and imaginative play, a need for physical activity breaks, difficulties following complex instructions and a strong tendency towards creativity. The study's key finding was that students frequently learnt more from their peers than from their teachers, suggesting the possibility for peer-based learning methodologies. These observations provide useful knowledge that will help improve instructional strategies and build an inclusive and engaging learning environment.

2.3 *Finding a Research Gap in the Domain of Aesthetics Development*

Young children are becoming increasingly disconnected from nature due to the effects of urbanization and modern lifestyle choices. This disconnection from nature can lead to negative impacts on children's physical and mental health, as well as hinder their ability to appreciate and understand the natural world. Aesthetic development is crucial for children's ability to appreciate and engage with beauty and creativity in the world around them, which has long-term implications for their overall well-being and ability to navigate complex social and cultural environments. As such, further research is needed to explore the potential of nature-based experiences to support children's aesthetic development, as well as the specific mechanisms through which these experiences can promote creativity, imagination, and a sense of wonder in children.

3 Design Ideation

A research gap was identified and a brainstorming phase was initiated to generate solutions, including an outdoor nature curriculum, school garden, field trips, nature, art and music activities. The brainstorming phase evaluated feasibility, impact, and resources for implementation, resulting in a well-rounded plan to enhance children's aesthetic appreciation for nature. This problem can be addressed by incorporating natural tools into nature-based activities in schools which can enhance sensory experiences, encourage creativity, and deepen children's connection with nature. The ideation stage involved creating activities to enhance children's aesthetic development through nature interactions. These activities aimed to foster empathy, kindness, and positive traits. Examples included nature walks, gardening, art sessions, storytelling, and eco-friendly projects. The goal was to promote sustainable practices and a sense of responsibility towards the environment. A total of 51 activities were developed for children's aesthetic development through nature-based experiences. A voting system was used to select the top six activities, which were deemed feasible, effective, and engaging for the target age group.

3.1 Testing of Harvest Pattern Play and Shape Matching Activity

Research on children's aesthetic development through nature-based activities identified six potential activities out of 51 activities. These activities were reviewed and evaluated by a panel of experts in the field of child development, using the Likert scale. The highest-scoring activity was "Harvest Pattern Play and Shape Matching," which uses nature tools to explore textures and colours, fostering creativity and imagination. Experts found this engaging and beneficial for children's aesthetic development. The Harvest Pattern Play and Shape Matching activity was tested at CSC Bal Vidhyalaya School, allowing children to engage in sensory and tactile experiences while learning about nature. The activity promotes social skills, aesthetic development, creativity, and environmental respect. The testing phase assessed the success and identified areas for improvement. The activity testing involved 25 children aged 3–6 from different socioeconomic backgrounds, displaying varying moods. The materials used were vegetables, colours, and paper. The testing was conducted at CSC Bal Vidhyalaya school in Azadpur, Delhi.

Table 2 includes testing of Harvest Pattern Play and Shape Matching Activity in 4 stages:

3.2 Terms of Measuring

The behaviour of the children during the exercise was evaluated based on their engagement, creativity, and feedback to determine the activity's effectiveness [4]. The youngster's attention and interest levels were observed while they were engaged in the activity to gauge their level of participation. Along with originality, the use of the various shapes, colours, and textures of the vegetables to create distinctive designs, as well as their willingness to take risks, were all factors that were considered

Table 2 Analysing and interpreting insights from the four stages of activity testing

Stages	Instructions	Artworks	Insights
Initial stage	The children were educated about vegetables and their health benefits, and were encouraged to recognize the vegetables they were given. They were then given various vegetables and asked to experiment with different prints. They were allowed to choose as many vegetables and colours as they wanted, and they enthusiastically requested more sheets to try out different designs and textures		• The children were really enthusiastic and engaged in the exercise • Before beginning the activity, the children took some time to select the vegetable • The children used too much paint when they were printing, so they were instructed to use the least amount of colour to create a texture that could be seen • The children were smelling the vegetables and feeling the texture of vegetable • Most children found the texture to be soft while others found its coarse • Some children conducted experiments using just one type of vegetable • The majority of children explored with several vegetables • While some children experimented with multiple colours, the majority of the children only used one colour

(continued)

Table 2 (continued)

Stages	Instructions	Artworks	Insights
Intermediate stage	The youngsters were then instructed to match the basic forms that were hung throughout the room with the artwork they had produced. Square, circle, triangle, pentagon, and hexagon were the basic shapes that were shown throughout the classroom		• According to their artwork, 13 out of 25 children were able to identify the right shape • While some children were able to accurately resemble the shapes • Most children only had a basic understanding of forms like circles, triangles, and squares • Children could recognize patterns that resembled pentagons and hexagons while not knowing what they were • Children enjoyed shape matching and weren't frightened to make mistakes • Before making a choice, kids tried matching the shapes several times

(continued)

Table 2 (continued)

Stages	Instructions	Artworks	Insights
Advanced stage	The children were then given a layout made up of a pattern. The children were then asked once more to print the pattern using vegetables. The children were allowed to experiment with various vegetables and colours. The selection of vegetables and combination of colour provided was not constrained		• In this stage, children were more talkative and the majority were experimenting with various vegetables • Children at this stage were also utilizing multiple colours • The majority of children used one vegetable for one colour and didn't dip it in a different colour • The children were instructed to print it inside the box. A number of them, though, were stamping the vegetables onto the pattern line • Children were lively and frequently requested additional paper • Children filled in the artwork quickly because they were antsy, but some children filled it carefully while deliberating where to place which colours

(continued)

Table 2 (continued)

Stages	Instructions	Artworks	Insights
Final stage	The last stages of the investigation once more involve children printing on plain paper. Once more, they are free to use as many vegetables as they like when printing, and the use of numerous colours or colour combinations is not constrained	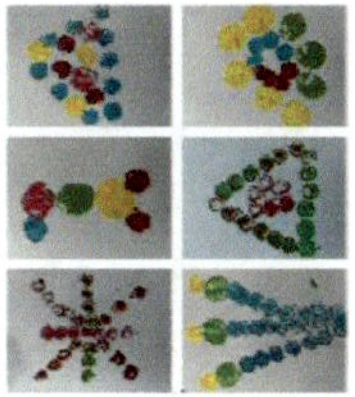	• The children at this stage did not print as restlessly as they did in the previous stage and were careful about what they wanted to create • A select few of the children produced artwork that was particularly meaningful to them • While the majority of the children attempted to create compositions • Some of the children were making symmetrical artwork • A handful of the children continued to make flowers • The last level was considerably cleaner because children had mastered vegetables and were much more open to experimenting with colours • The children at this age were far more willing to take chances by utilizing a variety of vegetables and colours

when rating their level of creativity in producing the vegetable prints. After the activity, children rated their experience using a 5-point Likert scale, enhanced with emojis to represent emotional states. Emojis, which are visual symbols conveying emotions, were used to make the scale more accessible to young children who may not fully understand numerical scales. The "angry" emoji represented a low rating—1, while the "thrilled" or "happy" emoji indicated a high rating—5. This method facilitated effective communication of the children's emotional responses, as emojis are universally understood and can bridge language or literacy gaps. The activity's success was evaluated through observations of engagement and social interaction, alongside children's feedback via an emoji-based Likert scale. The analysis revealed areas for improvement, with emojis aiding non-verbal communication.

4 Conclusion

The results suggest that nature-based learning experiences can positively impact children's aesthetic development by promoting creativity, imagination, and appreciation for the natural world. Children can explore different textures, colours, and forms found in the natural world, using their senses to appreciate the beauty of nature. The use of organic materials in art-making activities fosters imagination and creativity, as children develop an appreciation for the distinctive patterns, shapes, and colours found in nature. A sense of surprise and curiosity is fostered in youngsters through nature-based activities, inspiring them to investigate their environment and think creatively. The findings have significant implications for early childhood educators and policymakers to boost children's aesthetic and cultural development and general well-being by integrating nature-based learning experiences into early childhood education. However, the small sample size and short research duration limit the study's applicability to a wider population and the long-term impact of outdoor activities on young children's aesthetic development. Future research should address these limitations by increasing the sample size to a larger, more diverse population and conducting longitudinal studies to examine the long-term effects of nature-based activities on aesthetic development. Additionally, further research should explore the effects of cultural background and socioeconomic status on children's participation in outdoor activities and their aesthetic growth.

References

1. Häikiö TK (2020) Aesthetic learning processes in early childhood and early years education
2. Elliott S, Young T (2016) Nature by default in early childhood education for sustainability. Aust J Environ Educ 32(1):57–64. https://doi.org/10.1017/aee.2015.44

3. Read M (2019) Designing with color in the early childhood education classroom: a theoretical perspective. Creat Educ 10(06):1070–1079. https://doi.org/10.4236/ce.2019.106080
4. Dweck CS (2008) Brainology transforming students' motivation to learn. https://api.semanticscholar.org/CorpusID:146701907

Evaluation of the User Interface of Slicing Software for DLP 3D Printing Process

Rahul D. Ukey, Sangeeta Pandit, Avinash Sahu, Rajat Kamble, Sai Prakash, and Pankaj Kumar

1 Introduction

3D printing also known as Additive Manufacturing (AM) refers to a broad collection of technologies that operate on the principal of creating three dimensional objects from the bottom-up, bit-by-bit or layer-by-layer process, rather than removing the material until the final desired output is achieved [1]. It is utilised in a wide range of industrial, educational as well as medical sectors, such as automobile, engineering, and construction, dentistry, etc. [2]. Stereolithography (SLA), Digital Light Processing (DLP), Fused Deposit Material (FDM), Selective Laser Sintering (SLS), and other technologies are examples of 3d manufacturing techniques [3]. Digital Light Processing (DLP) is classified as vat photopolymerization, in which liquid photopolymer is selectively cured by light-activated polymerization in a vat. DLP employs a digital light projector screen to simultaneously flash a single picture of each layer (or many flashes for bigger portions). Therefore, the print can be achieved very fast in this method [4] as shown in Fig. 1.

The process consists of creating a CAD model using 3d modelling software like CATIA, SolidWorks, etc. After creating the CAD model, it is converted into the STL file (Stereolithography) which is sent to the slicing software, where the model is sliced, and g-codes are generated for the printer. Then, this desired output undergoes post-processing techniques, where it is cleaned, cured, and supports are removed from the model, as shown in Fig. 2.

The slicing software is critical in the 3D printing process. It slices the 3D model into thin layers and determines how each layer should be printed to achieve the best results. The software then creates G-codes with respect to the process parameters defined by the users, resulting in high-quality prints. The slicing software's precision is critical in deciding the print quality of 3D objects [6]. The user interface of the

R. D. Ukey · S. Pandit (✉) · A. Sahu · R. Kamble · S. Prakash · P. Kumar
PDPM Indian Institute of Information Technology Design & Manufacturing, Jabalpur, India
e-mail: s.pandit@iiitdmj.ac.in

© The Author(s), under exclusive license to Springer Nature Singapore Pte Ltd. 2025

S. Das et al. (eds.), *Human-Centered Digitalization, Volume 3*, Design Science and Innovation, https://doi.org/10.1007/978-981-96-4383-7_26

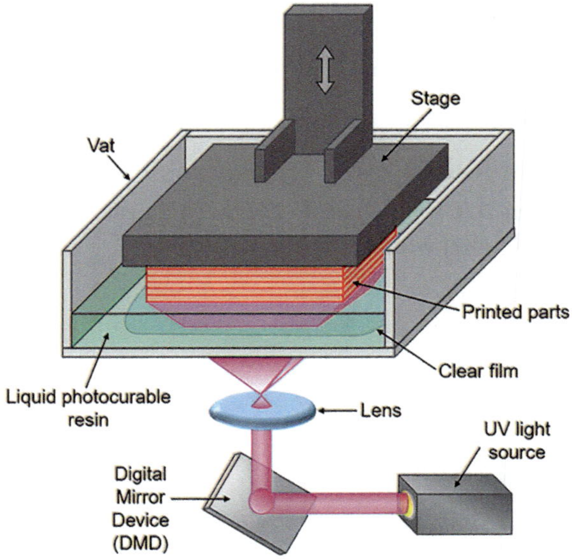

Fig. 1 Schematic diagram of DLP 3D Printing [5]

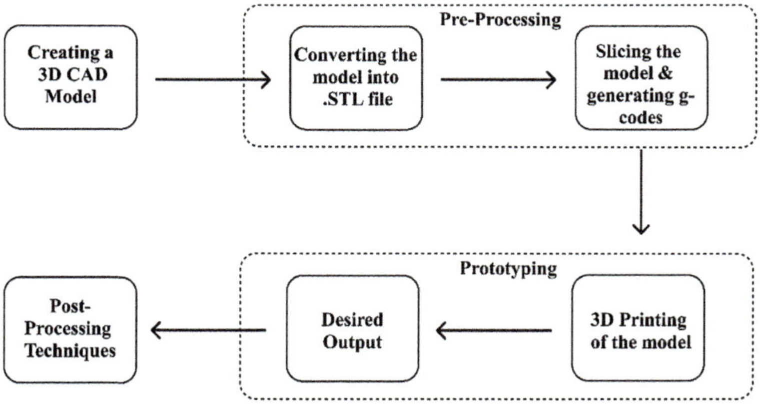

Fig. 2 3D printing process chain

software should be optimised in such a way that users with different backgrounds can navigate through the software and can use it with ease [7].

The slicer process parameter selection (layer thickness, print speed, exposure time, etc.) can significantly impact the 3D printing process parameters. It can produce different results if not selected properly, leading to poor print quality, more print time, and high material usage [8]. Poor user interface of the software can lead to difficulty in understanding, confusion among the users and dissatisfaction [9]. Therefore, the aim of the paper is to evaluate the different types of slicing software used for DLP

3D Printing to enhance the user experience which can lead to the high quality of the outcome or printed part.

2 Methodology

2.1 *Participants*

For this research paper, a small group of 10 evaluators with expertise in usability and user interface design were selected to perform the heuristic evaluation (HE). Also, 60 users with knowledge of DLP 3D Printing were selected randomly (30 beginners and 30 experts) to evaluate the user interface of the slicing software.

2.2 *Heuristic Evaluation (HE)*

Usability experts did heuristic evaluation (HE) to inspect the user interface against a set of heuristics or design principles [10]. The set of heuristic principles used to evaluate the user interface are shown in Table 1. For this research, the most popular slicing software over the past few years like Chitubox, Prusa Slicer, and Lychee [11] were selected and evaluated (Fig. 3).

Table 1 Heuristic principles

Code	Heuristic principles
P1	Visibility of the system
P2	Match between system and real world
P3	User control and freedom
P4	Consistency and standard
P5	Error prevention
P6	Recognition rather than recall
P7	Flexibility and efficiency of use
P8	Aesthetic and minimalist design
P9	Help and documentation
P10	Help users recognise, diagnose, and recover from errors

[A] Chitubox Slicer [B] Prusa Slicer [C] Lychee Slicer

Fig. 3 Interface of different slicing software

2.3 Quantitative Study

After heuristic evaluation (HE), a quantitative study was conducted with a sample size of 60 users. 30 beginner-level users and 30 expert DLP 3d Printing slicing software users. To analyse the existing problems identified in the heuristic evaluation and to validate the findings, a modified set of user interface evaluation questionnaires was developed based on four common evaluation heuristics: Nielsen's Attributes of Usability, Lewis' works on IBM Computer Usability Satisfaction Questionnaires, Schneiderman's eight golden rules of Interface Design and Chin's Questionnaire for User Interface Satisfaction [12–15] and were asked to rate their satisfaction with various aspects of the slicing software user interfaces.

The modified questionnaire based on the four principles mentioned above for evaluating user satisfaction of the slicing software was categorised into:

(A1) Learnability/Memorability: (1) How easy was learning how to use the slicer? (2) How easy was it to find the settings and features you needed? (3) How well did the documentation and tutorials help you learn how to use the slicer? (4) How easy was it to remember how to use the slicer after not using it for a while? (5) How consistent was the slicer's interface and terminology?

(A2) Efficiency/Speed: (1) How quickly could you complete the tasks using the slicer? (2) How responsive was the slicer to your input? (3) How fast was the slicer to slice models? (4) How efficient was the slicer's use of system resources? (5) How well did the slicer handle complex models?

(A3) Error Prevention/Forgiveness: (1) How often did you make errors while using the slicer? (2) How well did the slicer prevent errors? (3) How helpful were the slicer's error messages? (4) How easy was it to recover from errors?

(A4) Flexibility/Controllability: (1) How flexible was the slicer in meeting your needs? (2) How much control did you feel you had over the slicing process? (3) How customisable were the slicer's settings? (4) How easy was it to save and load custom slicing profiles? (5) How well did the slicer support different types of 3D printers and materials?

(A5) Helpfulness/Support: (1) How helpful was the slicer's documentation? (2) How helpful were the slicer's error messages? (3) How easy was it to find help online or from the slicer's community? (4) How well did the slicer's documentation and support keep up with new features and changes?

(A6) Overall Satisfaction: (1) How satisfied were you overall with the slicer? (2) Would you recommend the slicer to others? (3) Did the slicer have all the features you needed? (4) Was the slicer well-documented? (5) How likely are you to continue using the slicer in the future?

2.4 Statistical Analysis

To analyse the data, basic statistics like mean, min, max, and standard deviation were used. Similarly, percentage was used for the user's level of satisfaction.

3 Result

3.1 Participants

It is observed in this research that the evaluators had a mean average experience of more than 7.5 years, from a minimum experience of 2 years to a maximum of 13 years in evaluating the usability and user interface design of different software (Table 2).

The participants for the quantitative study selected were divided into two groups (Beginners and Experts) based on their knowledge of the DLP 3D Printing. The expert users had a mean average experience of more than 6.8 years as compared to the beginner users which is 1.06 years (Refer Table 3).

Table 2 Demographics of the expert evaluators ($n = 10$)

Characteristics	Mean (standard deviation)	Min	Max
Age	42 (± 12.02)	24	57
Experience	7.5 (± 3.53)	2	13

Table 3 Demographics of the users ($n = 60$)

Characteristics	Beginner level of users			Expert level of users		
	Mean (standard deviation)	Min	Max	Mean (standard deviation)	Min	Max
Age	35 (± 6.95)	22	45	43 (± 7.69)	24	55
Experience	1.06 (± 1.08)	0	3	6.8 (± 2.72)	3	11

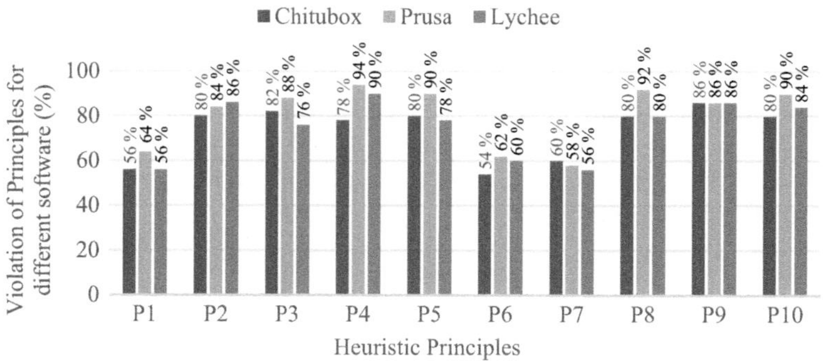

Fig. 4 Heuristic evaluation of different slicing software

3.2 *Heuristic Evaluation (HE)*

From the above study, it was revealed that there were several usability issues and violations in the user interfaces of the slicing software. The experts identified problems related to visual aesthetics, navigation, and the overall user experience. It was found that the visibility of the system (P1) (56, 64, 56%), recognition rather than recall (P6) (54, 62, 60%) and flexibility and efficiency of use (P7) (60, 58, 56%) were highly violated while evaluating the user interface of the different slicing software (Fig. 4).

3.3 *Quantitative Study*

The quantitative study provided important insights into the satisfaction levels of users with the slicing software user interfaces. It was found that the level of satisfaction for different slicing software among the users with beginner level of knowledge is less (40.42, 40.45, 39.64%) than the level of satisfaction among the users with expert level of knowledge (82.07, 90.29, 79.73%) as shown in Fig c. The beginner-level users found the interface to be of unclear labelling of icons and buttons and difficulties in locating certain features or functionalities (Fig. 5).

4 Discussion

Since its emergence, 3D printing technology has received considerable attention. In particular, the fast popularisation of DLP equipment has resulted in a surge of 3D printing technology [16]. Dozens of software companies provide open-source Additive Manufacturing build preparation software that includes slicing, resulting in a

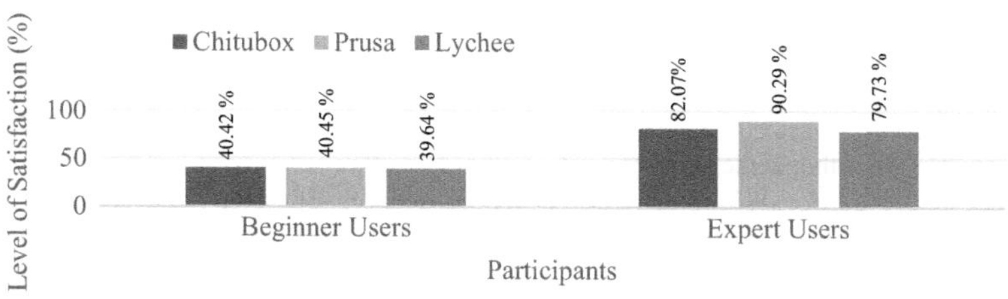

Fig. 5 Questionnaire result from different users

rapidly changing, volatile and dynamic market [3]. A study investigated the variation of the selected 3D printing process properties that originate from the choice of a slicer. The study found that different slicers can produce different results regarding print quality, print time, and material usage [8].

A study shows that the usability of user interfaces plays a crucial role in determining the quality and effectiveness of information systems [17]. Problems with the user interface can lead to various negative outcomes, including increased user errors, decreased user satisfaction, and reduced efficiency levels. Usability evaluation is an important issue in software development as it helps improve the product's overall quality [18]. Improving usability and functionality in user interfaces is essential for enhancing user satisfaction and promoting efficiency and minimising potential errors, particularly in critical settings such as intensive care units [19]. Moreover, usability evaluations provide a basis for improving the quality of a user interface design, ultimately enabling designers to make more effective decisions and perform their tasks more efficiently.

This study aimed to evaluate the user interfaces of slicing software for Digital Light Processing 3D Printing. The results clearly indicate a need for improvements to facilitate greater ease of use and more intuitive interaction, ultimately enhancing the overall user experience. To ensure these improvements, a future work should focus on a more detailed evaluation process, where the interface is evaluated based on more complex tasks given to the users of different level of knowledge. This further study can help provide a precise evaluation of the user interface for different slicing software.

5 Conclusion

Based on the results of the heuristic evaluation and quantitative study, it is evident that the user interfaces of slicing software for Digital Light Processing 3D Printing are in need of improvement to increase functionality and usability. The violations of flexibility and efficiency of use, recognition rather than recall, and visibility of the system highlight the areas that require attention in order to enhance the user experience and improve the quality of the product.

References

1. Jakus A (2019) Chapter 1: An introduction to 3D printing—past. Present, and future promise. 3D printing in orthopaedic surgery, 1–15. Author F, Author S Title of a proceedings paper. In: Editor F, Editor S (eds) Conference 2016, LNCS, vol 9999. Springer, Heidelberg, pp 1–13
2. Ahmed R, Hossan MM (2020) Construction of a low-cost 3D printer using Arduino (Doctoral dissertation, Sonargoan University). Author F Contribution title. In: 9th International proceedings on proceedings. Publisher, Location, pp 1–2 (2010)
3. Gibson I, Rosen DW, Stucker B, Khorasani M, Rosen D, Stucker B, Khorasani M (2021) Additive manufacturing technologies, vol 17. Springer, Cham, Switzerland, pp 160–186
4. Redwood B, Schöffer F, Garret B (2017) The 3D printing handbook: technologies, design and applications. 3D Hubs B.V.
5. Bae S-U, Kim B-J (2021) Effects of cellulose nanocrystal and inorganic nanofillers on the morphological and mechanical properties of digital light processing (DLP) 3D-printed photopolymer composites. Appl Sci 11(15):6835. MDPI AG. Retrieved from https://doi.org/10.3390/app11156835
6. Bryła J, Martowicz A (2021) Study on the importance of a slicer selection for the 3D printing process parameters via the investigation of G-code readings. Machines 9(8):163. MDPI AG. Retrieved from https://doi.org/10.3390/machines9080163
7. Bødker S (1991) Through the interface a human activity approach to user interface design. L. Erlbaum
8. Faruqi F, Friedman K, Cheng L, Wessely M, Subramanian S, Mueller S (2021) SliceHub: augmenting shared 3D model repositories with slicing results for 3D printing. arXiv preprint arXiv:2109.14722
9. Paluri SK (2020) Human computer interaction. In: Proceedings of 10th international conference on research methods and professional issues, United Kingdom, pp 1–7
10. Nielsen J, Molich R (1990) Heuristic evaluation of user interfaces. In: Proceedings of the SIGCHI conference on human factors in computing systems, pp 249–256
11. Manufactur3D (2023) 9 popular resin 3D printer slicer software. Manufactur3D. https://manufactur3dmag.com/9-popular-resin-3d-printer-slicer-software/
12. Yuniarto D, Marliana RR (2021) Evaluation of E-learning based on Nau (Nielsen's attributes of usability) according to students. J-Tin's—Jurnal Teknik Informatika 5(1). http://ejournal.stmiksumedang.ac.id/index.php/jtin/article/view/289
13. Lewis J, James R (1995) IBM computer usability satisfaction questionnaires: psychometric evaluation and instructions for use. Int J Hum-Comput Interact 7:57. https://doi.org/10.1080/10447319509526110
14. Faiza D, Farell G, Delianti VI, Thamrin, Rahmadika S (2022) Eight golden rules interface analysis for video conference information system. In: 2022 International conference on electrical engineering and informatics (ICELTICs), Banda Aceh, Indonesia, 2022, pp 131–135. https://doi.org/10.1109/ICELTICs56128.2022.9932115
15. Chin J, Diehl V, Norman K (1988) Development of an instrument measuring user satisfaction of the human-computer interface. In: ACM CHIi'
16. Haghsefat K, Tingting L (2020) 3D printing and traditional manufacturing technology analysis and comparison
17. Farzandipour M, Nabovati E, Zaeimi GH, Khajouei R (2018) Usability evaluation of three admission and medical records subsystems integrated into nationwide hospital information systems: heuristic evaluation. Acta Inf Med 26(2):133
18. De Castro A, Macías JA (2016) SUSApp: a mobile app for measuring and comparing questionnaire-based usability assessments. In: Proceedings of the XVII international conference on human computer interaction, pp 1–8
19. Barra DCC, Sasso GTMD, Almeida SRWD (2015) Usability of computerized nursing process from the ICNP® in intensive care units. Rev Esc Enferm USP 49:0326–0334

Work-Related Musculoskeletal Disorder Among the Phone Repair Occupation Group in India

Rajat Kamble, Sangeeta Pandit, Abhiram Vijayaraghavan, Sriya Kommu, Prisha Shriyan, Jyotsna Tripathi, Uday Tiwari, Shivansh Shukla, Anshul, Nived Prem, Swastik Bhowmik, and Nevin Jose

1 Introduction

In the current trends of digital India, there are millions of mobile phone users. To cater to these consumers, a mobile phone service and repair economy has emerged in almost every corner, spread across the town. A major portion of this belongs to the unorganised, self-trained Technicians working in small spaces, with makeshift workstations. These technicians work for prolonged hours being exposed to different work hazards such as biological hazards, chemicals, and carcinogenic fumes [1] and physiological factors such as repetitive work, awkward postures, and precision work. The work involves primary tasks such as opening the mobile, replacing screens which require use of tools requiring pinch grip and heating tools, operating in bent posture. Other intricate tasks involve repairing if the internal parts which requires much precision and involve the same tools along with soldering and use of different chemicals that emit fumes when burned.

Researchers have stated that working in poorly designed workstation environments results in extra effort, negatively affecting efficiency, productivity, and health [2, 3]. Technicians working in these setups have been observed to work continuously in static awkward posture for long durations. It has been found in many research that overall discomfort and pain in neck, back, and shoulders are common symptoms among the workers who sit for long hours for most of their workday [4–6]. Higher discomfort reports in fingers, hands, and eyes among workers involved in precision work.

R. Kamble · S. Pandit (✉) · A. Vijayaraghavan · S. Kommu · P. Shriyan · J. Tripathi · U. Tiwari · S. Shukla · Anshul · N. Prem · S. Bhowmik · N. Jose
Design Discipline, PDPM Indian Institute of Information Technology, Design and Manufacturing, Jabalpur, Madhya Pradesh, India
e-mail: s.pandit@iiitdmj.ac.in

© The Author(s), under exclusive license to Springer Nature Singapore Pte Ltd. 2025
S. Das et al. (eds.), *Human-Centered Digitalization, Volume 3*, Design Science and Innovation, https://doi.org/10.1007/978-981-96-4383-7_27

Though many research have been done on other precision-based prolonged sitting tasks, no such studies have been found that asses their work for ergonomic issues. Thus this study aims to find the WRMSDs among the repair technicians and to perform a postural analysis for the tasks involved and aid in development of better workstation.

2 Materials and Methods

Data collection and observation was done from 20 different repair service hubs in Jabalpur, Madhya Pradesh, 35 technicians were selected for the study and were presented with a Nordic questionnaire. All the technicians working in these repair service hubs were male. Interview, Direct observation, Photography, and Videography were done to understand the different tasks involved. Mobile dismantling/ assembling and Shouldering/repairing were the two major tasks that were focused in this study. Postural analysis was done on selected high-risk tasks with the help of Rapid Upper limb assessment sheet (RULA).

The data were analysed using simple descriptive statistics, Prevalence of MSDs and injuries concerning factors such as working hours and experience was found using the Chi-square test. p values 0.05 were considered statistically significant. Further, logistic regression was used, where the odds ratio (OR) with 95% confidence interval (CI) was calculated.

3 Results

3.1 Participants

The technicians had experience ranging from 1 month to 25 years. And worked for an average of 8.14 ± 2.4 Hours. Around 37% of the technicians had work experience of 0–4 years, 28% had experience of 5–10 years and 34% had experience of 11–25 years (Table 1).

3.2 Postural Analysis

Figure 1 shows the two tasks which are majorly performed during their work: Soldering components and dismantling phone and their components. Both of which are highly precision task and requires constant bent posture for long durations. As shown in Table 2 both the task reported high risk in the RULA analysis, which suggested investigation and implementation of changes in the task.

Table 1 Demographic characteristics of the repair technicians ($N = 35$)

Parameters	Categories	N (%)
Age	18–25	13 (37.1%)
	26–35	13 (37.1%)
	> 36	09 (25.7%)
Experience (years)	0–4	13 (37.1%)
	5–10	10 (28.6%)
	> 11	12 (34.3%)
Working hours (hours)	< 5	09 (25.7%)
	6–9	12 (34.3%)
	> 10	14 (40%)

(a) (b)

Fig. 1 Two major tasks involved, **a** soldering ICs and other components, **b** dismantling the phone components

Table 2 RULA assessment

Task	RULA score	Action category
Soldering	7 (high risk)	Investigate and apply changes
Dismantling	7 (high risk)	Investigate and apply changes

3.3 Self-reported Musculoskeletal Symptoms

Prevalence of discomfort in different body region concerning experience and working hours was found using $\chi 2$ test, almost all the technicians reported discomfort in at least one body region, the reports were considered in percentages and a p value below 0.05 was considered significant. Neck region had the highest report of discomfort followed by eye strain, lower back, and shoulders. We also found that discomfort in shoulder, lower back, and strain in eyes were significantly high among the technicians with experience more than 11 and those who worked for 10 + hours ($p < 0.05$) while injuries were found to occur significantly more among the technicians with less experience during the last 6 months (Table 3).

3.4 Association of Work Factors with Discomfort Symptoms

Binary logistic regression was performed for the analysis of the factors associated with discomfort in each body part. The results showed that the technicians with less experience were less likely to report discomfort in the shoulders, lower back and eye strain with OR 0.09 (95% CI 0.013, 0.607), OR 0.778 (95% CI 0.119, 5.1), and OR 0.2 (95% CI 0.002, 0.307), similar results was also noticed among the technicians wo worked for less hours, with OR 0.48 (95% CI 0.005, 0.417), OR 0.01 (95%CI 0.01, 0.176), and OR 0.048 (95%CI 0.005, 0.417) for the body parts shoulders, lower back and eye strain, respectively. Pain in neck was common across all experience (Table 4).

4 Discussion

The research performed considering MSDs among the technicians working at different mobile repair units of Jabalpur, Madhya Pradesh, has provided a brief knowledge about their working conditions and reports of MSDs in different body regions. The different tasks involved in their work require performing repeated precision tasks and working in an awkward posture for long durations, RULA analysis for the task has reported high risk and suggested immediate intervention to eliminate the risks. Research on similar working conditions involving sitting for long hours has been reported to develop MSDs in different body regions [7]. This study reports similar findings of high neck, shoulder, and lower back discomfort. This study revealed that the technicians who have worked for more years in the field were more likely to develop MSDs in the shoulder and lower back regions along with strain in the eyes.

This study also found that the technicians who worked for longer hours were also more likely to develop MSDs in the neck, shoulder and lower back regions, along

Table 3 Prevalence of discomfort with respect to the demographic factors

Parameters	Categories	Neck (%)	Shoulder (%)		Lower back (%)		Wrist (%)	Legs (%)	Injuries (%)		Eye strain	
Experience	0–4	53	30	< 0.05	23	< 0.05	30	53	77	< 0.05	23%	< 0.01
	5–10	60	50		70		10	50	60		70%	
	> 11	83	83		75		25	50	25		91%	
Work hours	> 5	55	22	< 0.01	11	< 0.01	44	44	66		22	< 0.01
	6–9	66	41		41		8	50	66		58	
	> 10	71	85		92		21	57	35		85	

Table 4 Statistical correlates of musculoskeletal disorders in different body regions

Parameters	Categories	Neck	Shoulder	Lower back	Legs	Injuries	Eye strain
Exp	0–4	0.23 (0.036, 1.5)	0.09 (0.013, 0.607)*	0.1 (0.16,0.627)	1.167 (0.242,5.6)	10 (1.5,62.7)	0.2 (0.002, 0.307)**
	5–10	0.3 (0.42, 2.1)	0.2 (0.228, 1.42)*	0.778 (0.119, 5.1)*	1 (0.187, 5.3)	4.5 (0.73,27.7)	0.21 (0.018, 2.4)
	> 11	1	1	1	1	1	1
Work hours	> 5	0.5 (0.087, 2.8)	0.48 (0.005, 0.417)**	0.01 (0.01, 0.176)**	0.6 (0.11, 3.2)	3.6 (0.616, 21.0)	0.048 (0.005, 0.417)*
	6–9	0.8 (0.15, 4.2)	0.12 (0.018, 0.785)*	0.055 (0.005, 0.568)*	0.75 (0.159, 3.5)	3.6 (0.71, 18.25)	0.233 (0.035, 1.5)
	> 10	1	1	1	1	1	1

* $p < 0.05$, ** $p < 0.01$, ** $p < 0.001$

with strain in the eyes. Many other research done in similar work environments has found similar findings, which resulted from repeated precision task, and working in awkward posture for long hours [6–10]. In this study, it was also found that the technician who had less experience had significantly higher reports on injuries such as cuts, burns incidents in the last 6 months. Similar findings were observed by research done in other fields where it was seen that workers with more experience worked in ergonomically safer ways than apprentices with less experience [11]. During our research at the field it was also observed that the units of technicians with more experience had better ergonomic equipment along with better workstations, whereas those with lower experience didn't have equipment such as magnifying tools. Reduced in eyesight was common among most of the experienced technicians, which led them to invest in magnifying units.

The limitations of this study are that an assessment of all the tasks involved is necessary to understand the problems effectively and to develop a workstation. A larger set of data will also be needed for better understanding.

5 Conclusion

The technicians involved in phone repair tasks are high risk of MSDs considering working for long hours in bend stooping posture on tasks that requires precision and constant strain in eyes. Working for many years in the field has resulted in MSDs in neck, shoulder, and lower back resulting poor livelihood. Strain in eyes have also let in reducing of eye sights which affects performing the daily life tasks. An ergonomic intervention is needed for a better workstation setup, to eliminate such awkward postures.

References

1. Montano D (2014) Chemical and biological work-related risks across occupations in Europe: a review. J Occup Med Toxicol 9(1):1–13
2. Chandrasekar K (2011) Workplace environment and its impact on organisational performance in public sector organisations. Int J Enterp Comput Bus Syst 1(1):1–19
3. Das S, Gangopadhyay S (2022) Effect of work environment noise on productivity, mental fatigue and blood pressure among Indian postal assistants. Age (years) 33(3.09):33–37
4. Arora SN, Khatri S (2022) Prevalence of work-related musculoskeletal disorder in sitting professionals. Int J Community Med Public Health 9(2):892–895
5. Anghel M, Argeanu V, Talpo C, Lungeanu D (2007) Músculoskeletal disorders (MSDS) consequences of prolonged static postures. J Exper Med Surg Res 4:167–172
6. Ghosh T, Das B, Gangopadhyay S (2010) Work-related musculoskeletal disorder: an occupational disorder of the goldsmiths in India. Indian J Community Med: Official Publ Assoc Prev Soc Med 35(2):321
7. Ariëns GA, Bongers PM, Douwes M, Miedema MC, Hoogendoorn WE, van der Wal G et al (2001) Are neck flexion, neck rotation, and sitting at work risk factors for neck pain? Results of a prospective cohort study. Occup Environ Med 58:200–207
8. Das D, Singh AK (2022) Risk factors associated with work-related musculoskeletal disorders among floor-sitting precision handicraft workers. Int Arch Occup Environ Health 95(5):1129–1145
9. Nimje AM, Kothare HG, Chaudhari SV, Sarath C, Vijay P (2022) Work-related musculoskeletal pain and the associated factors in mobile phone technicians: a cross-sectional study. J Datta Meghe Inst Med Sci Univ 17(2):219–222
10. Kamble R, Pandit S, Sahu A (2023) Occupational ergonomic assessment of MSDs among the artisans working in the Bagh hand block printing industry in Madhya Pradesh, India. Int J Occup Saf Ergon 29(3):963–969
11. Alwasel A, Abdel-Rahman EM, Haas CT, Lee S (2017) Experience, productivity, and musculoskeletal injury among masonry

Exploring the Latest Trends in Motion Sickness Research: A Network Analysis Approach

Yein Song, Cai Wang, Jungyeon Park, Jaesik Yang, Seonghyeon Kim, Sangwoo Bahn, and Myung Hwan Yun

1 Introduction

In recent years, significant advancements in technology coupled with increasing environmental concerns have catalyzed a transformation in the automotive industry, shifting from conventional Internal Combustion Vehicles (ICVs) to Electric Vehicles (EVs) and Autonomous Vehicles (AVs). This shift is expected to revolutionize transportation by enhancing safety, efficiency, and convenience. Despite ongoing advancements in the automotive sector, several hurdles must be overcome before achieving widespread commercialization and adoption of ICVs, EVs, and AVs. A pivotal issue to be addressed is motion sickness in transportation to ensure passenger comfort [1].

While motion sickness research has traditionally focused on ICVs, recent studies have extended to include EVs. These studies have found that EVs, unlike ICVs, do not experience noticeable gear shifting. However, the regenerative braking system in EVs creates significant deceleration forces that can cause motion sickness [2]. Thus, EVs and ICVs exhibit different acceleration and deceleration patterns that can induce motion sickness and must be analyzed separately.

Looking ahead, future AVs may utilize either ICV or EV technologies, depending on user needs and preferences [3]. It is critical to acknowledge the distinct characteristics of EVs and ICVs, while also considering the Non-Driving Related Tasks (NDRTs) that occupants are likely to engage in while traveling in AVs. Engagement

Y. Song · C. Wang · J. Park · M. H. Yun (✉)
Seoul National University, Gwanak-gu, Seoul, South Korea
e-mail: mhy@snu.ac.kr

J. Yang · S. Kim
Hyundai Motor Company, Bundang-gu, Seongnam-si, Gyeonggi-do, South Korea

S. Bahn
Kyung Hee University, Giheung-gu, Yongin-si, Gyeonggi-do, South Korea

© The Author(s), under exclusive license to Springer Nature Singapore Pte Ltd. 2025
S. Das et al. (eds.), *Human-Centered Digitalization, Volume 3*, Design Science and Innovation, https://doi.org/10.1007/978-981-96-4383-7_28

in NDRTs is anticipated to increase the likelihood of motion sickness [4]. Therefore, our research aimed to examine ICVs and EVs within the broader category of Human-Driven Vehicles (HDVs), as well as AVs as a separate category.

Consequently, this paper has gathered experimental research articles on motion sickness from the past five years to identify current trends and insights using network analysis. Moreover, it has compared HDVs, inclusive of both EVs and ICVs, with AVs. Additionally, a systematic review of these specific studies was carried out, although this paper will primarily discuss the findings from the network analysis.

2 Method

To discern recent trends in motion sickness research over the past five years, we employed the following methodology: (1) keyword selection through a preliminary search; (2) defining the search scope and conditions; (3) establishing inclusion and exclusion criteria; (4) evaluating relevance based on titles and abstracts; (5) conducting professional-based screening; (6) selecting the final collection of papers; and (7) identifying key trends using network analysis.

2.1 Data Acquisition and Selection

Five researchers began the process by reviewing two systematic literature reviews, one doctoral dissertation, and 18 experimental studies to construct a list of relevant search terms. These keywords were then combined to formulate search queries aimed at retrieving pertinent papers.

Motion sickness: "motion sickness" OR "car sickness" OR "induced motion sickness" OR "sickness symptoms".
Measure for subjective severity: "questionnaire" OR "MSQ" OR "SSQ" OR "MSSQ" OR "FMS" OR "rating" OR "survey" OR "scale" OR "assessment."
Measure for objective severity: "physiological" OR "physiology" OR "physical" OR "posture" OR "head" OR "postural".
Vehicle-oriented: "electric vehicle" OR "electric car" OR "autonomous vehicle" OR "automobile" OR "electric".
Considering driving environment: "real driving" OR "real car driving" OR "on-road" OR "driving."
Prediction of motion sickness: "model" OR "index" OR "machine learning" OR "deep learning" OR "regression" OR "multilayer perceptron" OR "algorithm" OR "estimate."

The search was conducted using electronic databases, targeting journal articles and conference papers, with the conditions set as follows: keywords must appear in the title, abstract, or keyword list; the fields of study must be engineering, humanities, or social science; the publications must be from the past five years (2018 to early 2023); and the papers must be in English.

Inclusion and exclusion criteria were defined by three experts who prioritized papers based on the relevance of their titles and abstracts to motion sickness research, such as the development and application of questionnaires, actual vehicle experiments, development of prediction and evaluation models, and comparisons of motion sickness models with vital signs. Studies that conducted human experiments were specifically selected. Papers that were inaccessible, unidentifiable, or focused solely on the development of hardware and software for mechanical and electrical technologies were excluded. Redundancies found across the databases were also removed.

An expert assessment was then conducted where five professionals rated each paper's relevance to the study's aims on a scale from 0 (no association) to 2 (high association). Papers averaging a score below 0.7 were excluded.

Following the relevance assessment, a full-text review of the selected papers was undertaken. During this phase, the studies were categorized based on who was studied, how the study was conducted, and what findings were sought.

2.2 *Network Analysis*

For the papers that met the selection criteria, a network analysis was performed on their keywords. The keywords were first collected and preprocessed, ensuring they were free of typographical errors. The process included tokenization, cleaning, normalization, and lemmatization. The data was then used to create a co-occurrence matrix based on the frequency of word pairs within the documents.

In the visualized networks of this paper, each word was represented as a node, with the relationships between words (co-occurrences) as edges. The thickness of these edges corresponded to the frequency of co-occurrence. Degree centrality and eigenvector centrality analyses helped determine each node's importance, and modularity analysis was employed to discern the network's structure and its clusters, or communities. A larger node size indicated higher eigenvector centrality and distinct colors coded each modularity class.

This analysis facilitated the identification of the primary topics within the entire dataset and subsets of human-driven vehicles (HDVs) and autonomous vehicles (AVs). Major topics were designated by words with an eigenvector centrality of 0.2 or greater. Words with very low centrality were disregarded. These topics were then compared by the type of driving entity.

Data preprocessing and co-occurrence matrix calculations were executed using Python. Network visualization and analysis were performed using Gephi 0.10.1.

3 Results

A total of 49 papers related to motion sickness were selected for this study. Among these, 17 studies focused on human driving vehicles (HV), including internal combustion engines and electric vehicles. The remaining 32 studies focused on Autonomous Vehicles (AVs).

After preprocessing, the total number of keywords extracted from all 49 papers was 443. Of these, 156 were found in HV studies, while 287 were associated with AV research. The network for the entire dataset comprised 127 nodes and 451 edges. The keyword network for HV consisted of 69 nodes and 198 edges, with 33 nodes corresponding to an eigenvector centrality of 0.2 or higher, representing 25.98% of the total. In other words, the keyword network for AV consisted of 87 nodes and 289 edges, with 36 nodes (52.17% of the total) corresponding to an eigenvector centrality of 0.2 or higher.

In the semantic network of keywords from HV paper, the top 10 eigenvector central nodes are sickness, motion, driving, vehicle, biosignal, head/body, simulator, cue, driver/passenger/user, model/optimization. The top 10 eigenvector central keywords in AV networks are sickness, motion, autonomous, model/optimization, vehicle, driving, comfort, driver, HCI/HMI, and control.

For all data, 13 communities were derived by the modularity class analysis on the entire dataset, with seven communities confirmed based on only 31 nodes displaying an eigenvector of 0.2 or higher. The first community consisted of nodes related to human perception, comfort, and interaction with the system. The second community contained nodes related to apparatuses that implement driving situations, real-time responses, and vibration. The third community was associated with mathematical and predictive models or data simulation. In addition, there are some keywords related to human factors or acceleration related to motion sickness and multisensory cues for prediction. Further, topics on head/body dynamics, autonomous driving, and biosignals were identified.

For HVs, There are 8 communities detected by Modularity class analysis in HV semantic network. The first community, labeled as "head/body tilt," consisted of nodes related to head/body, control, and tilt. The second community, named as "cue," comprised nodes for sensory stimulation and cues. The third one, interpreted as "simulator/real car," included nodes related to the apparatuses that implemented driving situations in the motion sickness study. Other communities were associated with the environment and occupants such as drivers and passengers. The community labeled as "model/optimization" included models that quantitatively predict motion sickness, such as subjective vertical theory, or mathematical approaches like optimization. Finally, the community named "biosignal" consisted of nodes related to gaze tracking, EEG, and biosignals.

For AVs, nine communities were found in the network for AV research. The community composed of acceleration, prediction, and cue nodes was named as a theme related to "cue." The "biosignal" community included eye (gaze) and biosignal nodes. Communities related to mathematical data simulation or predictive models were labeled as "model/optimization," though these differed from HV. Communities related to human–computer or machine interaction were discovered, along with topics on AV as a public service. The community associated with the perceived comfort of occupants was labeled "comfort." In the "Dynamics" theme, only nodes corresponding to an eigenvector centrality of 0.2 or higher were considered dynamics, while nodes below 0.2 were categorized as head/body and tilt. Additionally, topics related to seat arrangement and path planning were identified (Table 1).

4 Discussion

The topics commonly found in both HV and AV networks were biosignal, models or optimization, and cue. The "biosignal" topic refers to studies using biosignals as objective measures of motion sickness severity [5–8]. Both HV and AV biosignal communities included an "eye" node, suggesting studies that measure eye reflexes and gaze. The HV network also featured EEG, as evidenced by studies comparing the severity of motion sickness in drivers and passengers through electroencephalography [6]. Though not represented in the network, several papers included other physiological measures such as facial skin temperature changes, and electrodermal activity [8].

The "model/optimization" cluster in the HV network includes nodes for mathematical, evaluation, repeatability, susceptibility, SVC, theory, and objective, while in the AV network, it comprises control, predictive, simulation, intelligent, and risk. This topic is associated with mathematical approaches used in motion sickness research [9–11]. HV-related studies predicted motion sickness severity using established models like the Subjective Vertical Model (SVM) [12]. AV-related studies predicted motion sickness severity by simulating autonomous driving scenarios or by designing driving traffic systems optimized for motion sickness [8, 13, 14].

The "cue/anticipation" community relates to the theory that providing motion information, or cues, can mitigate the sensory mismatch that causes motion sickness. Studies aim to prevent or alleviate motion sickness by providing cues such as galvanic cutaneous stimulation, multisensory cues, visual and vibration cues [15–20].

The "head/body dynamics" topic aligns with studies based on the assumption that the occupants' body (particularly the head) shaking in response to driving conditions can contribute to motion sickness [21, 22]. To reduce lateral acceleration that contributes to motion sickness in AVs, a lateral control strategy was designed based on head tilt (Fig. 1).

The nodes "driver," "passenger," or "user" associated with "occupant" appeared in both HV and AV but were interpreted differently. In HVs, drivers anticipate changes potentially mitigating sensor conflicts [23]. AV studies related to comfort and attention suggest that the occupant's physical cognitive status may shift with the task changes, altering the trend of motion sickness.

Table 1 Networks and topics for the keywords from (a) All, (b) HV, and (c) AV

Class	Node	Deg.	Eig. cent	Class	ID	Deg.	Eig. cent
(a)				*(b)*			
0. Head and body tilt	Head/body	13	0.611	0. Cue	Sickness	30	1.000
	Control	4	0.236		Motion	30	0.987
	Tilt	6	0.217		Cue	7	0.252
1. Cue	Cue	12	0.471		Anticipation	7	0.302
	Galvanic	4	0.290		Acceleration	5	0.338
	Cutaneous	4	0.255	1. Comfort	Comfort	21	0.695
	Sensory	7	0.238		Occupants	16	0.654
	Aging	4	0.220		Real	4	0.219
2. Simulator or real car	Sickness	28	1.000		Time	7	0.308
	Motion	20	0.849		Attention/ perception	10	0.418
	Driving	14	0.664	2. Public service	Vehicle	29	0.854
	Vehicle	14	0.663		Autonomous	32	0.950
	Simulator	12	0.528		Service	4	0.248
	Real	6	0.343		Public	6	0.226
	Comfort	7	0.320	3. HCI	System	8	0.262
3. Environment	Environment	6	0.238		Multi	10	0.417
	Presence	4	0.228		Simulator	7	0.280
4. Occupants	Occupants	12	0.435		Driving	20	0.758
	Performance	8	0.287		HCI/HMI	13	0.629
	Balance	4	0.242		Integration	5	0.206
	Independent	4	0.224		Vibration	7	0.286
	Platform	4	0.207		Sensory	6	0.276
5. Model and optimization	Model/ optimization	12	0.412	4. Model and optimization	Control	14	0.491
	Mathematical	6	0.331		Predictive	8	0.337
	Evaluation	8	0.291		Model/ optimization	34	0.862
	Repeatability	4	0.259		Simulation	7	0.357
	Susceptibility	4	0.249		Intelligent	4	0.232
	SVC	4	0.233		Risk	6	0.211
	Theory	4	0.210	5. Seat arrangement	Seat	4	0.241
	Objective	4	0.209		Arrangement	4	0.232
6. Biosignal	Biosignal	17	0.641	6. Biosignal	Eye	8	0.289
	Eye	11	0.366		Biosignal	8	0.381
	EEG	3	0.227	7. Dynamics	Dynamics	11	0.451

(continued)

Table 1 (continued)

Class	Node	Deg.	Eig. cent	Class	ID	Deg.	Eig. cent
				8. Path planning	Path	6	0.220
					HF	9	0.413
					Planning	6	0.218

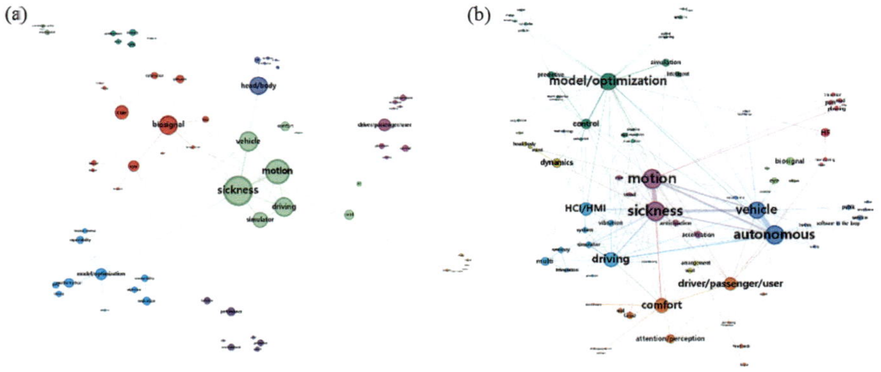

Fig. 1 Keyword network for **a** HV and **b** AV

The "simulator/real driving" community was unique to the HV network. Studies not specifically targeting AVs can still inform AV design if the subject participated as a passenger.

Seat arrangement and path planning are topics unique to AV, addressing motion sickness from seating configurations and optimizing the trajectory and motion of a vehicle to minimize sensory conflicts.

Public service in the AV network indicates automated driving technology's potential to revolutionize public service, with different considerations for motion sickness compared to private vehicles.

5 Conclusion

This research identified major topics and crucial factors of motion sickness influenced by engine types and driving entities in both human-driven and automated vehicles. Through a literature review and network analysis, we determined the trends and topical structures in existing research and noted differences based on the engine type and driving entity. Further studies may expand this work by including non-English research and covering longer periods. Findings contribute to understanding motion sickness in-vehicle environments and are expected to inform future research and practical improvements.

Acknowledgements The authors thank the anonymous reviewers and HIS laboratory members for their feedback and support. This research was supported by the National Research Foundation of Korea (NRF), which is funded by the Korea government (MSIT & MOE) (RS-2023-00222172, RS-2024-00409458, RS-2025-00558315). Additionally, this work was supported by Hyundai Motor Company.

References

1. Iskander J, Attia M, Saleh K, Nahavandi D, Abobakr A, Mohamed S, Asadi H, Khosravi A, Lim CP, Hossny M (2019) From car sickness to autonomous car sickness: a review. Transp Res Part F: Traffic Psychol Behav 62:716–726. https://doi.org/10.1016/j.trf.2019.02.020
2. Salter S, Diels C, Herriotts P, Kanarachos S, Thake D (2019) Motion sickness in automated vehicles with forward and rearward facing seating orientations. Appl Ergon 78:54–61. https://doi.org/10.1016/j.apergo.2019.02.001
3. Singh V, Singh V, Vaibhav S (2020) A review and simple meta-analysis of factors influencing adoption of electric vehicles. Transp Res Part D: Transp Environ 86:102436
4. Smyth J, Jennings P, Mouzakitis A, Birrell S (2018) Too sick to drive: how motion sickness severity impacts human performance. In: 2018 21st international conference on intelligent transportation systems (ITSC). IEEE, pp 1787–1793
5. Gálvez-García G, Albayay J, Fonseca F, Bascour-Sandoval C (2020) Joint and individual effectiveness of galvanic cutaneous stimulation and tactile stimulation at decreasing simulator adaptation syndrome. PLoS ONE 15(10):e0240627
6. Huang KC, John AR, Jung TP, Tsai WF, Yu YH, Lin CT (2021) Comparing the differences in brain activities and neural comodulations associated with motion sickness between drivers and passengers. IEEE Trans Neural Syst Rehabil Eng 29:1259–1267
7. Li C, Zhang Z, Liu Y, Zhang T, Zhang X, Wang H, Wang X (2022) Multi-dimensional and objective assessment of motion sickness susceptibility based on machine learning. Front Neurol 13. https://doi.org/10.3389/fneur.2022.824670
8. Xuan RP, Xiong Y, Brietzke A, Marker S (2021) Thermal infrared imaging based facial temperature in comparison to ear temperature during a real-driving scenario. J Therm Biol 96:102806
9. Jones ML, Le VC, Ebert SM, Sienko KH, Reed MP, Sayer JR (2019) Motion sickness in passenger vehicles during test track operations. Ergonomics 62(10):1357–1371
10. Wang J, Zhang L, Huang Y, Zhao J (2020) Safety of autonomous vehicles. J Adv Transport 2020:13 p, Article ID 8867757. https://doi.org/10.1155/2020/8867757
11. Li D, Hu J (2021) Mitigating motion sickness in automated vehicles with frequency-shaping approach to motion planning. IEEE Robot Autom Lett 6(4):7714–7720
12. Liu H, Inoue S, Wada T (2022) Motion sickness modeling with visual vertical estimation and its application to autonomous personal mobility vehicles. In: 2022 IEEE Intelligent vehicles symposium (IV). IEEE, pp 1415–1422
13. Siddiqi MR, Milani S, Jazar RN, Marzbani H (2021) Ergonomic path planning for autonomous vehicles-an investigation on the effect of transition curves on motion sickness. IEEE Trans Intell Transp Syst 23(7):7258–7269
14. Talsma TMW, Hassanain O, Happee R, de Winkel KN (2023) Validation of a moving base driving simulator for motion sickness research. Appl Ergon 106:103897. https://doi.org/10.1016/j.apergo.2022.103897
15. Almallah M, Hussain Q, Reinolsmann N, Alhajyaseen WK (2021) Driving simulation sickness and the sense of presence: correlation and contributing factors. Transport Res F: Traffic Psychol Behav 78:180–193

16. Bohrmann D, Bruder A, Bengler K (2022) Effects of dynamic visual stimuli on the development of carsickness in real driving. IEEE Trans Intell Transp Syst 23(5):4833–4842
17. DiZio P, Ekchian J, Kaplan J, Ventura J, Graves W, Giovanardi M, Lackner JR (2018) An active suspension system for mitigating motion sickness and enabling reading in a car. Aerospace Med Hum Perform 89(9):822–829
18. El Hamdani S, Bouchner P, Kunclova T, Lehet D (2022) The impact of physical motion cues on driver braking performance: a clinical study using driving simulator and eye tracker. Sensors 23(1):42
19. Htike Z, Papaioannou G, Siampis E, Velenis E, Longo S (2022) Fundamentals of motion planning for mitigating motion sickness in automated vehicles. IEEE Trans Veh Technol 71(3):2375–2384. https://doi.org/10.1109/tvt.2021.3138722
20. Kuiper OX, Bos JE, Diels C, Cammaerts K (2019) Moving base driving simulators' potential for carsickness research. Appl Ergon 81:102889. https://doi.org/10.1016/j.apergo.2019.102889
21. Saruchi SA, Mohammed Ariff MH, Zamzuri H, Amer NH, Wahid N, Hassan N, Abdul Kadir Z (2020) Lateral control strategy based on head movement responses for motion sickness mitigation in autonomous vehicle. J Braz Soc Mech Sci Eng 42:1–14
22. Wada T, Fujisawa S, Doi S (2018) Analysis of driver's head tilt using a mathematical model of motion sickness. Int J Ind Ergon 63:89–97
23. Rolnick A, Lubow RE (1991) Why is the driver rarely motion sick? The role of controllability in motion sickness. Ergonomics 34(7):867–879
24. Zheng Y, Shyrokau B, Keviczky T (2023) Mitigating motion sickness with optimization-based motion planning. Arxiv Preprint

Severity Assessment of the Wayfinding Problems in Mumbai's Local Train Commuting System

Shikha Agarwal and Prasad Bokil

1 Introduction

Wayfinding pertains to an individual's capacity to navigate from a reference point to a desired destination, encompassing both cognitive and behavioral aspects [1]. An individual who is well-oriented possesses a keen awareness of their current location, their desired destination, and the necessary steps to bridge the gap between them. Vision plays a pivotal role in furnishing crucial information, making it an essential element for wayfinding. Visual impairments constitute the most evident and severe perceptual hindrances to successful wayfinding [1, 2].

India, home to an estimated 25% of the world's population with visual impairments [3], presents a challenging context. Our understanding of wayfinding challenges faced by people with vision impairments (hereafter referred to as PVIs) primarily relies on research conducted in economically developed nations. However, exploring the real-world wayfinding challenges PVIs encounter in the context of rail transit environments in India provides fresh and invaluable insights that supplement the global knowledge base on this topic.

Mumbai local train commuting system (hereafter referred to as MLTCS) serves the daily mobility needs of 7.4 million commuters, including those with disabilities, in Mumbai, India's financial capital and most populous city, earning it the distinction of having the world's highest passenger density among urban railway systems [4]. Despite being the world's most crowded public transit system, it remains the most reliable and cost-effective mode of transportation in both the city and its suburbs [5]. As Mumbai's population continued to grow, the problem of frequent overcrowding

S. Agarwal (✉) · P. Bokil
School of Design, Indian Institute of Technology, Industrial Design Centre (IDC), Bombay, Mumbai, India
e-mail: shikha.agarwal@iitb.ac.in

P. Bokil
e-mail: prasad.bokil@iitb.ac.in

© The Author(s), under exclusive license to Springer Nature Singapore Pte Ltd. 2025

S. Das et al. (eds.), *Human-Centered Digitalization, Volume 3*, Design Science and Innovation, https://doi.org/10.1007/978-981-96-4383-7_29

has escalated, giving rise to numerous safety concerns that have been raised over the years. The unsafe conditions in trains and stations result in approximately 3500 fatalities annually, equating to an alarming average of 10 deaths per day [6]. It is not widely recognized that even the present wayfinding design can be a factor in compromising usability and safety. The research presented by Agarwal et al. [7] has shed light on the complete absence of social awareness regarding the wayfinding difficulties experienced by PVIs in MLTCS. The paper delves into the specific wayfinding challenges encountered by PVIs in MLTCS. To bring about positive change, it is essential to identify and prioritize issues negatively impacting the wayfinding of PVIs.

Severity assessment, a specific step within the process of usability evaluation [8, 9], is used in this study to categorize identified wayfinding problems by their impact level (e.g., critical, moderate, minor) to determine the priority of resource allocation to make MLTCS wayfinding friendly for PVIs. The severity assessment study reported in this paper builds upon the research conducted on PVIs by Agarwal et al. [7]. The study results presented in this paper are derived from the experiences and perceptions of train commuter subgroups categorized according to three levels of vision: No Vision, Low Vision, and Sighted. The field research aims to comprehend how the level of vision influences the severity assessment of wayfinding problems.

2 Severity Assessment Studies

As per Nielsen [9], the severity of a usability problem is typically assessed as a composite of three key factors: Impact-measuring the extent of trouble experienced by affected users. Persistence-Determining how frequently a user encounters the problem. Frequency-assessing the number of users affected by the problem. Further, Sauro [10] clearly distinguished between problem frequency and severity. According to him, problem frequency refers to the number of participants who encounter a problem divided by the total number of participants. Problem severity relates to the rating users give based on their perception of the problem's impact. The literature on usability evaluation features a significant debate regarding whether problem severity is correlated with the number of users encountering a problem [8, 10]. However, most literature concludes that the severity of a problem remains independent of the number of users it is likely to affect (i.e., frequency) [10]. To prioritize issues, the frequency and severity of problems should be considered [10]. Benini [11] elucidates the difference between severity and priority. He explains that the degree of severity is an inherent property of every unmet need and, statistically, is unrelated to the severity of other needs. In contrast, priorities emerge from comparisons among two or more needs. A strong positive correlation between the severity and priority of a specific need across the assessed areas enhances confidence in the assessment.

There is also an ongoing debate about the involvement of user evaluators or both in the usability test. Hertzum [8] emphasized that the severity assessments have probably been influenced by a subset of users who closely resemble the individual evaluators, making them unrepresentative of the entire user population. Enhanced results could be obtained by basing severity assessments on user-generated data or having users make these assessments instead of relying solely on evaluators. Hassenzahl [12], Law and Hvannberg [13] and Hertzum [8] noted that the involvement of users in assessing problem severity has remained largely unexplored thus far, highlighting a significant demand for discovering methods to obtain more dependable and democratic severity assessments than those performed by usability experts. Furthermore, when multiple users assess the severity of reported problems, it is likely to enhance the reliability of severity assessments compared to assessments made by a single evaluator [10, 14]. Caulton [15] emphasized that in the context of heterogeneous user subgroups, some problems may be encountered with equal probability across subgroups, while others may be unique to specific subgroups. Therefore, it is essential to identify a strong correlation between the severity and frequency of specific problems within a particular subgroup and across different subgroups. Ratings of usability problems are one of the most used methods to categorize usability problems based on their severity of impact [8–10, 12].

From the review of the existing literature, we understood that a dearth of studies engages users in severity assessments. Further, past studies fail to explain the criteria, reasoning, and experiences that govern user or usability expert decisions while assigning any value to the *p* item under assessment. The novelty of this study is the involvement of PVIs in the severity assessment process of wayfinding.

3 Methodology

Structured interviews were employed as the data collection method, with thirty participants recruited through the snowball sampling technique. Participant subgroups were established based on three levels of vision: No Vision, Low Vision, and Sighted, with ten participants selected for each subgroup in this study (Table 1).

Table 1 Age, gender, and familiarity with MLTCS of participants' subgroups

Groups	Age	Gender		Familiarity level		
		Male	Female	Rare	Occasional	Regular
NV	35.8	7	3	2	3	5
LV	29.2	5	5	1	2	7
S	29.1	5	5	5	2	3

3.1 Data Collection

A structured questionnaire was created in Hindi and English, covering seven broad wayfinding task categories: 1. Station entry and exit 2. Ticket purchase 3. Use of station facilities and services 4. Foot over bridge uses 5. Platform use 6. Train boarding and disembarking 7. Nonspecific or General (wayfinding concerns not specific to a single journey stage). These tasks encompassed many wayfinding problems [7]. In total, 70 wayfinding problems were evaluated for their severity impact, relevant to various participant subgroups. Among 70 wayfinding problems, there were common and unique concerns among participant subgroups [15]. Out of 70 problems, 42 (60%) were common to all three subgroups, 17 (24.28%) were common among the VI group (no vision and low vision), and 11 (15.71%) problems were common among participants with low vision and sighted individuals (Table 2).

Face-to-face interviews were conducted at each participant's home or workplace, with all interviews being voice recorded and lasting from 1 to 1:30 h. Wayfinding issues and task categories were verbally described in a predetermined order. The study employed the following data collection process.

Severity Rating: In the first step, participants were asked to rate the severity of wayfinding problems within each task category. The rating technique enabled an understanding of the absolute severity impact of each problem within its specific context, which was considered suitable for this study due to a large *p*-value of 70, making it less time-consuming and mentally taxing [11, 16]. They were likely given a 4-point Likert scale to assign severity scores to these problems. This Likert scale is modified from measurement scales developed by Neilsen [9] and Sauro [10].

0—Not a problem for me.

1—Minor: Causes hesitation in decision-making, low priority for fixing.

2—Moderate: Leads to time wastage and frustration; medium priority for fixing.

Table 2 *p* items in each task category for all users and subgroups no vision (NV), low vison (LV), sighted (S)

No.	Task category	No. of *p* items			
		Total	NV	LV	S
1	Station entry and exit	8	8	8	6
2	Ticket purchase	6	6	6	5
3	Use of station facilities and services	11	8	7	8
4	Foot over bridge use	5	5	5	2
5	Platform use	13	11	13	8
6	Train to board and disembark	15	12	11	12
7	Nonspecific/general	12	7	12	12
	Total	70	57	62	53

3—Critical: This results in wrong decisions, anger, and high priority for fixing.

This Likert scale is borrowed from the scale developed by Sauro in 2013, including 0-Not a problem for me, from the scale developed by Nielsen [14]

Qualitative Feedback: Participants provided qualitative feedback or explanations for their responses throughout the rating. This feedback offers valuable insights into the participants' thought processes, providing context for their ratings [10, 17].

The frequency of participants who did not encounter problems was documented, in line with Nielsen (1995). This suggests that specific wayfinding issues covered in the questionnaire did not negatively affect specific participants in this study. Furthermore, the frequency of participants unaware of specific wayfinding issues was also considered. Respondents were encouraged to share new wayfinding concerns outside the questionnaire [11]. Positive responses regarding currently functioning wayfinding aids and services in the MLTCS were also welcomed during the assessment [17]. Overall, this data collection process combines quantitative (severity ratings) and qualitative (participant feedback) methods to assess and prioritize wayfinding problems in different task categories comprehensively. This approach allows researchers to understand better the issues and the rationale behind the participants' assessments.

3.2 *Data Analysis*

Subgroups organized rating data, and the percentages of commuters assigning ratings on a scale from 0 to 3 were sorted from largest to smallest. Wayfinding concerns were examined based on three key parameters.

Severity Impact Factor (IF): This factor gauges how significantly commuters are affected by a particular wayfinding concern. It was calculated for each concern using the formula $(0 * n0 + 1 * n1 + 2 * n2 + 3 * n3)/(n0 + n1 + n2 + n3)$, n stands for no. of participants affected by the problem. This approach is consistent with Sauro [10], Neilsen [9], and Hertzum [8]. A predefined range for the IF value was used to categorize concerns as mild (0–1), moderate (1.01–2), or critical (2.01–3) in terms of their impact.

Problem Frequency (PF): This metric assesses the number of participants affected by a specific wayfinding concern, following the methodology proposed by Neilsen [9]. In this case, the number of participants aware of the problem, rated between 0–1, was considered in calculating problem frequency.

Pearson correlation coefficient test was performed to measure the linear correlation between problem frequency and severity impact. The same test was also performed to find the correlation between the number of usability problems and the severity ratings. The inter-rater reliability test, Fleiss Kappa [18] was performed to find the level of agreement between the raters for each wayfinding problem and the task category. Additionally, qualitative responses from participants were transcribed to understand the criteria employed for rating.

4 Results

4.1 Rating Results

Out of the 70 wayfinding problems, 29 were rated as critical, 37 had a moderate impact, and 4 were deemed to have a minor impact. Among the 70 issues, only one was "not validated" by 23 participants, concerning the absence of audio, visual signage, and help desk services in three languages: Hindi, English, and Marathi. Table 3 displays the ratings of the subgroups, the problem frequency (PF), and the severity impact factor (IF) for each problem within each task category. The table presents only the highest-rated problems in each task category. No participants added further wayfinding problems to the list discussed in the questionnaire, indicating that the concerns addressed in this study were comprehensive.

Unexpected patterns emerged in the severity ratings for specific items. Surprisingly, there was a positive correlation between severity ratings and the level of vision for certain wayfinding problems. Sighted individuals perceived limited availability of provisions to identify local trains based on the platforms they stop at and the train route as having a more severe impact on their wayfinding experience (2.10) compared to the low-vision group (1.90) and the no-vision group (1.10). Unlike sighted individuals, those with VI pay more attention to public announcements and rely heavily on public assistance for wayfinding decisions.

The Pearson correlation coefficient results, when performed between no. of issues in all task categories and average severity ratings given by subgroups to each task category, are the following: The Pearson Correlation coefficients for all 30 participants: – 0.24; No Vision: – 0.01, Low Vision: – 0.12 indicate that despite differences in the number of wayfinding problems in each task category, there is a weak and negligible linear relationship with the average severity ratings given by each participant subgroup. This means that differences in the number of wayfinding problems in each task category do not influence the severity ratings. However, a relatively strong and positive correlation was observed between no. the problems in each task category and the severity ratings for the sighted group (0.71). Further, the Pearson correlation coefficient results when performed between severity and frequency for different subgroups and all participants, we found the following results: Like previous studies conducted by Law and Hvannberg [13, 0.10], Hertzum [8, 0.29], and Sauro [10, 0.06] this study also identified a weak correlation between frequency and severity for all participants (0.35). Additionally, as Caulton [15] pointed out, severity and frequency may be strongly correlated within individual subgroups with unique problems compared to the overall correlation performed on all problems for all user groups. A reasonably strong positive correlation is observed in this study between problem frequency and severity for the No Vision (0.74) and Sighted (0.82) subgroups, with a negligible correlation for the low-vision group (0.29). It's important to note that correlation does not imply causation. These values only describe the linear relationship between the two variables. Other factors not considered in this analysis may influence the observed patterns.

Table 3 *p* items with the participant frequency (PF) and the highest severity impact factor in each task category (IF)

	Task categories and related problems	NV	LV	S	PF	IF
1	**Entry and exit from the station** The inability to see vehicles at the station entrance leads to chances of collision	10	10	0	20	2.75
2	**Ticket purchase** There are no direction signs to reach the nearest ticket hall from the platform, station entry, and foot over the bridge	10	10	10	30	2.23
3	**Use of station facilities** Lack of information to make new commuters aware of what type of facilities are available at each station and where they are located	10	9	9	28	2.46
4	**Foot over bridge use** There is a lack of information on what parts of the station each bridge connects before climbing it. As a result, people use the wrong bridge	10	10	10	30	2.77
5	**Platform use** Beepers are not located on all platforms. Even if they are present, they are often not working or are unheard of during peak hours. All these factors create difficulty in the platform's waiting area for people with disabilities to access a coach easily	10	9	0	19	2.58
6	**Train boarding and disembarking** Overcrowding of general commuters in coach for people with disabilities makes train travel difficult for PVIs	10	10	0	20	2.95
7	**General/non-task specific** Lack of awareness of provisions for instant help in case of emergency and security measures in running trains and on the station	9	4	9	22	2.64

Table 4 Kappa values for participant subgroups for each task category

No	Task category	NV	LV	S	ALL
1	Station entry and exit	0.01	0.06	0.13	0.15
2	Ticket purchase	0.03	0.11	0.09	0.16
3	Use of station facilities and services	0.34	0.38	0.10	0.23
4	Foot over bridge use	0.19	0.10	0.14	0.30
5	Platform use	0.22	0.12	0.15	0.20
6	Train to board and disembark	0.16	0.22	0.17	0.27
7	Nonspecific/General	0.17	0.24	0.08	0.21

Table 4 shows the results of a Fleiss' Kappa test for assessing inter-rater reliability among the three participant subgroups in the context of various task categories. Here's an interpretation of Table 4: For some task categories, there is relatively low agreement among the raters (e.g., Station entry and exit, with Kappa = 0.05; Ticket purchase, with Kappa = 0.16). For other task categories, there is moderate to substantial agreement among the raters (e.g., Use of station facilities and services, with Kappa = 0.23; Foot over bridge use, with Kappa = 0.30; Train Boarding and Disembarking, with Kappa = 0.27; General category, with Kappa = 0.21).

The table provides an overview of the inter-rater reliability for different task categories. It suggests that the level of agreement varies across the tasks, with some showing stronger agreement among the raters and others demonstrating weaker agreement.

4.2 Criteria Used by the Participants for Rating

Two factors influenced participant criteria for rating: challenges experienced without wayfinding support, limited self-reliant coping mechanisms or both. Problems that can be alleviated with coping mechanisms or wayfinding support are rated as less severe. Major criteria for evaluating more challenging wayfinding problems include task accomplishment difficulties, the experience of using station and train infrastructure, information comprehension, staff, service, signage deployment sufficiency and consistency, discomfort levels, problem occurrence frequency, particularly when wayfinding support and self-devised coping mechanisms are limited. On the other hand, certain problems are considered "manageable" with social support, wayfinding aids like public announcements, the M-indicator mobile app for local train routes, tactile floor indicators, and audio beepers on platforms to designate reserved coach waiting for areas for people with disabilities. Additionally, self-reliant coping mechanisms, such as using multiple senses, developing directional awareness, and employing preference and avoidance strategies, influence responses indicating lower severity.

5 Conclusion

The study successfully validated real-life wayfinding problems experienced by individuals who encounter them. Among the 70 concerns, 69 problems were validated. A "Severity Rating Index" is provided through this study that quantifies the degree of severity or impact of wayfinding problems on the wayfinding of three participant subgroups. The rating results provide designers, planners, and policymakers with a strategic framework for systematically addressing these wayfinding problems, enabling gradual yet significant improvements. To enhance local train usability, particularly for PVIs, addressing wayfinding concerns related to task categories entering and exiting the station building (NV: 2.35, LV: 2.20) and foot over bridge is crucial (IF: 2.08 respectively). Mechanisms should be designed to encourage access and use of station facilities by commuters with no vision (IF: 2.20). To address the most significant wayfinding challenges, improvements should commence at large-scale stations characterized by complex layouts, extensive scale, and overcrowding. This study is one of its kind to involve PVIs visually in severity assessment, providing a unique perspective and possibility of their participation in wayfinding evaluation studies. Unlike other severity assessments, this research offers criteria used by participants for their responses.

This study has a few limitations. Despite the high number of assessment items, all participants patiently completed the study. The limited sample size means the results may not be generalizable to a larger population, but they do offer valuable insights into issue criticality. Another limitation is that the rating technique used in the study often led to most wayfinding problems being perceived as having moderate or severe impact, making it challenging to discriminate between item scores.

There is a scope to extend the analysis to the familiarity variable. This study plays a crucial role in guiding the designing of wayfinding considerations for the three commuter groups: no-vision, low-vision, and sighted group, as part of the future scope of the research.

References

1. Passini R, Proulx G (1988) Wayfinding without vision: an experiment with congenitally totally blind people. Environ Behav 20(2):227–252. https://doi.org/10.1177/0013916588202006
2. Golledge RG, Klatzky RL, Loomis JM (1996) Cognitive mapping and wayfinding by adults without vision. In: Portugali J (ed) The construction of cognitive maps. Geo Journal Library, vol 32. Springer, Dordrecht. https://doi.org/10.1007/978-0-585-33485-1_10
3. Institute for Health Metrics and Evaluation (IHME) (2020) GBD compare data visualization. IHME, University of Washington, Seattle, WA. Available from http://vizhub.healthdata.org/gbd-compare. (Accessed 30-10-2021)
4. Gardas BB, Shimpi NR, Mahajan SB (2013) Value analysis for a Mumbai Local Train: A case study. Int J Sci Res Publ 3(6), ISSN: 2250-3153. http://www.ijsrp.org/research-paper-0613.php?rp=P181281
5. Kadam S (2015) Apprehension study of passengers travelling by Mumbai Suburban Trains. AADYA—Natl J Manag Technol 5:16–23, ISSN 2319-264X

6. Mehta KM (2012) Mumbai's lifeline has claimed 36,000 lives in 10 years. The Times of India. Retrieved from https://timesofindia.indiatimes.com/city/mumbai/mumbais-lifeline-has-claimed-36000-lives-in-10-years/articleshow/12739124.cms
7. Agarwal S, Mullick A, Ray GG (2017) Wayfinding challenges and strategies in Mumbai Suburban Railway. In: Research into design for communities, volume 1: proceedings of ICoRD. Springer Singapore, pp 161–172
8. Hertzum M (2006) Problem prioritization in usability evaluation: from severity assessments toward impact on design. Int J Hum-Comput Interact 21:125–146
9. Neilsen J (1994) Severity ratings for usability problems. Retrieved from https://www.nngroup.com/articles/how-to-rate-the-severity-of-usability-problems/
10. Sauro J (2014) The relationship between problem frequency and problem severity in usability evaluations. J Usability Stud Archive 10:17–25
11. Benini A (2013) Severity, and priority—their measurement in rapid needs assessments
12. Hassenzahl M (2000) Prioritizing usability problems: data-driven and judgement-driven severity estimates. Behav Inf Technol 19(1):29–42. https://doi.org/10.1080/014492900118777
13. Law ELC, Hvannberg ET (2004) Analysis of combinatorial user effect in international usability tests. In: Proceedings of the CHI 2004 conference on human factors in computing systems, 9–16
14. Nielsen J (1992) Reliability of severity estimates for usability problems found by heuristic evaluation. In: Conference on human factors in computing systems (CHI '92). Association for Computing Machinery, New York, NY, USA, pp 129–130 https://doi.org/10.1145/1125021.1125117
15. Caulton DA (2001) Relaxing the homogeneity assumption in usability testing. Behav Inf Technol 20:1–7
16. Fabbris L (2013) Measurement scales for scoring or ranking sets of interrelated items
17. Dumas JS, Molich R, Jeffries R (2004) Describing usability problems: are we sending the right message? ACM Interact 11(4):24–29
18. Fliess JL (1971) Measuring nominal scale agreement among many raters. Psychol Bull 76:378–382

In-Depth Review of Crafts from Natural Fibres: Traditional Techniques, Innovations, and Sustainability

Akanksha Dobriyal and Swati Sarkar

1 Introduction

Traditional fibre crafts, steeped in the cultural heritage of practically every civilisation on the planet, meld tradition, aesthetic appeal, and sustainability in a harmonious balance. Mostly sourced from plants and animals, these fibres-which include cotton, linen, wool, and silk-have been converted with great skill into artefacts and textiles since time immemorial [1]. Weaving and felting epitomize the ingenuity of human beings passed on through generations. The natural fibre crafts in this eco-friendly trend take prominence to bring the past into the present for a sustainable future [2].

1.1 Background on Natural Fibre Crafts

Natural fibre crafts originate deeply from ancient civilizations, when early human beings used the versatile properties of plant and animal fibres to make some very essential items in daily life [3]. Ancient crafts like India's cotton and China's silk symbolize cultural identity. Fibres like jute have deep historical roots, with modern crafts reflecting their cultural significance [4].

A. Dobriyal (✉) · S. Sarkar
School of Design, University of Petroleum and Energy Studies, Dehradun, India
e-mail: akankshadobriyal28@gmail.com

S. Sarkar
e-mail: swati.sarkar@alumni.iitg.ac.in

© The Author(s), under exclusive license to Springer Nature Singapore Pte Ltd. 2025

S. Das et al. (eds.), *Human-Centered Digitalization, Volume 3*, Design Science and Innovation, https://doi.org/10.1007/978-981-96-4383-7_30

1.2 Importance and Relevance in Modern Times

In today's era, natural fibre crafts address sustainability, ethical consumerism, and cultural preservation. As synthetic material concerns grow, natural fibres provide an eco-friendly alternative, often requiring less chemicals and energy [5]. Handcrafted natural fibre items preserve tradition and support artisans in a mass-produced age, blending modernity with sustainable craftsmanship [6].

1.3 Traditional Techniques in Natural Fibre Crafts

Traditional natural fibre crafts merge ancestral wisdom with environmental adaptability and cultural expression. The patterns over the years have been woven, raw fibres were spun into yarn, and handmade garments and decor. Felting is the process whereby fibres are amalgamated to create durable items [7]. Different cultures have unique practices, such as the Japanese art of 'Shibori' for dyeing [8] or the intricate 'Ikat' weaving popular in several Asian and Central American traditions [9]. Age-old techniques, refined over generations, guarantee product durability and authenticity, showcasing human creativity and our enduring bond with nature.

1.4 Introduction to Traditional Crafting Techniques

Traditional craftsmanship brings to light the deep connection between man and nature, and that inborn desire in man to create. Methods passed on through millenniums develop from basic survival tools into an intricate art, fusing function and beauty [10]. Weaving, the interlacement of yarns; spinning, which turns fibres into threads; felting, a process making dense fabric; and dyeing with designs that involve organic pigments-all techniques cultural in nature [11]. Each region has its specialty in craftsmanship, as dictated by environment, history, and needs of the people. Each of these techniques allows one to trace the past and showcases artisans' ingenuity, which is actually at the core of today's natural fibre crafts [12].

1.5 Range of Materials Used in Various Cultures

Traditional fibres and other materials in handicrafts have always shown cultural variety and regional availability. For example, the Asian cultures weave silk threads from silkworms into textiles such as brocades of China or sarees of India [13]. Africa, with its rich tapestry of tribes and traditions, boasts unique handicrafts using materials like raffia from palm trees or cotton spun into the colourful kente cloth of Ghana

[14]. The Navajo and other indigenous peoples made iconic wool patterns, while in South America, cotton, agave, and alpaca fibres are used. Pacific islands such as Fiji and Samoa tell stories using tapa cloth made from the bark of a mulberry tree [15]. European tapestries, especially of medieval France and Belgium, mix wool with threads of precious metals. Such material diversity tells of the ingenuity of artisans from around the world and narrates a story of sustainability, creativity, and pride in culture through their craft [16].

1.6 Asian Techniques and Materials

Various Asian cultures are rich in their histories and have mastered different unique fibre handicraft techniques, which are admired in various parts of the world. India is good at traditional methods of weaving and dyeing like block printing, bandhani, and ikat [17]. Chinese sericulture gave the world brocades, embellished with embroidery techniques such as Suzani. Japan introduced Sashiko stitching and the shibori dyeing method [18]. In Indonesia, it is batik where wax is painted onto fabric in incredibly intricate patterns and then dyed to spectacular results [19]. Materials such as Thai silk, Indian jute, and Philippine abaca represent regional flora. These techniques passed on for thousands of years epitomize both Asian skill and the spirit of their various homelands.

1.7 African Techniques and Materials

Natural fibre crafts maintain a rich heritage in diverse cultural tapestry across Africa, and specifically at the centre of it all for many cultures is weaving. Places like Ghana are famous for kente cloth, a vibrant silk, and cotton fabric with symbolic patterns [20]. The Yorubas are also renowned for their adire cloth, done by resistant dyeing with cassava paste to make such patterns. Central and West Africa use the raffia fibre taken from palm trees to create items like mats. The biodiversity of Madagascar yielded a different type of raffia textiles with geometric designs [21]. Agave sisal plants were grown to create durable baskets, rugs, and other items by East Africans; marrying utility with artistry [22]. Ugandan bark cloth from the inner bark of the Mutuba tree is a testament to sustainable craftsmanship. Techniques and materials such as these on this vast continent underline the genius in adaptation and creativity on the part of the African artisans while echoing the deep-seated relationship of people with their natural environment.

1.8 Native American Techniques and Materials

The closeness of nature to Native American tribes all over the Americas is reflected within various natural fibre arts. For instance, the Pomo in California create intricately complicated baskets with willow and redbud [23]. The Navajo people, indigenous to the American Southwest, have become synonymous with their detailed wool rugs and blankets. These textiles, dyed with native plants and minerals, showcase geometric and symbolic designs [24]. The Tlingit and Haida tribes weave Chilkat blankets of mountain goat wool and cedar bark in the Pacific Northwest. Plant fibres like yucca and bear grass are common in vessels, typically coloured with berry juice and crushed bark for natural dyes [25]. These natural materials and the construction methods used by Native American craftsmen betray their remarkable abilities and disclose an intimate spiritual and cultural association with the land, telling stories about ancestry, mythology, and the beauty of nature in cohesion.

1.9 European Techniques and Materials

Europe's varied climates and cultures fostered diverse natural fibre crafts. Belgium's Flanders is famed for wool tapestries, Ireland and France for linen and lace, and the British Isles for distinctive wool crafts like Aran knitting and Fair Isle patterns [26]. Countries like Spain and Italy have long histories of silk production in the Mediterranean, where techniques like ikat weaving, known locally as 'telas de lenguas' or 'rago di seta', were developed [27]. European dyeing traditions rely on natural dyes from sources like woad and madder, giving textiles unique hues. European fibre handicrafts have conveyed identity, tradition, and regional pride for centuries [28].

2 Modern Innovations in Natural Fibre Crafts

In modern times, natural fibre crafts blend tradition with innovation for today's eco-conscious world. A key trend includes using sustainable and recycled materials, like turning pineapple leaves into 'Piñatex', a leather substitute [29] Or the innovative use of spider silk and mycelium for creating fabrics and accessories [30]. Digital technology, including CAD systems, enhances weaving and knitting patterns, while 3D printing provides precise craft design. Evolved natural dyes now offer durability and sustainability without sacrificing colourfastness [31]. Bio fabrication, the process of cultivating fibres in labs from microbial sources, offers a promising horizon for crafting materials with reduced ecological footprints [32]. The global shift towards artisanal goods has spurred collaborations between artisans and designers, modernizing ancient techniques. Today's innovations in natural fibre crafts blend tradition and technology, envisioning a sustainable future.

2.1 Role of Technology in Enhancing Natural Fibre Crafts

Technology has revolutionized natural fibre crafts, linking traditions with modern needs. Advanced machinery improves fibre processing, while CAD tools help artisans visualize designs, enhancing precision and reducing waste [33]. Digital looms allow intricate weaving faster than by hand. Innovations like 3D printing introduce novel ways for fibre integration, merging with traditional materials [34]. Technology has also enhanced the longevity and vibrancy of natural dyes, making them more resilient to wear and environmental factors. While technology has introduced efficiency and expanded possibilities, it has also fortified the essence of natural fibre crafts, ensuring their relevance and appeal in the modern world.

2.2 Case Studies

Studies on natural fibre crafts emphasize tradition, sustainability, and innovation. The revival of nearly-extinct Muslin in Bangladesh, historically termed 'woven wind', demonstrates this, as collaborations between artisans, historians, and NGOs breathe life back using age-old methods [35]. India's Kerala 'Banana Sari' utilizes banana plant fibres for sustainable, luxurious textiles, blending ancient techniques with modern sustainability. In Peru's highlands, traditional alpaca wool crafts embrace technology to improve quality and align with global fashion. This exemplifies the blend of heritage, innovation, and market demand in fibre crafts.

2.3 Enhanced Durability Through Modern Practices

Modern natural fibre craft techniques prioritize aesthetics, durability, and sustainability. Textile innovations bolster fibre strength, with methods like mercerization enhancing cotton. Improved mordants ensure natural dyes now resist fading better [36]. Innovative weaving and knitting patterns, aided by computer software, enhance durability. Natural and advanced finishes now protect fibres from stains, moisture, and UV damage [37]. Collectively, these modern practices signify a thoughtful evolution in natural fibre handicrafts, emphasizing longevity and functionality alongside traditional beauty and craftsmanship.

2.4 Innovations in Aesthetic Appeal

Natural fibre crafts, merging traditional motifs with modern designs, highlight human creativity. Updated dyeing with organic pigments produces vibrant, eco-friendly shades. Combining fibres like silk-linen or bamboo-cotton introduces unique textures and finishes [38]. Digital design tools aid in the visualization and creation of intricate patterns, offering a fresh spin on age-old designs [39]. Furthermore, collaborations between global designers and local artisans bring diverse perspectives to the fore, infusing natural fibre handicrafts with a cosmopolitan touch [40]. This reimagined aesthetic appeal celebrates the material's inherent beauty and ensures that these handicrafts are versatile, contemporary, and relevant to modern tastes and styles.

3 Sustainability of Natural Fibre Crafts

Natural fibre crafts, rooted in renewable and biodegradable resources, epitomize sustainability in an environmentally-conscious era. Cultivating flax for linen, bamboo for its fibres, or sheep for wool, these crafts have a smaller carbon footprint than synthetic alternatives when responsibly managed [41]. Traditional artisans often use organic farming, avoiding harmful chemicals, benefiting both the environment and users. Natural dyes reduce water pollution, and as global focus shifts to sustainability, handicrafts that support local economies and cultural heritage gain prominence. Natural fibre crafts embody the harmony of man, material, and nature, promoting sustainability.

3.1 Challenges and Complexities

While natural fibre crafts offer myriad benefits and remain a testament to the delicate balance between nature and craftsmanship, their creation is laden with challenges and complexities. Cultivating natural fibres is vulnerable to environmental factors, including droughts, pests, and diseases, which can impact yield and quality. This unpredictability often leads to price volatility, making it difficult for artisans to maintain consistent profit margins [42]. Traditional methods of processing and weaving can be labour-intensive and time-consuming, limiting the scale of production and competing with mass-produced synthetics on price points [43]. Natural dyes, while eco-friendly, may lack the vibrancy and durability of chemical ones. Transferring ancient techniques to urban-focused youth and maintaining authenticity amid global standards is challenging. Though valued for sustainability, natural fibre crafts confront preservation and promotion hurdles.

3.2 Comparing Natural Fibres to Synthetic Materials

Natural fibres like cotton and wool come from plants and animals, making them renewable and biodegradable. In contrast, synthetics like polyester derive from petroleum, often having a larger carbon footprint and taking longer to decompose [44]. Natural fibres breathe well and are preferred for many garments, but degrade faster and can be costlier. Synthetics are customizable, often more durable, and economical but are being made eco-friendlier with recycling innovations. The choice hinges on purpose, desired features, and values. Table 1 shows findings of art and crafts from different countries.

Table 1 Findings of art and crafts from different countries

S. No.	Techniques used	Country	Findings	References
1	Block printing, bandhani (tie-dye), and Ikat	India	The Bagru prints of Rajasthan, the vibrant swirls of Gujarat's bandhani, and the mesmerizing weaves of Pochampally ikat represent India's diverse and timeless textile artistry	[17]
2	Suzani and Chikan	China	Tribal textiles in which brocades and Damask is created	[18]
3	Sashiko	Japan	Embroidery technique, that involves intricate, repetitive stitching patterns, often used for both decorative and functional purposes	[8]
4	Batik	Indonesia	A centuries-old art of wax-resist dyeing, producing intricate patterns and designs deeply rooted in cultural traditions and symbolism	[19]
5	Adire	Nigeria	Resist-dyeing technique, where patterns are crafted on textiles using indigo dye and various resist methods, reflecting the rich cultural tapestry of the Yoruba people	[21]
6	Hand Weaving	East Africa	The kikoi and kente cloths of East Africa are vibrant testaments to the region's hand weaving mastery, intertwining threads of tradition and identity	[22]
7	Aran knitting	Ireland	It weaves intricate cable patterns inspired by the island life, encapsulating tales of fishermen, families, and the rugged Celtic coastline	[26]
8	Weaving Technique-lenguas' or 'rago di seta'	Spain and Italy	Traditional, vibrant textiles characterized by flame-like patterns, celebrated in both Spanish and Italian textile heritage	[27]

4 Discussion and Conclusion

The crafts involving natural fibres are a combination of tradition, sustainability, and modern innovation. Inspired by ancient civilizations, these crafts reveal the timeless bond between the people and nature, brought to life through thoughtful, imaginative use of the resources the environment provides. In the modern world, beset with problems brought about by synthetic materials, this craft represents a whole range of sustainable solutions from ancient times to modern problems. More than being beautiful, they are representatives of the marriage of human imagination, cultural stories, and the gifts of nature. In modern times, these handicrafts face threats emanating from globalization and automation. This balance is complex to achieve between authenticity and environmental sustainability with economic viability, due to natural events and market fluctuations. As opposed to synthetics, the advantages of natural fibres entail a choice on environmental, economic, and social consequences. These crafts represent something more than aesthetic appeal; they are tales of human ingenuity and the benevolence of nature, presenting eco-friendly alternatives in an age where sustainability is key. It looks to the past as these crafts seamlessly merge age-old techniques with modern ones. The traditional methods are revered, while the integration of state-of-the-art technologies upgrades the craftsmanship and makes it relevant. Such collaboration between traditional artisans and contemporary designers gives way to items which are deeply rooted in history, yet tuned to modern tastes.

These crafts, being of renewable sources, would further the cause of a sustainable model wherein quality comes first over quantity and where consumption is further backed by awareness.

However, further research needs to be done to optimize their use in applications to meet today's demand. There should also be emphasis on the support of artisans because this is a heritage that needs preservation. This will marry their particular skills with modern design and hence come up with resonating artefacts. Finally, innovation and research in this area have the capacity to bridge historic wisdom with modern solution crafting of a sustainable path forward.

References

1. Srivastava N (2009) Natural fibers. Int J Fiber Text Res. Natl Inst Sci Commun Inf Resour
2. Thondhlana G, Pullanikkatil D, Shackleton CM (2020) Plant fibre crafts production, trade and income in Eswatini, Malawi and Zimbabwe. Forests
3. Reddy RM (2009) Innovative and multidirectional applications of natural fibre, silk-a review. Acad J Entomol, 71–75
4. International Workshop on Silk Handcrafts Cottage Industries and Silk Enterprises Development in Africa, Europe, Central Asia and the Near East, & Second Executive Meeting of Black, Caspian seas and Central Asia Silk Association (BACSA) Proceedings Black, Caspian Seas and Central Asia Silk Association (BACSA). Kozabirlik Sericultural Cooperative. Online. Available: www.bacsa-silk.org
5. Chakravorty R, Dutta P, Ghose J (2010) Sericulture and traditional craft of silk weaving in Assam

6. Shah A, Patel R (2016) Rural handicrafts
7. Affinito L, Maria Conti G, Motta M (2017) New vision on Knitwear design. How the traditional craft methodologies are evolving into fashion international scenarios. Des J
8. Pujeeb N (2020) The development of natural dyed fabric product with Shibori fabric dyeing technique. Palarch's
9. Bar G, Amrutha A (2005) Contemporisation of traditional Odishi Ikat
10. Yang Y, Shafi M, Song X, Yang R (2014) Preservation of cultural heritage embodied in traditional crafts in the developing countries. A case study of Pakistani handicraft industry. Sustainability Switzerland
11. Vakirtzi S, Georma F, Karnava A (2017) Beyond textiles: alternative uses of twisted fibres and evidence from akrotiri, Thera
12. Davico A, Mendonca P (2010) The use of natural fibres on architecture : the local economy and the arts and crafts. Universidade do Minho
13. Anderson JA (2009) China's Southwestern Silk Road in World History
14. Rice E (2020) The pattern of modernity: textiles in art, fashion, and cultural memory in Nigeria since 1960. Geschichts- und Kulturwissenschaften
15. Richards R, Matthews PJ (2021) Barkcloth from the Solomon Islands in the George Brown Collection. In: From field to museum—studies from Melanesia in Honour of Robin Torrence, ed. Jim Specht, Val Attenbrow, and Jim Allen. Technical reports of the Australian Museum, pp 245–258
16. Projects H, Missaggia C (2013) DigitalCommons@URI DigitalCommons@URI European Tapestries: history, conservation, and creation
17. Hann MA (2005) Patterns of culture: techniques of decoration and coloration. University of Leeds
18. Shaver C (1992) Sashiko: a Stitchery of Japan
19. Selamet J, Indonesian Batik Translation: a case study. Online. Available: http://eprints.binadarma.ac.id/2113/
20. Tyler MS (2016) Meanings of kente cloth among self-described American and Caribbean students of African descent
21. Zanne Aroa (2023) Evolution of Adire textile design of Nigeria_Diffusion & impact on globalization
22. Mwaniki AM (2018) Factors affecting sisal cultivation and adoption in Kiomo Division, Kitui County, Kenya
23. Kroeber AL (1909) American Anthropologist: California Basketry and the Pomo. JSTOR
24. Jett SC (1990) Culture and tourism in the Navajo country. J Cult Geogr
25. Gunther E (2022) Indian Life on the Northwest coast of North America as seen by the early explorers and fur traders during the last decades of the eighteenth century
26. Chisamore_Nora (2021) A history of hand Knitting in the British Isles and my contemporary adaptation
27. Chartier R (2008) Inscription and erasure: literature and written culture from the eleventh to the eighteenth century
28. Partarakis N, Zabulis X (2023) Safeguarding traditional crafts in Europe. Encyclopedia, 244–1261
29. Ashley V (2021) Knowledge of leather alternatives: an exploratory study: implications for education. J Text Sci Fashion Technol
30. Akosua Mawuse Amankwah., Edward Appiah., Charles Frimpong., Anthony Kentthe.: Structure of the Ghana Fashion Industry in the Implementation of Sustainable Strategies.In 24th IFFTI Conference, (2022).
31. Bunnell K (2004) Craft and digital technology
32. Collet C (2023) Explore the intersection of biology and design to develop new protocols for sustainable design
33. Jin H, Yang J (2022) Using computer-aided design software in teaching environmental art design. Comput Aided Des Appl, 173–183
34. Walters K (2021) A role for industrial looms in craft research. FormAkademisk

35. Sharmin F, Tousif Hossain S (2020) Revitalization of handloom communities for preserving the craft heritage of Bangladesh. Urban Reg Plann
36. Patil S, Mahapatra A, Gotmare VD, Patil PG, Bharimalla AK, Arputharaj A (2019) Effect of different mercerization techniques on tactile comfort of cotton fabric
37. Paul R (2015) Functional finishes for textiles: an overview
38. Kashyap M, Hooda A (2008) Amalgamation the Beauty of India. Int J Adv Eng Manag
39. Zoran A (2015) Hybrid craft: showcase of physical and digital integration of design and craft skills
40. Barker T, Hall A (2020) E-ARTISANS contemporary design for the global market
41. Thyavihalli Girijappa YG, Mavinkere Rangappa S, Parameswaranpillai J, Siengchin S (2019) Natural fibers as sustainable and renewable resource for development of eco-friendly composites: a comprehensive review. Front Mater
42. Simelton E, Fraser EDG, Termansen M (2009) Typologies of crop-drought vulnerability: an empirical analysis of the socio-economic factors that influence the sensitivity and resilience to drought of three
43. Meng S, Pan R, Gao W, Yan B, Peng Y (2022) Automatic recognition of woven fabric structural parameters: a review. Springer Nature
44. Cassie ABD (2009) Natural fibres versus man-made fibres. natural fibres. J Textile Inst Proc, pp 556–561

Color Psychology in Cause-Related Marketing: Unravelling the Impact of Color Choices on Consumer Perceptions and Behavior

Sonam Harsh Hardaha, Sarvesh Tripathi, and Tripti Singh

1 Introduction

Color plays an important role in marketing by evoking emotions, conveying messages, and influencing consumer behavior [1]. Marketing has long recognized the impact of color choice on consumer behavior and mood [1]. Colors are more than visuals; they carry emotional and psychological weight and are subtle but powerful instruments of consumer behavior [2]. Cause-oriented marketing is a concept that combines marketing strategies with social aspects and emphasizes the role of companies in making a positive contribution to society [3]. In recent years, cause-related marketing (CRM) has gained prominence as companies and organizations seek social or environmental involvement [4, 5]. Key elements of CRM campaigns include the strategic choice of colors [6], as colors have the ability to convey support for a product and engage consumers emotionally The field of cause-related marketing is where brands align with social or environmental causes to achieve authenticity and purpose [5]. Layers are added the choice of colors in these campaigns can have a significant impact on consumers' perceptions of the brand and the causes they support [7].

Although a lot of research has examined the concept of color, its application to cause-related marketing remains an underexplored area this literature review seeks to address this gap through knowledge of existing data on color perception, consumer behavior, and the formal use of color in marketing. Examine how it affects This study examines how two distinct palettes—brand display and perception of causes—influence consumer behavior. Our goal is to provide colors that match brand identity, such as McDonald's red and yellow, for social or environmental reasons, whether their effects differ from palettes attached to our study uses a comprehensive survey

S. H. Hardaha (✉) · S. Tripathi · T. Singh
Design Discipline, PDPM Indian Institute of Information Technology Design & Manufacturing, Jabalpur, Madhya Pradesh, India
e-mail: 22mds010@iiitdmj.ac.in

© The Author(s), under exclusive license to Springer Nature Singapore Pte Ltd. 2025

S. Das et al. (eds.), *Human-Centered Digitalization, Volume 3*, Design Science and Innovation, https://doi.org/10.1007/978-981-96-4383-7_31

questionnaire that combines quantitative and qualitative elements. Not only will the findings guide marketers, but they will also contribute to a broader discourse about the strategic use of color in marketing.

2 Methodology

2.1 Participants and Materials

The participants in this study for data collection and observation consist of 47 students from the Indian Institute of Information Technology, Design, and Manufacturing (IIITDM) in Jabalpur. The choice of this sample is based on convenience and accessibility. It reflects a diverse range of academic backgrounds and ages. No specific demographic restrictions were imposed, allowing for a broad representation of people.

In our study, we investigate the impact of color palettes on consumer perception in the context of cause-related marketing. To achieve this goal, we have selected McDonald's brand identity, characterized by the iconic red and yellow colors, and an environmental conservation cause represented by the color green. We created two posters with identical content, such as slogans, images, and messaging, but with different color schemes. Poster A (Fig. 1) features McDonald's brand colors, incorporating red and yellow. McDonald's, a globally renowned fast-food chain, has intentionally chosen its iconic color palette of red and yellow to evoke specific emotions and consumer associations. Red signifies energy, excitement, and appetite stimulation, while yellow is associated with happiness and friendliness. These colors have significantly contributed to McDonald's brand identity and are extensively studied in marketing and consumer psychology. Therefore, investigating the impact of these colors in our research is relevant and timely.

Poster B (Fig. 2), on the other hand, employed the color green associated with environmental conservation. Green is commonly linked to environmentalism and conservation, evoking feelings of eco-friendliness and responsibility. It's an ideal color for cause-related marketing research, given its emotional resonance and message alignment. Additionally, the growing importance of environmental issues makes it a socially relevant topic for our study.

We collected data through primary quantitative research using the survey. The questionnaire has been thoughtfully designed to uncover important insights. It enables the identification of key factors: brand recognition, emotional engagement, purchase intent, and overall poster preference. By asking participants about their prior knowledge of the brand, emotional connection to the poster, inclination to purchase based on the advertisement, and their preference between Poster A (Cause-Related Colors) and Poster B (Brand Identity Colors), we gain valuable data for our study.

To verify the theoretical insights, two hypotheses were developed and tested:

Fig. 1 Poster A–Uses the brand identity *colors*

Fig. 2 Poster B–Uses the cause-related colors

Null Hypothesis (H0): The selection of color palettes in cause-related marketing materials has no significant impact on consumer perceptions, emotional engagement, or purchase intent.

Alternative Hypothesis (H1): The selection of color palettes in cause-related marketing materials significantly influences consumer perceptions, emotional engagement, and purchase intent.

Table 1 Comparison between brand identity palette and cause palette

Dependent variable	Brand identity palette	Cause palette	*F*	*p*-values
Visual appeal	3.9149	3.9787	0.167	–
Excitement	3.7872	4.0000	2.231	–
Trust	3.8298	3.3617	6.446	$p < 0.05$
Empathy	3.4255	3.8085	6.781	$p < 0.05$

2.2 Data Collection

To gain a more accurate understanding of the participants in the survey questionnaire, we collected their demographic information. Following that, the visual appeal of each poster was then evaluated using a Likert scale (Table 1), and the emotions evoked by each were rated (including interest, trust, and empathy) (Table 2). Further, through answering op-ed questions and sharing their preferences, feelings, and opinions on how color choices affected the participants, participants had a way to contribute qualitative insights. The purpose of incorporating structured and open-ended questions was to gain an accurate understanding of how the participants perceived the marketing materials.

2.3 Statistical Analysis

Basic statistics were used to calculate the average values of dependent variables such as visual attractiveness and emotional involvement. Comparisons between experiments were performed using ANOVA. In all cases, a p value < 0.05 was considered significant.

3 Result and Discussion

The participants used a 5-point Likert scale to rate the visual appeal of the "brand identity colors" and the "cause-related colors," with higher scores signifying stronger appeal. The "brand identity color" received an average rating of around 4.1, which was higher than the "cause-related colors," as indicated by the data displayed in Table 1. It may be concluded that the established brand color scheme has a stronger appeal based on the respondents' initial visual appeal (Fig. 3).

The study also delved into emotional engagement, including interest, trust, and empathy evoked by the posters. Participants provided ratings on a Likert scale, which allowed for microscopic assessments of their emotional reactions. Table 2 provides a deeper understanding of emotional engagement. It is clear that "brand identity colors" motivated the respondents more, with an average of about 4.0, compared

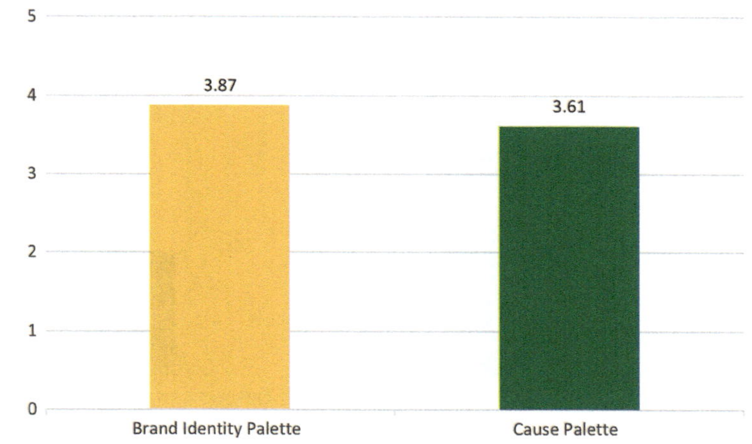

Fig. 3 Visual appeal ratings

to "cause-related colors," which received an average motivation number about 3.5. Comparable confidence levels for the two palettes Available, which evolves about 3.5. Notably, "cause-related colors" outperform "brand-related colors" in arousing empathy, with an average rating of about 3.7, while the latter score around 3.2. This interesting result suggests that cause-related colors dominate in evoking emotional empathy (see Fig. 4).

The study examined the effect of posters on purchase intention. Respondents were categorized as "yes" or "no." Table 3 reveals that the majority of respondents expressed the intention to purchase both the "brand identity colors" and the "cause-related colors," with approximately 70% and 65% indicating "yes" for purchase

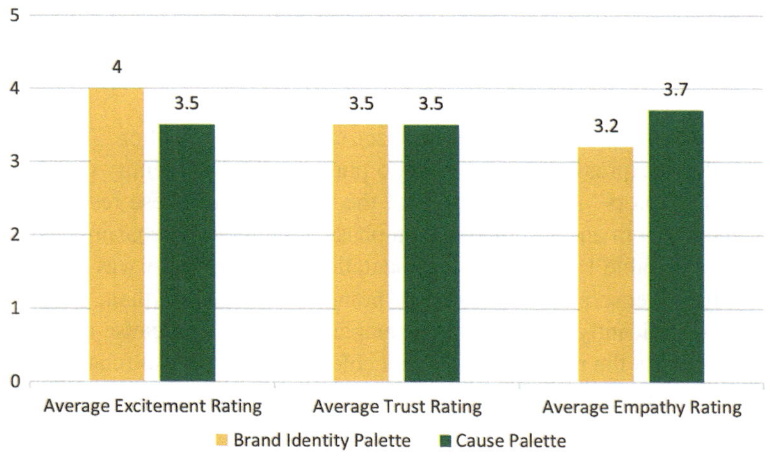

Fig. 4 Emotional engagement ratings (Excitement, Trust, and Empathy)

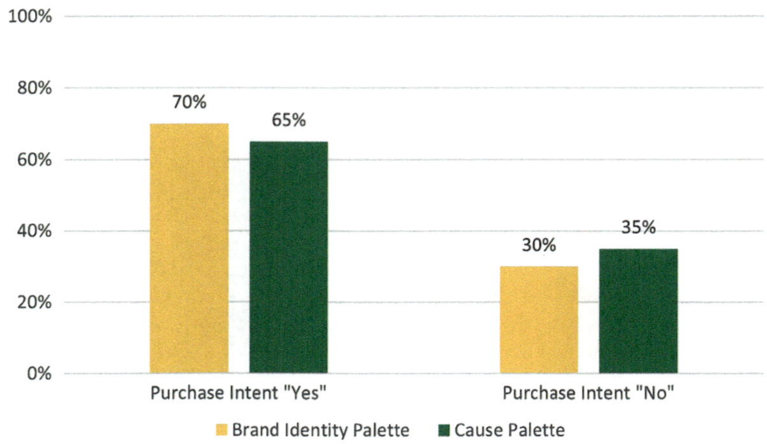

Fig. 5 Purchase intent

intention, respectively, thus causing both to be taken into consideration in purchasing decisions the effect is highlighted (see Fig. 5).

The results of this study provide valuable insight into the complex relationship between color and consumer behavior. They emphasize the importance of color choice in marketing strategies, especially in the context of brand recognition and cause-related marketing.

"Brand identity colors" excel in initial visual appeal, typically scoring 4.1. But "cause-related color" is no slouch, with an average visual pleasure rating of around 3.7. This indicates that a well-established brand color scheme initially has an advantage in terms of tangible impact, but cause-related colors are also appealing to respondents in terms of emotional engagement. "Brand identity colors" generate more excitement, with an average rating of 4.0, indicating that they can evoke positive emotions and happiness, while "cause-related colors" tend to exceed empathy, with an average rating of about 3.7. This indicates that the palette of causes resonates more and is more sympathetic among respondents.

Moreover, both have a significant impact on purchase intention. About 70% of the respondents expressed the intention to purchase "brand identity colors," while "cause-related colors" influenced 65% of the respondents. These results highlight the influence of both and stimulate the emphasis on both in purchasing decisions.

As shown in Table 1, emotional representation trustworthiness was found to have significantly higher scores ($F = 6.44$) for "brand identity colors" than "brand-related colors" and significantly higher scores for empathy ($F = 6.78$) in cause-related colors. The findings reject the null hypothesis that color choices in cause-related marketing have little effect on consumer mood, emotional involvement, or purchase intention.

The findings highlight the complexity of color selection in marketing. While the basic visual appeal of a brand's established colors may be strong, cause-related colors excel in eliciting empathy and continue to make a big impact on purchase.

4 Conclusion

In this study, we examined the effectiveness of color selection in cause-related marketing campaigns, using McDonald's as a case study. The findings show that when assessing visual appearance, brand identity colors received slightly higher average scores compared to cause-related colors. This suggests that colors associated with brand recognition are more visually appealing to consumers. However, emotional engagement showed differences in emotional domains. The cause-related colors scored highly on trustworthiness, indicating that colors matching a meaning can evoke a sense of trust and confidence. On the other hand, the brand identity colors scored higher on appeal, indicating that they can elicit a more evocative and engaging emotional response. Interestingly, empathy ratings were quite similar between the two palettes.

An interesting finding of this study is the effect of colors on purchase intention. The results showed that the choice of colors, whether symbolic or semantic, had little effect on purchase intention. That means that, while color can have a visual impact and evoke emotions, it can't be the determining factor in purchasing decisions when it comes to cause-related marketing.

This study emphasizes the importance of color in marketing and the need for marketers to understand the subtle relationship between color and consumer behavior. It provides a foundation for future research and practical applications in dynamic marketing. Brands should consider the aesthetic and emotional impact of color when developing their marketing strategies. The study encourages marketers to carefully select colors that are associated with their brand and causes. Effective presentations should use color to create visual appeal and emotional engagement, fostering a stronger connection with customers. Overall, this study encourages further exploration of the strategic use of color in marketing, considering not only visual aesthetics but also emotional and psychological aspects. Future research can provide valuable insights into explicit and emotionally appealing marketing strategies through the examination of the complex relationship between color and consumer behavior.

References

1. Hunjet A, Vuk S (2017) The psychological impact of colors in marketing. Int J Vallis Aurea 3(2):42–54
2. Eiseman L (2000) Guide to communicating with color. Grafix Press, U.S.A
3. Sattar MM, Barr JC, Sfodera F (2023) Cause-related marketing. In Singh A, Sharma S, Singh A, Unanoglu M, Taneja S (eds.), Cultural marketing and metaverse for consumer engagement, pp 71–95. IGI Global. https://doi.org/10.4018/978-1-6684-8312-1.ch006
4. Samu S, Wymer W (2014) Cause marketing communications: consumer inference on attitudes towards brand and cause. Eur J Mark 48(7/8):1333–1353
5. Brønn PS, Vrioni AB (2001) Corporate social responsibility and cause-related marketing: an overview. Int J Advert 20(2):207–222

6. Sliburyte L, Skeryte I (2014) What we know about consumers' colour perception. Procedia Soc Behav Sci 156:468–472
7. Lim D, Baek TH, Yoon S, Kim Y (2020) Colour effects in green advertising. Int J Consumer Stud 44(6):552–562. Chicago

Exploring the Impact of Additive Manufacturing on Personalized Footwear

Avik Sarkar and Sagar Ghuge

1 Introduction

Additive Manufacturing (AM), also called 3D printing, is one of the technological breakthroughs becoming a game changer due to its unique layer-by-layer manufacturing approach from digital design files to create a tangible object without tooling and product-specific setup [1]. AM offers several key benefits compared to conventional manufacturing (CM), including low lead time, cost, unparalleled product customization, on-demand manufacturing, no constraint of economies of scale, and reduced time and cost, especially for products with small volumes and complex geometry [2, 3]. Because of this, AM technology has surged in popularity, finding applications across various industries, such as aerospace, automotive, robotics, food production. A notable application is the custom manufacturing of footwear, enabling rapid prototyping, novel designs, and greater personalization. Despite its prominent potential in this domain, AM is still in the early stage of delivering personalized footwear according to requirements [4, 5].

As a consequence of the increasing prevalence of sedentary lifestyles and poor health habits, more people are experiencing health problems that result in higher body mass indexes (BMIs). This often leads to foot issues, such as ulcers and pain, particularly during prolonged periods of physical activity like standing and walking

A. Sarkar
Department of Mechanical Engineering, Swami Vivekananda School of Diploma, Malandighi, Durgapur, West Bengal, India

S. Ghuge (✉)
Department of Operations and Decision Science, T A Pai Management Institute, Manipal Academy of Higher Education, Manipal, Karnataka, India
e-mail: sagar.ghuge@manipal.edu

A. Sarkar
Department of Mechanical Engineering, yadavrao Tasgaonkar College of Engineering and Management, Karjat, Raigad, India

© The Author(s), under exclusive license to Springer Nature Singapore Pte Ltd. 2025
S. Das et al. (eds.), *Human-Centered Digitalization, Volume 3*, Design Science and Innovation, https://doi.org/10.1007/978-981-96-4383-7_32

[6]. This has prompted healthcare experts to emphasize the importance of comfortable footwear that can adapt to daily activities and aid in the rehabilitation of lower limbs. Given the foot structure variances across individuals, the custom-made 3D-printed soles emerge as a promising solution. These soles and shoes, engineered using 3D scans of an individual's foot, cater specifically to the user's plantar structure. This level of customization goes beyond merely achieving a good fit; it allows the insoles to be tailored to specific activities, such as walking, running, or jumping. Achieving such a high level of complexity and precision with increased flexibility in customized footwear is a challenge when using CM. Thus, AM is extensively employed in personalized footwear due to its economic attractiveness with state-of-the-art technical possibilities, offering supply chain agility while attaining sustainability [6, 7]. Thus, the exploration of AM for personalized footwear is gaining traction and has caught the attention of practitioners and researchers. To the best of the authors' knowledge, there is a scarcity of literature review that contributes to the domain AM in personalized footwear. Therefore, the present research work is carried out to fill this void in the existing body of literature and offer an overview of AM in the personalized footwear industry.

2 Objective and Methodology

This research aims to demonstrate an extensive overview of the available scientific literature on AM in the personalized footwear domain. To attain this, the authors conducted an in-depth literature review of scientific publications and white papers offered by Industry leaders. The review is formed on a rigorous, reproducible, scientific, and transparent process that enables the discovery of essential trends related to the chosen study topic. Applying systematic review principles will, therefore, assist in eliminating mistakes, lessen random effects, enhance the legitimacy of the resultant evidence, and give more reliable outcomes in drawing conclusions and offering future research directions.

2.1 Carrying Out the Literature Review

At the outset, the authors compiled the list of scientific publications gathered through the WoS and Scopus databases published in English. The authors utilized the keywords "3D print," or "Additive Manufacturing," and "customized footwear," or "personalized footwear," along with their synonyms, which are included in the search query of Wos and Scopus database. While examining WoS resources, the authors only found 10 articles. Furthermore, the Scopus database yields 2 results. In the next stage, the authors eliminated duplicate articles by complying with the WoS and Scopus outcomes, restraining the final count to 10 Nos. In the third stage, the authors read the abstract to exclude the publications which are not pertinent to the specified

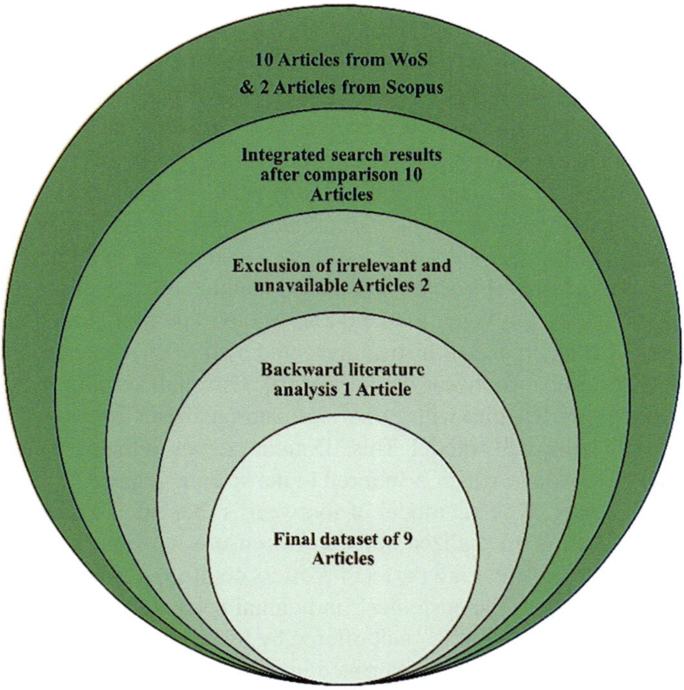

Fig. 1 Systematic literature review process

research area. In the fourth stage, a full assessment is carried out, which results in qualifying all eight articles. Further, in the fifth stage, one article is captured through backward literature analysis, resulting in the final count of the research article to 9 Nos. Further, the scarcity of scientific literature in the field motivates the authors to explore the white papers, market research reports and allied documents to understand the AM in personalized footwear from an industrial context (Fig. 1).

3 Content Analysis

The content analysis of the systematic literature review is classified into four segments: Evolution of personalized footwear, Type AM technologies and materials and Recent Landscape, Role of AM in mass customization of footwear, and Personalized footwear industry outlook through the lens of AM technology.

3.1 Evolution of Personalized Footwear

In today's world, the purpose of footwear has evolved beyond simply providing protection for the feet. Shoes now serve various functions, such as symbolizing elegance for women, facilitating horseback riding for men, concealing imperfections, and reflecting one's social status. Unlike in the past, where there was no distinction between the design of the right and left shoe until the nineteenth century, the current market and its customers demand more personalized and unique footwear [8]. AM technology has evolved as an alternative manufacturing technology for mass customizations due to the benefits of no-constrained complexity and economical attractiveness. This emergence is also underscored by the 0.30% contribution of AM technology in the global footwear market [9, 10]. Personalized foot manufacturing involves several steps. It begins with gathering anatomical data from a customer's feet and transforming it into a 3D model. This 3D model serves as the basis for designing the 3D model of footwear, which is tailored to the specific requirements and preferences of the customer. This 3D model of footwear is then subjected to simulations and various tests using specialized software to ensure its robustness [10]. These steps ensure that the footwear is a perfect fit, offers comfort, and provides the necessary support according to the customer's individual needs and preferences. A very renowned example is the "Phits™" suit offered by Materialize©, which assesses the dynamic data of users to develop 3D printed insoles [11].

3.2 Type AM Technologies and Materials

There are seven distinct types of AM technologies, but it's important to note that not all of these technologies are suitable for producing personalized footwear. Various constraints, such as material compatibility, accuracy, and size, play a significant role in determining their suitability [1]. To provide an overview of this field, the authors have compiled a summary in Table 1 that focuses on the most commonly used AM technologies and materials for personalized footwear. For instance, powder bed fusion (PBF) stands out as one of the most extensively employed AM technologies in the manufacturing of footwear, particularly for mass customization. It plays a pivotal role in producing a significant portion, often ranging from 80 to 85%, of the final components within an advanced and integrated business model [10]. In terms of materials, personalized footwear companies predominantly rely on thermoplastic polymers and their various iterations. These materials are favored for their durability, strength, and flexibility [14]. Notably, water-soluble polyvinyl alcohol (PVA), acrylonitrile butadiene styrene (ABS), polylactic acid (PLA), and polycarbonate (PC) are also among the top choices in the footwear industry for achieving personalization through AM [12].

Table 1 AM technologies and materials used for personalized footwear [12–16]

Process categories	AM technologies	Materials
Material extrusion	FDM, FFF	Thermoplastics filament (TPU, TPR, TPE, ABS, PLA, PC, Nylon)
Vat photopolymerization	SLA, DLP	Photosensitive resins (TPU, TPR, TPE, Nylon, Silicone)
Powder bed fusion (Polymers)	SLS	Polymer and composite powder (TPU, TPR, TPE, Nylon, Carbon Composite)
Material jetting	Material jetting DOD (Drop on demand	Photopolymer resins (TPU, TPR, TPE, Nylon)
Binder jetting	Binder jetting	Polymer powder
Haptic coating	Haptic coating	Polyurethane dispersion-PUD resin
Special process	DLS	Photosensitive resins (TPU)
Special process	RLP, LP	BASF liquid polyurethane

* SLA–Stereolithography, SLS–Selective Laser Sintering, FDM–Fused Deposition Modelling, FFF–Fused Filament Fabrication, MJF–MultiJet Fusion 3D Printing, RLP–Rapid Liquid Printing, LD–Liquid Drawing (Pneumatic Material Extrusion), DLS–Digital Light Synthesis

3.3 *Role of AM in Mass Customization of Footwear*

Mass customization will play a significant role in the forthcoming footwear industry, mainly driven by 3D printing capable of quickly and precisely acquiring everyone's distinctive geometric and anatomy data. AM technology utilizes individual footwear 3D scanning equipment and 3D printers, which can be accessed physically or online via web portals or applications which can be easily integrated through mobile devices [10]. These technologies are in the initial research stage and will evolve over the years. For example, SmarTech anticipates that the footwear-specific 3D scanning equipment market will bolster customized product digitization and will generate over $130 million in annual sales by 2029 [10]. The advent of 3D footwear printing techniques holds immense promise for enhancing the functional capabilities and traditional production processes of customized foot orthotics and personalized shoes for an individual. It is considered a significant technological advancement in this field. To illustrate, each HP Multi Jet 3D printer can produce up to 21,000 pairs of foot orthotics or personalized shoes annually. This translates to saving over 40 tons of material waste per year when compared to conventional CNC polypropylene milling [17]. Consequently, the adoption of such innovative technology stands to accelerate market growth substantially and cost-effectively. In 2018, the worldwide footwear business earned over $260 billion in revenue [9].

The market is mature and increasing slowly since hundreds of millions of people worldwide, particularly in emerging nations, can now buy high-quality footwear. While customized 3D-printed footwear is now most typically outsourced to service bureaus for prototyping and production services, SmarTech anticipates that the bulk of AM usage in the footwear sector will take place in-house over time [9]. The inclination toward broad use of AM in the footwear sector is linked to a macro trend of growing personalization of consumer items, which can even now be observed in everything from automobiles to jewelry. Currently, most customers prefer mass-produced brand products more than custom-made ones, which implies that product customization is still in its early stages. However, younger generations are growing to favor custom-made products, particularly as they become more widely available and enhance comfort. Simultaneously, AM is driving further mass customization in the footwear business by digitizing the traditional footwear manufacturing process. Thus, the AM mass customization era is evolving with the convergence of personalization and digital automation. Footwear is predicted to be one of the first consumer product groups to undergo this transformation.

3.4 *Personalized Footwear Industry Outlook Through the Lens of AM Technology*

Many organizations are embracing AM to produce personalized footwear, gaining a competitive edge in the market. They are attracted to the advantages of rapid prototyping and customization, all at a significantly lower cost compared to CM. Global footwear giants like Nike, Adidas, and Reebok have been at the forefront of 3D printing technology for at least five years [7].

These companies are not only using AM to manufacture entire shoes but also individual components of footwear. Figure 2 provides an overview of these footwear components. Additionally, the authors have listed these components along with the companies utilizing AM technology to produce them in Table 2.

Let's take a closer look at some noteworthy examples. For instance, New Balance has team up with 3D Systems to create 3D-printed midsoles for running shoes. These midsoles are flexible, offering adequate support. They are made using SLS technology with DuraForm Flex [18], a thermoplastic elastomer material. This material

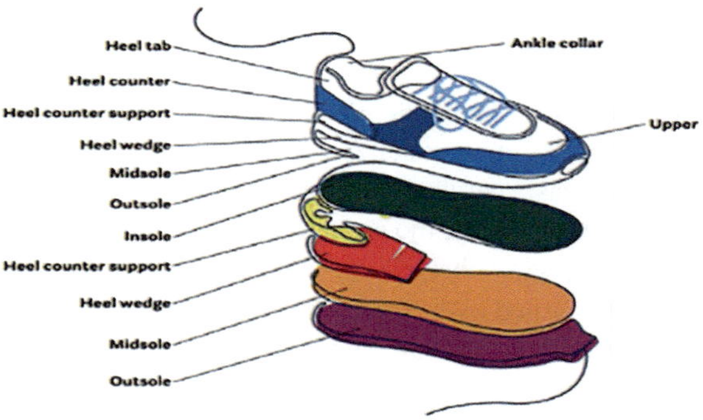

Fig. 2 Footwear components

Table 2 Comprehensive mapping of available personalized footwear parts producible though AM technologies [12–16]

Footwear	Footwear parts and applications	Suited for	Technology	Materials
Zeller feld	Entire shoe	Customized and sustainable comfort on a budget	FDM, FFF	TPU filament
Dior derby	Entire shoe (Robust but weight reduced)	High fashion meets 3D innovation	SLS	TPU powder
Botter x reebok	Entire shoe	A blend of high-end design and practicality	SLS	TPU powder
Nike fly print	Entire shoe, mass-customized uppers, custom football cleats	The marathoner's dream shoe	FDM, SLS	TPU & nylon filament, TPU & Nylon powder
Adidas x Carbon 3D	Lattice mass produced midsoles	The future of performance footwear	DLS	Photosensitive resins (TPU)
Reebok liquid series x BASF	Weight reduced and high-rebound outsoles and lacing	Revolutionizing shoe design with liquid laces	RLP, LD	BASF liquid polyurethane
New Balance X 3D Systems	Flexible and shock absorbent midsoles, Limited edition midsoles, customized spike plates	Pushing boundaries with rebound resin	SLS, SLA	TPU & Nylon powder, photosensitive resins (TPU & Nylon)
FUSED footwear	Entire shoe (Recyclable)	Eco-conscious Footwear for the Future	FDM	TPE filament
Under Armor	Limited edition midsoles	Advanced cushioning and weight-saving features	SLS	TPU Powder
Ecco	Highly customized midsoles	Aimed at increasing the wearer's comfort	SLA	Photosensitive resins (Silicone)
Hilos	Limited edition outsoles and heel platform (Recyclable)	The Sustainable 3D Printed Shoe	SLS	TPU Powder
Athos	Highly customized shoes	3D Printed Climbing Shoes	MJF	TPU

(continued)

Table 2 (continued)

Footwear	Footwear parts and applications	Suited for	Technology	Materials
The Parametriks Print 001	Entire Shoe (Parametric Design)	A shoe that stands out from the crowd for its comfort	SLA	Photosensitive resins (TPU & Nylon)
Pleko	Insole, outsole, spikes, and ribs	The Carbon Fiber Shoe for Running	SLS	Carbon Fiber Composite Powder

* SLA–Stereolithography, SLS–Selective Laser Sintering, FDM–Fused Deposition Modelling, FFF–Fused Filament Fabrication, MJF–MultiJet Fusion 3D Printing, RLP–Rapid Liquid Printing, LD–Liquid Drawing (Pneumatic Material Extrusion), DLS–Digital Light Synthesis

combines strength and flexibility to provide exceptional comfort for sports shoes. AM technology facilitates the creation of intricate structures that enhance shock absorption and reduce the footwear's weight. Subsequently, New Balance partnered with Formlabs to leverage their stereolithography technology to develop the Triple Cell platform for the FuelCell Echo. Another example is ECCO, which introduced its Quant-U service in 2019, allowing customers to personalize their shoes through 3D printing. This service employs a 3D scanning procedure to determine the orthopaedic requirements of each individual's feet. ECCO utilizes this technology to craft 3D-printed insoles tailored to its customers' specific needs. These insoles are made of silicone, a material that offers both stability and cushioning. Additionally, ECCO claims that these insoles are easily replaceable and can be cleaned [19]. According to the recent report of SmartTech [10], the worldwide footwear business is expected to grow $400 billion by 2029, in which the contribution of AM technology will be 1.5%, which is whooping $5.9 billion compared to the recent contribution of 0.3%.

4 Discussion and Future Research Direction

The personalized footwear industry is known for its focus on catering to the unique needs and preferences of individual customers through customization. In recent years, this sector has experienced substantial growth and popularity, driven by cutting-edge technologies like AM, which allows for the creation of high-quality, one-of-a-kind products that stand out in the market. AM is a groundbreaking manufacturing method that is disrupting the industry due to its unlimited manufacturing possibilities, rapid customization, freedom from constraints related to economies of scale, on-demand production, and sustainability.

Key characteristics of the personalized footwear industry include mass customization, make-to-order production, pricing considerations, a strong emphasis on achieving a perfect fit and maximum comfort through personalization, diverse applications ranging from sports and orthopedics to military and fashion, brand differentiation, a culture of continuous innovation, and a growing focus on sustainable practices.

Major players in the industry, such as Nike, Adidas, and Reebok, have widely adopted AM technology to diversify their product offerings and offer tailored solutions that cater to individual customer preferences. Notable AM technologies used in the industry include vat polymerization, filament extrusion, material jetting and powder bed fusion. In 2018, AM accounted for only 0.3% of the total footwear market, but this is projected to increase significantly to 1.5%, equivalent to $5.9 billion by 2029. Furthermore, the revenue generated by AM in the footwear industry is experiencing an impressive year-on-year growth rate of 19.5%. This signifies that the contribution of the AM industry to the personalized footwear sector is on track to represent around 30% of the total AM revenue.

However, there is a notable research gap when it comes to selecting the most suitable AM technology for personalized footwear according to requirements. Currently, there is a lack of comprehensive understanding of the criteria for selecting AM technology and an assessment of the technical capabilities of available options with regard to specific customer preferences and needs. The absence of relevant criteria or attributes and a methodical, data-centric approach for technology selection can lead to the oversight of crucial factors and result in wasted efforts. Unfortunately, there is scarcity of research in this domain. Therefore, this systematic literature review is valuable and specifically, it has set the stage for identifying missing links and promising avenues for future research. These future research scopes are outlined beneath:

Research Direction 1: Evaluating AM technology for personalized footwear when data is limited.

Research Direction 2: Developing a framework for selecting the appropriate AM technology for personalized footwear.

Research Direction 3: Understanding the characteristics of personalized footwear that are well-suited for AM technology.

Research Direction 4: Exploring the impact of using design for AM in creating personalized footwear.

Research Direction 5: Investigating how AM affects the modularity and integrity of personalized footwear.

5 Conclusion

The review aims to provide an overview of AM technology for the personalization of footwear and to create the foundation that researchers and practitioners can utilize to develop the framework for identifying compatible AM technology according to customer needs and preferences. The authors carried out the systematic literature review with the following specific aims. At the outset, the author demonstrates the evolution of personalized footwear to understand its current status. Second, to catalog the various AM technologies and materials employed in creating personalized footwear, which can serve as valuable information for choosing the right AM technology process in line with customer preferences. Third, this review outlines the steps involved in achieving mass customization through AM and the growing trend toward using AM in the realm of personalized footwear. Lastly, the outlook on personalized footwear provides an in-depth assessment of AM technology and its real-world applications in the context of mass customization.

In line with these objectives, this review makes two significant contributions. Firstly, it reveals that there is limited research available on the selection of suitable AM technologies for personalized footwear. Furthermore, the feasibility of personalized footwear depends on the level of customization and the potential for redesign based on user applications, which is influenced by technical compatibility and economic viability. Second, the review identifies five noteworthy areas for future research opportunities within the emerging field of AM in personalized footwear.

References

1. Ghuge S, Dohale V, Akarte M (2022) Spare part segmentation for additive manufacturing—a framework. Comput Ind Eng 169:108277. https://doi.org/10.1016/j.cie.2022.108277
2. Ghuge S, Akarte M (2024) Additive manufacturing service bureau selection: a Bayesian network integrated framework. Int J Prod Econ 276:109348. https://doi.org/10.1016/j.ijpe.2024.109348
3. Ghuge S, Akarte M, Raut R (2024) Decision-making frameworks in additive manufacturing management: mapping present landscape and establishing future research avenues. Benchmarking An Int J. https://doi.org/10.1108/BIJ-12-2023-0845
4. Ghuge S, Parhi S (2023) Additive manufacturing service provider selection using a neutrosophic best worst method. Procedia Comput Sci 217:1550–1559. https://doi.org/10.1016/j.procs.2022.12.355
5. Ghuge S, Akarte M, Pandey A (2022) Determining and validating the spare parts selection criteria for additive manufacturing using Delphi technique. In: 2022 IEEE International conference on industrial engineering and engineering management (IEEM). IEEE, pp 1247–1251
6. Gong T, Kang L (2021) Application Analysis of 3D Printing Technology in Design Field: Taking Shoe Design as an Example. Sci Program 2021:1–8. https://doi.org/10.1155/2021/5662460
7. Ukobitz D, Faullant R (2021) Leveraging 3D printing technologies: the case of Mexico's footwear industry. Res Manag 64:20–30. https://doi.org/10.1080/08956308.2021.1864919
8. Stoica M, Crisan N, Cucu L, Prisecaru DA (2022) Study of customized plastic high heels footwear. Mater Plast 59:252–260. https://doi.org/10.37358/MP.22.1.5576
9. Additive manufacturing research previously known as smartech analysis (2019) 3D-printed footwear 2019–2029, an analysis of the market potential of 3D printing in the footwear industry
10. Smartech analysis (2019) 3D printing in footwear
11. Materialise phits: The world's first dynamic 3D-printed orthotic
12. Sher D (2020) Five footwear industry leaders using 3D printing for production today. In: voxelmatters
13. Schwaar C (2023) Off to the Laces! 3D printed shoes: the cream of the crop in 2023. In: all3dp.com

14. Printermods 3D printed shoes in 2020: big brands are on board
15. 3D Sourced (2023) best 3D printed shoes in 2023 (Sneakers, Heels & More)
16. Carlota V (2022) 3D printing news rankings 3D printed shoes: what's available on the market today? In: 3d natives
17. Verified market research (2023) Insoles market size and forecast. In: VMR Sept 2023. https://www.verifiedmarketresearch.com/product/insoles-market/
18. Adrienne SO (2019) New balance's latest shoes come with 3D-printed soles. In: GEAR. https://www.wired.com/story/new-balance-triplecell-3d-printed-shoe/
19. Wheatley T (2019) New balance launches 3D printed shoe platform. In: GET run. https://www.getsweatgo.com/running/reviews/new-balance-launches-3d-printed-shoes-platform-95

Workspace Ergonomics: An Integrated Approach Toward Sustainable Sitting and Seating

Avik Sarkar, Chetan Deshmukh, and Sagar Ghuge

1 Introduction

The inevitable popularity of paperless offices in India under the context of the fastest-growing IT sector across the globe is increasing the number of individuals who use sitting positions as their primary and prolonged office posture. However, this practice comes with several downsides due to the inadequate implementation of proper seating biomechanics, ergonomics, and dimensional analysis, along with a lack of anthropometric data for a generic population [1]. However, using standard ergonomic chairs has several drawbacks. As a result, the analysis of sitting posture and the design of working seats based on the ergonomics idea for prolonged sitting with posture correction have steadily gained attention from academics and designers [2]. This article focuses on examining and validating the construction and precise dimensions of a dual-purpose ergonomic office chair, particularly emphasizing Anthropometrics and Ergonomics for extended sitting with posture correction within the unique context of anthropometric data for the entire Indian population.

A. Sarkar (✉)
Department of Mechanical Engineering (Product Design and Development), Yadavrao Tasgaonkar College of Engineering and Management, Karjat, Raigad, Maharashtra, India
e-mail: 989ovik@gmail.com

Department of Mechanical Engineering, Swami Vivekananda School of Diploma, Malandighi, Durgapur, West Bengal, India

C. Deshmukh
Department of Mechanical Engineering, Terna Engineering College, Nerul, Navi Mumbai, Maharashtra, India

S. Ghuge
Department of Operations and Decision Science, T A Pai Management Institute, Manipal Academy of Higher Education, Manipal, Karnataka, India

© The Author(s), under exclusive license to Springer Nature Singapore Pte Ltd. 2025
S. Das et al. (eds.), *Human-Centered Digitalization, Volume 3*, Design Science and Innovation, https://doi.org/10.1007/978-981-96-4383-7_33

Fig. 1 Poor sitting posture after prolonged sitting [3]

2 The Lack of a Generic and Ergonomic Office Chair

- The backrest of most of the common office chairs adequately supports the waist, where users must consciously align their back and waist while seated. However, owing to prolonged sitting and attention to an electronic screen, individuals often unconsciously lean forward because of hazy vision and waist tiredness. This leads to poor sitting posture, which in turn causes further waist and back strain, according to Fig. 1 (marked in red color).
- People might become bored and operate less effectively if they maintain the same position for an extended period.
- When people need to bend forward to draw or write, the angle created by the trunk and thigh narrows, which raises pressure on the pelvic and abdomen regions of the body. The upper body leans forward at the same moment, which wears out the neck.
- Heart disease, diabetes, hypertension, depression, and anxiety are just a few of the ailments that are more common in those who lead sedentary lifestyles [4].

3 Introduction to Usage Scenarios and Structure

The different types of seating angles must be swapped to maintain a stabilized upper body since kneeling restricts the movement of the lower legs. As a result, the switching device's controller is located beneath the seat. An attachment in the form of a rotating mechanism to the seat post tilts the seat surface forward [5].

When assuming a kneeling position, the knee rest is very crucial. To offer enough support for the knees and lower legs, the knee supporter should be at right angles to the Femur (thigh bones). The angle and height of the supported knee can be adjusted simultaneously; please see Fig. 2. Providing sufficient provision and movement space for the knee during the kneeling posture, the column of the knee rest is connected to the chair foot and the column, and the base plate of the knee rest is optimally designed so that it does not affect the leg space of the legs while sitting [5].

Fig. 2 Seat and knee rest adjusting mechanism [6]

4 Ergonomics

4.1 Force Analysis of the Lower Limb

Analyzing the physiological composition of the human body, the fibula and tibia collectively form the human calf. According to the figure, the tibia is a moderately big bone that is situated in the front of the lower leg. When knelt, the lower leg's major stress point is the top section of the tibia.

According to Fig. 3, the majority of the calf's nerves and blood arteries are found in the muscles on the rear of the calf. The tibia, i.e., shinbone (frontal part of the calf muscle), bears a specific amount of pressure when the contacting area for the calf to the knee support is broad and soft, without placing excessive strain on nerves and blood vessels, which would cause poor circulation of blood and impassiveness of calf.

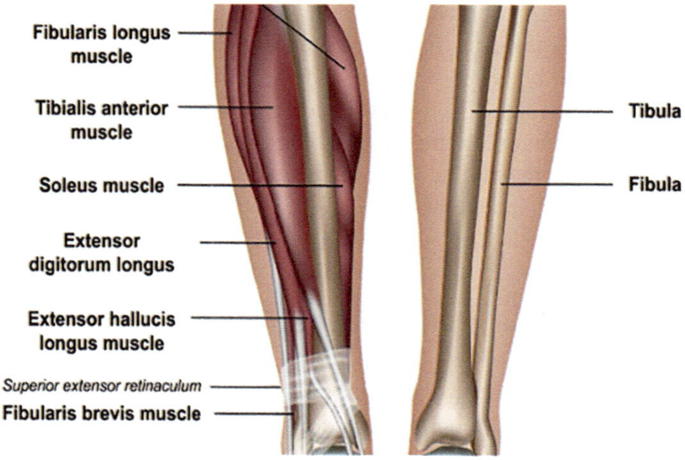

Fig. 3 Lower leg anatomy [7]

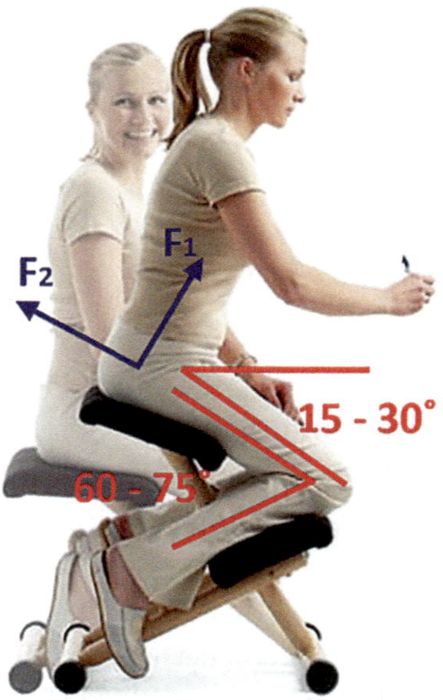

Fig. 4 Force analysis in kneeling position [6]

4.2 *Balance Analysis*

The stability of humans, while seated, is one of the fundamental needs for seat design as depicted in Fig. 4. The upper limb is sustained by forces F1 along the surface of the seat to the hip area and F2 along the thigh area. The two forces are acting upwards along the direction of the upper body [8].

The seating posture, along with the tilting seating pan, can offer some support when the individual leans back, but simultaneously, pressure on the quadriceps will grow, and the human body tends to curve down along the surface for a forward-leaning position. When someone leans forward, their center of gravity shifts forward, increasing the angle involving the thigh and body, and decreasing the torque from the thigh to body. This means that to maintain the body's balance, the person cannot lean forward too much, and they must keep their upper body upright [8].

4.3 *Analysis of Spine Curvature in a Kneeling Position*

Standing posture is more likely to result in waist muscular exhaustion than sitting, according to the electromyographic examination of the waist in these two situations. Since the majority of the weight of the body is supported by the feet, knees, and thighs in standing posture, lumbar muscle fatigue is challenging to develop. The waist is in the finest shape right now [9]. Similar to standing, when kneeling, some of the weight is supported by the calf, which relieves strain on the waist.

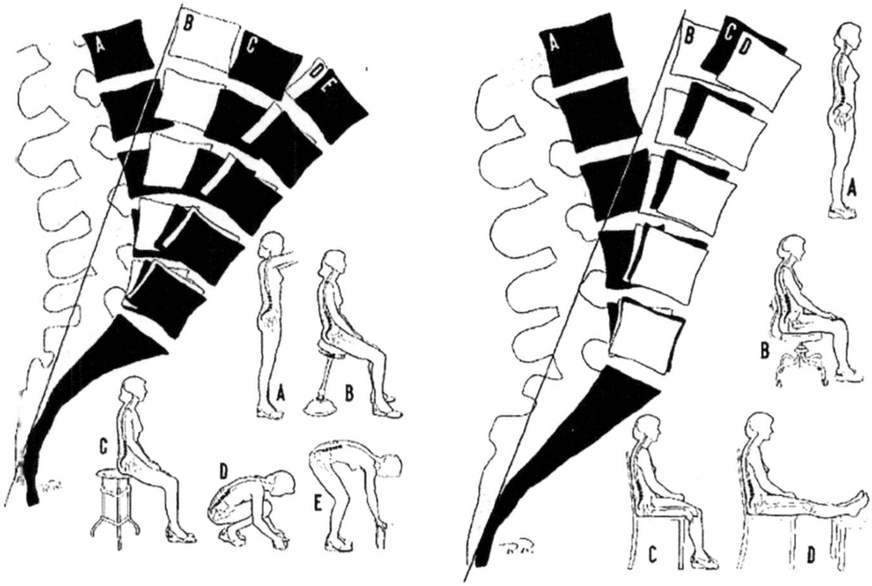

Fig. 5 Alignment of the spine in sitting and during various activities [10]

Slightly tilting the forward seat surface reduces kyphosis compared to a slightly tilting back seat. A slightly tilting forward seat tends to reduce kyphosis more than a slightly tilting rear seat [10]. A similar conclusion can be made from Fig. 5. The degree of the lumbar posterior is inversely proportional to the angle of the truck between the thigh, and the process becomes more nominal.

5 Selection and Analysis of Anthropometrics for Better Ergonomic Dimensions

Since this seat's kneeling position more closely resembles that of standing, the spine curvature remains more normal [9]. Methodically, the basic statistical anthropometric source for the design of this ergonomic chair is Indian Anthropometric Dimensions for Ergonomic Design Practice by the National Institute of Design as the research is primitively focused on the Indian Anthropometrics and Ergonomics context. The primary parametric considerable dataset is shown in bold in Table 1 [10, 11].

Table 1 Sitting posture dimensions [10, 11]

Percentile		Male (mm)						Female (mm)					
Anthropometric nomenclature	Chair dimensions	5th	25th	50th	75th	95th	Mean	5th	25th	50th	75th	95th	Mean
Thigh clearance height with raised knee	Seat height (kneeling)	530	564	587	610	**644**	588	500	525	540	555	596	542
Popliteal length	Seat height (normal sitting)	380	406	425	443	**471**	426	365	386	399	417	441	399
Buttock to knee length (normal sitting)	The gap between the seat and knee support	489	529	558	581	615	**556**	459	503	526	553	585	527
Buttock to popliteal length (normal sitting)	Seat depth	399	431	455	477	**512**	456	384	417	439	468	494	442
Buttock to leg length (normal sitting)	Overall depth (normal sitting)	640	654	719	779	**779**	669	540	579	639	729	759	651
Buttock to leg length (raised on toe)	Overall depth (kneeling)	569	599	659	699	**769**	658	559	639	669	689	719	655
Knee-to-knee (relaxed)	Knee rest width	349	409	449	484	**559**	455	259	329	394	439	549	394
Hip breadth	Seat width	272	309	331	355	405	334	259	296	314	341	**429**	323

5.1 Seat Height

To afford the lower portion of legs ample space over the surface of the seat while implementing the kneeling posture, the kneeling position seat is a tiny bit taller than the seated position. Therefore, the maximum height of the seat is 644 mm for kneeling posture and 471 mm for normal sitting. As a reference, the 95th percentile of the thigh clearance height with raised knee and popliteal length for males is used. Furthermore, the gap between seat and knee support is considered to be 556 mm to accommodate most of the users' population by taking into account the mean value of men's buttock to knee length for normal sitting posture.

5.2 Chair Width

The breadth of the hips, while seated, serves as the primary benchmark for chair surface width. To lessen the strain brought on by tight seats, the chair's width should be able to guarantee that individuals have sufficient space to relocate about as well as take into account the amount of clothing adjustment and psychological adjustment. The chair's width is 429 mm on the basis of the 95th percentile for hip breadth for females in seating posture. Additionally, knee rest width is 559 mm based on the 95th percentile of knee-to-knee distance in a relaxed posture for males.

5.3 Sitting Depth

Considering sitting posture, the prime stresses of the human body are centric to the back of the thighs and the hips; considering the 5th percentile for female buttock to popliteal length for normal sitting posture as a reference, a seat depth of 512 mm can provide corrective support to the lumbar spine and the hips. Simultaneously, an overall depth of 779 mm can provide enough space for movement at the knees, based on the 95th percentile of buttock to leg length in normal sitting as well as raised toe posture for males.

5.4 Sitting Angle

During the experimentation of varying seat and knee angles for prolonged sitting, Dr. A. C. Mandal found that the lumbar support loses effectiveness when the sitter moves forward [12]. Furthermore, he concluded that the sitting posture is at its best when the seat angle is at $+ 15°$ and the knees at $70°$ [13].Thus, analyzing the graph

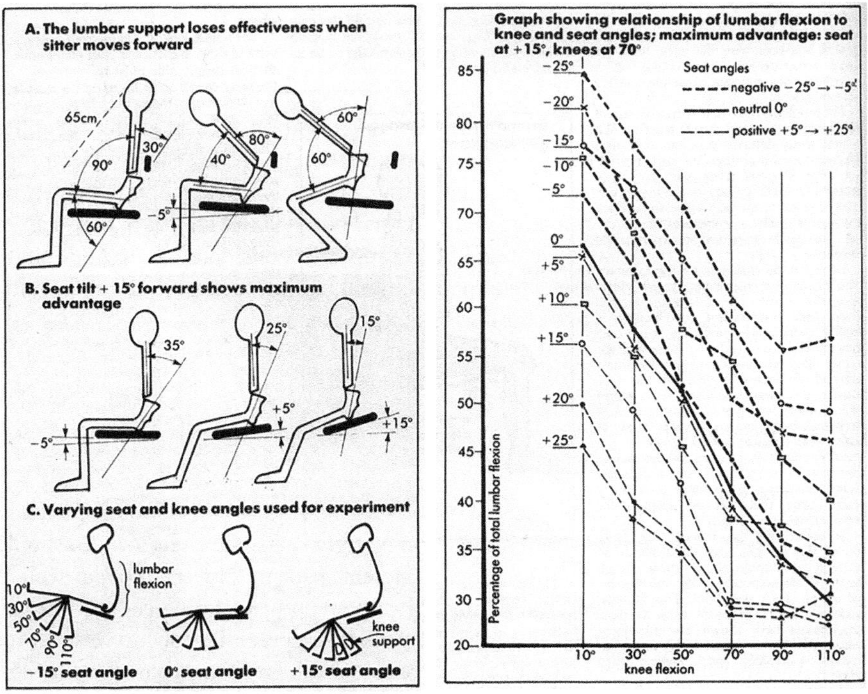

Fig. 6 Sitting angle guide [16]

showing the relationship of lumber flexion to knee and seat angle, the range of 15–30° for seat angle and the range of 60–75° find the best relevance to help the human body's spine in a sitting posture to be closer to the natural state[14, 15] (Fig. 6).

5.5 *Material of Chair Surface*

Kneeling requires increasing the person's friction with the seat surface while maintaining balance. The friction coefficient of leather is the highest [17] when compared to popular chair upholstery covering materials such as leather, nylon cloth, and textile fabric. As a result, the upholstery of the seat for the chair will be made of leather.

6 Conclusion

This research focuses on the Anthropometrics and Ergonomics of the ergonomic office chair for both kneeling and sitting posture in the novel context of dimensional anthropometric data for the entire Indian population by analyzing the optimal aspect of a healthy and prolonged sitting posture. This ergonomic office chair offers superior ergonomics compared to the traditional seat, demonstrating a novel kind of human–machine interaction and posture correction for prolonged sitting. However, there are still a lot of flaws, and more advancements are required in areas like aesthetics, features, construction, materials, and modularity in the near future. To sum up, this article caters to a basic but novel aspect of ergonomics and anthropometric data for the entire Indian population through dimensional analysis of the ergonomic office chair and to make sustainable sitting and seating highly compatible with workspace ergonomics.

References

1. Fang YW, Pan ST, Duan JW (2014) Health hazards and countermeasures of musculoskeletal diseases in office sitting postures. J Heal Educ Heal Promot 9:326–329
2. Yan Q, Qin S (2007) New seat design based on Ergonomics. J Ergon 13:44–47
3. Saraswat A (2020) What kind of effects can I expect from having a good posture while sitting? In: Quora. https://www.quora.com/What-kind-of-effects-can-I-expect-from-having-a-good-posture-while-sitting
4. Departmental news (2002) Physical inactivity a leading cause of disease and disability, warns WHO. In: World Heal. Organ. https://www.who.int/mediacentre/news/releases/release23/en/
5. Tang Z, Bao W, Chen X et al (2022) Comfort kneeling seat for school-age children. J Shanghai Jiaotong Univ. https://doi.org/10.1007/s12204-022-2463-0
6. Opsvik P. Varier Multi In: Varier. https://www.varierfurniture.com/en/multi
7. Murray EA (2023) Muscles in the Anterior Compartment of the Leg. In: Lect Ser How We Move Gross Anat. Motion. https://www.wondriumdaily.com/muscles-in-the-anterior-compartment-of-the-leg/
8. Cao C, Wei Y, Tang J (2020) Design of kneeling-sitting dual-use office chair based on ergonomics. E3S Web Conf 179:02009. https://doi.org/10.1051/e3sconf/202017902009
9. Hou JJ (2009) Study on kneeling sitting position and design of new office chair. D. Nanjing Forestry University, Nanjing
10. Chakrabarti D (1997) Indian anthropometric dimensions for ergonomic design practice. National Institute of Design, Paldi, Ahmedabad, India
11. Roebuck JA (1999) Indian anthropometric dimensions for ergonomic design practice by debkumar chakrabarti 1997, 161 pages, Paldi, Ahmedabad, India: National Institute of Design ISBN 81-86199-15-0
12. MANDAL ÅC (1976) Work-chair with tilting seat. Ergonomics 19:157–164. https://doi.org/10.1080/00140137608931528
13. Mandal AC (1986) An investigation of the lumbar flexion of office workers. In: Corlett N, Wilson J MI (ed) Ergonomics Of working postures, proceeding. Taylor & Francis;, London, pp 345–354
14. Mandal AC (1981) The seated man (Homo Sedens) the seated work position. Theory Practice Appl Ergon 12:19–26. https://doi.org/10.1016/0003-6870(81)90089-2
15. Mandal AC (1991) Investigation of the lumbar flexion of the seated man. Int J Ind Ergon 8:75–87. https://doi.org/10.1016/0169-8141(91)90027-J
16. Wright FE (2012) The problem with sitting. In: Auton Kneeling Chair. https://fewrightautonomy.wordpress.com/page/2/
17. Wang HQ, Tong HT, Wang ZB, Feng CQJ (2016) Study on the friction coefficient between synthetic leather and leather and cloth. J China Leather 45:8–11

Easing the Process of Breastfeeding for Mothers in Public Space

Sakshi Shirsekar and Arundhati Guha Thakurta

1 Introduction

Breastfeeding is one of the most beautiful phenomena for a mother and her child. Numerous medical organizations like WHO and Gynecologists suggest solely nursing for initial six months (juice, no formula or water). And immediately after the introduction of additional meals, it is recommended to continue breastfeeding throughout the first year of a child's birth [1].

Infants get the optimal nourishment from breast milk. It has an absolutely ideal balance of vitamins, protein, and fat. It comprises everything that a baby needs to develop. All of this is delivered in a condition that is more digestible than baby formula or any other form of supplements. Breast milk incorporates antibodies that aid in the infant's defense against viruses and germs [1]. Breastfeeding reduces a likelihood of developing asthma or allergies in babies. In addition, infants who are solely breastfed for the first six months had fewer respiratory diseases, ear infections, and episodes of diarrhea. Furthermore, they experience lesser hospitalization and medical visits [2].

Several research has connected breastfeeding to better IQ scores in late adolescence. It has not only benefitted infants, but also nursing moms in several ways. For instance, physical proximity, skin-to-skin contact, and eye contact help mothers and infant connect with each other and feel secure. As they develop, breastfed babies have a higher probability to acquire the ideal weight rather than becoming overweight. However, it is believed to reduce the risk of obesity, diabetes, and some tumors [1].

S. Shirsekar · A. G. Thakurta (✉)
School of Design, NMIMS Deemed-to-be-University, Mumbai, India
e-mail: arundhati.thakurta@nmims.edu

S. Shirsekar
e-mail: sakshi.shirsekar20@nmims.edu.in

© The Author(s), under exclusive license to Springer Nature Singapore Pte Ltd. 2025

S. Das et al. (eds.), *Human-Centered Digitalization, Volume 3*, Design Science and Innovation, https://doi.org/10.1007/978-981-96-4383-7_34

However, despite breastmilk and nursing being advantageous for both mothers and children, women are discouraged from applying these practices since they are required to breastfeed not only in their houses but publicly too.

2 Literature Review

2.1 What Inhibits Mothers from Breastfeeding in Public?

Despite nursing being an essential and natural aspect of motherhood, there is a serious paucity of public facilities that allow mothers to do so discreetly. Mothers who breastfeed publicly are also subjected to the gaze of strangers, which may lead to discomfort and embarrassment [3]. Breastfeeding in public is highly deemed to be a taboo. Cultural belief that considers breasts to be a sexual object, the sight of an infant breastfeeding seems indecent. However, the act of nursing has become sexualized because of its association with intimacy [4]. Many mothers cover their breasts while breastfeeding in public, but this does not prevent strangers from gazing and causing discomfort. People's curiosity and embarrassment make it difficult for nursing mothers to contentedly nurse in public spaces.

2.2 India Mothers Do Not Want to Breastfeed in Public!

Regarding the Indian setting, in order to comprehend the problem better, a survey was conducted using a web-based questionnaire and targeted on 900 mothers between the ages of 25 and 45 from metropolitan and nonmetropolitan areas. According to the survey, 93% of moms found breastfeeding in public to be very challenging and uncomfortable due to the shortage of nursing rooms. Only 6% of respondents reported finding separate breastfeeding and nursing rooms. Furthermore, 83% of working women reported breastfeeding in public. The most frequent locations where moms breastfeed their infants were their private vehicles, public transportation, restaurants, parking spaces, and mall trial rooms [5] In accordance with the analysis, lack of appropriate and hygienic/safe areas to nurse and discomfort due to others staring causing loss of privacy are the two primary hurdles to breastfeeding in public. On the other hand, when it comes to nursing in public, the majority of Indian moms seek seclusion and intimate contact with their babies. This is because their cultural background prohibits nursing in public without covering.

2.3 Existing Solutions Trying to Assist Breastfeeding in Public Setup

Lactation Rooms—A lactation room (lactorium) is a private area for breastfeeding mothers to nurse their children. Access to these rooms in India is restricted to certain locations. These rooms may be seldom accessible during routine commutes. In addition, the accessible ones are not well maintained for nursing mothers to utilize them [6].

Nursing Bras—This product was designed for lite group and has a zipper/velcro in the bra to facilitate breastfeeding in public. Low- and middle-income individuals may not be able to afford these items.

Covering Cloths—The majority of traveling women use a dupatta, shawl, or other garment to conceal their breasts when breastfeeding their infants. However, this product tends to fall off the body, causing inconvenience. Additionally, it becomes difficult to wrap the clothing when both infants and belongings must be managed. It also strains the mother's hand since she must elevate the infant to nurse.

While interviewing mothers and people, some observations were noted. One of the most vital findings was that, even in the worst-case situation, working mothers do not want to breastfeed in public. Instead, then nursing in public, mothers prefer to flee to their homes or restrooms. This is due to the fact that they have seen individuals looking out of curiosity. There is a significant gap here since most individuals do not have adequate exposure to nursing in public settings via media or other means. This shapes their perception of what is unacceptable and what should not be done in public. Nursing in public places makes mothers feel humiliated and uncomfortable [7]. Another finding is that mothers are expecting to have that isolated time with their child while she is nursing in public as it is so personal to both of them, but withholding feeding the hungry child for personal reasons is not a solution. This void must be addressed by ***encouraging and assisting mothers to breastfeed in public spaces by enhancing her nursing experience, assuring comfort to them and avoiding social hesitation."***

3 Methodology

3.1 Participants

The research comprised of 42 individuals from Mumbai, India. To understand the perspectives of diverse participants, 15 low, 14 middle, 2 high-income mothers, 2 professionals, and 9 commuting men and women were surveyed. With interview responses, the primary stakeholder group was narrowed down to lower- and middle-income group mothers and travelers. This was centered on the insights that working mothers spend less time with their babies, resulting in decreased breastmilk consumption. Low and medium socio-economic group because they face a lot of problems as they lack accessibility and are tied with social constraints.

3.2 Data Collection and Synthesis

Secondary data was elucidated from reliable sources and sites. Perhaps more than twenty-five research papers and publications were considered. This information was derived from authoritative medical sources and other health specialists. This data assisted in developing a comprehensive understanding of this. The primary data was referenced with various personas. Data from several personas were interconnected to identify patterns and insights. Using persona-specific surveys, interviews, and comprehensive observation, the nature of the issue in the problem area was determined after a thorough investigation. A variety of tools were employed throughout the synthesizing process. The use of empathy maps yielded unique insights into the all the 4 personas' environment. Affinity maps were essential for segmenting problem areas and assessing data trends. Journey maps aided in the development of empathy by depicting the user's overall itinerary of the process of breastfeeding. The last phase included converting information into a how and why laddering framework to get a fundamental understanding of the challenges faced by mothers.

3.3 Conceptualization

Brainstorming—During the brainstorming phase, raw ideas were documented based on problem-solving keywords. Consideration was given to a mother's desires and needs in terms of privacy, which was the primary approach adopted. In addition to raising awareness of the topic's relevance, several campaigns were devised.

Low Fidelity Prototypes—The 3 prospective concepts were shortlisted from a wide range of options after the idea was validated by the users. To better grasp a product's functionality and design, minimal viable products were developed. In consideration of user feedback, a final solution was determined.

User Validation—MVPs were used to assess concepts with users in the field. A brief questionnaire was designed to identify what aspects of the concept worked and did not work. Ultimately, based on user feedback and insights, the solution was limited down to a single product and a collaborative effort to promote awareness.

4 Results and Discussion

4.1 *Result of Stage 1: Key Observation and Insight*

The primary research delved further into the issues mothers confront in public spaces. All the interviews ($N = 31$) were divided into three personas: Low-income mothers ($N = 10$), middle-income mothers ($N = 11$), medical officials ($N = 2$), and general commuters ($N = 8$) (Fig. 1).

Despite using a shawl, dupatta, or stoll, only 27% of mothers from middle-income group consented to breastfeeding in public. The other 72% (Fig. 2) of mothers stated that they would still attempt to establish a secure, clean bathroom or a location to breastfeed their babies in public areas. This is mostly attributed to the fact that many commuters see nursing women in public with curiosity. On the other hand, among local computers, 67% (Fig. 3) of both males and females believed it was uncomfortable to witness a nursing mother in public. This was a situation in which women did cover their babies and nurse them. This hesitancy was mostly since these individuals lacked sufficient experience to standardize the procedure.

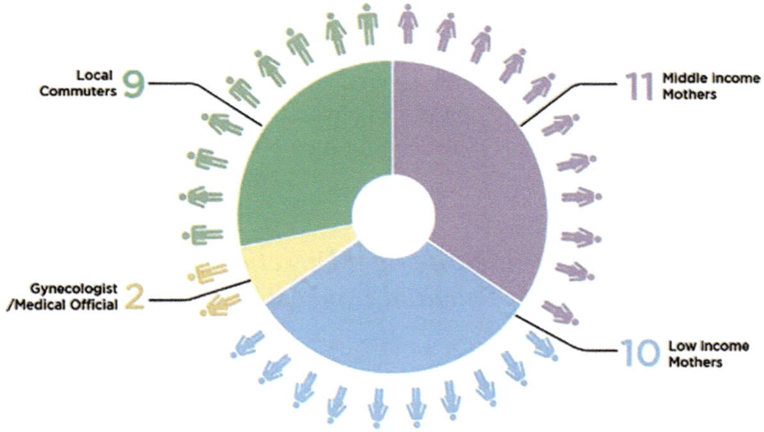

Fig. 1 Stakeholder demographics

Fig. 2 Percentage of mother's breastfeeding

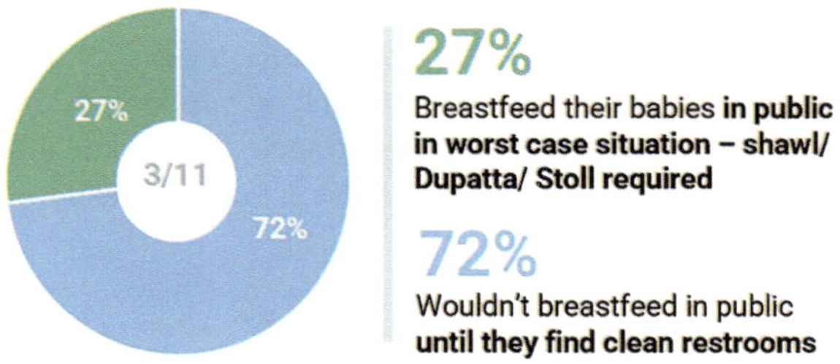

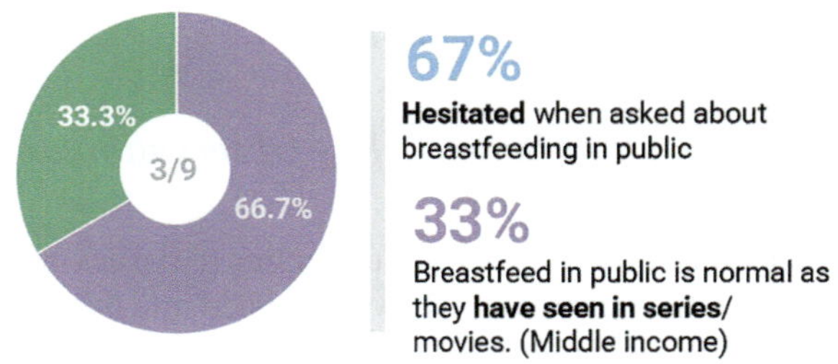

Fig. 3 Percentage of commuters being hesitant

The impact on mothers who were unable to breastfeed as a result of internal and environmental circumstances was enormous. Primarily, this was establishing a poor analogy for society by demonstrating that nursing in any fashion is immoral and should never be done. Similarly, it damages a mother's mental and emotional health and stresses her out as she becomes trapped in an associated taboo that prevents her from nursing even if the baby is wailing. In addition, moms who were originally encouraged to breastfeed in public lack the confidence to do it again after encountering these obstacles. Despite this, the baby's appetite reduction remains a worry in light of the distance issue.

In the primary research, both women from low and middle-income strata ($N = 21$) highlighted several problems in addition to emotional challenges. These physical challenges and concerns made it difficult for breastfeeding women to nurse in public. This resulted in a persistent strain on the mother's hand when she lifted the child to breastfeed. The fact that in addition to this physical strain, dupatta or cover do not provide complete privacy made it difficult for mothers to continue the practice.

4.2 Result of Stage 2: Identifying a Direction and Segregation of Potential Concept

After defining all the gaps and insights, the most important intervening directions were, to retain privacy, provide safety and comfort, and eliminate social reluctance. Given that the solution must be mother-centric, ensuring privacy was a vital factor. In primary study, mothers acknowledged that they want a private moment with their babies when breastfeeding, since it facilitates an emotional connection.

Multiple approaches in terms of products and campaigns were developed with the main objective of assisting mothers to nurse in public space and normalizing public breastfeeding. In this segment, the developed notions were whittled down to three potential concepts with the assistance of a social campaign that may aid in giving a comprehensive solution to the issue. For the same, low-fidelity prototypes were built (Fig. 4).

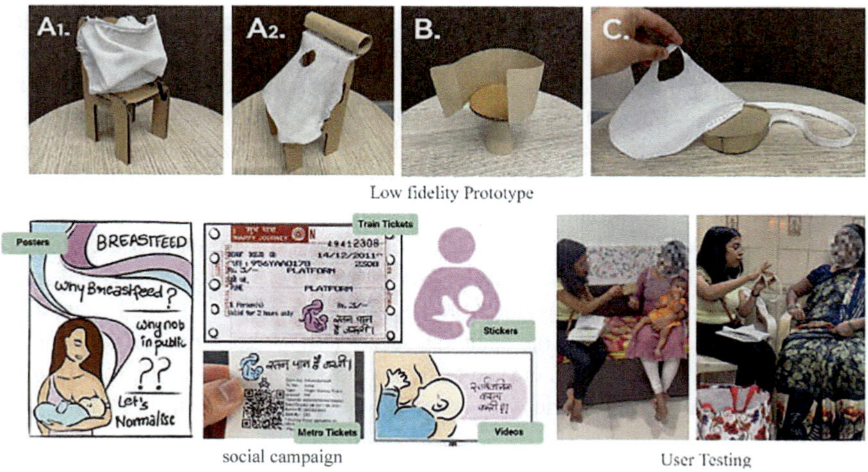

Fig. 4 Prototype and user testing

4.3 *Result of Stage 2: Final Concept*

After defining all the gaps and insights, the most important intervening directions were, to retain privacy, provide safety and comfort, and eliminate social reluctance. Given that the solution must be mother-centric, ensuring privacy was a vital factor. In primary study, mothers acknowledged that they want a private moment with their babies when breastfeeding, since it facilitates an emotional connection. An optimal solution to the issue encountered by mothers was determined. Potentially bridging the gap was a comprehensive approach that incorporates a bag design and social campaign. "***DearMaa***" is a bag developed for breastfeeding women to facilitate a more convenient nursing experience in public. The bag assists mothers to breastfeed in public when they have to do so in front of a large crowd (Fig. 5).

Fig. 5 Final product, a sling nursing bag

Features of the Bag—The product is a zippered bag in which both mother and child's belongings can be accommodated. The bag prevents the need for additional luggage or belongings for the mother. The distinctive selling feature of the bag is that it has two hooks on each side and a side zipper on the back, with a nursing apron attached. When mothers desire to nurse their children, they may pull the apron from the zipper, lock it to the bag's side, and wrap it around their neck. Within the cover, mother may nurse her child in solitude and without obstruction. This ensures locking of the cover from both sides, keeping the chest region covered even if the infant becomes agitated or moves. As highlighted in user testing, lifting the baby to breastfeed causes severe physical strain on the arms. To address this issue, the product is an oval-shaped bag with a soft cover depression at the back that serves as a backrest for the baby to sleep on while being fed. This bag's curved cushiony design allows women to raise their babies in accurate feeding angle while nursing without exerting strain on their arms (Fig. 6).

About Campaign—The campaign included promoting posters, nursing illustrations on transport tickets and reserved seats in public transport for nursing mothers. The purpose of the campaign is to raise awareness about the significance of breastmilk and nursing, demonstrate the necessity for it in public spaces, and prevent mothers from feeling socially embarrassed. The objective is to normalize breastfeeding in

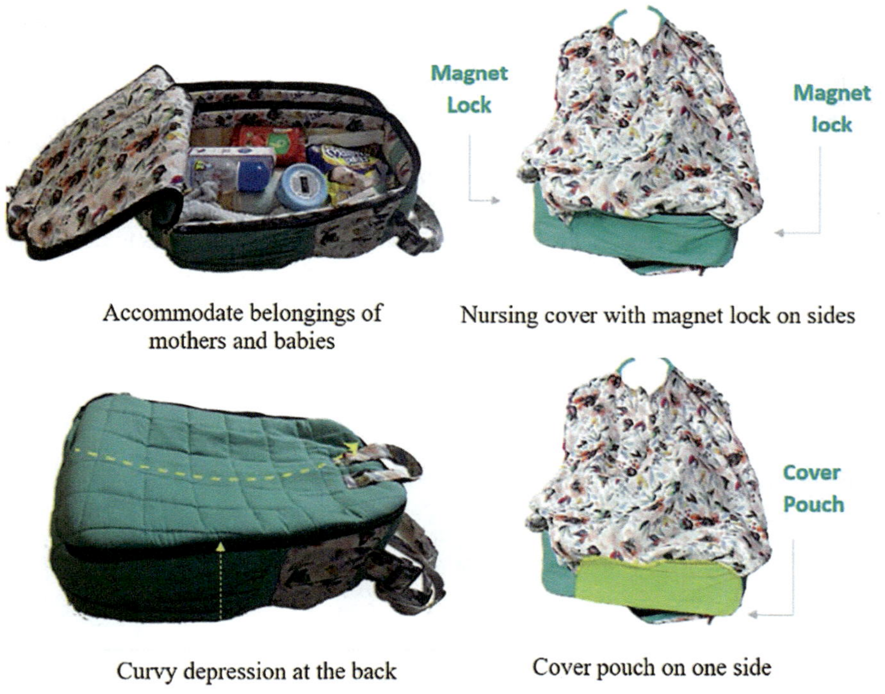

Fig. 6 Features of bag

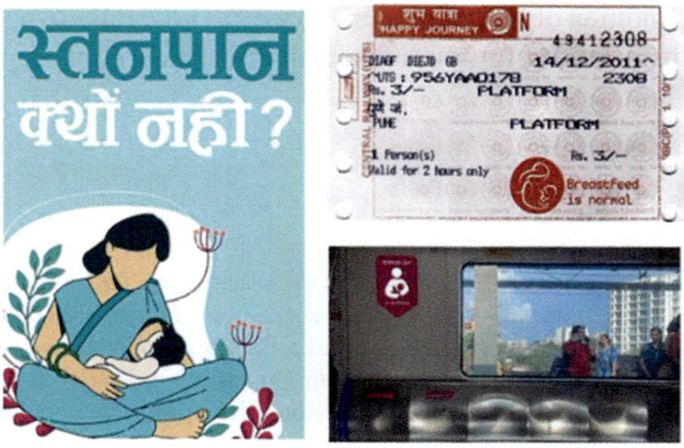

Fig. 7 High fidelity of social campaign

public and raise awareness about the necessity of feeding the child whenever needed in order to increase mothers' confidence in nursing in public when necessary.

Overall User Feedback

After building a solution with high fidelity, users evaluated the prototype. User testing was conducted in two phases. In Phase 1, 14 individuals, 11 mothers, and 3 nursing mothers were evaluated. Phase 1 improvements were implemented in Phase 2 as a result of all of their feedback. In this phase, there were 12 participants that were assessed. It was noticed that mothers valued the curve since it made it simpler for them to feed their infants. 24 out of 26 mothers stated that the device was simple to use since they could comprehend its functionality without any instruction. They appreciated the cover hook's 100% cover-locking assurance. Mothers acknowledged that babies could sit comfortably in the depression created by the curvature as that becomes a motivating influence for mothers to nurse in public while needed (Fig. 7).

5 Conclusion

Considering the findings and insights of both secondary and primary research on nursing in public, it is immediately evident that this is a genuine issue that countless mothers experience. Due to a number of difficulties, traveling mothers have chosen alternatives to breastfeeding instead. However, they are oblivious of the potential problems they may encounter in the future. There will always be uncertainty about the baby's needs. But why should a woman be constantly condemned and be under guilt for being unable to nurse her kid in public when she needs to? There is an urgent need to address the problem and find a satisfactory resolution. Following the current research, a nursing bag was designed with the mother's requirements and desires in mind in order to make it user centric. The designed product is the ideal solution for the issue face in the current condition. The campaign and product designed and demonstrated in the paper is plausible, feasible, and also has the ability to alter the condition of Indian mothers, as well as urge folks to make this a reality and normalize the necessity for breastfeeding in public.

References

1. Taylor RB (n.d.) The benefits of breastfeeding for both mother and baby. WebMD. Retrieved September 18, 2022. https://www.webmd.com/parenting/baby/nursing-basics
2. World health organization. (n.d.). Babies and mothers worldwide failed by lack of investment in breastfeeding. World Health Organization. Retrieved 18 Sept 2022. https://www.who.int/news/item/01-08-2017-babies-and-mothers-worldwide-failed-by-lack-of-investment-in-breastfeeding
3. Sheehan A, Gribble K, Schmied V (2019) It's okay to breastfeed in public but–international breastfeeding journal. BioMed Central. Retrieved 18 Sept 2022.
4. Roberts A (nd) Advantages of breastfeeding over bottlefeeding. Retrieved from Health and Wellness Love to Know: https://www.lovetoknowhealth.com/pregnancy/advantages-of-breastfeeding
5. % of Indian moms feel uncomfortable breastfeeding in public places. Health Vision (2020) Retrieved 25 July 2022. https://healthvision.in/93-of-indian-moms-feel-uncomfortable-breastfeeding-in-public-places/#:~:text=93%25%20of%20Indian%20moms%20feel%20uncomfortable%20breastfeeding%20in%20public%20places
6. van Dellen SA, Wisse B, Mobach MP (2022) Effects of lactation room quality on working mothers' feelings and thoughts related to breastfeeding and work: a randomized controlled trial and a field experiment. Int Breastfeed J 17:57. https://doi.org/10.1186/s13006-022-00499-0
7. Amir LH (2014) Breastfeeding in public: You can do it?. Int Breastfeed J. Retrieved 18 Sept 2022. http://www.ncbi.nlm.nih.gov/pmc/articles/PMC4298082/

Do Indian Audiences Prefer Banking Print Advertisement Visual Metaphor?

Partha Das, Manoj Majhi, Gurdeep Singh, and Gobinda Banik

1 Introduction

The Banking system is the backbone of any country's economic system. Banking is a pivotal component of the financial system that involves a range of operations, including accepting deposits, lending money, and providing financial services. It is vital in enabling economic expansion, fostering financial stability, and bolstering the overall economy [1]. The relationship between banks and customers is an essential component of this ecosystem. Advertisements can help spread information among many populations [2]; hence, the importance of banking advertisements is a crucial domain of concern. In India, the importance of print advertisement is paramount [3, 4]. In most parts of rural India, the popularity of print advertisements over other mediums is undeniable. The total percentage of print advertising expenditure (adex) in India last year was 22 percent, the largest in the world (the global share is 5 percent), according to Pitch Madison Advertising Outlook Report 2022 [5]. Though digitalization is happening rapidly in banking advertisements, Indian print advertisements still have an important hold in the Indian market.

In banking advertisements, visual symbols and metaphors are commonly employed. A visual symbol, such as an image or icon, conveys clear messages quickly, transcending language barriers and serving as a universal communication tool widely used in branding and signage [6, 7]. Visual or pictorial metaphors use imagery to represent ideas beyond their literal meaning, effectively conveying complex concepts and evoking emotions [8, 9]. In the last 25 years, metaphor research has changed its direction. Now, it is not confined only to the linguistic domain; it has spread its arms

P. Das (✉) · M. Majhi · G. Singh
Department of Design, Indian Institute of Technology Guwahati, Guwahati, Assam, India
e-mail: parthadasiitg@gmail.com

G. Banik
Department of English, Basirhat College, Basirhat, West Bengal, India

© The Author(s), under exclusive license to Springer Nature Singapore Pte Ltd. 2025

S. Das et al. (eds.), *Human-Centered Digitalization, Volume 3*, Design Science and Innovation, https://doi.org/10.1007/978-981-96-4383-7_35

in every field, especially in advertising. The use of visual metaphors in advertisements is one of the popular fields of study in advertisement research. Some academic works can be found where researchers claimed that visual metaphor effectively grabs the audience's attention [10, 11]. However, such research studies and experiments were not conducted on any particular product category. Recent studies show that the impact of visual metaphors relies on product categories [12–14].

In the wake of the paucity of literature in the banking advertisement domain, it is vital to explore whether visual metaphors are effective in banking advertisements, and rigorous studies are deemed necessary. The present research, directed in such context, aims to understand the audience's ad-liking toward 'with visual metaphor' and 'without visual metaphor' advertisements. It critically examines the gender differences in ad-liking 'with visual metaphor' and 'without visual metaphor' advertisements in the context of Indian banking print advertisements. In principle, it attempts to answer the unexplored research question in the present context that narrates as RQ 1: 'How do Indian audiences respond toward Indian banking print advertisements possessing 'visual metaphor' and 'without visual metaphor,' and RQ 2: Does any difference lies between Indian male and female audiences responses toward advertisements 'with visual metaphor' and 'without visual metaphor,' especially in the context of Indian banking print advertisements?

2 Literature Review

The literature review below describes the relationship between visual metaphors in advertising, gender roles in advertising interpretation, and existing visual analysis of banking advertisements research. The subsequent sections of this paper assist as a foundation for the study, elucidating the present knowledge scenario in the domain.

2.1 Visual Metaphors in Advertising

The existing literature hints that the 'Visual Metaphors' help substantially enhance the effectiveness of advertisements. Research by Ang and Lim indicates that pictorial metaphors generate more favorable attitudes toward advertisements and increase purchase intentions compared to non-metaphorical representations [15]. These visual metaphors function as an indirect form of persuasion, subtly conveying advertising appeals that resonate with consumers [16]. Moreover, Cao et al. highlight that metaphors' visual structure can capture attention and enhance the message's credibility, reinforcing the notion that well-crafted visual metaphors can lead to a more positive reception of advertisements [17]. Visual metaphors' usefulness is also dependent on their cognitive processing. Studies show that visual pictures are processed more intuitively than verbal statements, helping customers to create faster emotional associations with the promoted items [18]. Such a cognitive advantage is reinforced by the findings of the research study conducted by Mothersbaugh et al., who claim that the stylistic focus of visual components in advertising might boost memory and brand identification [19]. Myers & Jung further explored the interplay between consumer self-view and cognitive style. Their research studies illustrated how individual differences in processing visual metaphors can shape advertising effectiveness [10]. Several other researchers interpreted that the complexity of visual metaphors can also influence their effectiveness. They discovered elaborate visual metaphors might convey rich ideas but can also generate ambiguity, complicating consumer understanding [20]. Such complexity can deter engagement if consumers lack the requisite knowledge or time to process the used metaphor. Conversely, more straightforward visual metaphors are more accessible and can lead to more robust consumer responses, as indicated by research on metaphorical versus straightforward advertising approaches [21].

2.2 Gender Roles in Advertising Interpretation

Gender roles in advertising have been extensively researched. Previous studies have found substantial gender disparities in how people perceive information and respond to adverts [22, 23]. Although both men and women can grasp social cues, disparities develop in their reactions to them, making women more vulnerable to their influence [24]. Various researchers have argued that women are more inclined to systematically examine advertisement content when supplied with precise information, whereas males tend to use heuristics in their analysis [25]. Researchers have explored gender differences in language usage and demonstrated that employing gender-fair language can effectively reduce gender stereotyping and discrimination [26].

Especially in the Indian context, gender interpretation of advertisements is shaped/influenced by traditional media stereotypes [27]. Indian researchers, particularly Pathak and Barthwal, have observed a shift toward empowered portrayals of women. Such a shifting landscape of advertising highlights the intricate and ever-changing ways in which gender is perceived in media [28, 29]. Despite this, research on interpreting visual metaphors from a gender perspective, especially in the Indian context, remains limited. This study seeks to fill this gap by examining how men and women prefer banking print advertisements, featuring 'with visual metaphor' and 'without visual metaphor' differently.

2.3 Visual Analysis of Banking Advertisements Research

The visual component of a banking print advertisement assumes a crucial function in captivating the audience's attention and effectively communicating the bank's brand, values, and services. The selection of the visual component is conducted with meticulous attention to convey a profound sense of confidence, security, and financial well-being. Generally, these advertisements showcase a prominent depiction of a bank building, producing an impression of stability and enduring nature.

Limited scholarly literature has attempted to explore the visual appeal of banking advertisements. Foroudi et al. found that consumers perceive a corporate logo's name, design, and typography as pivotal for a successful image. A well-crafted logo significantly shapes clients' perceptions and their responses to marketing efforts [30]. Guo studied Citibank's use of visual symbols to convey its corporate culture, revealing the potential of localized cognitive frameworks and contemporary advertising in disseminating corporate ideas [31]. Leong et al. delved into value communication's efficacy in enhancing service visualization, highlighting the importance of cognitive linkage in forming positive opinions [32]. Their research focused on financial security and social recognition. Song et al. explored the impact of font case and color temperature on investment decisions and trust assessments, indicating that lowercase letters foster more favorable reactions and influence trustworthiness, indirectly affecting investment intentions. In their study, the warm color temperatures enhanced the influence of all capitalized terms [33].

Visualizing the limited existing research in this field, investigating the utilization of visual metaphors in banking advertising remains a pertinent area for further exploration. Thus, it is worth investigating/examining the use of visual metaphors in banking advertisements.

3 Methodology

The study was conducted in three stages: (1) selecting stimuli for the experiment, (2) participant selection and sample collection, and (3) data analysis. In the first stage, advertisements (stimuli) were selected from published magazine ads. The second stage was participant selection and sample collection from all over India. The final stage was data analysis through statistical software, and afterward, study results were interpreted to develop a research outcome.

3.1 Selection of Stimuli for the Experiment

The advertisements (stimuli) were chosen from widely circulated general interest print magazines such as India Today, Outlook, The Week. The study focused on banking commercials spanning a period of five years (2017–2022). Two experts were engaged to select the advertisements/stimuli. One of the experts involved has been an ad director of a leading Indian ad agency for the last seven years; the other expert works as an assistant professor at a central university in India and has been teaching and researching in the field of visual communication for over six years.

The selection parameters emphasized the strength of metaphors in the advertisements, assessing whether they contained metaphorical clues and to what extent these clues were present. For ads without metaphors, the clarity of the message became the primary consideration. The experts were shown 30 banking advertisements: 15 'with visual metaphors' (WVM) and 15 'without visual metaphors' (WOVM). Through iterative review, experts selected six ads for the experiment: three ads 'with visual metaphors' (WVM) and three other ads 'without visual metaphors' (WOVM) related to Car Loans (CL), Education Loans (EL), and Fixed Deposit (FD) banking print advertisements.

3.2 Participant Selection and Sample Collection

The snowball sampling technique was applied to collect data. Snowball sampling is a non-probability sampling technique in which existing participants recruit new participants, expanding the sample [34]. An online survey was conducted through Google Forms, and the survey forms were distributed via Facebook Messenger and WhatsApp to get audience responses. University students from all corners of India participated in the experiment, each bringing unique perspectives and experiences from their diverse regional backgrounds, enriching the overall participation. A total of 109 participants responded. After screening the data, 102 responses were selected for further final study, with 51 males and 51 females among them. The respondents were between 20 and 40 years old. The mean age of participants was 28.4 years. The chosen age range was intentionally designed to encompass a wide range of opinions and experiences, mirroring the demographics of younger and more senior students. All participants' minimum qualification was a bachelor's degree or the equivalent of 15 years of education. The participants must be a readers of at least one Indian magazine—India Today, Outlook, or The Week regularly.

3.3 Data Analysis

The IBM SPSS Statistics 22 version was used to conduct a paired t-test and an independent t-test. The statistical test measured significant differences between the means of independent groups. Here, the independent variables considered were ads 'with visual metaphor,' 'without visual metaphor,' and gender; the dependent variable was ad-liking. The ad-liking was measured with a 7-point Likert (where '1' indicates—least like, and '7' indicates—most like) scale for collecting participants' responses (Fig. 1).

Fig. 1 Examples of ads 'with visual metaphor,' 'without visual metaphor' where **a** Car Loan ad 'with visual metaphor' (CLWM), **b** Car Loan ad 'without visual metaphor' (CLWOM), **c** Education Loan ad 'with visual metaphor' (ELWM), (d) Education Loan ad 'without visual metaphor' (ELWOM), **e** Fixed Deposit ad 'with visual metaphor' (FDWM), and (f) Fixed Deposit ad 'without visual metaphor' (FDWOM)

4 Result and Discussion

The statistical analysis was conducted to compare two sample groups: advertisements 'with visual metaphor' (WVM) and advertisements 'without visual metaphor' (WOVM). A paired samples t-test was used to assess the differences in preference between Car Loan ad 'with visual metaphor' (CLWM) and Car Loan ad 'without visual metaphor' (CLWOM), Education Loan ad 'with visual metaphor' (ELWM), and Education Loan ad 'without visual metaphor' (ELWOM), and Fixed Deposit ad 'with visual metaphor' (FDWM), and Fixed Deposit ad 'without visual metaphor' (FDWOM) as illustrated in Table 1.

From the values from Table 1, it can be observed that, for CLWVM and CLWOVM, the mean difference was 0.725 ($M = 5.372$, $SD = 1.802$, $SEM = 0.178$), with a 95% confidence interval (CI) ranging from 0.371 to 1.079. In the t-test ($n = 102$, $df = 101$), this produced a t-statistic of 4.064 that proves the two groups are significantly different ($p < 0.001$). Likewise, the comparison between ELWVM and ELWOVM showed a mean difference of 0.686 ($M = 5.246$, $SD = 1.116$, $SEM = 0.110$), with a 95% CI between 0.466 and 0.905. The t-test statistics for 101 df produced $t = 6.206$, confirming a statistically significant difference of the groups $p < 0.001$. For FDWVM and FDWOVM, the mean difference was 0.598 ($M = 4.896$, $SD = 1.083$, $SEM =$

Table 1 Paired samples test

Advertisement comparison WVM VS WOVM	Paired differences					t-value	df	Sig. (2-tailed)
	Mean difference (M)	Standard deviation (SD)	Standard error of the mean (SEM)	95% Confidence interval (CI) of the difference				
				Lower	Upper			
Car loan	0.725	1.802	0.178	0.371	1.079	4.064	101	0.001*
Education loan	0.686	1.116	0.110	0.466	0.905	6.206	101	0.001*
Fixed deposit	0.598	1.083	0.107	0.385	0.810	5.576	101	0.001*

0.107), with a 95% CI ranging from 0.385 to 0.810. The t-test, conducted with 101 df, yielded a t-statistic of 5.576, demonstrating a statistically significant difference ($p < 0.001$).

The results indicate that the WVM group's mean scores were higher than WOVM's. These findings indicate that the use of visual metaphors in advertisements can be very useful in improving the effectiveness of the advertisements as compared to the ads 'without visual metaphors' in the context of print advertisements for various banking advertisements. The data shows that ads containing visual metaphors achieve a high rate of strong ad-liking compared to those 'without visual metaphors' across various products advertising categories, viz., car loans, fixed deposits, and education loans. These results are in corroboration with earlier studies in this domain [35].

Simultaneously, Fig. 2 depicts that visual metaphor-based advertisements (CLWVM, FDWVM, and ELWVM) were rated higher than their non-visual equivalents (CLWOVM, FDWOVM, and ELWOVM) for ad-liking. These findings suggest the advantage of including visual metaphors in advertisements regarding financial products of the banking sector. Incorporating these will make their ads even more effective and understandable.

Table 2 depicts the statistical comparison of data differences between two sex groups, male and female, as observed for the six advertisements, including those 'with visual metaphor' and 'without visual metaphor.'

An exploratory study of CLWVM characterized by the performance of both genders shows significant performance differences in female and male test participants even when the hypothesis of equal variances is not held. Compared to females, male participants ($M = 5.372$, $SD = 0.915$, $SEM = 0.128$) scored, on average, 0.352 units higher ($M = 5.019$, $SD = 1.655$, $SEM = 0.231$). The difference in mean values was statistically significant with $p < 0.005$ with both t-test comparisons, indicating the possibility that gender is a moderating variable in the test scores of male and female participants. In contrast, for CLWOVM, the test results analysis reveals no statistically significant gender disparity, irrespective of whether the assumption of equal variances is met. Male participants ($M = 4.607$, $SD = 1.844$, $SEM = 0.258$) exhibited a slightly higher mean score than female participants ($M = 4.274$, $SD = 1.898$, $SEM = 0.265$), with a modest mean difference of approximately 0.371 units.

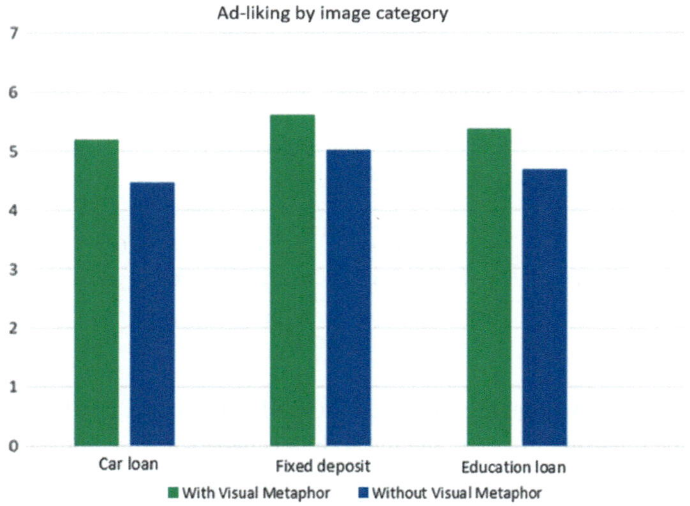

Fig. 2 Analysis of ad-liking of the ads 'with visual metaphor'

Table 2 Descriptive statistics of test scores by gender

Ad category		Gender	N	M	SD	SEM
Car loan	WVM	Male	51	5.372	0.915	0.128
		Female	51	5.019	1.655	0.231
	WOVM	Male	51	4.607	1.844	0.258
		Female	51	4.274	1.898	0.265
Education loan	WVM	Male	51	5.294	1.540	0.210
		Female	51	5.470	1.653	0.231
	WOVM	Male	51	4.823	1.492	0.209
		Female	51	4.568	1.565	0.219
Fixed deposit	WVM	Male	51	5.294	1.540	0.164
		Female	51	5.470	1.653	0.175
	WOVM	Male	51	4.823	1.492	0.202
		Female	51	4.568	1.565	0.197

However, this difference did not reach statistical significance, as suggested by the p-values ($p > 0.005$) for both t-tests, indicating that gender does not seem to influence the variability in test scores among the participants significantly.

In the case of ELWVM, the gender gap in test results was not statistically significant, irrespective of whether the assumption of equal variances was met. Male participants ($M = 5.294$, $SD = 1.540$, $SEM = 0.210$) and female participants ($M = 5.470$, $SD = 1.653$, $SEM = 0.231$) showed similar mean test scores, with a slight mean difference of approximately 0.578 units favoring female participants. However, this difference did not achieve statistical significance, as indicated by the p-values ($p > 0.005$) for both t-tests, suggesting that gender is not a significant factor influencing the variability in test scores among participants. Likewise, the analysis of the ELWOVM did not find any statistically significant gender disparities in the results, regardless of the assertions made and whether equal variances held. Male (M = 4.823, SD = 1.492, SEM = 0.209) and female participants (M = 4.568, SD = 1.565, SEM = 0.219) exhibited comparable mean test scores, with a modest mean difference of approximately 0.402 units favoring male participants. However, this finding fell short statistically as both t-tests had p-values equal to or ($p > 0.005$), thus explaining that gender does not contribute significantly to the variability of test scores among the participants.

In the case of FDWVM, the study suggests no gender difference in the test scores, irrespective of whether the assumption of equal variances is met. Male participants (M = 5.294, SD = 1.540, SEM = 0.164) and female participants (M = 5.470, SD = 1.653, SEM = 0.175) showed similar mean test scores, with a small mean difference of approximately 0.685 units favoring male participants. However, this difference was not statistically significant, as evidenced by the p-values ($p > 0.005$) obtained from both t-tests, indicating that gender does not emerge as a statistically significant factor in explaining the variation in test scores among the participants. Lastly, in the context of FDWOVM, there was no statistically significant difference in test scores across genders, irrespective of whether the assumption of equal variances was satisfied. Male participants (M = 4.823, SD = 1.492, SEM = 0.202) and female participants (M = 4.568, SD = 1.565, SEM = 0.197) displayed similar mean test scores, with a modest mean difference of approximately 0.782 units favoring male participants. However, this difference did not reach statistical significance in either case, as indicated by the p-values ($p > 0.005$) for both t-tests, suggesting that gender does not substantially impact the range of test scores observed among the participants.

5 Conclusion

The statistical analysis demonstrates that advertisements incorporating visual metaphors consistently achieve higher mean scores than those without visual metaphors across different financial products, including car loans, education loans, and fixed deposits. The t-tests carried out indicate that there are significant differences due to the effectiveness of the visual metaphors in increasing the scale and response to such advertisements. These results propose that integrating visual metaphors into advertising strategies can be a powerful tool to advance consumer engagement and efficiency in financial product advertising.

The statistical analysis of the six advertisements displays that gender does not consistently appear as a significant factor influencing test performance across different ad types. Although a statistically significant gender disparity was examined in CLWVM, where male participants scored higher on average than female participants, this pattern was not evident in the other ad categories. In the case of

CLWOVM, ELWVM and ELWOVM, and FDWVM and FDWOVM, no significant gender differences were observed in test scores. These findings hint that, except for CLWVM, gender does not play a considerable role in influencing the variability of responses to these financial advertisements. Therefore, the impact of gender on ad effectiveness appears to be context-dependent and may vary based on the presence or absence of visual metaphors.

6 Limitations and Future Scope

Despite the current study's thorough examination, it has limitations, such as not taking into account class inequalities based on family income and educational skills. The participants were master's or Ph.D. students, which might induce bias. Replicating the study with participants from various socioeconomic classes, academic backgrounds, nationalities, and ages might result in more thorough replies and interpretations. Moreover, in the future, researchers may take up studies to conduct experiments on print banking advertisements from a design perspective, exploring different visual metaphor typologies to understand how these metaphors enhance ad quality and audience reception in India. This approach could become a vital tool for designers to create more effective banking advertisements for Indian users. The current study suggests two main paths for future research: qualitative analysis through in-depth user interviews and quantitative analysis through administering surveys with higher sample numbers.

Acknowledgements The authors would like to extend their heartfelt gratitude to all the participants who generously contributed their time and insights to this study.

References

1. Rakshit B, Bardhan S (2019) Does bank competition promote economic growth? empirical evidence from selected South Asian countries. South Asian J Bus Stud 8(2):201–223. https://doi.org/10.1108/SAJBS-07-2018-0079
2. Lin LF, Li YM, Wu WH (2015) A social endorsing mechanism for target advertisement diffusion. Inf Manag 52(8):982–997. https://doi.org/10.1016/j.im.2015.07.004
3. Kalro AD, Sivakumaran B, Marathe RR (2010) Comparative advertising in India: A content analysis of English print advertisements. J Int Consum Mark 22(4):377–394. https://doi.org/10.1080/08961530.2010.505887
4. Chattopadhyay T (2010) The effects of message strategy and execution framework on teenage boy's processing of print advertisements in India. Int J Bus Sci & Appl Manag (IJBSAM) 5(2):17–28
5. Anand S (2024) Ad spends on print media in India rose to 39% last year. https://economictimes.indiatimes.com/industry/services/advertising/ad-spends-on-print-media-in-india-rose-to-39-last-year/articleshow/89625530.cms?from=mdr , last accessed 2024/09/08

6. Latała-Matysiak D, Marciniak M (2023) The influence of the lotus flower theme on the perception of contemporary urban architecture. Space Cult 26(4):647–655. https://doi.org/10.1177/120633122110323
7. García-Peñalv FJ, Ordonez de Pablos P, García J, Therón R (2014) Using OWL-VisMod through a decision-making process for reusing OWL ontologies. Behaviour & Information Technology, 33(5):426–442. https://doi.org/10.1080/0144929X.2012.709538
8. Das P, Singh G, Banik G, Majhi M (2023) Creating a proxy advertisement for cognitive response evaluation in visual metaphor studies under advertisement research. In: International conference on design and digital communication, pp 765–776. Springer Nature Switzerland, Cham. https://doi.org/10.1007/978-3-031-47281-7_62
9. Indurkhya B, Ojha A (2017) Interpreting visual metaphors: asymmetry and reversibility. Poet Today 38(1):93–121. https://doi.org/10.1215/03335372-3716240
10. Myers J, Jung JM (2019) The interplay between consumer self-view, cognitive style, and creative visual metaphors in print advertising. J Mark Commun 25(3):229–246. https://doi.org/10.1080/13527266.2016.1197296
11. Phillips BJ (2003) Understanding visual metaphor in advertising. In: Scott LM, Batra R (eds.), Persuasive imagery: A consumer response perspective, pp. 297–310. Routledge, London
12. Margariti K, Boutsouki C, Hatzithomas L, Zotos Y (2019) Visual metaphors in food advertising: A cross-cultural study. Food Res Int 115:338–351. https://doi.org/10.1016/j.foodres.2018.11.030
13. Chang CT, Wu YC, Lee YK, Chu XY (2018) Right metaphor, right place: choosing a visual metaphor based on product type and consumer differences. Int J Advert 37(2):309–336. https://doi.org/10.1080/02650487.2016.1240468
14. Chang CT, Yen CT (2013) Missing ingredients in metaphor advertising: The right formula of metaphor type, product type, and need for cognition. J Advert 42(1):80–94. https://doi.org/10.1080/00913367.2012.749090
15. Ang SH, Lim EAC (2006) The influence of metaphors and product type on brand personality perceptions and attitudes. J Advert 35(2):39–53. https://doi.org/10.1080/00913367.2006.10639226
16. Zhao H, Lin X (2019) A review of the effect of visual metaphor on advertising response. In: The 4th international conference on economy, Judicature, Administration and Humanitarian Projects (JAHP), pp. 29–34. Atlantis Press. https://doi.org/10.2991/jahp-19.2019.7
17. Cao S, Wang H, Zou X (2018) The effect of visual structure of pictorial metaphors on advertisement attitudes. Int J Mark Stud 10(4):60. https://doi.org/10.5539/ijms.v10n4p60
18. Fenko A, De Vries R, Van Rompay T (2018) How strong is your coffee? The influence of visual metaphors and textual claims on consumers' flavor perception and product evaluation. Front Psychol 9:53. https://doi.org/10.3389/fpsyg.2018.00053
19. Mothersbaugh DL, Huhmann BA, Franke GR (2002) Combinatory and separative effects of rhetorical figures on consumers' effort and focus in ad processing. J Consum Res 28(4):589–602. https://doi.org/10.1086/338211
20. Djafarova E (2016) Why do advertisers use visual metaphors? A Pragmatic Perspect Mark Rev 16(2):203–214. https://doi.org/10.1362/146934716X14636478977593
21. Ding X, Feng P, Wang J, Lin M (2022) Metaphorical or straightforward? comparing the effectiveness of different types of social media advertising. Front Neurosci 16:851729. https://doi.org/10.3389/fnins.2022.851729
22. Das P, Majhi M (2021) Gender role portrayal in indian advertisement: a review. In: International conference of the indian society of ergonomics, pp 461–471. Springer International Publishing, Cham. https://doi.org/10.1007/978-3-030-94277-9_40
23. Booth A, Nolen P (2012) Salience, risky choices and gender. Econ Lett 117(2):517–520. https://doi.org/10.1016/j.econlet.2012.06.046
24. Venkatesh V, Morris MG, Ackerman PL (2000) A longitudinal field investigation of gender differences in individual technology adoption decision-making processes. Organ Behav Hum Decis Process 83(1):33–60. https://doi.org/10.1006/obhd.2000.2896

25. Boshoff C, Toerien L (2017) Subconscious responses to fear-appeal health warnings: An exploratory study of cigarette packaging. South Afr J Econ Manage Sci 20(1):1–13. https://doi.org/10.4102/sajems.v20i1.1630
26. Sczesny S, Formanowicz M, Moser F (2016) Can gender-fair language reduce gender stereotyping and discrimination? Front Psychol 7:154379. https://doi.org/10.3389/fpsyg.2016.00025
27. Kumari S, Shivani S (2014) Female portrayals in advertising and its impact on marketing communication—pieces of evidence from India. Manage Labour Stud 39(4):438–448. https://doi.org/10.1177/0258042X1557802
28. Pathak S, Mondal I, Mishra S (2024) Unravelling the linguistic tapestry: a discursive study of gender portrayal in select indian electronic advertisement. Rupkatha J Interdiscip Stud Humlties 16(1). https://doi.org/10.21659/rupkatha.v16n1.02g
29. Barthwal S (2024) Gender portrayals and perceptions in the new age society of India. Indian J Gend Stud 31(1):102–121. https://doi.org/10.1177/0971521523121053
30. Foroudi P, Melewar TC, Gupta S (2014) Linking corporate logo, corporate image, and reputation: An examination of consumer perceptions in the financial setting. J Bus Res 67(11):2269–2281. https://doi.org/10.1016/j.jbusres.2014.06.015
31. Guo L (2017) Symbol analysis of financial enterprises' advertisements–a case study of citibank–. Cultura 14(1):81–87. https://doi.org/10.3726/CUL.2017.01.08
32. Leong VS, Hibbert S, Ennew C (2018) Communicating value to enhance service visualization. J Serv Mark 32(6):645–656. https://doi.org/10.1108/JSM-04-2017-0114
33. Song Y, Luximon Y, Luo J (2020) A moderated mediation analysis of the effect of lettering case and color temperature on trustworthiness perceptions and investment decisions. Int J Bank Mark 38(4):987–1005. https://doi.org/10.1108/IJBM-09-2019-0315
34. Naderifar M, Goli H, Ghaljaie F (2017) Snowball sampling: A purposeful method of sampling in qualitative research. Strides in development of medical education, 14(3). https://doi.org/10.5812/sdme.67670
35. Das P, Basak A, Vaidya G, Singh G, Banik G, Majhi M (2024) Unveiling gender differences in visual metaphor interpretation in indian banking print advertising. Arch Des Res 37(4):99–116. https://doi.org/10.15187/adr.2024.08.37.4.99

Study on Military Combat Training Using Head-Mounted Devices Through Mixed Reality

Pranav Chaturvedi, Nikhil R. Banbe, and Nilakshi Yein

1 Introduction

Military operations are one of the most important fields that use mixed reality nowadays. It has affected almost every practical field on the globe [1]. The military is increasingly using virtual reality in addition to traditional combat tactics [2]. The objective of this research is to analyse military training that employs mixed reality and head-mounted devices.

Military performance and capability are critically dependent on combat training. In the past, military personnel have been trained using conventional means to get them ready for battle. On the other hand, there's growing curiosity about how new technology, like mixed reality (MR) simulations, can enhance military training [4]. With mixed reality, consumers can have an immersive experience by combining elements of virtual and augmented reality. In military training, MR simulations can simulate combat conditions in a secure and controlled manner, giving soldiers a realistic environment in which to hone their skills. By making combat training more interesting and stimulating for trainees, this technology has the potential to enhance combat training [3].

In spite of earlier research, additional research is required to fully understand the impact of MR simulations on military training. Examining how well MR simulations work to improve military personnel's combat performance and training satisfaction is particularly important.

P. Chaturvedi (✉) · N. R. Banbe · N. Yein
School of Design, UPES, Dehradun, India
e-mail: pranavchaturvedi.design@gmail.com

© The Author(s), under exclusive license to Springer Nature Singapore Pte Ltd. 2025

S. Das et al. (eds.), *Human-Centered Digitalization, Volume 3*, Design Science and Innovation, https://doi.org/10.1007/978-981-96-4383-7_36

2 Literature Review

2.1 *Mixed Reality (MR) Based Simulations*

Virtual reality (VR) and augmented reality (AR) are components of mixed reality (MR) systems used for training in combat. The MR system is a wearable that falls within the "Head-mounted device" (HMD) category. As a result, the soldier using this system—referred to as a VR user—can identify items in 3D throughout their surroundings [2].

Head-mounted devices, or HMDs for short, are small display or projection technologies installed on helmets, hats, or glasses. Heads-up displays (HUDs) are a kind of head-mounted display (HMD) that overlays an image on the user's perspective of the actual world in AR applications and the virtual world in VR cases, without obstructing their vision [5].

It is essential to first examine how military training is now conducted in order to assess the potential benefits of this technology for HMDs [7]. The most effective and accessible virtual shooting simulators employ large projection screens and can have a high degree of accuracy, graphics, and artificial intelligence (AI), but they don't require the user to move, use cover wisely, or engage with the environment [6]. Their ability to maintain the visual appeal of a video game is especially noteworthy, as in multiuser simulation, the user can still distinguish between a flat screen and the real floor, roof, and nearby objects or companions [8, 9].

2.2 *Combat Rehearsal Exercise via MR*

A method through which combatants could employ real-life scenarios in which they had to move quickly, run into cover, and dodge fire from simulated foes who indulged with them in the same way as they do in real-world training. The inclusion of virtuality into the actual environment would force the user to behave and move in the same manner as in a genuinely threatening circumstance during the training exercise. The VR user may locate friendly soldiers, hostile forces, and traps on the battlefield by utilising GPS-equipped devices. The system should have image processing capabilities to recognise them. To make such assessments, image processing can employ pattern recognition [10]. (Fig. 1).

Creative integration of various technologies might make this degree of simulation a reality. The best enterprises for computer engineering are creating:

- High-resolution augmented reality (AR) gadgets will view the actual environment with virtual enhancements like 3D CGI opponents, automobiles, people, and more. In the actual scenario, explosions, shootouts, and many other circumstances might be practically replicated. These AR viewing devices may be worn as AR glasses or HMDs, like Google Glass [11].

Fig. 1 The virtual environment was rendered for the HMD MR system [1]

- 3D mapping of the exact area where the simulation would take place using geographically referenced data may make it possible for electronically manufactured objects to interact with the real environment, resulting in a very realistic experience. Thus, just like an actual adversary, computer-generated elements may take advantage of the surroundings to further their objectives. This was done at the US Army Battle Lab at Fort Benning. [12].
- In addition to enhancing immersion, realistic sound effects would enable the user to react to virtual events like an explosive device going off or an airstrike supporting it. But using surround sound output alone won't cut it. The software's spatial sound must calculate "the correct volume, filtering, and reflections of the sound or reverberation of objects at distance" in order for it to frequently enhance the player's sensation of presence when playing a game. It needs to convince the user that the virtual world is genuine by offering "sensory proof".
- It may be possible to interact with virtual components without touching a screen or pressing a button by utilising specialised devices like gloves with sensors or

simply shooting at simulated AR using laser transmitters mounted on real firearms. Virtual enemies may shoot back and be able to inflict penalties on the player for being struck, as is the case with an iPhone/iPad game that currently uses this type of augmented reality interaction: It's a device created by Parrot Drone called the AR Hunter [13]. Using pressure points, tiny electrical discharges, or compressed air, a customised vest might be used to "stimulate the sense of strikes and impacts received" by the user [14].

All of these cutting-edge methods might provide military training with a simulation that is incredibly accurate and could replace live training exercises, which are more expensive since they require more staff, training ammunition, pyrotechnics, and other special efforts. Additionally, since actual ammunition is not required, such a mixed environment can be promising to reduce the number of injuries and fatalities in conventional training methods, and it will furthermore cut the expenses of training, much like employing virtual vehicle simulators does.

3 Aim and Objectives

Aim: To analyse the impact of mixed reality simulations on improving combat training.

Objectives:

1. To study the combat performance of participants who received mixed reality combat training versus standard instruction.
2. To assess the level of satisfaction with training between individuals who received mixed reality combat training and those who received standard training.
3. To identify the advantages and disadvantages of mixed reality combat training over traditional training and the potential benefits of introducing mixed reality simulations into military combat training programmes.

4 Methods and Materials

4.1 Experiment: Setting and Subject Profile

An experiment with 15 participants (8 males and 7 females, aged between 19–30 years old) was carried out for 2 days in an indoor space after explaining the research topic to them and obtaining their permission to use the data from the experiment in the research paper. The participants who were chosen for the experiment had no physical combat experience, were all physically healthy, and had no visual health issues; the youngest was a 19-year-old male and the oldest was a 30-year-old female. 5 of the 15 participants had prior experience of being in a virtual space using VR headgear;

7 including those 5 had prior experience of playing digital video games involving combat of any kind; and the remaining 8 had no prior experience of either being in the virtual space or playing combat-related video games. We divided the participants into two groups: Set 1 had the 7 participants with prior experience in video combat gaming, and the other 8 participants with little to no experience were grouped in Set 2

4.1.1 Procedure

The purpose of the experiment was to learn more about the effectiveness of virtual simulation training for a regular person who has never engaged in physical combat of any kind. The participants were instructed to don the Oculus headgear and play the Superhot VR game, which involves combating their way out of various levels, each of which features a different combat setting like hand-to-hand combat, shooting at approaching targets, and using the surroundings to defend themselves. Each participant received the same set of Oculus and played the same levels of the simulation. Each participant played all 6 levels an equal number of times (i.e., 2 times) at the same difficulty level. After a brief demonstration and explanation of the game, each participant played all 6 levels on the first day. Every participant was allowed to retry a level if they failed it and to play until they cleared all 6 levels successfully; there were no time constraints imposed by the experimenter. Each participant's time and number of attempts to successfully clear the levels were recorded. On the second day, the entire procedure of the first day was repeated, and new observations were made. Following the experiment, each participant was interviewed face to face about their experience and learnings from the experiment, and their insights were written down.

5 Results

The difference in performance between Sets 1 and 2, as well as the performance of Set 2 on the first day compared to their performance on the second day, show that training in a virtual environment and playing video games can improve reflexes in the real world and that competing in a virtual environment feels more rewarding and natural than competing in digital video games. Table 1 shows the result/outcome of the experiment discussed above.

5.1 Observations

Participants in Set 1, with prior experience of maneuvering in a virtual space and playing combat video games, needed less time to learn the physics of the game and

Table 1 Outcome of the conducted experiment

SETS	Number of attempts (Day 1)	Number of attempts (Day 2)
Set 1 (All 6 levels)	12	9
Set 2 (First 3 easier levels)	21	14
Set 2 (Later 3 difficult levels)	74	39

* Set 1 consists of participants with prior experience in VR and combat-related digital video gaming
*Set 2 consists of participants with little to no prior experience in either VR or video gaming

successfully complete the level than those in Set 2, with little to no prior experience in a virtual setting. On the first day, Set 1 participants cleared all 6 levels in an average of 1.7 attempts. The rest of Set 2 took an average of 2.65 attempts to clear the first 3 ßlevels, which is easier than the last 3 of the 6 levels. The Set 2 participants required an average of 9.25 tries to complete the final 3 levels with higher difficulty settings. On the second day, the participants in Set 1 made 1.2 attempts to clear all 6 levels, decreasing the total number of attempts taken by Set 1 from 12 on the first day to 9 on the second day. The Set 2 participants took on average 1.75 attempts to clear the first 3 levels, decreasing the number of attempts from 21 to 14. For the later 3 difficult levels, Set 2 participants took an average of 4.875 attempts, decreasing the number of attempts from 74 to 39.

5.2 Insights from Face-To-Face Interviews

The experimenters scheduled a face-to-face interview after the experiment was a success. Interviews with each participant yielded insights. Although most of the Set 1 participants were pleased with their performances on both days, 2 of those 7 said they could now complete all 6 levels in a single attempt. Due to their gaming background, the Set 1 participants believed they had quicker reflexes than the Set 2 participants. They asserted that because of their dependence on gaming, their reflexes had significantly improved in the real world. They claimed that, in comparison to digital video games, the virtual environment gave them a more realistic experience of combat-related games, and completing the levels felt more rewarding.

The Set 2 players found the game to be both frustrating and thrilling. 5 out of 8 participants said they would perform better if given the chance to play again and that they would like to play more games like Superhot VR because it exhilarated and challenged them.

5.3 *Findings from the Experiment*

Due to their extensive practice on the first day, it was found that both groups performed better on the second day. Set 2's performance showed that their reflexes substantially improved on the second day compared to the first, but little could be drawn from Set 1's outstanding performance on both days.

6 Conclusion

The study concludes that there is room for growth in the field of military training with regard to the impact of mixed reality on simulation-based military combat training. Through more realistic and immersive training, military personnel can better prepare for real-world combat situations with the aid of mixed reality simulations. In addition to increasing training frequency and cost-effectiveness, mixed reality simulations may enhance decision-making, precise shooting, and overall combat performance. There are several disadvantages to take into account despite the possible advantages, including technical problems, adapting to new environments, and user discomfort.

These challenges might be addressed and solved, though, as technology develops and becomes more widely available.

7 Discussion and Future-Cone

Following its examination of the effects of mixed reality on military combat training through simulations, the paper concludes that it is a subject of growing importance within the military training community. Through more realistic and immersive training, military forces can enhance their decision-making, marksmanship, and combat effectiveness with the use of mixed-reality simulations. Mixed reality simulations can also offer a more economical and effective training option than conventional ones. On the other hand, there are certain possible issues that need to be resolved, like technical issues, situational adaptation, and user discomfort. The potential advantages of mixed reality simulations in military combat training are a topic of much discussion, notwithstanding these obstacles.

Military personnel may ultimately receive their primary training from mixed reality simulations rather than more conventional techniques. Improved operational outcomes could result from having military personnel who are better equipped and trained.

Future research could also focus on how mixed-reality simulations can be integrated with other emerging technologies like robots and artificial intelligence. This could result in training simulations that are more effective and realistic, allowing military personnel to practise intricate situations that were previously hard or impossible to recreate.

References

1. Ter Haar R (2005) Virtual reality in the military: present and future. In: 3rd Twente Student Conf. IT. Citeseer
2. Rathnayake WP (2018). Usage of mixed reality for military simulations. In 2018 International Conference on Current Trends towards Converging Technologies (ICCTCT), pp 1–5. IEEE
3. Boren T, Provost L, Hendrix R, DeLeo G (2014) Simulation fidelity and training effectiveness in military decision-making
4. Bohannon RW, Litts K, Grier R, Willeford B (2019) Mixed reality training in marksmanship: An assessment of user satisfaction and training effectiveness
5. Bimber O, Raskar R (2005) Spatial augmented reality: merging real and virtual worlds. Wellesley: A K Peters
6. CIOpGLO (2013) Virtual shooting simulators: fundamentals, applicability for military training, and a comparative study of the Brazilian Army. Campinas, Brazil
7. Amorim JA, Matos C, Ana R.M. Cuperschmid, Per M. Gustavsson, Cesar Pozzer (2013) Augmented reality and mixed reality technologies: enhancing training and mission preparation with simulations
8. Laser Shots (2013) Real combat: firearms training system. Product reference sheet. Orlando, United States
9. Elbit Systems (2012) Land forces training systems: platforms and weapon operators training systems. Product reference sheet. Haifa, Israel
10. King-sun Fu, fellow, ice, and Azrael resented, fellow. IEEE, Pattern recognition and image processing
11. Starner T, IEEE (2013) Project glass: an extension of the self
12. MetaVR (2013) Virtual Ft. Benning urban training site. Retrieved August 19, 2013 from http://www.metavr.com/downloads/MetaVR_ FtBenning-Brochure.pdf
13. Kammerer T (2012) TargetHunter for AR.Drone. Retrieved August 19, 2013 from https://itunes.apple.com/us/app/targethunter-for-ar-drone/id416855825?mt=8&ign-mpt=uo=
14. Simulogica (2012) SimIR: Interactive police and security training system. Product reference sheet. Florianópolis, Brazil

Enhancing Parking Accessibility in Shopping Malls: A Comprehensive Study on Challenges and Solutions with License Plate Recognition and NFC Integration

Soumitra Bhoir, Skandabhirama Nadimpalli, Shraddha Golhe, Ria Raina, Sahil Tahlani, and Arunita Paul

1 Introduction

Parking management is an important issue due to the growing number of cars worldwide.

In parking management, parking guidance in order to facilitate the search for vacant parking spaces is a fundamental service, car searching is also an essential service when car owners attempt to find their parked cars [1]. As the parking layout of parking facilities is often quite uniform in the indoor parking garages and may consist of hundreds or thousands of parking spaces, it puts a cognitive load on the car owner's minds to memorize and identify their parking spaces [2]. Meanwhile the management of parking spaces are willing to help customers to find their vehicles as soon as possible to improve customer satisfaction [1]. When parking lots are cluttered with too much or poorly placed signage, it can lead to confusion and lack of clarity. Hence it is of great importance in researching parking navigation for the entire transit

S. Bhoir (✉)
Thane, Maharashtra, India
e-mail: soumitra.bhoir@gmail.com

S. Nadimpalli
Nagpur, Maharashtra, India

S. Golhe
Hyderabad, Telangana, India

R. Raina
Bangalore, Karnataka, India

S. Tahlani · A. Paul
Kolkata, India
e-mail: arunita@karnavatiuniversity.edu.in

© The Author(s), under exclusive license to Springer Nature Singapore Pte Ltd. 2025

S. Das et al. (eds.), *Human-Centered Digitalization, Volume 3*, Design Science and Innovation, https://doi.org/10.1007/978-981-96-4383-7_37

of entering and finding a vacant spot and searching the car while exiting. Over the years, technology has been harnessed to enhance the parking experience in shopping malls. Some of the technologies that have been used include: Utilizing License Plate Recognition or RFID tags to allow seamless entry and exit [3, 4]. Mobile applications provide real-time information on parking availability, helping visitors plan their visit. The global positioning system (GPS) is a conventional and widely used approach for user localization but may not be accurate for indoor environments [2]. In response to the accelerating shift from wired to wireless technology, our study proposes a parking navigation system for indoor parking. Our approach involves the integration of Near Field Communication (NFC) readers for secure user data transmissions and License Plate Recognition (LPR) systems to enable real-time vehicle tracking. By merging NFC readers and LPR systems, we have proposed a system that tackles the cognitive load imposed by the activity of tracking and navigating inside parking spaces, thereby enhancing overall parking experience [3, 5].

2 Methodology

The research paper employs a mixed methods approach to thoroughly explore the parking experience of drivers within designated parking areas at NEXUS Ahmedabad One Mall and Palladium Mall. This investigation utilizes a variety of data collection techniques to offer a well-rounded comprehension of the parking process and its associated determinants. The study meticulously examined various aspects of parking, such as the availability of vacant parking spaces on both weekdays and weekends. Additionally, the research focused on the challenges drivers face in locating their parked cars when exiting the parking facility during both weekdays and weekends. We used the following processes and methods in our mixed methods approach.

Direct observations for understanding real-time parking behavior, questionnaires, semi-structured interviews, customer journey mapping, and visual representation of data. For analysis of data, the SMART method was utilized to set clear and well-defined objectives that guided the direction of our study. The SMART criteria was instrumental in ensuring that our research goals were Specific, Measurable, Relevant, and Time-related.

The combination of quantitative and qualitative data collected through various methods allows for a comprehensive analysis. The findings from this study aim to provide valuable recommendations for optimizing parking facilities, enhancing customer satisfaction, and improving the overall parking journey within the observed shopping malls in Ahmedabad.

3 Existing Systems (Observations)

Understanding the operation of the existing system plays the role of an essential cornerstone in highlighting the loopholes. Specific touchpoints were shortlisted which required special emphasis, the targets are namely:

- The navigation signages in operation
- Effectiveness of alphanumeric allotment
- Token system
- Segmentation in parking.

These specific touchpoints were finalized as it invites a mixture of multiple functions to be performed at a singular time. The inflow of entries was constructed into graphs and pie charts for better understanding and gauging the effectiveness of the existing systems and to forecast a path of action by highlighting areas of discomfort.

The navigation signages in operation: The portion attempts to identify how effective the signages are cognitively to ensure seamless flow of traffic through the layout of the parking.

The recordings were classified under five categories, namely:

- Very ineffective
- Hard to follow
- Satisfactory
- Effective
- Very effective (Fig. 1).

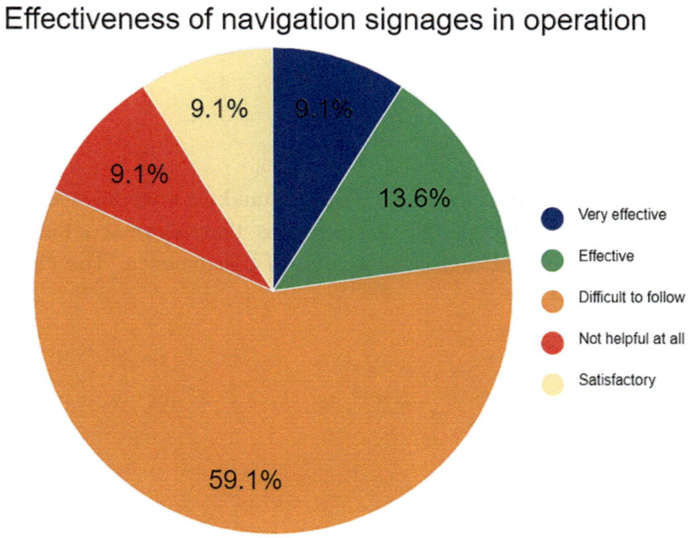

Fig. 1 Effectiveness of navigation signages in operation

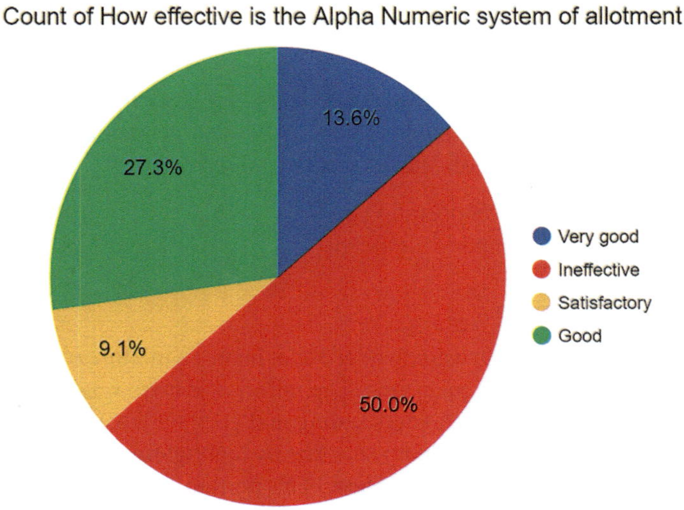

Fig. 2 Count of effectiveness of alphanumeric system

Effectiveness of the Alphanumeric System: A portion of the questionnaire to understand the effectiveness of the alphanumeric system and to uncover points on the cognitive load experienced by users who are subjected to utilizing the system. The survey more importantly focused on understanding how well the system aided people in remembering the numeric in the scenario of retrieving the vehicle.

The questionnaire aimed at segregating experiences under specific options which are as follows:

- Ineffective
- Satisfactory
- Good
- Very Good (Fig. 2).

Effectiveness of the Token System: To understand the convenience of the token, the questionnaire subjected people to questions that addressed the query. The responses were categorized under the degree of convenience. The categorization was as follows:

- Very inconvenient
- Inconvenient
- Satisfactory
- Convenient
- Very convenient (Fig. 3).

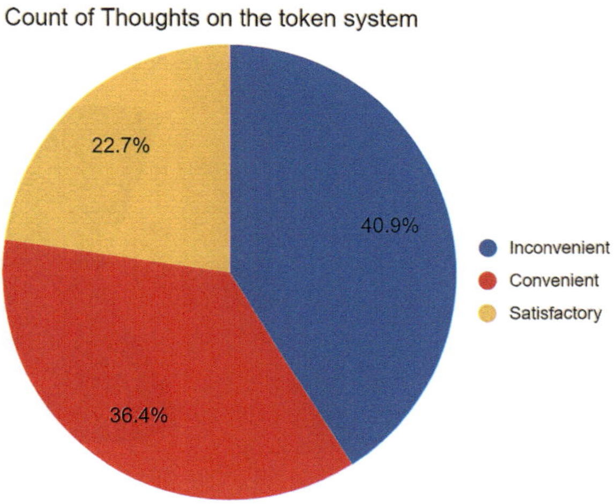

Fig. 3 Count of thoughts on the token system

3.1 User Insights

This section elaborates on the independent variables that were finalized for clearer analysis of the data collected. They are namely:

- Age of the user
- Frequency of visit to the mall
- Mode of transport used to visit the mall.

These variables were selected keeping **time and feasibility** of the research in mind based on the data acquired from the structured interview and observations of the users in the mall, the insights of all 3 variables are as follows:

Age of the Users: The data above helped us in identifying a potential trend of the age group visiting malls during weekdays (Fig. 4).

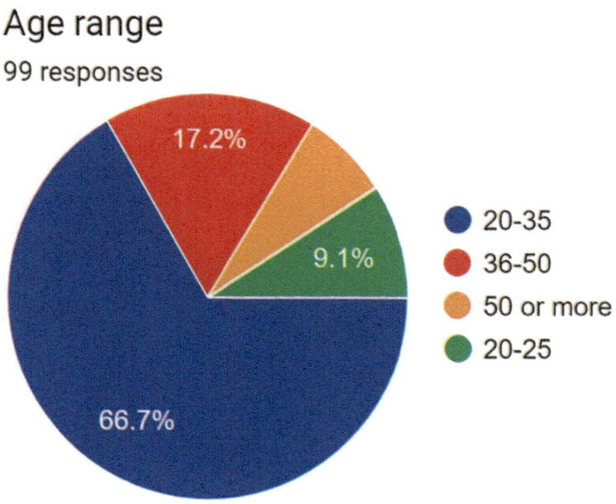

Fig. 4 Age range

Fig. 5 Frequency of visiting shopping malls

Frequency of Visit: These insights derived give us a trend from regular visitors to irregular visitors helping in analyzing the later data (Fig. 5).

Mode of Transportation: The variable mode of transportation gave us insights about the preferred mode among visitors visiting the mall. According to data represented in 1.1(c) 52% of visitors prefer the car as a mode of transportation implying that the majority of the respondents are mall parking users. Additionally 39% of the respondents prefer motorcycles implying that approximately 91% of visitors in total are parking users. The remaining 9% is for public transportation (8%) and other (1%). These findings provide valuable insights into the transportation choices made by the parking users.

Human Presence

During the interview session conducted by us, we got the insight about how human presence makes them feel comfortable and secure. During the observation stage, the key insight derived was that the majority of the users seeked confidence and trust with the presence of the staff (Fig. 6).

Fig. 6 Graph of seek for assistance

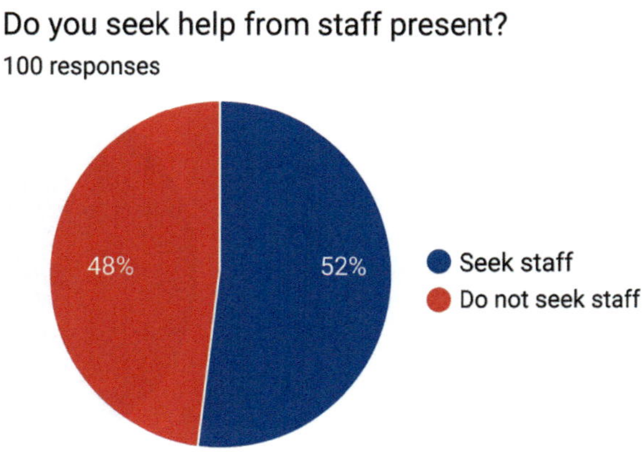

3.2 User Insights

This section provides insights about the 3 dependent variables namely,

- Satisfaction level from parking facilities
- Ease of finding a parking spot
- Preference of parking spot.

The data collected from both the malls is represented through a pie chart combined and individually through line graphs.

Satisfaction Level from Parking Facilities

This variable tells us about the level of satisfaction achieved from the parking facilities offered by the mall to the users. Five categories of responses were recorded namely:

- Very Dissatisfied
- Dissatisfied
- Neutral
- Satisfied
- Very Satisfied (Fig. 7).

Ease of Finding a Parking Spot

These insights provide valuable information about the overall ease of parking within the mall premises (Fig. 8).

Preference of Parking Spot

These findings offer insights into the factors that influence parking spot choices among the users (Fig. 9).

In summary, significant insights on perception and preference were derived from plotting the graphs of the three variables above. Line graphs 1.2(d) and 1.2(e) of palladium and Nexus one were prepared for a better understanding of the impact of the three variables in each mall. The data collected contributed toward

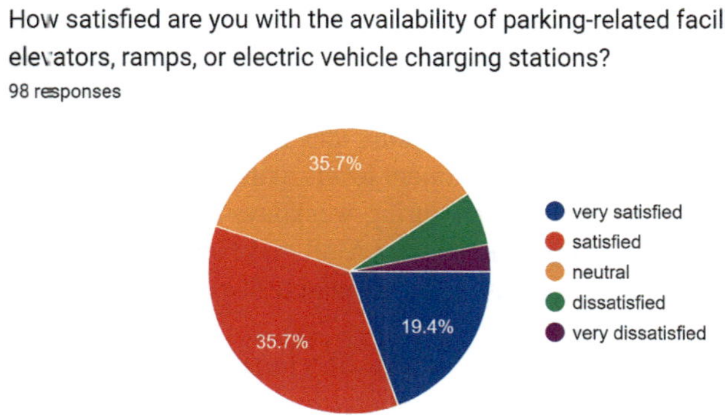

Fig. 7 Satisfaction level graph

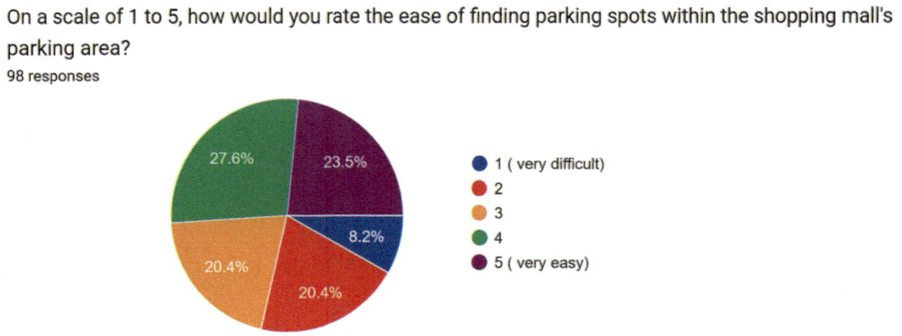

Fig. 8 Rating of the ease of finding parking spots

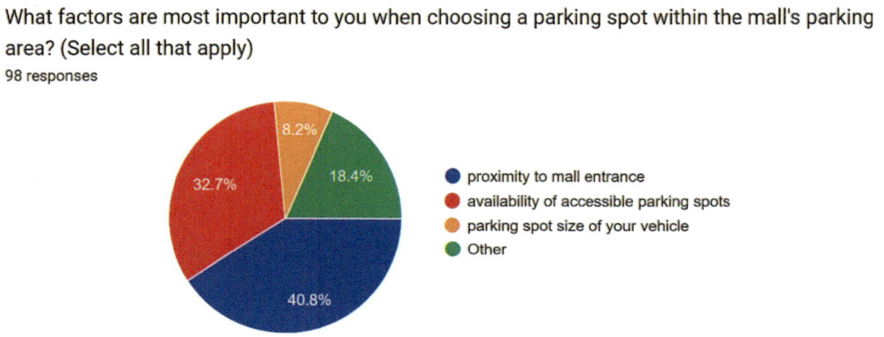

Fig. 9 Important factors while choosing a parking spot

exploratory research and a better understanding of parking experience creating potential implications for further analysis.

3.3 *Weekend Analysis*

The study delves primarily into visitors' evaluations and inclinations, aiming to acquire comprehensive insights into the contentment levels, convenience of discovering parking spaces, and parking preferences during weekdays and an attempt at putting forth a forecast was made keeping Palladium Mall and Nexus Ahmedabad One as subjects. The findings from the forecast revealed notable disparities in visitors' contentment with the parking amenities.

Examining the ease of discovering parking spaces, the forecast unveils diverse experiences among the drivers, keeping the conditions for observations constant (Figs. 10, 11, and 12).

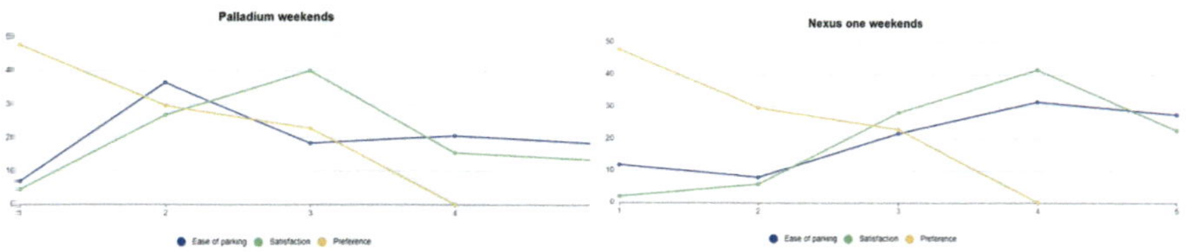

Fig. 10 Palladium and Nexus Mall weekend analysis

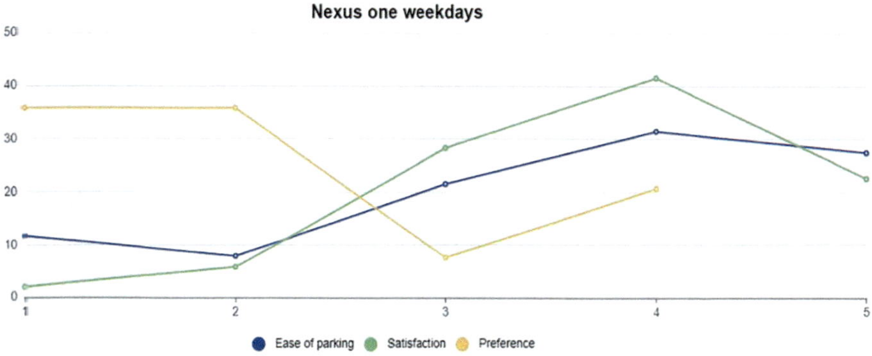

Fig. 11 Nexus Mall weekday analysis

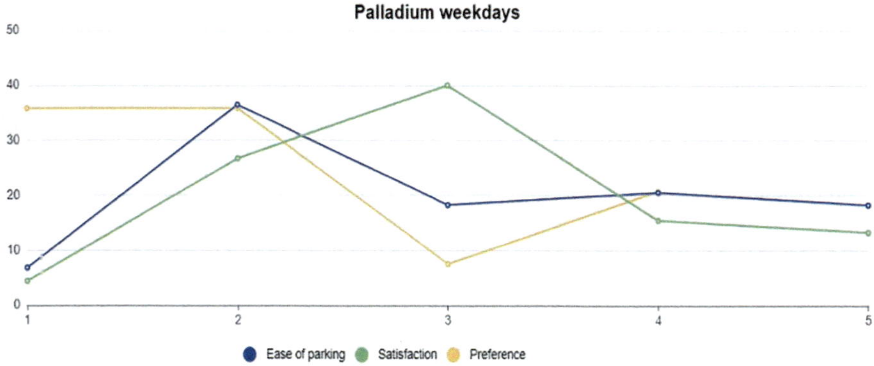

Fig. 12 Palladium Mall weekday analysis

4 System Design

4.1 System Overview

The proposed system procedure can be divided into three processes:

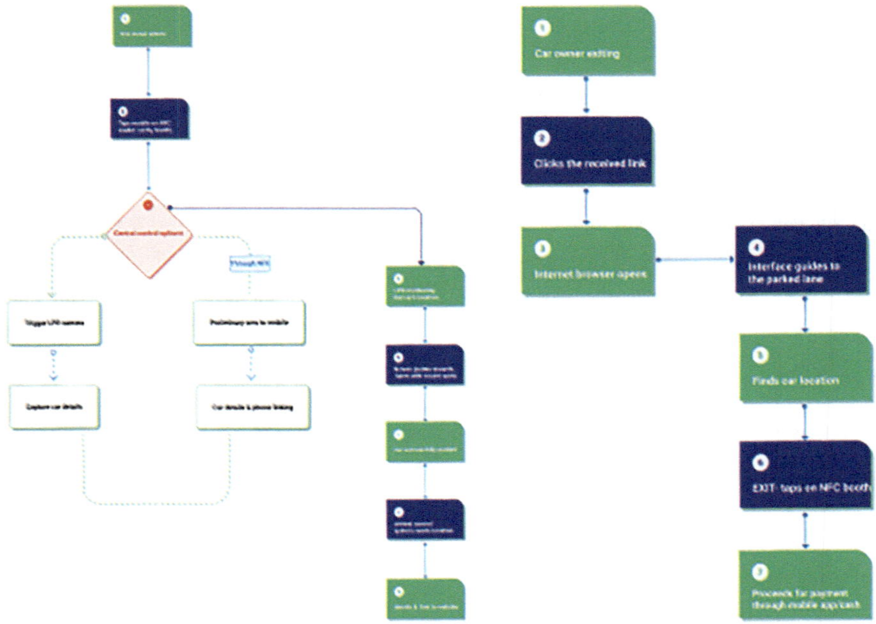

Fig. 13 System design

1. User taps the NFC reader on the entry booth, then the LPR camera identifies the car and links it to the mobile tapped.
2. At the time of return a link is provided to the user which directs them to the web browser showing the fastest route to the lane that the car is parked in.
3. At exit the car owner can tap on the NFC reader on the exit booth that would send them to a payment portal to choose to pay in cash or to their preferred online payment app (Fig. 13).

4.2 NFC

Near Field Communication (NFC) is a short-range wireless protocol with applications in secure mobile payments, data exchange, and IoT connectivity. Objects with NFC tags connect directly to smartphones, tablets, or intermediary readers, all linked to the Internet. High-end items use silicon CMOS NFC tags, while low-cost products opt for thin-film transistor technology. In NFC systems, a reader device powers NFC tags without batteries, using inductive coupling to transmit data via a 13.56 MHz carrier wave, facilitated by a rectifier and a load modulator for amplitude shift keying (ASK) modulation [6].

4.3 License Plate Reader (LPR)

A License Plate Reader (LPR), utilizes optical character recognition to automatically capture and interpret license plates on vehicles. LPR systems can extract alphanumeric characters from plate images, transforming them into machine-readable text. These systems find extensive use in security and law enforcement for identifying stolen vehicles, apprehending suspects, and monitoring vehicles linked to criminal activities. LPR technology also streamlines parking management, traffic monitoring, border control, and custom inspections, enhances vehicle tracking in logistics, and assists in parking enforcement. It relies on specialized cameras and real-time image recognition software, significantly contributing to security and efficiency across various sectors.

4.4 Workflow

Cognitive load refers to the mental effort and capacity required to process information, solve problems, make decisions, or perform various cognitive tasks. There is a lot of cognitive load created by the whole parking and exiting process. When the car enters the parking facility they have to take a token. It creates a cognitive load to keep it safe and remembering where the token was kept. The token system increases wait time obstructing flow of traffic at the exit. It also wastes paper which is not sustainable. In the proposed navigation system users tap their mobile phone on a NFC reader placed on the entrance booth. The NFC communicates information about the phone to the central control system and sends a preliminary SMS to the mobile device saying welcome to our parking facility [7]. After the tap the central control system triggers the LPR camera which captures the car details. The central control system hence identifies the car and links the data to the mobile device that was tapped. The LPR system then tracks the cars movements where it is traveling in the facilities which turn it is taking and which lane it is parked in. As the car moves ahead from the entrance booth there would be panels on each turn that would display the number of vacant spots available on that lane after taking that turn. Hence there would be no slow moving cars congesting the traffic wondering where to park looking for a vacant parking spot. After the car is parked the LPR notifies the system parked location and sends a message on the mobile device linked to that car [3, 6]. The message contains information as in date, time, parking spot number, lane number and a Clickable Link. After coming back to the parking facility while exiting, the car owners can click on this link and view the fastest route recommended on the interface on the web browser. Once the owner finds the car they could just tap on the NFC reader on the exit booth that would send them to a payment portal to choose to pay in cash or to their preferred online payment app [5].

5 Discussion

Mall systems often cause frustration due to improper signage and long exit waits. With the introduction of NFC readers, the token system gets eliminated thereby reducing wait time making the experience better. High mall traffic also made parking accessibility a problem which has been resolved through the solution proposed by us. We did SWOT analysis of the proposed concept to get a detailed analysis of the solution.

SWOT analysis of the existing system was performed which helped us identify the limitations and scalability of the proposed idea. The SWOT analysis is mentioned below:

The proposed system streamlines transit, reducing wait time and boosting efficiency.

Based on the survey collected, our solution proposed enhances convenience, customer satisfaction and accessibility providing accurate navigation and accessibility to their vehicles.

With the rapid advancement in technology, integration of NFC readers and LPR sensors for innovative user experience. The conventional token system has been eliminated thereby reducing the cognitive load of the user, providing them with a hassle free experience. Technical failures in the system could lead to disruption and dissatisfaction among the users.

With the rise in cybertheft, data sharing becomes a serious concern among the users.

Resistance to using newer technology due to lack of trust and comfort is still prevalent in the market. The solution could help boost revenue, engage brands and attract customers.

The stored data enhances understanding of user behavior for improved consumer service.

Custom options help in building trust, comfort thereby enhancing engagement and revenue. Shifting data and security regulations may raise costs. Relying on NFC readers and LPR sensors poses a disruption risk. The link provided would help users reach their parking lane easily but on arrival they would have to navigate their own vehicle. For example—The lane H2 has 5 cars as the maximum capacity so the user would be guided to H5 and then will have to navigate their vehicle. The data analysis during weekends has been done on an approximation which might change the capacity of cars in a single lane and create problems in navigation.

6 Conclusions

In this study, we've thoroughly examined the challenges surrounding parking accessibility in shopping malls. A thorough investigation was done regarding Issues such as inefficiency in finding vacant parking spots, delay due to inefficiency of the token system, the cognitive load on visitors while finding the parked car and congestion at entry and exit points have been identified as key pain points. To address these challenges, we've proposed a practical solution that integrates License Plate Recognition (LPR) technology and Near Field Communication (NFC) systems, aimed at enhancing the parking journey. In conclusion, our research and proposed solution are designed to enhance parking accessibility in shopping malls, leading to improved user satisfaction and a more enjoyable shopping experience, making the process more user-friendly and less mentally draining for both drivers and mall management. In the future, this solution could help increase potential revenue, involvement of brand, and customer engagement.

Acknowledgements The authors thank Dr. Arunachalam Muthiah, Prof. Shyam Prajapati, who provided expertise that greatly assisted the research, and we also thank our colleagues T Sai Aayush and Pranav Manikanandan for their contributions.

References

1 Sheng SW (2018) A BLE-based Pedestrian navigation system for carsearching in indoor parking garages. MDPI, Hsinchu 30401
2. Chen Z, Fu J, Sun R, Han H (2014) QR code location based reverse car-searching route recommendation model. In: 2014 IEEE international conference on vehicular electronics and safety (ICVES). Jinan, China
3. Rahman CA, Badawy W, Radmanesh A (2023) A real time vehicle's license plate recognition system. In: IEEE conference. Calgary, Alberta, Canada
4. Rana ME, Shanmugam K, Thian YY (2021) RFID based on-the-spot navigation system for car parks. J Appl Technol Innov 5(4). e-ISSN: 2600-7304
5. Juntunen A, Luukkainen S, Tuunainen VK (2010) Deploying NFC technology for mobile ticketing services. In: Ninth international conference on mobile business. Espoo, Finland
6. Myny K, Tripathi AK, van der Steen J-L, Cobb B (2015) Flexible thin-film NFC tags. In: IEEE communications magazine
7. Chang S-L, Chen L-S, Chung Y-C, Chen S-W (2023) Automatic license plate recognition. IEEE
8. Baglane SB, Kulkarni MS, Raut SS, Khatavkar TS (2014) Parking management system. IJMER, Pune, India
9. Agrawal KG, Sadhani S, Khandelwal S, Koul S (2015) Parking navigation and payment system using IR sensors and RFID technology. Int J Comput Appl 111(15):0975–8887
10. Naick M, Siridha N, Divya P, Madhuri P (2021) Parking space detection using image processing. IJSRSET 8(4):211–215
11. Du H (2013) NFC technology: today and tomorrow. Int J Futur Comput Commun 2(4)
12. Noto M, Sato H (2000) A method for the shortest path search by extended Dijkstra Algorithm. Yokohama, Japan 221-8686
13. Paidi V, Fleyeh H, Håkansson J, Nyberg RG (2018) Smart parking sensors, technologies and applications for open parking lots: a review. IET J 12(8):735–741
14. Montañés JAP, Adriana, Rodríguez M, Prieto IS (2013) Smart indoor positioning/location and navigation: a lightweight approach. Int J Artif Intell Interact Multimed 2(2)

Enhancing Campus Mobility: Integrating Sustainable Transport Solutions for University Environments

Sabyasachi Biswas, Uma Adhyaru, and Sandipan Bhattacharjee

1 Introduction

Newer universities are emerging in the regions outside the main city in India. Out of the 42,343 colleges in India, 60.56% are in the rural areas, stated the All India Survey on Higher Education (AISHE) report 2019–20 released on June 10. Due to oversaturation in urban areas, it is obvious that newer colleges and schools are built in the rural areas, moreover majority of Indian population lives in rural areas. Consequently, students and faculty members at these universities encounter connectivity problems. Typically, these institutions are located on the outskirts and have extensive campuses due to the affordability and accessibility of land. As a result, they feature a diverse array of buildings and present challenges in navigating the vast college grounds. In such scenarios, there is often a considerable distance between campus structures and dormitories, leading to a sense of detachment from the city itself. This can result in tiresome and costly commutes, especially due to limited transportation infrastructure in these areas. Additionally, the presence of narrow walkways and scattered facilities forces students to cover long distances on foot frequently. Furthermore, we have recognized the need to establish better connections between the university and other parts of the city to reduce transportation expenses and minimize the need for walking for errands.

Indian Tier 1 cities are well connected with the increase of public transport and metros, India still lacks proper public transport facilities in the outskirts of cities or

S. Biswas (✉)
School of Architecture and Planning, DIT University, Dehradun, India
e-mail: sabyasachi.biswas@ditiuniversity.edu.in

U. Adhyaru
Department of Industrial Design, Karnavati University, Gandhinagar, India

S. Bhattacharjee
Department of Communication Design, Karnavati University, Gandhinagar, India
e-mail: sandipan@karnavatiuniversity.edu.in

© The Author(s), under exclusive license to Springer Nature Singapore Pte Ltd. 2025

S. Das et al. (eds.), *Human-Centered Digitalization, Volume 3*, Design Science and Innovation, https://doi.org/10.1007/978-981-96-4383-7_38

rural areas. Rural population has always relied on private transport as their primary mode of travel. The considerable distances between destinations, the distributed local population, and the irregular visits of migrating residents have posed difficulties in establishing an effective public transportation system as a reliable substitute for personal vehicles. Nevertheless, given the pressing environmental concerns of today's climate crisis, it is imperative to discover and put into action sustainable rural mobility solutions that are environmentally friendly, inclusive, and feasible [1].

Another determinant of rural–urban transport is that the residents of higher-income districts tend to opt for private motorized transportation, such as cars, for their daily commutes. Conversely, individuals in poorer regions rely more on non-motorized modes and public transportation, often facing the necessity of longer commutes [2, 98–106]. Private transport dominates the rural areas, but problems are faced by those without access to them. Therefore, service provision should focus on making the best use of personal vehicles, along with exploring creative ways to promote public transport. In rural areas, along with the challenge of having to travel long distances to access jobs, other social factors also make it difficult to get the same opportunities. This research paper aims to integrate sustainable transportation planning into expansive university campuses in remote areas. Its primary goal is to discover ways to connect remote areas within the campus and establish eco-friendly transportation links to off-campus destinations like residential areas, bus stops, or taxi stands. Addressing this issue necessitates a comprehensive understanding of traffic planning and land use, adopting a holistic approach. The objective is to reduce the reliance on single-occupancy vehicles, as students frequently use personal cars for commuting, and instead, introduce an environmentally friendly and sustainable transportation network that facilitates convenient travel for both short and long distances.

2 Design Process

We employed the "Double-diamond" design process to develop solutions for this issue. The first step was to understand the problems people are facing. That's when we realized that even though having a huge campus is a blessing, it is getting tiring to cover the distance between buildings. Therefore, initially, we conducted secondary research to gather information regarding the problem, including its causes, potential solutions, and relevant data, through research papers and articles. Using this data as a foundation, we devised questionnaires and surveys to ascertain whether users genuinely perceived this as a problem and, if so, to gather statistical insights. Based on the answers we received, we created user personas signifying various groups of users. Personas were used to support the user journey maps and empathy maps. These tools helped us empathize with the users and understand their pain points and gain points, which served as the basis of the ideation. Many ideas emerged in the ideation stage, which were determined through trial and error methods.

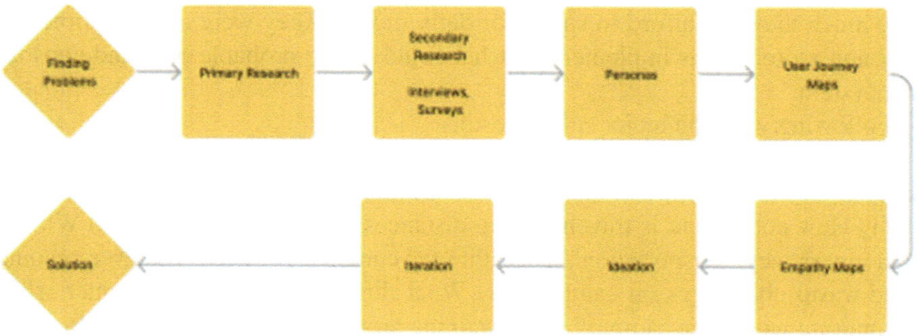

3 Literature Review

We began our research by learning about how different universities manage this problem of distance. While researching we found that UCLA offers various transportation options for its students, including biking, UCLA SafeRide, and the BruinBus service. Riding a bicycle is encouraged with over 3500 free bike parking spaces and newly added bike lanes around campus. UCLA SafeRide provides safe, reliable evening transportation via vans and buses, and the Safe Loop service runs approximately every 30–55 min. The BruinBus is a year-round bus service serving the UCLA campus and Wilshire Center. The TripShot app allows us to request "on-demand" rides from Monday to Thursday, between 7:00 p.m. and 12:00 a.m.

Purdue University offers free rides for its students, faculty, staff, and retirees on all CityBus routes throughout Lafayette and West Lafayette.

The University of Kansas provides free bus rides in Lawrence for students, faculty, and staff with their KU Card, creating a "fare-free" zone. Services are coordinated through a single Transit Guide, website, and customer service contact. Harvard University promotes biking and walking with bike corridors, Bluebikes stations, and an interactive map for bicycle facilities on campus. They also offer shuttle services, with real-time information available through the Passio Go! App.

4 Methodology

As mentioned earlier, the first step was to understand the problems people are facing. Therefore, after we received the bigger picture of the problem and various measures taken by other universities around the globe, we devised questionnaires and surveys to build a foundation for our research. The survey form included questions regarding campus mobility like the mode of transport chosen to travel to various parts of the campus, and if cycle/non-motorized mode of transport is a viable option. They were also questioned about their preferred mode of transport for long-distance travel and how much they can afford to spend for daily travels. They were also enquired if a new non-motorized is implemented, what could be the probable pros and cons for that system.

Following were the major survey results.

The purpose of the interviews was to understand the authentic problems of the users in the campus without bias. Some of the questions asked were

(1) How convenient is traveling long distances on campus for you? (2) What is your emergency transportation plan within the campus? (3) Have you ever encountered travel challenges on campus? (4) What difficulties do you encounter while commuting within the campus? (5) Is walking around campus often hectic for you? (6) How would you travel to the extended campus in an immediate situation? (7) Does weather conditions impact your campus commuting experience, and how? (8) How can we improve the campus commuting experience? (9) Could golf carts be a useful addition to campus transportation? (Fig. 1).

From the list of user problems, we sorted out common ones:

(A) Weather conditions (B) Infrastructure (C) Parking's are away (D) Have to walk a lot (E) Carrying weight and too many items while walking becomes difficult.

Through surveys we also understood that students face problems while traveling to nearby spots like cafes and shops, and transport hubs like bus stands or auto stands, because they are distant and tiring if you walk and expensive if we take an auto. Hence there was also a need to establish connections between these nearby stations and the college campuses.

User personas were created signifying various user groups. Personas were used to support the user journey maps and empathy maps. These tools helped us empathize with the users and understand their pain points and gain points, which served as the basis of the ideation. Many ideas emerged in the ideation stage, which were determined through card sorting, iceberg model, and other trial and error methods (Figs. 2 and 3).

After brainstorming for possible solutions on problems, it is hinted that there is a need for a vehicle or a transport system inside and between two campuses and around.

5 Probable Solution

Notably, 86.2% of the respondents expressed a need for a transportation system as their preferred solution. They believed that having a vehicle would also be advantageous in extreme weather conditions, such as intense heat, dust, and rain. We utilized tools like empathy maps, consumer journey maps, and personas to shape designs that are centered around the needs of the users. Given that our campus, like many other large campuses, is situated in a remote location, there arose a necessity to establish a link between the college and transportation hubs like bus stops and auto shuttle stands, which, in turn, connect to other parts of the city. Consequently, students can easily access bicycles from designated stands when required, enabling them to

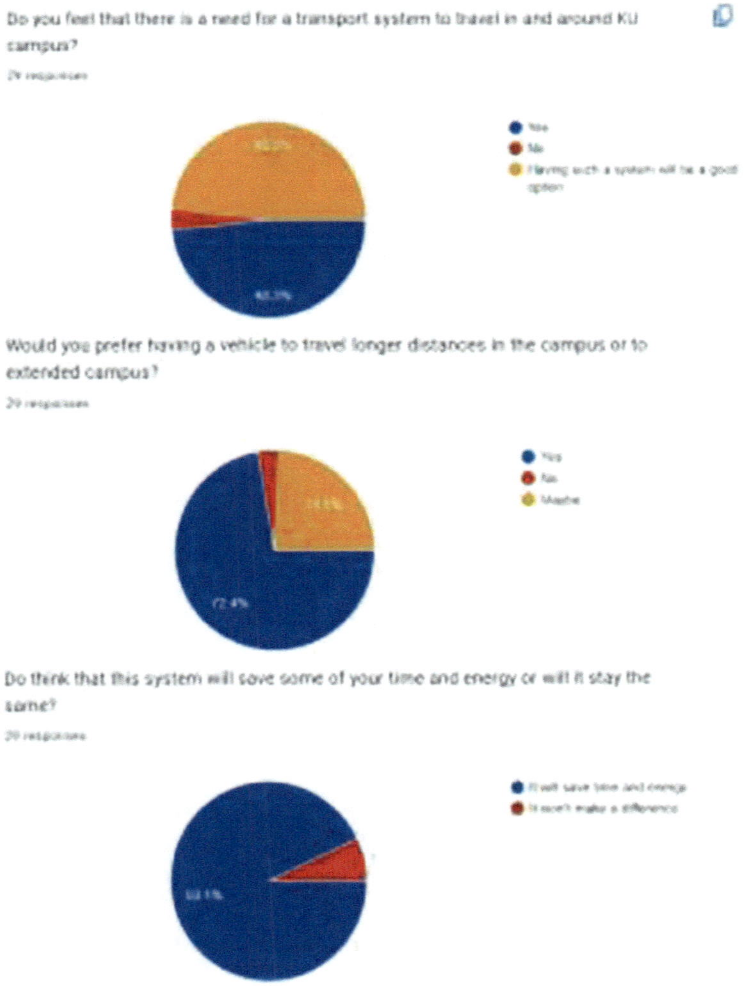

Fig. 1 Pie chart showing survey results

complete errands and travel within the campus. The choice of vehicle centers around the concept of a non-motorized, environmentally friendly mode of transportation.

The initial study included surveying 33 surveys and 10 interviews. The survey showed that 100% of people take multiple rounds around campus of which 86.2% of people have experienced exhaustion and fatigue from walking. Therefore, 75.5% of people opted for an autorickshaw and paid ₹20–₹30 for one way travel for a distance of less than 500 m. 48.3% completely agreed that there is a need for a travel system on campus, another 48.3% felt that it would be a good option, the rest 3.4% (one person) felt there was not. 93.1% of respondents think that it will save their time and energy.

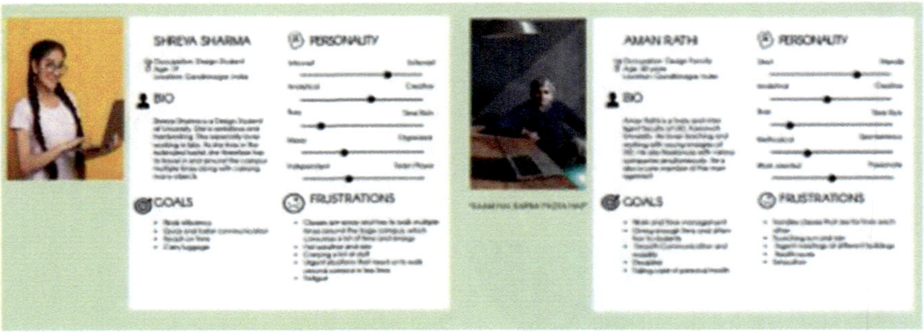

Fig. 2 User persona

Fig. 3 Ideations

To solve this issue of constant commutation in the campus and around, we propose the idea of creating a transportation network. This network is also referred to as "*Last-mile Transport services*." "Last mile transportation refers to the provision of transport services from mass transit hubs to the final destinations around them [1], [2]." Pengfeng et al. [3] would consist of a vehicle system network which has stands placed at regular intervals in the entire campus, which will be used by students to travel around the campus when in need. They can park it in the nearest stand to the building. Lock-unlock works by scanning the student-faculty ID card. Considering the narrow roads of the campus and individual traveling, a bicycle or an e-scooter was the initial

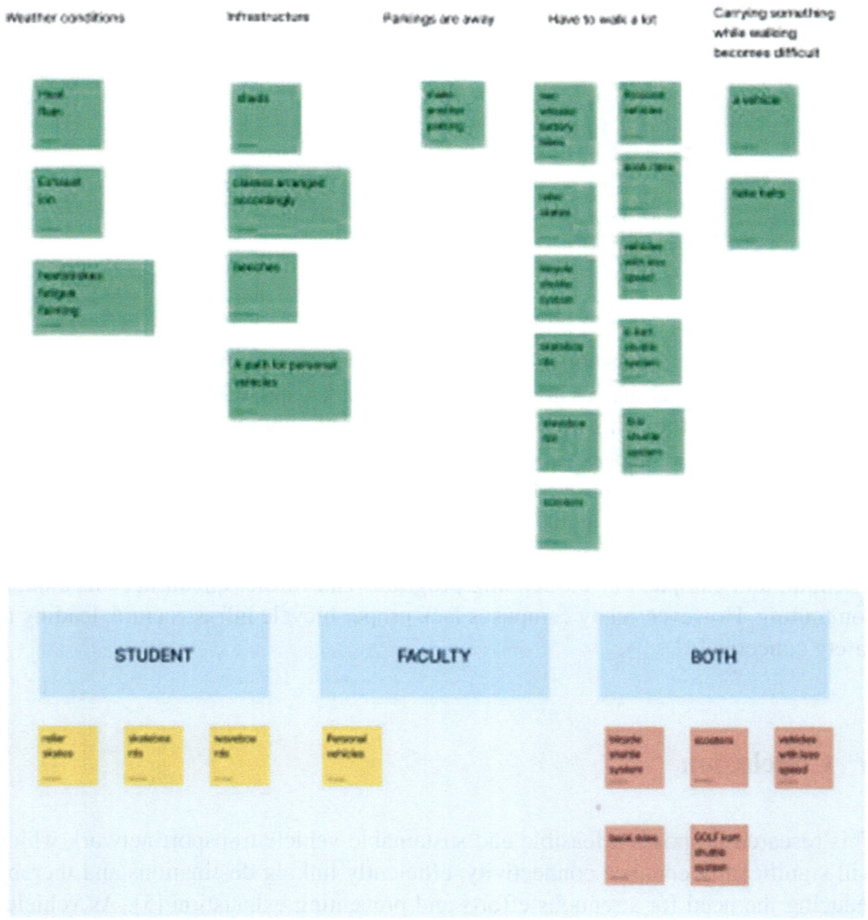

Fig. 3 (continued)

idea. But, with more research, we can come up with a non-motorized, sustainable vehicle for last mile connectivity. For better experience an app was thought of and medium fidelity user interfaces were made:

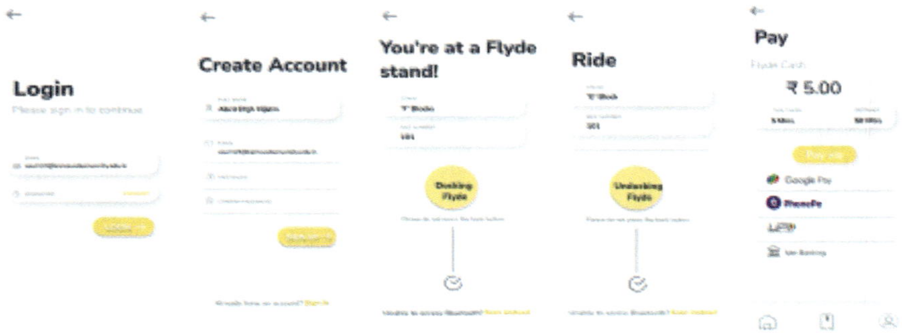

The reflections on sustainable transportation planning on campuses extend to various environments such as hospitals, research parks, and office complexes, aiming to reduce exhaustion. This approach involves encouraging biking and ridesharing. Bicycles are highlighted as a pollution-free, space-efficient, and accessible mode of transportation on campuses. UC boulder serves as an example with its comprehensive bicycle network, including multi-use paths and dismount zones to ensure safety. To promote cycling, secure bike parking, and bike racks on buses are essential, and campus bicycle-lending programs can further encourage sustainable commuting. However, many campuses lack proper bicycle infrastructure, leading to safety concerns [4].

6 Conclusion

This research proposes a feasible and sustainable vehicle transport network which will significantly enhance connectivity, efficiently linking destinations and thereby reducing the need for strenuous efforts and preventing exhaustion [5]. As vehicles like bicycles and e-scooters work the best and are the preferred options for this network, it creates a sustainable and environment friendly ecosystem, which is fun to use. It also reduces single-occupancy vehicle use and is pocket friendly.

This research paper aims to integrate sustainable transport planning within and around expansive university campuses. The primary objective is to establish efficient connections between remote points within the campus as well as to external locations such as residential areas, bus stations or auto stands—through a sustainable vehicle network. University campuses, often situated on the outskirts of a city, present challenges due to vast distances between buildings and limited connectivity to urban centers. This results in inconvenient, tiresome, and costly commutes, worsened by the inadequate transport infrastructure. To address this, the research would undergo an intensive understanding of traffic planning and land use as a cohesive approach. This reduces single-occupancy vehicle use as students prefer using personal vehicles to travel around; and introduces an environmentally friendly, sustainable transportation network and helps people travel to near-far places easefully. E-scooters can be another proposal for a transport vehicle. According to Rebecca Sanders and team,

"E-scooters are primarily valued for their convenience and the ability to get somewhere faster than walking while not having to drive; they are also generally considered fun to ride. Few of the perceived benefits differed significantly by sociodemographic, suggesting that e-scooters may help fill an urban transportation niche. This niche may be particularly relevant for areas where cars are discouraged or impractical, such as on or around university campuses, and in congested urban areas" [6, 217–227].

When the vehicle travels to other places out of campus, there needs to be proper arrangement of bike lanes that do not hamper and expose threat to the driver. According to Dehghanmongabadi A. and Hoşkara Ş., "Based on findings of the questionnaire survey and visual analysis carried out on the site, there are no appropriate motivating factors to encourage cycling among EMU commuters. Besides the fact that bike lanes are considered to be inappropriately-shared space on vehicular roads, bicycle parking and racks are efficiently located around campus and their conditions are not always suitable for use. There is also a lack of security for bike stations, shading elements, signage for cyclists, showering facilities, and ancillary services" [5].

Also from the survey it is understood that users feel a need to incorporate a vehicle, which not only helps them save energy in the exhausting heat, but also helps them travel a little faster than walking and carrying their belongings.

References

1. Poltimäe H, Rehema M, Raun J et al (2022) In search of sustainable and inclusive mobility solutions for rural areas. Eur Transp Res Rev 14(13). https://doi.org/10.1186/s12544-022-00536-3
2. Jain M, Korzhenevych A, Hecht R (2018) Determinants of commuting patterns in a rural-urban megaregion of India. Transp Policy 68:98–106. https://doi.org/10.1016/j.tranpol.2018.04.018

3. Pengfeng S, Sun Y, Xie B, Xiu Xu S, Xu G (2021) Data-driven shuttle service design for sustainable last mile transportation. Adv Eng Inf 49. https://doi.org/10.1016/j.aei.2021.101344
4. Balsas CJL (2003) Sustainable transportation planning on college campuses, vol 10. Elsevier. https://doi.org/10.1016/S0967-070X(02)00028-8
5. Dehghanmongabadi A, Hoşkara Ş (2018) Challenges of promoting sustainable mobility on university campuses: the case of eastern Mediterranean University, vol 10. MDPI. https://doi.org/10.3390/su10124842
6. Sanders RL, Branion-Calles M, Nelson TA (2020) To scoot or not to scoot: findings from a recent survey about the benefits and barriers of using E-scooters for riders and non-riders. Transp Res Part Policy Pract 139:217–227. https://doi.org/10.1016/j.tra.2020.07.009
7. Mathew JK, Liu M, Seeder S, Li H, Bullock DM (2019) Analysis of E-scooter trips and their temporal usage patterns. Sch J 89(6):44–49. https://www.proquest.com/scholarly-journals/analysis-e-scooter-trips-their-temporal-usage/docview/2251991862/se-2
8. Papantoniou P, Yannis G, Vlahogianni E, Attard M, Regattieri A, Piana F, Pilati F (2020) Developing a sustainable mobility action plan for university campuses. Transp Res Procedia 48:1908–1917. https://doi.org/10.1016/j.trpro.2020.08.223
9. Toor W, Havlick SW (2004) Transportation and sustainable campus communities: issues, examples, solutions. Island Press, United Kingdom
10. Mathew J, Liu M, Li H, Seeder S, Bullock D (2019) Analysis of E-scooter trips and their temporal usage patterns. ITE J 89:44–49
11. SafeRide, NYU. https://www.nyu.edu/life/travel-and-transportation/university-transportation/safe-ride-van-service.html
12. Tripshot on-demand. https://www.tripshot.com/on-demand/. https://www.tripshot.com/resources/?category=Shared%20Shuttles
13. https://www.micro-scooters.co.uk/
14. Kost C, Mwaura N, Jani A, Van Eyken C (nd) Streets for walking & cycling. UN-Habitat & Institute for Transportation and Development Policy. https://www.transformative-mobility.org/wp-content/uploads/2023/03/Streets-for-walking-and-cycling-Web-Version-uCC0Dx.pdf
15. Wang H (2012) Approximating the performance of a last mile transportation system [Submitted to Sloan School of Management, and Department of Civil and Environmental Engineering, in partial fulfillment of the requirements for the degrees of master of science in operations research and master of science in transportation at the Massachusetts]. Massachusetts Institute of Technology. https://dspace.mit.edu/bitstream/handle/1721.1/78148/829296935-MIT.pdf?sequence=2
16. Verma A, Harsha V, Subramanian GH (2021) Evolution of urban transportation policies in India: a review and analysis. Transp Dev Econ 7:25. https://doi.org/10.1007/s40890-021-00136-1
17. Qwe'Vontae Eure (2020) Home overview highlights factors that influence city micro-mobility: an understanding of campus e-scooter use [Poster Session]. Learning Commons, Atrium. https://digitalcommons.odu.edu/undergradsymposium/2020/postersession/
18. Ramaswamy NS (1998) Rural transport in India. Curr Sci 75(8):800–803. http://www.jstor.org/stable/24101627
19. Nadimuthu LPR, Victor K (2021) Performance analysis and optimization of solar-powered E-rickshaw for environmental sustainability in rural transportation. Environ Sci Pollut Res 28:34278–34289. https://doi.org/10.1007/s11356-021-12894-x
20. Singh SK (2005) Review of urban transportation in India. J Public Transp 8(1):79–97. https://doi.org/10.5038/2375-0901.8.1.5
21. AISHE Final Report 2019–20—English, Government of India, Ministry of Education. www.aishe.gov.in
22. https://www.niche.com/blog/5-advantages-of-going-to-college-in-a-rural-area/#:~:text=Rural%20areas%20typically%20have%20low,Bown%20College
23. Transportation and sustainable campus communities: issues, examples, solutions, Will Toor, Spenser Woodworth Havlick, Island Press, 2004, 293 pp. https://books.google.co.in/books?id=YTFPnuCuLjwC

State-of-the-Art Knowledge in Industry 4.0-Enabling Technology Adoption Challenges in Assembly-Line Manufacturing: A Cognitive Ergonomics Perspective

Vidyotama Sandal, Urmi R. Salve, and Sougata Karmakar

1 Introduction

The development of Industry 4.0-enabled advanced digital technologies and their application within various industrial sectors, especially manufacturing, has been one of the most popular and extensive topics of interest in the scientific research community. The Industry 4.0 concept integrates cyber-physical systems (CPS) and technologies, such as the Internet of Things (IoT), cloud computing, and big data analytics.

Adopting and implementing I4.0-enabled technology in manufacturing has significant implications for human operators. An ideal combination of these technologies is required to fully exploit their potential and deliver significant impacts on the business. [1] Integrating I4.0 solutions and practices can improve resilience and responsiveness, mitigating disruptive operational, and financial performance effects. [2] However, adopting I4.0 technologies requires workers to adapt to new roles and acquire the necessary knowledge and training. [3]

Industry 4.0 requires a different type of worker capable of performing cognitive work and participating in decision-making. [4] The design of I4.0 technologies should also consider ergonomic and flexible aspects to support operators in complex tasks. [5] Adopting I4.0 technologies in the manufacturing industry has territorial effects, with a primal influence of geographical location and regional characteristics on technology adoption and implementation. The contributing factors include the availability of skilled labour, proximity to technology providers, quality of regional infrastructure, and presence of innovation ecosystems among the many [6].

V. Sandal · U. R. Salve · S. Karmakar (✉)
Department of Design, Indian Institute of Technology, Guwahati, India
e-mail: karmakar.sougata@gmail.com

© The Author(s), under exclusive license to Springer Nature Singapore Pte Ltd. 2025
S. Das et al. (eds.), *Human-Centered Digitalization, Volume 3*, Design Science and Innovation, https://doi.org/10.1007/978-981-96-4383-7_39

Existing literature allows the researchers to assume that technology adoption is lower than expected, which can be perceived as an implication of lower contextual rates, mostly due to proximity constraints. Geographical factors like distance between manufacturing facilities, technology hubs, and support institutes also hinder these adoption rates. This likelihood further decreases as distance from available infrastructure and research support increases. Lower adoption rates exist with fewer resources and lower access to technological support.

The study aims to investigate the available scientific literature on the effect of technology adoption barriers and implementation patterns on the human operators involved in manufacturing systems equipped with Industry 4.0-enabled technologies. A better understanding of the factors that influence and require skills for successful technology adoption by manufacturing firms will allow and improve the survival of the workforce in the industrial environment. Knowledge from this review can help guide the future research agenda and direct the community towards human-centric and sustainable technology adoption in Industry 4.0 and, consequently, Industry 5.0 scenarios. The identified research gaps are expected to assist the establishment of the 'Smart Factory' concept with higher effectiveness and efficiency in the manufacturing sector.

Following the framework, a scoping review was initiated, identifying relevant research questions and setting the required parameters [7]. The literature must observe the assembly-line manufacturing system as a whole and not as the individual components while also considering the significant involvement of human operators as an essential resource within the system. For experimental studies, the manufacturing system must be studied as an entity, with the literature including participant characteristics and objectives of the study. The following research questions were formed to guide the review.

1. What are the relevant contributions of existing literature towards Industry 4.0 technology adoption regarding assembly-line manufacturing systems?
2. What are the identified adoption barriers and implementation patterns for Industry 4.0-enabling technologies for manufacturing?
3. What are the influencing factors and required skills that impact the cognitive capabilities of the human operators within Industry 4.0-enabled manufacturing systems?

2 Methodology

The authors performed a scoping review to address the research questions and map the concepts concerning areas of interest. The review methodology was selected due to the high count of available publications studying and discussing the implications of the Fourth Industrial Revolution in various areas of implementation. The leniency in not requiring well-structured research questions or hypothesis testing also encouraged the authors to pursue a scoping review over a systematic literature review, as the methodology provides an overview of the research domain and opportunities for open-ended discussions for the research community.

The literature search was performed in March 2023 and was refreshed with a follow-up attempt in September 2023 in the SCOPUS database using the keywords mentioned below. The search string sets and keywords were categorised per the formulated research questions.

The keywords "Industry 4.0", "Assembly 4.0", and "Operator 4.0" were used to search for articles addressing the Fourth Industrial Revolution. "Technology Adoption" and "Technology Implementation" were used to cross-check studies on the phenomenon. The authors used the strings "adoption barriers", "implementation barriers", "adoption patterns", and "implementation patterns" to address RQ2. Consideration for RQ3 included the following keywords and strings, "critical success factors", "influencing factors", and "required skills". Finally, the keywords "human factors" and "cognitive capabilities" were also included to find studies focusing on the ergonomics aspect of Industry 4.0-enabled technology implementation. TABLE provides details of the search strings and their yielded search results from the SCOPUS database (Table 1).

The search was limited to journal articles published in English between 2013 and 2023, with further restrictions to 2018 as the lower limit. The relevance of the literature was assessed based on the titles and abstracts to identify articles that focused on Industry 4.0 implementation within the manufacturing sector and addressed at least one of the established research questions. Review studies and studies not conducted within manufacturing practices were excluded (Table 2).

Snowball sampling methodology was used, along with forward citation, to ensure no relevant articles were excluded during the search. The initial and final screening resulted in 42 research articles, which were then read through. The full-text analysis of the selected articles focused on identifying adoption barriers and enablers for Industry 4.0-enabling technologies in manufacturing while also identifying factors that influenced the cognitive performance of human operators involved in such environments.

Table 1 Strings and keywords used for database search

TITLE-ABS-KEY	Article count (Post-exclusion)
("Industry 4.0" OR "Assembly 4.0" OR "Operator 4.0") AND ("Technology Adoption" OR "Technology Implementation") AND ("adoption barriers" OR "implementation patterns")	8
("Industry 4.0" OR "Assembly 4.0" OR "Operator 4.0") AND ("Technology Adoption" OR "Technology Implementation") AND ("critical success factors" OR "influencing factors" OR "required skills" OR "factors" OR "skills")	94
("Industry 4.0" OR "Assembly 4.0" OR "Operator 4.0") AND ("Technology Adoption" OR "Technology Implementation") AND (cognitive* OR "human factors")	6

Table 2 Inclusion and exclusion criteria for selecting articles

Exclusion Criteria	Inclusion
• Not published in English • Not published between 2018 and 2023 • Not published as a journal article • Published as a review study	• Must refer to manufacturing practices • Must refer to Industry 4.0 implementation within manufacturing • Must address at least one RQ

Information about the focus of the study, participant details (demographics, industrial sector, technology types, adoption levels, etc.), variables, and investigation parameters relevant to technology adoption were selected from the full-text analysis.

3 Results

The Industry 4.0 concept has gained significant and consistent traction in both industrial and academic research fields, with the phenomenon being developed and examined by researchers from different fields of study. Initial search and finalised articles for this review exhibit a steady incline in the interest of researchers regarding the topic, as the landscape is currently populated with abundant literature on the Industry 4.0 paradigm and its related aspects. Relevance in literature can be marked back to 2011, when the term was coined. Since then, the research community has directed their attention towards the digital transformation of industries towards the 'Smart' or 'Industry 4.0' level. The publications selected for this review are between 2018 and 2023, with an increment in the article count as the developments persevered.

Manufacturing firms from across the globe are currently implementing Industry 4.0-enabled technologies to survive in highly competitive global markets and sustain their businesses. Selected articles, conducted surveys, and studies in countries have been listed at the end of this paper. Economic stability in the world is maintained via multiple trade and economic unions to ensure efficient import–export practices. The authors observed higher adoption of digital technologies in Brazil, Italy, China, Germany, India, USA, Spain, Mexico, and Japan (from highest to lowest), an interesting mix of developing and developed countries. The varied economies present an opportunity to analyse the impact and correlation between technology adoption levels and economic scenarios of the concerned countries (Fig. 1).

3.1 I4.0-enabled Technologies and Manufacturing

The selected literature contributes towards understanding the impact and potential of Industry 4.0-enabled technologies in various aspects of manufacturing. The aim

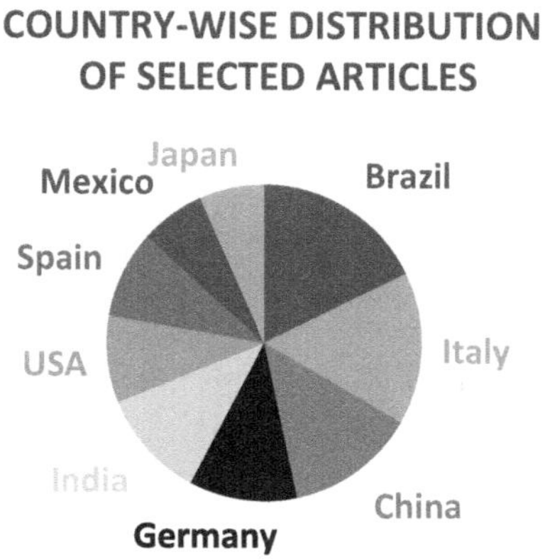

Fig. 1 Country-wise distribution of selected articles

was to address the knowledge gaps and provide theoretical perspectives on the adoption of digital technologies. Aspects such as shorter innovation-development cycles, resource efficiency, on-demand individualisation, faster decision-making via decentralisation, and higher production flexibility have been highlighted as facilitators towards their adoption [8].

The trade-offs between different production targets and the adoption of specific I4.0-enabled technologies have been investigated, highlighting the need for flexible operations. Explored the same and observed that manufacturing firms can adopt to achieve higher productivity, quality, and operational flexibility [9]. Also examined the relationship between manufacturing strategies, critical success factors (CSFs) and technology adoption scenarios in the digital transformation process of manufacturing firms [4]. Many studies, therefore, contribute towards providing managerial insights for anticipating potential issues and defining coherent benchmarks before, during and after the implementation (Treviño-Elizondo & García-Reyes, 2023).

Some have worked on understanding the extent to which I4.0-enabling technologies have been adopted across countries and industrial sectors, filling the gap by providing quantitative analysis of manufacturing systems. The literature highlights interconnectivity, interoperability, virtualisation and information transparency as reliable and consistent themes for investment and implementation preferences [10].

Assembly-Line Manufacturing

The literature focused on the potential benefits and impacts of the Industry 4.0 concept on manufacturing operations, including assembly-line manufacturing [1]. I4.0-enabled technologies exhibit different performance outcomes in the assembly line with different combinations. The enterprise's manufacturing performance is also impacted by implementing digital technologies as stand-alone and bundled identities.

Tables 3 and 4 provide details on the application domains and performance impacts of stand-alone and bundled technologies upon their implementation in manufacturing.

Table 3 Application domains and performance impacts for stand-along technologies

Application domain	Performance impact	Technology (Stand-alone)
Production process	• Reduced quality control and product development costs • Higher production efficiency and effectiveness • Reduced throughput and lead times	• Artificial Intelligence • Robots • Co-bots • Big Data and Analytics • Additive Manufacturing
Management and control	• Improved process quality • Reduced plant productivity	• Big Data and Analytics • Robots

Table 4 Application domains and performance impacts for bundled technologies

Application domain	Performance impact	Technologies (Bundled)
Maintenance	• Reduced maintenance costs and machine downtimes • Higher plant productivity and product quality • Improved service effectiveness and product customisation	• Network and sharing • Network and sharing + data processing • Computing + network and sharing + data processing
Production process	• Increased production flexibility, efficiency, and effectiveness • Improved product and production quality	• Network and sharing + data processing • Digital production + network and sharing • Digital production + data processing • Digital production + network and sharing + data processing • Visualisation + data processing • Computing + network and sharing • Computing + network and sharing + data processing
Management and control	• Increased production flexibility and efficiency • Reduced manufacturing and delivery times • Increased overall equipment effectiveness (OEE) • Reduced operational, transportation and storage costs • Improved energy consumption efficiency	• Network and sharing + data processing • Computing + network and sharing • Computing + data processing

The Industry 4.0 concept is based on cyber-physical systems (CPS) principles and future-oriented technologies that enable communication via interactions between human and machine components in assembly-line manufacturing [11]. The adoption of I4.0-enabling technologies in assembly-line manufacturing has been observed by integrating various technologies, such as sensors, additive manufacturing, augmented reality, and rapid prototyping using 3D printing, among others [12].

Different sectors are changing within the industrial environment and the manufacturing and assembly-line paradigms due to the introduction and establishment of concepts like Smart Factory and Smart Manufacturing during the Fourth Industrial Revolution [13]. Simultaneously, the literature has identified a need for advanced research and examination to understand the preparedness and acceptance levels of I4.0-enabled technologies in assembly-line manufacturing [12]. Employee acceptance refers to the willingness of workers to adopt and integrate advanced technologies into their daily work. Acceptance is crucial because the success of technologies depends on the understanding, trust, and comfort among workers despite their advanced front. [14] also discuss that employee acceptance is about ensuring confidence and support within the workforce, as well as their impact on boosting productivity and improving organisational performance. This has been suggested following the significant role of managers in convincing the assembly-line operators to embrace this industrial change due to digitalisation [15–17].

Manufacturing Flexibility

The literature on manufacturing flexibility about I4.0-enabled technologies has made significant contributions. It has provided a comprehensive understanding of the potential impact of different I4.0 technological concepts on manufacturing flexibility, including their influence on available and needed flexibility dimensions [8]. The need for manufacturing companies to implement a synergic bundle of I4.0 technologies rather than adopting scattered technological solutions has been emphasised in the literature, as bundles of technologies have a broader effect on performance than single technologies adopted in isolation [18].

Additionally, the role of I4.0 solutions in fostering shorter innovation/development cycles, resource efficiency, individualization on demand, faster decision-making, and higher flexibility in production have been discussed [18]. The literature also emphasised the importance of combining I4.0-enabled technologies to fully exploit their potentialities and deliver major impacts on business, as the most tangible benefits come from the synergy associated with their combined use [19]. Furthermore, it discusses the necessity of I4.0 base technologies, such as machine connectivity and real-time information systems, in achieving different types of internal flexibility on the shop floor [20].

3.2 Adoption-Implementation of I4.0-Enabled Technologies

The Industry 4.0 concept stands for the integration of information and communication technologies (ICT) and industrial technologies [21]. However, there is no definitive list of I4.0-enabled technologies yet, as the usage is context-based within the industrial sector. Clarity must also be established when using the terms 'adoption' and 'implementation', as the two are consecutive stages within the technology assimilation process and must not be used synonymously (Raj, Dwivedi, Sharma, de Sousa Jabbour, & Rajak, 2020). The decision-making and introduction step of a new technology is the 'adoption', while the actual deployment and utilisation of the solution is its 'implementation'.

Various frameworks have been developed to study the factors affecting the adoption and implementation of these technologies, with the TOE framework being the most preferred due to its theoretical and empirical strengths [6]. The framework allows for a broader perspective by taking the technological, organisational, and environmental dimensions into consideration during the adoption-implementation of I4.0-enabled technologies, as mentioned in Table 5.

Adoption Enablers

The integration of digital technologies and automation facilitates the transformation of traditional assembly-line manufacturing into the modern Assembly 4.0 concept. Technologies like IoT, cloud and big data allow assembly-line manufacturers to gather real-time data for monitoring machine performance, production efficiency and product quality.

Various enablers influence the adoption of digital technologies within the I4.0 paradigm in manufacturing. These include top management support, a skilled workforce, a robust information technology infrastructure, collaboration and partnerships, a supportive regulatory and policy environment, sufficient financial resources, effective change management practices, scalability and interoperability of technologies, cybersecurity measures, and a continuous learning and improvement culture (Maretto, Faccio, & Battini, 2023). These enablers play a crucial role in facilitating the successful adoption and implementation of I4.0 technologies, driving transformation, and ensuring the integration and connectivity of these technologies within the manufacturing environment.

Implementation Patterns

The literature recognised and highlighted the importance of implementing synergic solutions of I4.0-enabled technologies, instead of opting for scattered technology adoption practices. Complete and efficient exploitation of potential benefits of the Industry 4.0 concept can be ensured in this manner [1]. The implementation of I4.0-enabled technologies requires improved internet system facility, enhanced technological knowledge, and enabled interconnectivity of machines and devices capable

Table 5 Technological, organisational, and environmental dimensions for I4.0-enabled technology adoption

Technological dimensions	Organisational dimensions	Environmental dimensions
• Competitive advantage • Complexity • Compatibility • Cost	• Top management support • Technology competence • Technological infrastructure • Employee skills • Firm size	• Government support • Competitive pressure • Labour shortage

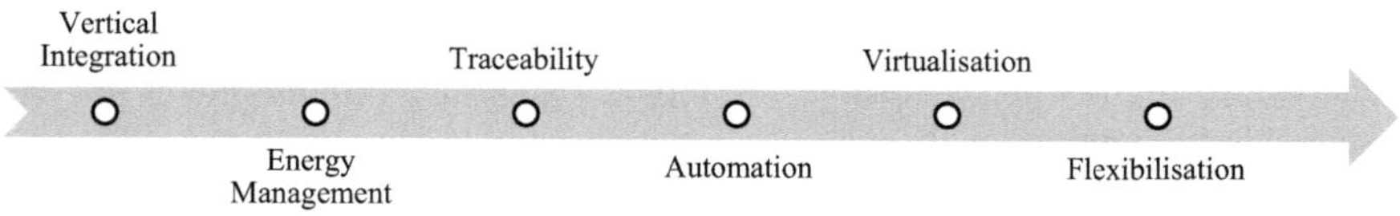

Fig. 2 Framework adaptation for adoption-implementation patterns of Smart Manufacturing

of managing big data, facilitating data sharing, and developing data security systems [22].

Technologies and practices such as vertical integration, automation, robotics, human–machine interaction, and additive manufacturing are encompassed within the 'Smart Manufacturing' dimension of the Industry 4.0 concept [23]. Different technology arrangements can be implemented simultaneously to achieve specific production targets, such as flexibility, process quality, and productivity [1].

The implementation patterns of I4.0-enabled technologies have been observed to vary between small- and medium-sized enterprises (SMEs) and larger enterprises, establishing proportionality between technology implementation and company size [24]. The current degree of implementation varies across different technologies, industrial sectors and economies, with some technologies exhibiting higher levels of adoption-implementation than others [11].

Figure 2 explains an adaption of the framework summarising adoption-implementation patterns of the Smart Manufacturing dimension of the Industry 4.0 concept.

Adoption-Implementation Barriers

Manufacturing firms are expected to overcome various barriers to effectively implementing I4.0-enabled technologies. These barriers impact the integration of advanced digital technologies and hinder the potential benefits of the technology advancement process. One of the main barriers is the lack of access to advanced technologies and the high cost of implementation, which can hinder inclusive participation in the Industry 4.0 paradigm. Additionally, there are several internal barriers within organisations, such as a lack of competence, employee resistance to change, and unclear benefits to workers, which can impede the adoption of I4.0-enabled technologies [5].

Furthermore, inadequate infrastructure, including a lack of reliable electricity supply and ICT challenges, poses a significant barrier to adopting I4.0 technologies [19]. The complexity of Industry 4.0, high implementation costs, market uncertainty, security concerns, lack of expertise, and regulatory challenges have also been identified as barriers to the successful implementation of I4.0 technologies [18]. These barriers can hinder organisations from fully embracing the potential benefits of I4.0 and require strategic interventions to overcome them.

Moreover, the sustainable adoption of I4.0-enabled technologies in manufacturing faces additional challenges. Inequality, fragmented task environments, insufficient regulatory and policy frameworks, and human capital challenges can hinder the widespread and equitable adoption of these technologies in the manufacturing sector

[19]. Therefore, it is crucial to address these barriers to ensure that adopting I4.0 technologies aligns with sustainable development goals and promotes inclusive and sustainable economic growth.

3.3 Factors Affecting Human Operators in I4.0-Enabled Manufacturing Systems

The cognitive capabilities of human operators in 4.0-enabled manufacturing systems are influenced by factors such as the design of cognitive automation interfaces, individuals' learning curves, and the incorporation of analytical information. Required skills include interdisciplinary knowledge, IT competencies, adaptability, communication skills, and understanding of process relations. Upskilling programs are needed to address the scarcity of Industry 4.0-related skills. Considering technology, organisation, and human factors, the socio-technical approach can enhance cognitive capabilities in Industry 4.0-enabled manufacturing systems.

Influencing Factors for Operators

The design of cognitive automation solution interfaces influences the cognitive capabilities of human operators in I4.0-enabled manufacturing systems. Understanding the individual learning curves and cognitive processes have also been recognised as influencing factors within the literature. The incorporation of analytical information provided by novel software systems has also been observed to influence the cognitive capabilities of the manufacturing workforce as well [25]. Factors identified for the successful implementation of technologies include top management support, employee engagement, organisational infrastructure, alignment with organisational strategies, and change management [4].

Operator Skills for I4.0-Enabled Manufacturing

In the context of I4.0-enabled manufacturing systems, the adoption and use of machine flexibility technologies demand workers to acquire new digital skills [8]. Workers in these systems need to be capable of performing cognitive work, including data processing, interpreting information, and decision-making. The impact and effects of co-bot utilisation on workers' skills in different manufacturing activities have been analysed using the Labour Process Theory (LPT) framework [18].

The literature emphasises the need for workers to adapt to new technologies and acquire the necessary knowledge and training to operate them effectively [8]. The design of I4.0 technologies, such as co-bots, must consider ergonomic, and flexible features to assist operators with more complex tasks and enhance their skills. Overall, the implementation of advanced technologies in I4.0 manufacturing systems requires workers with flexible skills to adapt to changing production demands [26].

4 Discussion

The manufacturing sector has experienced a significant shift in recent years with the advent of the Fourth Industrial Revolution. This new era of digitalisation has revolutionised various aspects of manufacturing by integrating advanced digital technologies within the working environment. Technologies such as Artificial Intelligence (AI) possess the potential to achieve optimised production processes, increased efficiency and enhanced overall competitiveness. With IoT, physical components have been connected and synchronised within the system. The connectivity levels allow for seamless and real-time communication and exchange of information between machines, systems, and humans in the system. The integration of CPS, involving human operators and machines, has also been another crucial aspect of Industry 4.0. The concept allows for intelligent communication and collaboration between humans and automated systems, achieving higher efficiency and effectiveness in manufacturing operations.

For assembly-line manufacturing, the I4.0 concept enabled the acceleration of the development and implementation of advanced technologies to achieve significantly reduced cycle times and streamlined processes, allowing for a quicker response to consumer and market demands. Additionally, the concept also brought about increased productivity and flexibility in the assembly line. Real-time data and analytics allow manufacturers to monitor the production processes more clearly, providing improved opportunities to identify and rectify breakdowns. Shorter setup times, reduced errors, and equipment downtimes have maximised the production output, allowing firms to utilise their resources efficiently. Increased mass customisation and broader product variety are evident customer demands in the current global market.

This industrial revolution has entirely shifted the industrial scenario and is still in the adoption-implementation stage in various sectors and economies. Therefore, the need to observe and identify enablers and barriers to technology adoption and implementation patterns has become an extensive multidisciplinary research paradigm. With this, many researchers highlighted the societal impact of Industry 4.0 and the potential economic shifts the I4.0 concept is expected to bring about. Some researchers have also highlighted the need to study the impact of digital technologies on the workforce involved in these manufacturing systems. Existing research on the aspect is currently limited, providing opportunities for researchers from different fields. The suggestion is to explore the idea of the human worker as a resource component within the system that can be improved and not replaced.

5 Conclusion

The authors investigated existing literature on I4.0-enabled technology implementation in assembly-line manufacturing to provide readers with a comprehensive overview of the field. Literature has enticed the development of novel concepts, such as Operator 4.0, Work 4.0, and even Industry 5.0, as research interest in the ongoing industrial revolution has widened. Unlike prior revolutions, Industry 4.0 and Industry 5.0 paradigms have been observed and developed about their impact on society and the environment. The reviewed literature highlights the lack of definition in context-based technology implementation in assembly-line manufacturing. Some challenges manufacturing firms face have been discussed, such as the lack of industrial standards for implementing and applying these technologies in the presence of human operators within the manufacturing assembly. The need to examine the I4.0-enabled technology adoption-implementation per contextual parameters such as sector, economy, readiness, and preparedness levels against the impact on human operators within the manufacturing systems has been identified.

References

1. Battaglia D, Galati F, Molinaro M, Pessot E (2023) Full, hybrid and platform complementarity: Exploring the industry 4.0 technology-performance link. Int J Prod Econ 263 p 108949. https://doi.org/10.1016/j.ijpe.2023.108949
2. Enrique DV, Marcon É, Charrua-Santos F, Frank AG (2022) Industry 4.0 enabling manufacturing flexibility: Technology contributions to individual resource and shop floor flexibility. J Manuf Technol Manag 33(5):853–875. https://doi.org/10.1108/JMTM-08-2021-0312
3. Bilbao-Ubillos J, Camino-Beldarrain V, Intxaurburu-Clemente G (2021) Industry 4.0, proximity constraints and new challenges for industrial policy. Eur Plan Stud 29(2):329–345. https://doi.org/10.1080/09654313.2020.1753660
4. Tortorella G, Kurnia S, Trentin M, Oliveira GA, Setti D (2022) Industry 4.0: What is the relationship between manufacturing strategies, critical success factors and technology adoption? J Manuf Technol Manag 33(8):1407–1428. https://doi.org/10.1108/JMTM-02-2022-0088
5. Maisiri W, Van Dyk L, Coeztee R (2021) Factors that inhibit sustainable adoption of industry 4.0 in the south african manufacturing industry. Sustainability 13(3):1013. https://doi.org/10.3390/su13031013
6. Kinkel S, Baumgartner M, Cherubini E (2022) Prerequisites for the adoption of AI technologies in manufacturing—Evidence from a worldwide sample of manufacturing companies. Technovation 110:102375. https://doi.org/10.1016/j.technovation.2021.102375
7. Arksey H, O'Malley L (2005) Scoping studies: Towards a methodological framework. Int J Soc Res Methodol 8(1):19–32. https://doi.org/10.1080/1364557032000119616
8. Höse K, Amaral A, Götze U, Peças P (2023) Manufacturing flexibility through Industry 4.0 Technological Concepts—Impact and Assessment. Glob J Flex Syst Manag 24(2):271–289. https://doi.org/10.1007/s40171-023-00339-y
9. Enrique DV, Marodin GA, Santos FBC, Frank AG (2023). Implementing industry 4.0 for flexibility, quality, and productivity improvement: Technology arrangements for different purposes. Int J Prod Res 61(20), 7001–7026. https://doi.org/10.1080/00207543.2022.2142689
10. Castelo-Branco I, Cruz-Jesus F, Oliveira T (2019) Assessing Industry 4.0 readiness in manufacturing: Evidence for the European Union. Comput Ind 107:22–32. https://doi.org/10.1016/j.compind.2019.01.007

11. Yu F, Schweisfurth T (2020) Industry 4.0 technology implementation in SMEs—A survey in the Danish-German border region. Int J Innov Stud 4(3):76–84. https://doi.org/10.1016/j.ijis.2020.05.001
12. Van Dun DH, Kumar M (2023) Social enablers of Industry 4.0 technology adoption: Transformational leadership and emotional intelligence. Int J Oper & Prod Manag 43(13):152–182. https://doi.org/10.1108/IJOPM-06-2022-0370
13. Bortolini M, Faccio M, Galizia FG, Gamberi M, Pilati F (2020) Design, engineering and testing of an innovative adaptive automation assembly system. Assem Autom 40(3):531–540. https://doi.org/10.1108/AA-06-2019-0103
14. Cordero D, Altamirano KL, Parra JO, Espinoza WS (2023) Intention to Adopt Industry 4.0 by Organizations in Colombia, Ecuador, Mexico, Panama, and Peru. IEEE Access 11:8362–8386. https://doi.org/10.1109/ACCESS.2023.3238384
15. Chumnumporn K, Jeenanunta C, Simpan S, Srivat K, Sanprasert V (2022) The Role of leader and the effect of customer's smart factory investment on firm's industry 4.0 technology adoption in Thailand. Int J Technol 13(1):26. https://doi.org/10.14716/ijtech.v13i1.4814
16. Dabić M, Maley JF, Črešnar R, Nedelko Z (2023) Unappreciated channel of manufacturing productivity under industry 4.0: Leadership values and capabilities. J Bus Res 162 p 113900. https://doi.org/10.1016/j.jbusres.2023.113900
17. Tortorella G, Prashar A, Antony J, Mac Cawley AF, Vassolo R, Sony M (2023) Role of leadership in the digitalisation of manufacturing organisations. J Manuf Technol Manag 34(2):315–336. https://doi.org/10.1108/JMTM-09-2022-0312
18. Kamp B, Gibaja JJ (2021) Adoption of digital technologies and backshoring decisions: Is there a link? Oper Manag Res 14(3–4):380–402. https://doi.org/10.1007/s12063-021-00202-2
19. Cui Y, Liu W, Rani P, Alrasheedi M (2021) Internet of Things (IoT) adoption barriers for the circular economy using Pythagorean fuzzy SWARA-CoCoSo decision-making approach in the manufacturing sector. Technol Forecast Soc Chang 171:120951. https://doi.org/10.1016/j.techfore.2021.120951
20. Cirillo V, Fanti L, Mina A, Ricci A (2023) The adoption of digital technologies: Investment, skills, work organisation. Struct Chang Econ Dyn 66:89–105. https://doi.org/10.1016/j.strueco.2023.04.011
21. Zhou B, Zheng L (2023) Technology-pushed, market-pulled, or government-driven? The adoption of industry 4.0 technologies in a developing economy. J Manuf Technol Manag 34(9), 115–138. https://doi.org/10.1108/JMTM-09-2022-0313
22. Adebanjo D, Laosirihongthong T, Samaranayake P, The PL (2023) Key enablers of industry 4.0 development at firm level: findings from an emerging economy. IEEE Trans Eng Manag 70(2):400–416. https://doi.org/10.1109/TEM.2020.3046764
23. Marcon É, Soliman M, Gerstlberger W, Frank AG (2022) Sociotechnical factors and Industry 4.0: An integrative perspective for the adoption of smart manufacturing technologies. J Manuf Technol Manag 33(2): 259–286. https://doi.org/10.1108/JMTM-01-2021-0017
24. Roth K, Farahmand K (2023). A socio-technical study of industry 4.0 and SMEs: recent insights from the upper midwest. Sustainability 15(16):12559. https://doi.org/10.3390/su151612559
25. Khourshed NF, Elbarky SS, Elgamal S (2023) Investigating the readiness factors for industry 4.0 implementation for manufacturing industry in Egypt. Sustainability, 15(12):9641. https://doi.org/10.3390/su15129641
26. Pedota M, Grilli L, Piscitello L (2023) Technology adoption and upskilling in the wake of Industry 4.0. Technol Forecast Soc Chang. 187 p 122085. https://doi.org/10.1016/j.techfore.2022.122085

Challenges of ATM Usage by Wheelchair Users—A Simulation Study

Avika Mondal, Vansh Kumar Singh, Srijan Singh, and Gourab Kar

1 Introduction

Automatic Teller Machines (ATMs) have become ubiquitous in the urban landscape of India. They offer fast and easy banking services for users as well as reduce operational expenses for banks. There are about 267,000 ATMs in India with an average of 18.5 million transactions-per-day [1]. With financial inclusion a key priority for India's economic growth, annual ATM transactions are projected to increase from six billion in 2023 to eight billion in 2027 [2]. While ATM use has become a daily necessity [3, 4], usage of an ATM is not necessarily simple and intuitive. To operate an ATM, the user has to respond to a series of sequential prompts displayed on the ATM interface, which demand considerable cognitive and physical abilities [5]. Studies have revealed that older adults and people-with-disabilities encounter greater challenges in the use of ATMs compared to younger adults [1, 6–8]. These challenges could be related to recalling the personal identification number, getting the required amount of money, or completing all steps required for the transaction [9]. The root cause for these outcomes could be attributed to the decline in cognitive and physical abilities for older adults as well as the psychological obstacles in ATM use—such as feelings of mistrust, lack of privacy, and a sense of insecurity.

Researchers and designers have used principles of universal design and participatory approaches to develop inclusive ATMs that accommodate the needs of older adults and people-with-disabilities. Changes to an ATM's interface such as reduced sequential instructions, fewer recall items, minimal user inputs and enhanced on-screen graphics can improve ATM usability for older adults [10]. Rearranging the display-and-control and providing adequate knee-and-leg space can enable older adults and wheelchair users to access ATMs [11]. Research on wheelchair accessibility of ATMs in India suggests that ramp slopes at the entrance to ATM-kiosks are

A. Mondal · V. K. Singh · S. Singh · G. Kar (✉)
Department of Design, Indian Institute of Technology Delhi, New Delhi, India
e-mail: gourab@iitd.ac.in

© The Author(s), under exclusive license to Springer Nature Singapore Pte Ltd. 2025
S. Das et al. (eds.), *Human-Centered Digitalization, Volume 3*, Design Science and Innovation, https://doi.org/10.1007/978-981-96-4383-7_40

steeper than what the accessibility guidelines mandate, location of number pad and screen is higher than the optimal reach envelope for wheelchair users, clearance for the knee and leg is missing and there is lack of adequate horizontal surface near the entrance [12]. Therefore, this study investigated ATM-kiosk usage by wheelchair users to understand their accessibility challenges. Research questions focused on: (1) What are the environmental design challenges for wheelchairs users in use of an ATM-kiosk? (2) How can environmental design challenges be ranked in terms of a hierarchy of needs? (3) What are layouts of ATM-kiosks that offer improved environmental access for wheelchair users?

2 Methods

Two ATM-kiosks on the campus of a public-funded research university in New Delhi, India were selected for the study. Both ATM-kiosks are located in the academic area on campus and witness a steady footfall of users throughout the day. The first ATM-kiosk (Kiosk #1), located adjacent to the branch office of a major public-sector bank has three ATMs (1, 2, and 4), a cash deposit machine (3) and passbook printing machine (5). The second ATM-kiosk (Kiosk #2) belongs to another public-sector bank, is located opposite the Central Library, and houses a single ATM (see Fig. 1).

In both locations, researchers measured dimensions of the ATM-kiosk and created floor plans. Next, researchers conducted a simulation of an ATM transaction and video-recorded the sequence of actions. In the simulation, a researcher assumed the role of the wheelchair user while another video-recorded the sequence of activities. A battery-powered wheelchair was used; researchers had no prior experience of using battery-powered wheelchairs. The protocol for the simulation had the following steps—go up the access ramp to the ATM-kiosk, reach and open the ATM-kiosk door

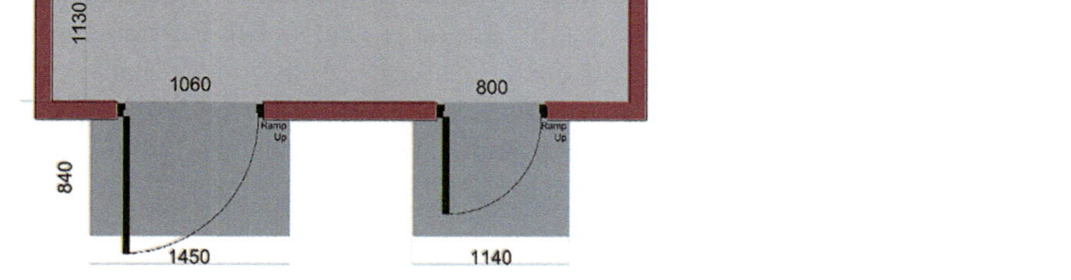

Fig. 1 Floor plan of Kiosk #1 (left) and Kiosk #2 (right)

from outside (if more than one door, try both), enter the ATM-kiosk, interact with the ATM (try side access and front access), turn around (if possible), open the ATM-kiosk door from inside, exit the ATM-kiosk, navigate down the access ramp. In both locations, simulations were conducted during daytime and at night. Three researchers took turns to assume the role of the wheelchair user (see Fig. 2).

Video recordings of the simulation exercise were reviewed by the research team as a group to identify key interaction points where the wheelchair users faced challenges. Keyframes of these interaction points were overlaid on the plan view of the ATM-kiosk. This analytical approach to review video data and map it to the spatial layout enabled researchers to systematically breakdown the ATM transaction process into a sequence of tasks and subtasks which could be analyzed individually. Keyframes of interaction points were analyzed in two ways—(a) a qualitative analysis using the seven fundamental principles of design [13]—*discoverability, affordances, signifiers, conceptual model, feedback, mapping, and constraints*; and (b) a quantitative analysis using the principles of universal design [14]—*equitable use, flexibility in use, simple and intuitive use, perceptible information, tolerance for error, low physical effort, and size and space for approach and use.*

The use of two methods of analysis enabled researchers to rank-order the severity of needs for wheelchair users. Based on findings from the needs analysis, the research

Fig. 2 Sequence of keyframes for ATM transactions in Kiosk #1 (top) and Kiosk # 2 (bottom)

team decided to focus on developing solutions for the seventh principle of universal design—size and space for approach and use [14]. The criteria for wheelchair maneuvering and clearances for side access, front access, T-turn, L-turn, and U-turn, were based on international standards and recommended wheelchair access guidelines [15, 16]. Using an inside-out approach to design, with unimpeded wheelchair movement being the key design goal, multiple layouts of the ATM-kiosk were generated. From these layouts, the most space efficient and practically feasible ATM-kiosk layouts were selected for design refinement. The final design consists of two ATM-kiosk layouts—one for a single-ATM scenario and another for multi-ATM scenario, both of which enable environmental access for wheelchair use.

3 Findings

3.1 Qualitative Analysis

Findings from the qualitative analysis are outlined below:

Discoverability. Kiosk #1 is 10 m away from the pavement and is approached by a series of ramps. The kiosk has two doors; only one of them is wide enough to enable wheelchair access (1060 mm). Kiosk #2 is located 12 m away from the pavement. The entrance to the kiosk is poorly lit, there is a 25 mm level difference between the floor level outside and inside the kiosk, and the door is barely wide enough to enable wheelchair access (900 mm).

Affordances. Kiosk #1 has two self-closing doors, one of which is wide enough to enable wheelchair access. The door swings outwards when pulled from the outside, allowing the wheelchair user to maneuver inside the kiosk and access the ATM in the side-orientation or front-orientation. Kiosk #2 has an inward opening door which allows wheelchair users to maneuver inside the kiosk and access the ATM in side-orientation. Neither ATM-kiosk has affordances to indicate direction of door swing.

Signifiers. Kiosk #1 lacks signifiers to specify which of the two doors is wheelchair accessible. None of the two ATM-kiosks offer any signifiers regarding the direction of door swing and the status of occupancy inside the kiosk.

Conceptual Model. From the wheelchair user's perspective, conceptual models about safe and optimal routes to access the ATM-kiosk may be incompatible with the reality of the environmental design. The lack of affordances and signifiers hinder wheelchair user's ability to pre-determine the route to take, identify the door to reach, anticipate door swing direction, and accurately estimate side access or front access to the ATM.

Feedback. Having entered Kiosk #2, the wheelchair user does not get any audible feedback about the self-closing door shutting by itself. This may be a safety hazard.

For Kiosk #1, there is lack of visual feedback for the wheelchair user about the possibility that other users may be near or behind the wheelchair user. Note that Kiosk #1 has three ATMs inside, so multiple users may be present inside the kiosk. In both kiosks, lack of auditory and visual feedback compromises the perception of safety.

Mapping. The absence of affordances and signifiers for wheelchair users results in a mismatch between *knowledge in the world* and *knowledge in the head*. Consequently, wheelchair users may have to undertake multiple attempts to access the ATM-kiosks. For example, while there are two doors in Kiosk #1, only one door is wheelchair accessible. This poses a mapping problem for the first-time wheelchair user and may lead to errors.

Constraints. The door swing direction and spatial layout of ATM-kiosks constrain the movement of the wheelchair user. In Kiosk #1 pulling the door outside imposes restrictions on ease of wheelchair ingress. In Kiosk #2 one has to push the door, hold it in place, then maneuver the wheelchair inside and turn right. To exit the kiosk, one has to drive the wheelchair in reverse, while keeping the door open with the other hand.

3.2 *Quantitative Analysis*

The keyframes for interactions points were rated on a scale of 0–10, with 0 being not at all inclusive and 10 being the most inclusive. Ten key interaction points were scored using the universal design checklist [14]. Scores for Kiosk #1 reveal that tolerance for error (18) and flexibility in use (24) had lowest scores for inclusion, followed by size and space for approach and use (30). The cumulative scores not only offer a quantitative measure of inclusion but also rank-order the hierarchy of needs (see Table 1).

3.3 *Redesigned Layouts*

In response to the need to reconsider the '*size and space for approach and use*,' two layouts of ATM-kiosks are proposed—one for a single-ATM scenario and another for multiple-ATM scenario (See Fig. 3). Both layouts feature a sliding door with 915 mm of clear doorway space for wheelchair access and an 800 mm turning circle radius for a wheelchair U-turn. The floor space for the single-ATM kiosk and the multiple-ATM kiosk are 3.72 m^2 and 5.60 m^2, respectively. These layouts of the ATM-kiosks enable a wheelchair user to perform a full-180°-turn inside and clearance on either of the ATM offers wheelchair users the flexibility to access the ATM from the side or the front orientations. These two ATM-kiosk layouts are optimized to have the lowest

Table 1 Scores for interaction keyframes based on the universal design checklist

	Ramp approach	Door access	Door #1 enter	Door #2 enter	Kiosk enter	Side access	Front access	Wheelchair rotate	Exit door	Ramp exit	Total
Equitable use	5	3	0	4	5	3	4	6	6	4	40
Flexibility in use	3	1	0	3	3	1	5	4	3	1	24
Simple and intuitive use	4	2	0	5	4	3	5	2	3	2	30
Perceptible information	3	2	0	5	3	5	4	3	4	3	32
Tolerance for error	1	1	0	3	3	4	3	1	1	1	18
Low physical effort	6	2	0	4	4	8	4	8	4	4	44
Size and space for approach and use	3	4	0	3	3	5	3	4	4	1	30

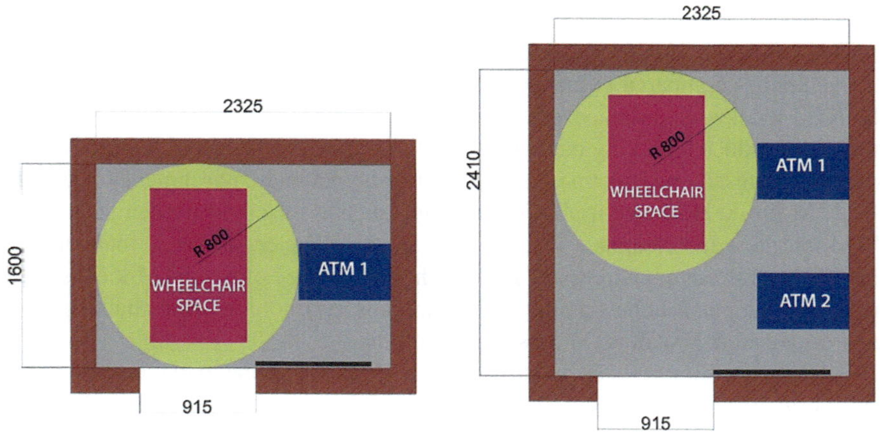

Fig. 3 Single-ATM layout (left) and multiple-ATM layout (right). All dimensions in mm

footprint while meeting minimum wheelchair accessibility needs for environmental access. The multiple-ATM layout can be extended to accommodate addition units.

4 Discussion

This study applied the seven fundamental principles of design [13] and the principles of universal design [14] to evaluate and highlight challenges for wheelchair access in ATM-kiosks. While previous studies have focused on redesigning the ATM interface and offering adequate knee-and-leg space to accommodate front access for wheelchair users [10–12], this study focused on need to optimize spatial layouts of ATM-kiosks to offer appropriate size and space for approach and use for wheelchair users

From a methodological perspective the study has several strengths. First, the use of full-scale simulation [17] and video-analysis enabled researchers to identify and classify the degree of human–environment fit [18] for all key interaction points. Second, keyframes for the interaction points were mapped to spatial layout to combine time and space parameters in the analysis. Third, use of qualitative and quantitative methods to analyze keyframes offered a robust and effective analytical framework to capture the instances of design exclusion. Fourth, adoption of universal design checklist provided an effective method to rank degree of exclusion from the perspective of wheelchair access. Finally, the research team was able to develop a systematic method to translate research evidence into environmental design guidelines for inclusive wheelchair access.

While findings from the study address a crucial gap in knowledge about wheelchair access for ATM-kiosks, there are several limitations that need to be mentioned. First, participants in the study were not regular wheelchair users; therefore, there may be differences in user behavior compared to habitual wheelchair users. Second, a battery-powered wheelchair with joystick controls was used, which has different mobility characteristics compared to manual wheelchairs. Third, while wheelchair access to the ATM-kiosk was evaluated, other challenges for older adults due to visual impairments and reduced working memory capacities were not considered. Fourth, design solutions focused on guidelines for the interior layout and environmental access to the ATM-kiosk. However, usability challenges such as the cognitive demands of the ATM interface and clearance requirements for the wheelchair user to interact with the ATM unit were not addressed. Finally, the sample size of two ATM-kiosks in an academic campus is at best a pilot study. Findings may not represent challenges for environmental access in ATM-kiosks across India.

Future work should focus on expanding the project's scope to encompass all the seven principles of universal design [14]. For ATM-kiosks in India to be truly inclusive, there needs to be adequate representation of a wide group of people across a range of ages and abilities. These groups would include children, older adults, and people-with-disabilities. A wider survey of ATM-kiosk typologies and interior layouts needs to be conducted to map out the challenges of human–environment fit [18] between diverse groups of people and a wide range of environmental designs. The layouts of ATM-kiosks proposed in this study are just one among many possibilities to create solutions that are *useable by all people, to the greatest extent possible, without the need for adaptation or specialized design* [19].

References

1. National Payments Corporation of India. https://www.npci.org.in/what-we-do/nfs/product-statistics/volumes. Last accessed 10 Nov 2023
2. The Indian Payments Handbook—2022–2027, PwC India. https://www.pwc.in/assets/pdfs/the-indian-payments-handbook-2022-2027.pdf. Last accessed 10 Nov 2023
3. Czaja SJ (1997) Computer Technology and the Older Adult. In: Helander MG, Landauer TK, Prabhu PV (eds) Handbook of Human-Computer Interaction. Elsevier Science B.V, Amsterdam, pp 797–812
4. Petrie H (2001) Accessibility and usability requirements for ICTs for disabled and elderly people: a functional classification approach. In: Nicolle C, Abascal J (eds) Inclusive Design Guidelines for HCI. Taylor & Francis, New York, pp 29–60
5. Jamieson BA, Rogers WA (2000) Age-related effects of blocked and random practice schedules on learning a new technology. J Gerontol 55 B(6):343–353
6. Rogers WA, Gilbert DK, Cabrera EF (1994) An in-depth analysis of automatic teller machine usage by older adults. In: Proceedings of the human factors and ergonomics society 38th annual meeting, pp 142–146
7. Rogers WA, Fisk AD (1997) ATM design and training issues. Ergon Des 5(1):4–9
8. Tarakanov-Plax A (2005) Design concept for ATM machine, accessible for the elderly users in Israel. In: Proceedings of the international conference on inclusive design
9. Rogers WA, Cabrera EF, Walker N, Gilbert DK, Fisk AD (1996) A survey of automatic teller machine usage across the adult life span. Hum Factors 38(1):156–166
10. Chan CC, Wong AW, Lee TM, Chi I (2009) Modified automatic teller machine prototype for older adults: a case study of participative approach to inclusive design. Appl Ergon 40(2):151–160

11. Jung H, Ko YJ (2017) ATM design applying the universal design concept. Arch Des Res 30(2):123–137
12. Kumar A, Sharad KS, Anwer S (2012) Difficulties faced by the wheelchair users in accessing automated teller machine (ATM): a case study. Indian J Physiother Occup Ther-Int J 6:142–144
13. Norman D (2013) The design of everyday things: revised and expanded edition. Basic books
14. Story MF (1998) Maximizing usability: the principles of universal design. Assist Technol 10(1):4–12
15. 2010 ADA Standards for Accessible Design. https://www.ada.gov/law-and-regs/design-standards/2010-stds/. last accessed 10 Nov 2023
16. D'Souza C, Steinfeld E, Paquet V, Feathers D (2010) Space requirements for wheeled mobility devices in public transportation: analysis of clear floor space requirements. Transp Res Rec 2145(1):66–71
17. Steinfeld E, Danford GS (Eds.) (1999) Enabling environments: measuring the impact of environment on disability and rehabilitation. Springer Science & Business Media
18. Iwarsson S, Ståhl A (2003) Accessibility, usability and universal design—positioning and definition of concepts describing person-environment relationships. Disabil Rehabil 25(2):57–66
19. Mace R (1997) What is universal design. The Center for Universal Design at North Carolina State University, vol 19. 2004

Roadeze—A Two-Wheeler Rider-Safety Intervention for Last-Mile Delivery Personnel

P. M. Arun and Gourab Kar

1 Introduction

1.1 Context

At least 1,68,491 people were killed on India's roads in 2022, with motorized two-wheeler (MTW) riders accounting for the highest share of fatalities (44.5%) [1]. This is not an isolated statistic, but part of a trend where in most less-motorized countries the vulnerable road users—pedestrians, bicyclists and MTW-riders—sustain a vast majority of fatalities and injuries due to road traffic accidents [2, 3]. The road safety situation is a particular concern in densely populated urban megacities like Delhi, Chennai, Mumbai and Bengaluru, which are also highly motorized [4]. Among the urban cities in India with a million-plus population, Delhi bags the top spot for road traffic fatalities [1], with 38.1% of fatalities and 43.7% of injured road crash victims belonging to the MTW category [5]. Despite this alarming scenario, safety research on MTW-riders in India is scarce [6].

In recent years, MTWs have gained popularity for last-mile delivery of goods and food delivery in low-and-middle-income countries [7, 8]. Studies suggest that trips made for delivering goods or food were often incentivized to complete within a certain duration, compelling the MTW-riders to be in time pressure and at the same time these riders frequently used their mobile phones to keep the end consumer updated about the delivery status [8]. Driving under time pressure may reduce a rider's focus which may negatively impact their driving performance and increase crash probability [9, 10]. In addition, MTW-riders' tendency to use mobile phones increases under time pressure [11]. The simultaneous occurrence of time pressure and increased use of mobile phones can drastically degrade driving performance with

P. M. Arun · G. Kar (✉)
Department of Design, Indian Institute of Technology Delhi, New Delhi, India
e-mail: gourab@iitd.ac.in

© The Author(s), under exclusive license to Springer Nature Singapore Pte Ltd. 2025

S. Das et al. (eds.), *Human-Centered Digitalization, Volume 3*, Design Science and Innovation, https://doi.org/10.1007/978-981-96-4383-7_41

serious consequences for MTW-rider safety [12]. Therefore, this research-informed-design study examined challenges for MTW-based last-mile delivery personnel in New Delhi, India with a goal to reduce on-road distractions and increase rider safety during the delivery process.

1.2 Research Questions

Research questions focused on: (1) What is the typical last-mile delivery workflow for the MTW-rider? (2) What are the human–machine interaction touchpoints for the MTW-rider in the delivery workflow? (3) What are sources of on-road distractions for the MTW-rider in the delivery workflow? (4) How can on-road distractions during delivery be reduced for the MTW-rider?

2 Methods

2.1 Process

This research-informed-design project adopted the design thinking process with five stages—empathize, define, ideate, prototype and test [13]. First, the design team conducted literature reviews, precedent analysis, ethnographic observations, and user interviews to understand the context. Second, data from the empathize phase was reframed into an actionable design brief. Third, ideations based on the design brief were developed into low- and medium-fidelity prototypes. Fourth, prototypes were tested with representative users to understand which of the design features worked and how the designs could be improved. Finally, a high-fidelity prototype was tested in a real-world environment with MTW-riders to validate the design solution. The study was conducted over a 14-week academic semester by the first author (see Fig. 1).

Empathize. Naturalistic observations and unstructured interviews with MTW-riders enabled the design team to understand the delivery workflow. The typical

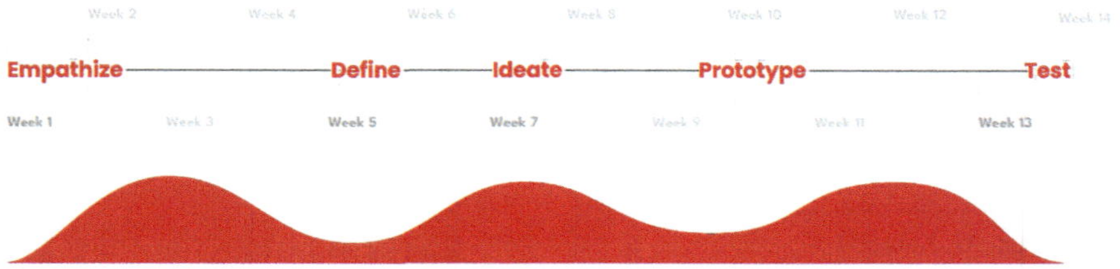

Fig. 1 Design thinking process—empathize, define, ideate, prototype and test

workflow comprises five phases—order receipt, order acceptance, transit, waiting, and delivery (see Fig. 2). Mapping touch points across the goods and/or food delivery workflow revealed three categories of human–machine interactions: (a) accepting/declining delivery request using delivery application, (b) voice communication through a phone call with vendor and/or customer for pickup and delivery and (c) route navigation using a separate application on the mobile phone. Informed consent was taken from the MTW-riders prior to the observations and interviews. The research revealed the dynamic nature of the last-mile delivery process—use of separate applications for order receipt, navigation and voice communication necessitates frequent switching between applications.

Define. Analysis of data from empathize phase revealed that MTW-riders frequently switch between applications on their mobile phone. To recreate task conditions in a controlled environment, a static simulation of the application switching task was conducted. Five participants performed a simulated driving task while concurrently switching applications on a mobile phone. Results of the simulation study suggest that participants frequently ended up in virtual traffic collisions as they were distracted in switching between applications. Therefore, the design team focused on how to

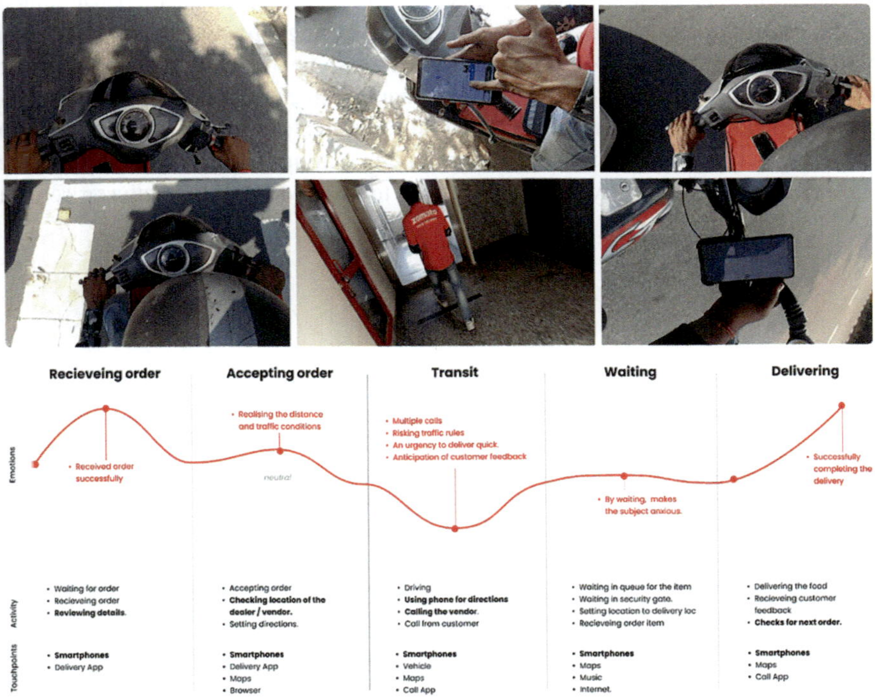

Fig. 2 Delivery workflow—order receipt, order accept, transit, waiting and order delivery

synthesize a solution that reduced or eliminated on-road distractions for MTW-riders. The redesigned interaction had to reduce need to switch between applications, improve tactile feedback and decrease instances of handlebar disengagement.

Ideate. Ideation sketches considered four concepts: (a) handlebar-based tactile mobile phone controls; (b) mobile phone holders with tactile button overlays; (c) mobile phone display mirrored to a MTW dashboard; (d) helmet-mounted projection display. Based on resources available the handlebar-based tactile mobile phone controls idea was taken forward. Sketches and paper-mockups considered different form factors for the hardware and information layouts for the user interface (see Fig. 3).

Prototype. Foamboard prototypes with swappable paper-based wireframes were created to simulate the user interaction. A static test-rig with a handlebar setup in a desktop environment was built to test how eye-gaze and hand positions were impacted by the design configurations (see Fig. 4).

Test. The final prototype featured handlebar-mounted tactile controls connected to the mobile phone through the USB-cable. The test was conducted in two phases—first, a control phase with the mobile phone screen as the interface device used during a typical delivery cycle. Second, an intervention phase where handlebar-mounted tactile controls were used as the interface device during a typical delivery cycle. In both phases, two cameras were used—one, to record the eye-gaze of the MTW-rider and the other to record actions on the handlebar. Parameters analyzed were: (a) eye-gaze, (b) handlebar disengagement, (c) mobile screen interactions, (d) navigation, (e) switching between applications and (f) receiving and making calls. The video data was reviewed and coded by the research team to reveal differences between control

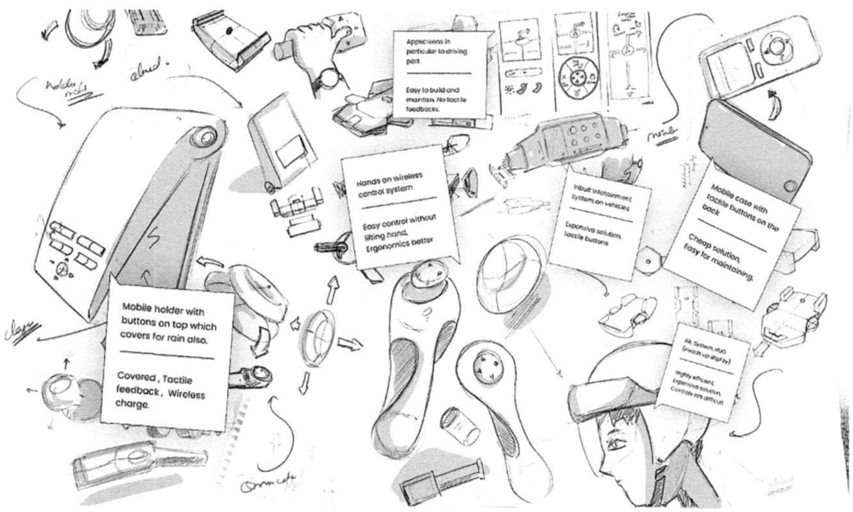

Fig. 3 Ideation sketches of different design concepts

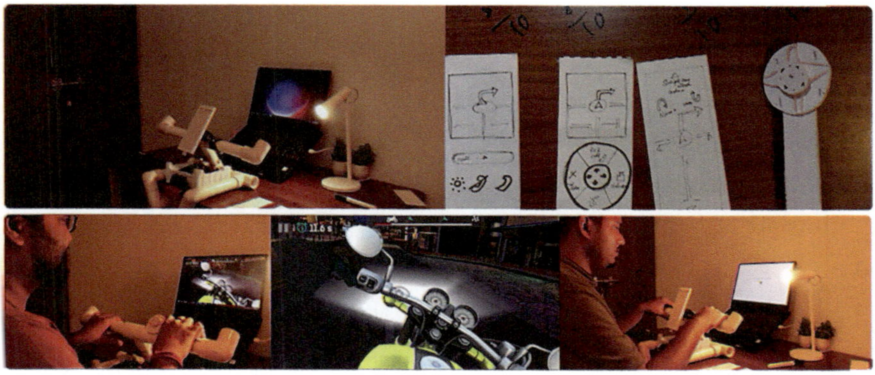

Fig. 4 Static test-rig with prototypes tested for eye-gaze and hand positions

and intervention conditions with respect to eye-gaze and handlebar disengagement (see Fig. 5).

3 Findings

Findings suggest that navigation tasks resulted in most instances of distracted vision for the control condition, followed by order receipt and voice calls, respectively. In comparison, the instances of distracted vision were reduced by 40% for the intervention condition compared to the control. In terms of handlebar disengagement, navigation tasks result in most instances of handlebar disengagement for control condition followed by order receipt and voice calls, respectively. However, for the intervention condition, instances of handlebar disengagement were reduced by 77% when compared to the control. The reduction in instances of distracted vision and handlebar disengagement for the intervention condition suggest that handlebar-based tactile controls may offer significant safety benefits compared to mobile phone touchscreen controls for delivery personnel (see Table 1).

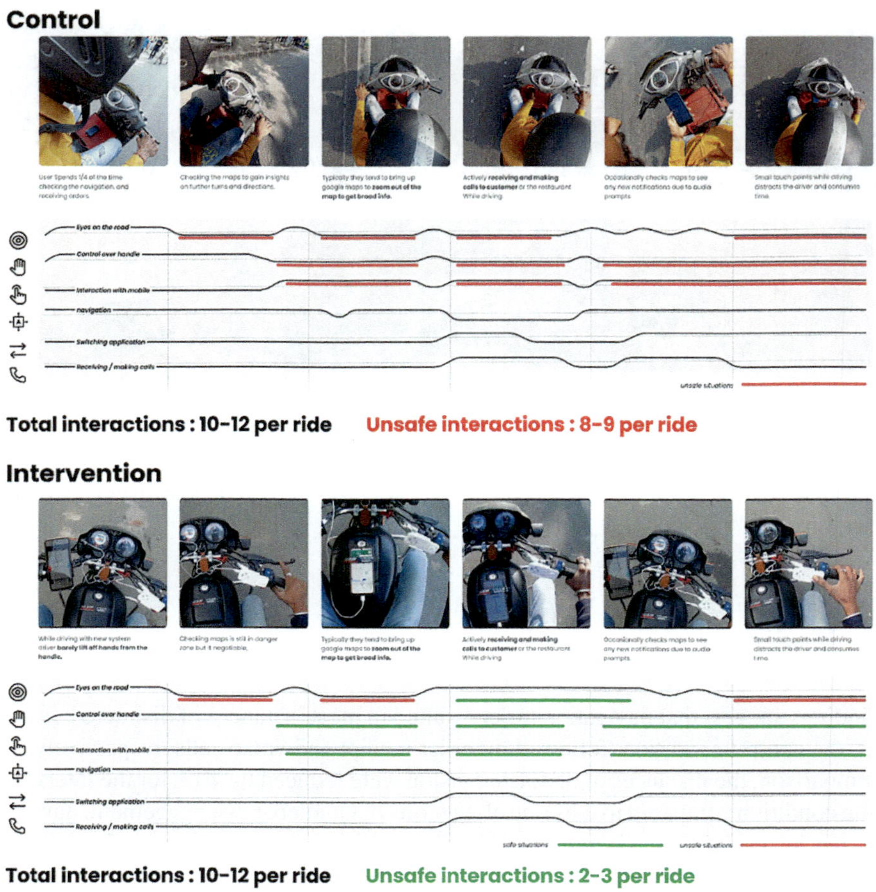

Fig. 5 Comparison of real-world task performance between control and intervention

Table 1 Comparison of MTW-rider performance between control and intervention conditions

Activity	Instances of distracted vision		Instances of handlebar disengagement	
	Control	Intervention	Control	Intervention
Order receipt	04	02	05	00
Navigation	15	09	06	01
Voice calls	03	02	02	00
Total	22	13	13	01

4 Discussion

Our findings revealed that a high number of mobile phone interactions in the control condition resulted in a significantly elevated risk of distracted driving during the delivery process, potentially leading to accidents. Use of mobile phone touchscreen interfaces for MTW-riders raises safety concerns as these interactions demand visual attention away from the riding task. Our intervention consisting of a handlebar-based tactile control mapped to the mobile phone application significantly reduced the instances of handlebar disengagement as well as instances of distracted vision. The strengths of the study include: (1) engagement with the real-world goods and food delivery context; (2) a human-centered-design process where physical and cognitive challenges of the MTW-rider were used as the evidence for the redesign; (3) iterative design and testing across the design process; (4) validation of final solution through rigorous field tests.

However, the study has the following limitations: (1) while tactile controls on the handlebar reduce instances of handlebar disengagement, they may not attenuate the risks of in-attention blindness in driving [14]; (2) the design of the handlebar-based tactile controller is only for motorcyclesthe design needs significant modification so that the controller can be fitted to handlebars of scooters; (3) sample size of participants and duration of field tests needs to be increased; (4) parameters for quantifying performance differences between control and intervention conditions needs fine-grained analysis. Future work includes iterative refinements in the hardware and interface design as well as longer duration tests with a larger sample of MTW-riders. Anticipated advancements in technology over the next decade may offer improved MTW-rider safety through voice-activated controls and helmet-based heads-up displays. The use of tactile controllers for mobile phone interactions by MTW-riders engaged in last-mile delivery can reduce instances of handlebar disengagement and decrease on-road distractions to improve rider safety.

References

1. Road Accidents in India—2022, Ministry of Highways & Road Transport, Govt. of India. https://morth.nic.in/sites/default/files/RA_2022_30_Oct.pdf. Last accessed 10 Nov 2023
2. Mohan D, Tiwari G (2000) Road safety in less motorized countries-relevance of international vehicle and highway safety standards. In: Vehicle safety 2000—proceedings of an international conference Held 7–9 June 2000, Institution of Mechanical Engineers, London, UK
3. Mohan D (2002) Road safety in less-motorized environments: future concerns. Int J Epidemiol 31(3):527–532
4. Thombre A, Ghosh I, Agarwal A (2023) Examining factors influencing the severity of motorized two-wheeler crashes in Delhi. Int J Inj Control Saf Promot 1–14
5. Delhi Traffic Police (2021) Road accidents in Delhi. Accident Research Cell, Delhi Police. https://traffic.delhipolice.gov.in/sites/default/files/uploads/2021/Delhi-Crash-Report-Final-2021.pdf. Last accessed 10 Nov 2023
6. Gutierrez MI, Mohan D (2020) Safety of motorized two-wheeler riders in the formal and informal transport sector. Int J Inj Contr Saf Promot 27(1):51–60

7. Kieling RR, Szobot CM, Matte B, Coelho RS, Kieling C, Pechansky F, Rohde LA (2011) Mental disorders and delivery motorcycle drivers (motoboys): a dangerous association. Eur Psychiatry 26(1):23–27
8. Zhang Y, Huang Y, Wang Y, Casey TW (2020) Who uses a mobile phone while driving for food delivery? The role of personality, risk perception, and driving self-efficacy. J Safety Res 73:69–80
9. Pawar NM, Khanuja RK, Choudhary P, Velaga NR (2020) Modelling braking behaviour and accident probability of drivers under increasing time pressure conditions. Accid Anal Prev 136:105401
10. Pawar NM, Velaga NR (2021) Effect of time pressure on steering control of the drivers in a car-following situation. Transport Res F: Traffic Psychol Behav 80:218–236
11. Lissy KS, Cohen JT, Park MY, Graham JD (2000) Cellular phone use while driving: Risks and benefits. Boston, MA: Harvard Center for Risk Analysis, Harvard School of Public Health. p 100
12. Gupta M, Pawar NM, Velaga NR, Mishra S (2022) Modeling distraction tendency of motorized two-wheeler drivers in time pressure situations. Saf Sci 154:105820
13. Brown T (2008) Design thinking. Harv Bus Rev 86(6):84
14. Strayer DL, Drews FA (2007) Cell-phone–induced driver distraction. Curr Dir Psychol Sci 16(3):128–131

Operational Requirements for Headway-based Public Transport System with Battery Electric Buses Under Swapping Strategy

S. Sajikumar and D. Bijulal

1 Introduction

Implementing electric vehicles as transportation units solves the pollution growth rate in line with the directions of the Paris Agreement 2015. Public bus transport organizations all over the world are now in the strategic stage of shifting their crude oil-based public transport units to Battery Electric Buses (BEB) to implement this policy. The related most common crucial drawbacks are capital investment, range anxiety, and charge replenishment [1].

The drawbacks of BEBs such as cost and range anxiety are being addressed by researchers around the world. The remaining most significant issue becomes the energy replenishment (charging). The depots can address it by overnight charging, fast charging at intermediate points, and battery swapping. Each of these alternatives has their pros and cons. Though most of the studies in the BEB public transportation system are related to charging, battery swapping remains an acceptable option. Battery swapping was introduced by Hartford Electric during the early 1900s [2].

Battery swapping is the process of replacing the used battery with a fully charged battery. It is fast and efficient, reduces waiting time, prolongs battery life, increases the overall efficiency of power systems, supports schedule adhesion, and help to achieve cost reduction [3].

S. Sajikumar (✉)
Research Scholar, Government Engineering College, Barton Hill, APJ Abdul Kalam Technological University, Thiruvananthapuram, Kerala, India
e-mail: sajikumars9928@gecbh.ac.in

D. Bijulal
Professor in Mechanical Engineering, Government Engineering College, Barton Hill, Thiruvananthapuram, Kerala, India

© The Author(s), under exclusive license to Springer Nature Singapore Pte Ltd. 2025

S. Das et al. (eds.), *Human-Centered Digitalization, Volume 3*, Design Science and Innovation, https://doi.org/10.1007/978-981-96-4383-7_42

2 Related Literature

The literature covers different aspects of the battery bus operations. Some relevant to this study include: mathematical framework to find optimal solutions for battery charging stations [4], the optimal overnight charging schedule to minimize the battery aging cost [5], methodology for developing a BEB overnight charging schedule at the depot with uncertain charging time [6]. Some literature addressed the issues such as: method to optimize the schedule of fast-charging in the BEB network to minimize the cost of the charging process and control the impact on the power grid [7] using a heuristic approach to determine the number of fast chargers required at intermediate points, a model to optimize the fast charging schedule with the minimum cost of operation [8], model to find the optimum number of charging stations and additional vehicles required so as to manage the range limitation of the BEB [9], holding-based corrective control strategy for avoiding headway deviation in electric bus operation [10].

2.1 Battery Swapping

Literature records the benefits of quick energy exchange and savings of peak power in a BEB operating in battery swapping mode [11]. In an attempt by [1] routing and scheduling strategies for the battery electric fleets based on swapping and charging demand analysis was developed. They have developed an algorithm for calculating the number of backup battery packs. Optimization of the design of a swapping–based public transport network considering the capacity of the battery module, rating of the charger, swapping schedule, charging schedule, and the required number of battery modules [12]. Models to find the number of swapping stations and the optimal location of the swapping station were proposed by [13]. Previous works in the energy management area of BEB operation did not consider the headway of operation and its effect. The objective of this work is to optimize the reserve battery inventory based on the fleet strength and headway of the transport network and to develop a model schedule for the swapping process in alignment with the bus operation schedule.

3 Energy Management in BEB Operation

Energy management is a complex activity that includes various battery-related activities, such as the purchase, storage, maintenance, charging, and recycling and disposal of batteries. In the case of depot charging, the BEB units are kept on charging in night charging schedules. Any early discharging of the battery is managed by an additional bus rather than an additional battery. In the case of fast charging, the battery is charged in an intermediate fast charging station(s) during the operation of the bus.

For these two methods, inventory of a sufficient number of chargers is required simultaneously for accommodating various recharging requirements. The battery units are integrated with the BEB units. In the case of the swapping method, the discharged battery is exchanged with a fully charged battery at a swapping center without affecting the bus schedule. To follow the swapping strategy, a reserve stock of charged batteries is required, which necessitates a basic model for BEB operation.

4 Assumptions

- There is a designed schedule for bus operation, with a fixed number of buses.
- A single route with one origin and one destination is assumed.
- Schedule deviations of operation are not considered.
- Operational characteristics of all the electric buses operating in the network and the chargers in the swapping station remain the same.
- Charging and discharging deviations among trips are not considered, thereby the range of BEB is assumed constant.

5 Base Model

The BEB operation is assumed to have separate bus operation management and energy management verticals (EMV). The battery inventory in this paper is assumed to have (a) battery in the bus (B_i) and (b) battery kept at the swapping center (R_j). The battery at the swapping center is either in a fully charged state or in a charging state. Management of all these categories of batteries is the aim of the energy management vertical.

EMV has a swapping schedule, a charging schedule, and a maintenance schedule. The fixed schedules of bus operation are the basis for the schedules of EMV. The stock of batteries consists of two categories: (a) swappable batteries in the bus (n) and (b) reserve batteries at the swapping center for swap (m). The number of swappable batteries shall be the same as that of the number of total buses in the freight [1]. The number of reserve batteries will vary according to various factors of operation. The complexities in optimization of reserve batteries and corresponding schedule management in Battery Electric Bus transport systems are discussed in this paper.

5.1 Swapping Flow

Figure 1 shows the swapping flow. An electric bus V_1 with a fully charged battery starts its operation from the depot at the scheduled departure time. This fully charged battery is from the general battery stock represented by B_i. Swapping affinity of a battery is a term used for representing the requirement generated by the battery to become fully charged, which starts from the beginning of the BEB operation.

Affinity to swap after a distance of x kilometers in time t can be estimated from various state of charge values (SoC).

$$\phi_{xt} = \frac{\mathrm{SoC}_{\max} - \mathrm{SoC}_{xt}}{\mathrm{SoC}_{\max} - \mathrm{SoC}_{\min}} \tag{1}$$

Affinity to swap before the commencement of the first trip is zero, where $x = 0$and t $=0$. After a certain operation time, it will reach its maximum value 1. The bus must be at the swapping center at this crucial level of affinity 1. The battery in the bus V_1 is swapped with a fully charged battery from the reserve battery stock in the swapping center R_j. The total number of batteries available at the swapping center is m. From this stock, R_1 is supplied against B_1, after which the inventory of reserve batteries reduces to R_{j-1}. The entry of B_1, into the swapping center gets added to the stock of battery on charging represented in the model as C_k until it becomes fully charged.

After charging, the battery gets added to the reserve stock of fully charged batteries R_j and enters into the swapping stock. This process of conversion needs a total time t_φ which includes the time for swapping activities, pre-charging activities, charging, and post-charging activities until it enters the stock.

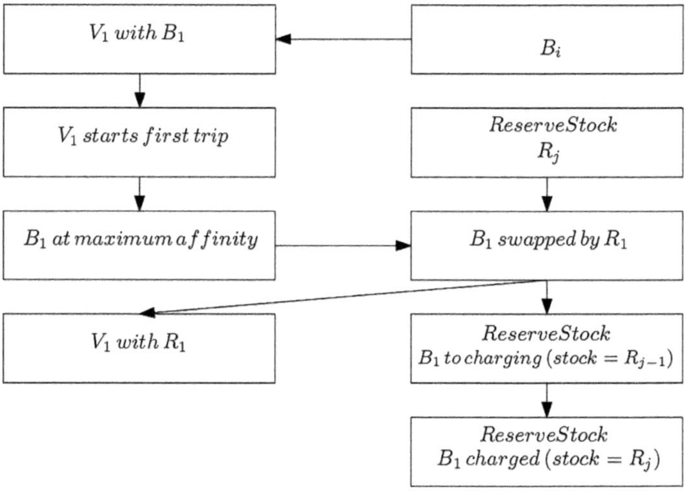

Fig. 1 Swapping flow

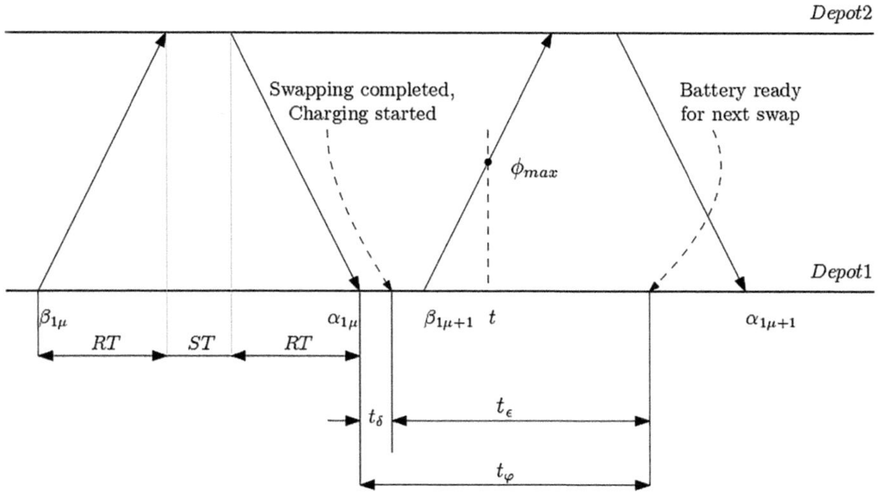

Fig. 2 Basic swapping model

Figure 2 shows a basic swapping model. A bus operation is planned between origin depot 1 and destination depot 2. A trip includes two running times (RT) and one stand time (ST).

If the charge of the battery during operation is expected to reach $\phi_{\max}$ after the completion of first trip, the battery needs to be swapped at the depot 1. The bus restarts its operation after the designed *ST* at depot 1. The battery taken for charging is ready to swap after the time t_φ, and can be used for satisfying the affinity generated by the battery of another bus. This time t_φ includes swapping time (t_δ), preparation time before charging (t_p), and charging time (t_ε). Time of scheduled arrival is represented by α and scheduled departure is represented by β. The schedule number is represented by μ.

5.2 Battery Requirement

In the swapping model, the number of additional battery requirements is labeled as the reserve stock. The maximum stock of batteries without considering the economy factor is twice that of the number of busses which leads to very high investment in the energy management vertical. At the maximum affinity point of a battery at a certain time point, the bus must be at the swapping center and the swapping center must have a minimum of one fully charged battery for the swapping process. The next requirement of a fully charged battery arises at the swapping center after the headway designed by the transport organization, for the next bus having battery at $\phi_{\max}$.

5.3 Effect of Headway

Now consider the next bus V_2 which starts with a headway of H from the first bus. After attaining its maximum affinity, bus V_2 reports at the swapping center and will continue operation after the stand time ST with the swapped fully charged battery. The reserve stock of the battery is related to the headway of operation. The effect of headway in determining the reserve battery strength can be categorized into two ways.

- Condition 1:$H > t_{\varphi}$, Here, the total number of batteries required at the swapping center, $R_j = 1$
- Condition 2:$H \leq t_{\varphi}$, As the battery in the first bus is in the charging mode at the time of maximum affinity of the battery of the second bus, the swapping center requires an additional battery as a reliever in the stock of the swapping center. The total number of batteries required at the swapping center,

$$R_j = \frac{t_{\varphi}}{H} \tag{2}$$

6 Swapping Schedule

The swapping schedule receives its input from the bus schedule. A swapping schedule contains the time of starting of swapping ($t_{\delta s}$), time of completion of swapping ($t_{\delta c}$), time of starting of charging (t_s), and time of completion of charging ($t_{\epsilon c}$).

$$t_{\delta s} = \alpha_{\gamma\mu} \tag{3}$$

$$t_{\delta c} = \alpha_{\gamma\mu} + t_{\delta} \tag{4}$$

$$t_{\epsilon s} = t_{\delta c} + t_{\rho} \tag{5}$$

$$t_{\epsilon c} = t_{\epsilon s} + t_{\epsilon} \tag{6}$$

7 Case Study

An intercity bus public transportation system considered in this article is operating with conventional diesel buses. The steps needed by the system for shifting to Battery Electric Buses under the swapping policy are: Purchase of BEB with swappable batteries, fixing the location of the swapping center, design factors such as stock of

Table 1 Schedule of the origin depot

Bus numbers	Departure from depot 1	Running time (hr)	Arrival at depot 2	Stand time (min)	Departure from Depot 2	Arrival at depot 1
1	5.00	2	7.00	30	7.30	9.30
2	5.30	2	7.30	30	8.00	10.00
3	6.00	2	8.00	30	8.30	10.30
4	6.30	2	8.30	30	9.00	11.00
5	7.00	2	9.00	30	9.30	11.30
6	7.30	2	9.30	30	10.00	12.00
7	8.00	2	10.00	30	10.30	12.30
8	8.30	2	10.30	30	11.00	13.00
9	9.00	2	11.00	30	11.30	13.30
10	9.30	2	11.30	30	12.00	14.00
1	10.00	2	12.00	30	12.30	14.30

battery, number of charging points, and swapping schedule. Stand time is equal to headway is the basic assumption taken in this case study. The length of the selected route is 60 kilometers and the running time required to complete the one-way trip is two hours. The designed stand time at the origin and destination is half an hour. The selected BEB has a range of 120 kilometer for a single charge, hence the swapping station can be set at the origin. The automatic swapping center is assumed with 4 minutes swapping time. The preparation time is assumed negligible. The time for charging the battery is three hours. The basic bus schedule shown in Table 1 is required to prepare the schedule for the swapping station.

7.1 Swapping Schedule

The swapping schedule is derived from the basic bus schedule using the equations explained in Sect. 7. The schedule prepared is shown in Table 2. The departure time of the first bus from depot 2, after its first one-way trip is 7.30 and it will report back at depot 1 by 9.30. This is the scheduled time for the starting of swapping process. After the completion of swapping process, this bus can be used for the next operation from 10.00 AM. The battery removed from the bus entered in to the charging center gets fully charged by 12.34. This battery can be used for satisfying the affinity of the seventh bus. Hence, the number of batteries required in the reserve stock is six. Total number of bus required for the operation is ten.

Table 2 Swapping schedule

Bus numbers	${}^{t}\delta s$	${}^{t}\delta c$ (min)	${}^{t}\varepsilon s$	t_{ε}	${}^{t}c$
1	9.30	4	9.34	3	12.34
2	10.00	4	10.04	3	13.04
3	10.30	4	10.34	3	13.34
4	11.00	4	11.04	3	14.04
5	11.30	4	11.34	3	14.34
6	12.00	4	12.04	3	15.04
7	12.30	4	12.34	3	15.34
8	13.00	4	13.04	3	16.04
9	13.30	4	13.34	3	16.34
10	14.00	4	14.04	3	17.04

8 Conclusions

The operational requirements of a swapping-based public transport are discussed in this work. The functions-based separation of transport operations into operation management and energy management clearly explains various operational requirements, existing in the swapping-based public transport. The basic characteristics of bus operation, the headway, and the total time spent by the battery at the swapping station are the determining factors behind the requirement of a battery in a swapping station. The requirement of battery as double the number of buses has come down with the proper use of available information. This simple case study reveals the importance of headway in the design of the stock of reserve battery. Properly maintained headway will not make an additional requirement for battery at the swapping station. A properly maintained swapping schedule will further help maintain the bus operation under the designed headway. The base model with constant headway and trip in a single route is considered in this work. Variable headway, multiple routes, inclusion of stand time variability, and running time variability are the future directions in this area.

References

1. Li W, Li Y, Deng H, Bao L (2018) Planning of electric public transport system under battery swap mode. Sustainability 10(7). https://doi.org/10.3390/su10072528
2. Ban M, Yu J, Li Z, Guo D, Ge J (2019) Battery swapping: an aggressive approach to transportation electrification. IEEE Electrif Mag 7(3):44–54. https://doi.org/10.1109/MELE.2019.2925762
3. Ahmad F, Alam MS, Alsaidan IS, Shariff SM (2020) Battery swapping station for electric vehicles: opportunities and challenges. IET Smart Grid 3(3):280–286. https://doi.org/10.1049/iet-stg.2019.0059

4. Leou RC, Hung JJ (2017) Optimal charging schedule planning and economic analysis for electric bus charging stations. Energies 10(4). https://doi.org/10.3390/en10040483
5. Houbbadi A, Trigui R, Pelissier S, Redondo-Iglesias E, Bouton T (2019) Optimal scheduling to manage an electric bus fleet overnight charging. Energies 12(14):1–17. https://doi.org/10.3390/en12142727
6. Zheng F, Wang Z, Liu M (2022) Overnight charging scheduling of battery electric buses with uncertain charging time, vol 22(5). Springer Berlin Heidelberg. https://doi.org/10.1007/s12351-022-00740-y
7. Abdelwahed A, van den Berg PL, Brandt T, Collins J, Ketter W (2020) Evaluating and optimizing opportunity fast-charging schedules in transit battery electric bus networks. Transp Sci 54(6):1601–1615. https://doi.org/10.1287/trsc.2020.0982
8. He Y, Liu Z, Song Z (2020) Optimal charging scheduling and management for a fast-charging battery electric bus system. Transp Res Part E Logist Transp Rev 142:0–36. https://doi.org/10.1016/j.tre.2020.102056
9. Estrada M, Mensión J, Salicrú M, Badia H (2021) Charging operations in battery electric bus systems considering fleet size variability along the service. Transp Res Part C Emerg Technol 138(Sep):2022. https://doi.org/10.1016/j.trc.2022.103609
10. Gkiotsalitis K (2021) Bus holding of electric buses with scheduled charging times. IEEE Trans Intell Transp Syst 22(11):6760–6771. https://doi.org/10.1109/TITS.2020.2994538
11. Kim J, Song I, Choi W (2015) An electric bus with a battery exchange system. Energies 8(7):6806–6819. https://doi.org/10.3390/en8076806
12. Ayad A, El-Taweel NA, Farag HEZ (2021) Optimal design of battery swapping-based electrified public bus transit systems. IEEE Trans Transp Electrif 7(4):2390–2401. https://doi.org/10.1109/TTE.2021.3083106
13. Moon J, Kim YJ, Cheong T, Song SH (2020) Locating battery swapping stations for a smart e-bus system. Sustain 12(3):1–21. https://doi.org/10.3390/su12031142

Process Safety Management in the Indian Steel Industry: A Comprehensive Evaluation

Shatrudhan Pandey, Abhishek Kumar Singh, and Shreyanshu Parhi

1 Introduction

Process Safety Management (PSM) ensures industries' success by creating safer and healthier industrial ecosystem. Reducing workplace accidents, injuries, near-miss incidents, and work-restricted cases correlates with increased productivity [1]. In today's highly competitive market, industries are focused on cutting costs while boosting profitability. Consequently, significant resources are allocated to process safety, crucial in minimizing losses caused by accidents and work restrictions related to manufacturing processes [2]. PSM practices are typically guided by regulatory frameworks such as those established by the US Occupational Safety and Health Administration (OSHA). These frameworks provide organizations with a comprehensive approach to managing process safety. However, in India, industries often overlook process safety and working conditions. Despite contributing approximately 3% to global manufacturing, India accounts for over 30% of industrial fatalities. The steel industry, a significant contributor to India's economic growth, mirrors this challenge, with a higher incidence of severe accidents than its international counterparts [3]. Many Indian industries lack robust safety management systems that align with global best practices. To mitigate hazardous incidents and reduce work-related restrictions, the steel industry in India needs to recognize the value of safe

S. Pandey (✉) · A. K. Singh
Department of Production and Industrial Engineering, Birla Institute of Technology, Mesra, Ranchi India
e-mail: er.shatrudhanp@gmail.com

A. K. Singh
e-mail: abhishekkumar@bitmesra.ac.in

S. Parhi
Department of Operations Management and Quantitative Techniques, IMI, Bhubaneswar, India
e-mail: shreyanshu.p@imibh.edu.in

© The Author(s), under exclusive license to Springer Nature Singapore Pte Ltd. 2025
S. Das et al. (eds.), *Human-Centered Digitalization, Volume 3*, Design Science and Innovation, https://doi.org/10.1007/978-981-96-4383-7_43

working environments and gradually integrate PSM practices [4]. Steel manufacturing involves inherent hazards that require careful and often complex management. PSM not only aims to protect workers but also safeguards the environment, assets, and surrounding communities [5]. Workers in the steel industry face several safety and health hazards due to the nature of their work [6]. In steel industries, workers face significant safety and health risks due to the hazardous nature of work. PSM is an analytical tool designed to prevent the release of highly hazardous chemicals, as defined by the Environmental Protection Agency (EPA) and OSHA. It encompasses a range of interrelated strategies to manage risks associated with the process industries, aiming to reduce the frequency and severity of chemical and energy release incidents. Industries must compile detailed Process Safety Information as part of these safety protocols before conducting any required process hazard analyses. The failure of PSM can result in devastating and costly consequences [7]. Figure 1 shows the various elements of PSM.

Significant industrial incidents have highlighted the critical need for Process Safety Management (PSM). One of the most notable PSM failures occurred in Bhopal, India, in 1984, when the release of methyl isocyanate resulted in over 2,000 deaths and thousands of injuries [8]. Another report emphasized the challenges of maintaining safety during project execution, particularly in Indian steel plants, where poor educational qualifications and inadequate training of contractual workers increase safety risks. In 2012, an explosion during the stabilization of a newly installed oxygen plant at the Visakhapatnam Steel Plant resulted in the loss of 11 lives [9]. A survey conducted at the Indian Iron and Steel Company (IISCO) in West Bengal revealed that workers' lack of Occupational Safety and Health (OSH) training, insufficient safety knowledge, and non-technical education were significant contributors to accidents [10]. Additionally, an industrial accident at the Bhilai Steel Plant in Chhattisgarh in 2018 claimed the lives of 14 workers, with 10 others suffering severe injuries. Investigations pointed to poor safety management, as gas remained in the Coke Oven Gas Pipeline, and a metal plate was removed from a flange joint without proper risk assessment. A resulting spark ignited the gas, leading to the fatal fire. The inquiry found that safety protocols were not followed, the fire brigade was improperly deployed, and no safe escape routes were arranged [11]. Despite being a flagship unit of the Steel Authority of India Limited (SAIL), Bhilai Steel Plant has struggled with workplace safety issues. Workers reported the absence of a safety committee for the past two years. In another incident on June 13, 2014, six workers were killed and over 40 others injured due to a methane gas leak at a water pump house in the same plant. These incidents underscore the urgent need for improved safety management practices in India's industrial sectors [12]. As per "Occupational Safety, Health and Working Conditions Code, 2020" the employer shall be responsible for maintaining in his establishment such health, safety, and working conditions for the employees as may be prescribed by the Central Government [13]. Today, many steel industries around the world have achieved records of no lost time injuries and zero fatalities. However, the Indian iron and steel industry is still a death trap. Poor health and safety conditions at workplaces also reflect the overall poor performance

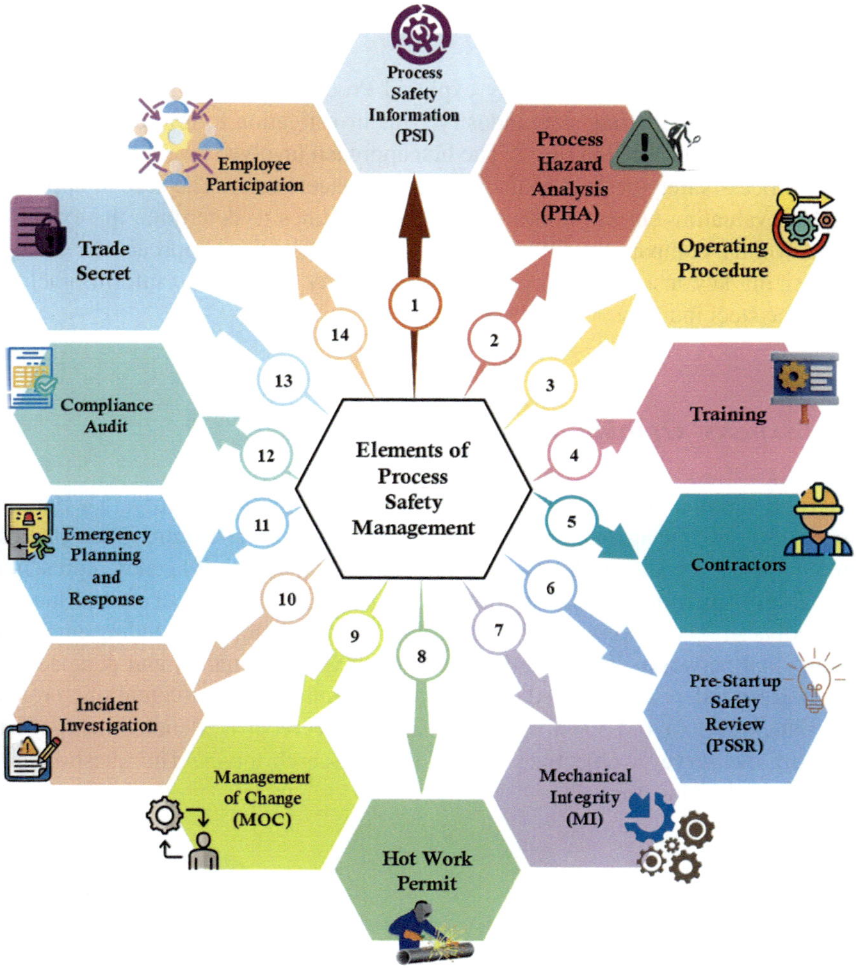

Fig. 1 Elements of Process Safety Management

of the sector [14]. Over the years, the country has witnessed a rise in process safety-related incidents. Factors such as industrial growth, the inherently hazardous nature of processes, and upward trends in incidents have highlighted the need to identify shortcomings in the current framework and propose measures for enhancing process safety. In light of challenges like restricted data availability, insufficient incident and near-miss reporting, and a lack of comprehension and application of Process Safety Management (PSM), the research examined the operational and safety practices within the steel industry in India. The steel industry was chosen due to its critical nature and susceptibility to severe events. It is important to note that this research was the first to be conducted in the Indian context.

2 Methodology

A qualitative research methodology explored Process Safety Management (PSM) practices within the Indian steel industry. This investigation followed three distinct approaches, as illustrated in Fig. 2. The first approach involved leveraging the expertise of process safety professionals in the Indian steel sector. The second approach entailed evaluating current relevant national guidelines to determine the extent of PSM coverage compared to OSHA PSM standards. The third approach focused on synthesizing key insights from global process safety literature to inform practices within the steel industry.

2.1 Experts' Opinion Survey

An expert opinion survey was conducted using a questionnaire to gather data on Process Safety Management (PSM). The survey collected insights from process safety experts, key steel industry stakeholders, and government agency representatives. Table 1 provides a summary of the respondents from the steel sector. Surveys and feedback requests were distributed through various channels, including email and professional networks, based on the relevance of the information and perspectives sought. Respondents were asked to offer candid opinions while maintaining professional integrity to avoid potential biases. Each expert received detailed instructions regarding the survey's objectives and relevant research topics. The questionnaire items are outlined in Table 2.

Fig. 2 Methodology

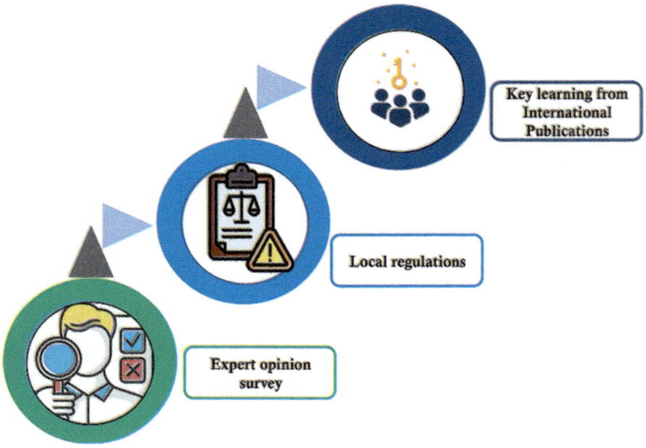

Table 1 Experts profile

Expert No	Profiles
1.	Associated with the major steel industry of India as a safety manager with 12–15 years of experience in process safety
2.	About 10–15 years of experience in the steel industry working as post of general manager—safety
3.	Associated with the labor department with experience of 10–15 years
4.	Has 40 years of working experience in safety and health in the safety and health information bureau
5.	Has over 35 years of experience as an assistant director—safety
6.	Has 09 years of experience as a safety officer in different steel plants in India
7.	Has diverse experience as a safety officer, mechanical engineer, and operation manager with 20 years of experience
8.	Has 30 years of experience as head of department- safety in various steel plants in India
9.	Has 22 years of experience as an environmental health and safety manager
10.	Has 30 years of total experience and 14 years of experience in occupational safety and health

Table 2 Questionnaire items

1.	Could you describe your background and experience within the steel industry?
2.	What are the current regulatory measures to prevent workplace accidents in the steel industry?
3.	From your perspective, what are the key lessons from past workplace accidents should steel industries adopt to prevent future incidents?
4.	What additional measures should be taken to prevent major accidents in the steel industry?
5.	Do you believe process safety implementation is deficient within the steel industry?
6.	What factors do you think contribute to this deficiency?
7.	Does your organization have a separate PSM program from the OSH program?

2.2 *Local Regulations*

In India, extant regulations dutifully address health, safety, and environmental concerns but notably lack explicit Process Safety Management (PSM) provisions. While dedicated PSM requirements are absent, some elements within the existing regulations indirectly contribute to PSM principles. Notably, the regulatory framework undergoes scrutiny, focusing on aligning standards with those set by the Occupational Safety and Health Administration (OSHA). This reflects a concerted effort to harmonize local regulations with globally recognized safety norms. By integrating international benchmarks, India aims to enhance industrial safety practices and align with established global standards, fostering a safer working environment and minimizing the risk of industrial accidents within the country.

2.3 *Key-Learning from International Publication*

This study harnessed valuable insights by examining international publications centered on process safety. The research deliberately referenced pivotal works that delve into strategies and measures adopted by developed countries. These references illuminated the country's various approaches to reducing the risk of recurrent industrial accidents within their regions. By drawing upon developed countries' experiences and best practices, the study aimed to contribute a nuanced understanding of successful strategies in Process Safety Management, potentially informing and improving safety protocols on a global scale. The focus on insights from developed nations provides a valuable benchmark for enhancing industrial safety practices and fostering a proactive approach to risk mitigation in various industrial settings.

3 Results and Discussions

3.1 *Experts' Opinion*

The survey responses indicated a notable absence of process safety-specific regulations in the country, leaving a significant gap in protection against incidents. While local regulations related to Occupational Safety and Health (OSH) offer some indirect process safety coverage, these are limited in scope. Despite the lack of mandatory requirements, some steel plants in India have proactively adopted Process Safety Management (PSM) practices. These plants conduct regular PSM audits to improve safety measures; however, there is a need for a more unified and coordinated effort to address the existing gaps comprehensively. Out of the ten respondents, seven (70%) underscored that the Bhopal disaster offered vital lessons in process safety. They identified several critical areas for improvement, such as safety management culture, mechanical integrity programs, training, change management, safe work practices, emergency response, and Process Hazard Analysis. On the other hand, three respondents (30%), all from the steel industry, noted that they have implemented PSM programs within their organizations and designated responsible individuals to oversee them. However, they also pointed out that the lack of regulatory compulsion hampers the broader implementation of PSM practices across the industry (Table 3).

Table 3 Comparison of PSM Coverage: Indian Local Regulations vs. OSHA PSM Elements

Local regulations and acts	PSM elements								
	Employee participation	Process Safety Information (PSI)	Process Hazard Analysis (PHA)	Operating procedure	Training	Contractors	Pre-Startup Safety Review (PSSR)	Mechanical Integrity (MI)	Hot work permit
The Factories Act, 1948	✓	✓	✗	✓	✓	✓	✓	✗	✗
The Mines Act, 1952	✓	✗	✗	✓	✓	✓	✗	✗	✗
The Dock Workers (Safety, Health and Welfare) Act, 1986	✓	✗	✗	✓	✓	✓	✗	✗	✓
The Indian Boilers Act, 1923 (amended2007)	✓	✗	✗	✓	✓	✓	✗	✗	✓
The Dangerous Machines (Regulation) Act, 1983	✓	✗	✗	✓	✓	✓	✗	✗	✓

(continued)

Table 3 (continued)

Local regulations and acts	PSM elements				
	Management of Change (MOC)	Incident investigation	Emergency planning and response	Compliance audit	Trade secret
The Factories Act, 1948	✗	✗	✓	✗	✗
The Mines Act, 1952	✗	✗	✗	✗	✗
The Dock Workers (Safety, Health and Welfare) Act, 1986	✗	✓	✓	✗	✗
The Indian Boilers Act, 1923 (amended2007)	✗	✓	✓	✗	✗
The Dangerous Machines (Regulation) Act, 1983	✗	✓	✓	✗	✗

3.2 *Local Regulations*

3.3 *Key Insights from International Publications*

The research titled "The Role of Leadership in Process Safety Management System" "No Process Safety Management System is an Island" by Levovnik et al. explores the relationship between Process Safety Management systems, factors affecting safety performance, and different leadership styles. Drawing from existing literature, it identifies the factors that wield the most significant influence on the successful implementation and performance of safety management systems [15]. Aziz et al. introduced a practical approach to implementing a Process Safety Information (PSI) system that aligns with the requirements. The focus is establishing structured strategies for effectively handling documentation, facilitating communication, and maintaining a written program to manage, revise, and update critical information about hazardous chemicals, equipment, and technology [16].

Goh et al. include interviews with government, academia, and private-sector industry experts. The goal was to learn from their insights about how the Singapore process industry can draw lessons from the Bhopal disaster. These interviews developed a systemic Process Safety Management model, focusing on key operational elements [17]. Zhao et al. stated that China has witnessed a series of chemical accidents with severe consequences for both people and the environment over the past two decades. The Chinese government has taken significant steps to strengthen legislation and regulatory standards in response to these accidents. Progress in this regard necessitates government support and a shared commitment across all hazardous industries to learn from past accidents [18]. Mannan & Waldram address a paradigm shift in process safety and the creation of a user-friendly, multinational accident database that is freely accessible and multilingual. It emphasizes the need for sharing comprehensive information, including causes and "lessons learned." The goal is to prompt action by authorities to establish the database and even consider compulsory use in certain circumstances. Ultimately, the aim is to enhance safety practices and prevent future accidents by learning from past incidents [19].

Research findings by Moura et al. introduce a novel approach for applying lessons from previous accidents to inform and enhance design reviews, offering a proactive strategy for preventing future incidents. MATA-D is a valuable resource for understanding the complexities of significant accidents in high-technology industries and provides insights into mitigating risks through improved design and accident prevention strategies [20]. Sarshar et al. have identified various challenges associated with the planning process that could impact significant accident risk. The research draws from investigation reports, interviews, and a workshop conducted within the industry. The challenges identified encompass various aspects of planning that can influence significant accident risk. The study proposes improvements designed to enhance the planning process in response to these challenges. These enhancements are targeted explicitly at better managing and mitigating significant accident risks, providing practical recommendations to bolster safety and reduce the likelihood of major incidents [21]. Shin analyzed significant industrial accidents in Korea from 2008 to 2012 and uncovered several key findings: Among all the accidents studied, 57.1% were attributed to human failure, while 32.1% occurred during the startup phase of operations. The analysis identifies deficiencies in preventive systems and the capacity to address human failures within chemical plants [22].

4 Conclusions

The Indian steel industry urgently needs to advance its process safety performance. Currently, Process Safety Management (PSM) in this sector is still in its nascent stages. Interviews with ten process safety experts and a thorough literature review have identified several key challenges: the lack of local PSM regulations, limited awareness of PSM principles, insufficient leadership engagement, absence of dedicated PSM system implementation, and ineffective near-miss and incident reporting coupled with poor database management. Despite these issues, some steel plants have voluntarily adopted PSM systems, demonstrating a proactive stance. Nevertheless, there is a pressing need for governmental support to develop PSM-specific regulations and a unified commitment across the industry to implement a comprehensive PSM program in India effectively.

References

1. International Labour Organization (2005) Sectoral Activities Programme Code of practice on safety and health in the iron and steel industry, pp 1–106
2. de Souza FHB, Pacheco RN (2018) Evaluation of Process Safety Management (PSM) Implantation and Estimate of Results : A Case Study at a Steel Making Industry in Minas Gerias. In: xxxviii encontro nacional de engenharia de producao A Engenharia de Produção e suas contribuições para o desenvolvimento do Brasil, Dec 2018 [Online]. https://steel.gov.in/sites/default/files/Framework Document for Safety Guidelines.pdf
3. Kartikay K, Bandi J (2019) The Indian steel industry: growth, challenges and digital disruption (PwC I.). https://www.pwc.in/assets/pdfs/consulting/technology/the-indian-steel-industry-growth-challenges-and-digital-disruption.pdf
4. Ministry of Steel G (2019) 25 Safety Guidelines for Iron & Steel Sector. Government of India. https://steel.gov.in/sites/default/files/Framework. Document for Safety Guidelines.pdf
5. World Steel Association (2021) Process Safety Management Fundamentals. [Online]. https://www.worldsteel.org/en/dam/jcr:eb3f7e7a-4f7a-4eef-b2bb-85ce7b2825ea/Process%2520Safety%2520Management.pdf
6. Siddesh P, Indrasen S (2019) Effective implementation strategy of safety practices in Indian steel industry using focused group discussion technique i-manager's. J Futur Eng Technol 15(2):19. https://doi.org/10.26634/jfet.15.2.16586
7. Occupational Safety and Health Administration O (2000) Process Safety Management. Process Saf Manag OSHA 3132: pp 1–54 [Online]. https://www.osha.gov/sites/default/files/publications/osha3132.pdf

8. Ness A (2015) Lessons learned from recent process safety incidents. Chem Eng Prog 111(3): 23–29 [Online]. https://www.aiche.org/sites/default/files/cep/20150323.pdf
9. Subhra Dhara SK (2016) Effective Project Management in Steel Industry Annu Tech Vol Metall Mater Eng Div Board, Inst Eng 2:43–61 [Online]. https://www.ieindia.org/webui/ajax/Downloads/WebUI_PDF/Technical_Volume_Pdf/ATV_MMDB.pdf
10. Rajak R, Chattopadhyay A, Maurya P (2021) Accidents and injuries in workers of iron and steel industry in West Bengal, India: prevalence and associated risk factors. Int J Occup Saf Ergon. https://doi.org/10.1080/10803548.2021.2012021
11. Bhilai Steel Plant Report (2019) Violation of operational Safety & health standards and industrial accidents in Bhilai steel plant. A Report to the Workers Mar 2019 [Online]. https://www.groundxero.in/wp-content/uploads/2019/04/compiled-BSP-report-pdf_pagenumber.pdf
12. Rai S (2021) Is safety compromised at SAIL's Bhilai Steel Plant ? Steel 360. https://www.steel-360.com/stories/steel/is-safety-compromised-at-sails-bhilai-steel-plant
13. M. of L. and Justice (2020) The occupational safety, health and working conditions code, vol 37. pp 1–86
14. Somnath Kumar BM, Keshari KK, Kumar V, Chatterjee A, Prasad A, Gupta SK (2018) Improving safety, health & environment in steel industry. Miner Met Rev
15. Levovnik D, Gerbec M, Dimovski V (2019) The role of leadership in process safety management system—no process safety management system is an Island. Chem Eng Trans 74:1375–1380
16. Aziz HA, Shariff AM, Rusli R (2014) Managing process safety information based on process safety management requirements. Process Saf Prog 33(1):41–48. https://doi.org/10.1002/prs.11610
17. Goh YM, Tan S, Lai KC (2015) Learning from the Bhopal disaster to improve process safety management in Singapore. Process Saf Environ Prot 97:102–108. https://doi.org/10.1016/j.psep.2015.02.004
18. Zhao J, Suikkanen J, Wood M (2014) Lessons learned for process safety management in China. J Loss Prev Process Ind 29:170–176. https://doi.org/10.1016/j.jlp.2014.02.010
19. Mannan MS, Waldram SP (2014) Learning lessons from incidents: a paradigm shift is overdue. Process Saf Environ Prot 92(6):760–765. https://doi.org/10.1016/j.psep.2014.02.001
20. Moura R, Beer M, Patelli E, Lewis J, Knoll F (2016) Learning from major accidents to improve system design. Saf Sci 84:37–45. https://doi.org/10.1016/j.ssci.2015.11.022
21. Sarshar S, Haugen S, Skjerve AB (2015) Challenges and proposals for managing major accident risk through the planning process. J Loss Prev Process Ind 39:93–105. https://doi.org/10.1016/j.jlp.2015.11.012
22. Shin IJ (2013) Loss prevention at the startup stage in process safety management: from distributed cognition perspective with an accident case study. J Loss Prev Process Ind 27:99–113. https://doi.org/10.1016/j.jlp.2013.11.003

Self-defense and the Youth

Haashika Shah and Siddhesh Shirsekar

1 Introduction

Self-defense is a practical countermeasure focused on protecting oneself from harm, distinct from martial arts. It emphasizes avoiding conflict and using mental strength rather than initiating attacks. In India, with a large youth population and rising crime rates, especially against women and children, self-defense education becomes crucial. Despite misconceptions, self-defense doesn't require uniforms or rituals; it's a simple skill accessible to all. Studies show that most attackers lack martial arts proficiency. Therefore, proper training provides a significant advantage in real-life situations. Self- defense classes not only teach fighting skills but also enhance situational awareness. This essential skill, like driving or CPR, is vital for individuals of all ages and genders (Prabha, R.; Kamath, K, D'Mello; Jayadevan, 2019).

2 Literature Review

What is self-defense?

Self-defense is a collection of abilities that includes perception, confidence, verbal confrontation, and physical and mental methods and skills, to effectively combat and withstand intense situations of danger. (Kamath, K, D'Mello.)

H. Shah · S. Shirsekar (✉)
School of Design, SVKM's Narsee Monjee Institute of Management Studies (NMIMS) Deemed-to-Be-University, Mumbai, India
e-mail: siddhesh.shirsekar@nmims.edu

H. Shah
e-mail: haashika.shah30@nmims.edu.in

© The Author(s), under exclusive license to Springer Nature Singapore Pte Ltd. 2025

S. Das et al. (eds.), *Human-Centered Digitalization, Volume 3*, Design Science and Innovation, https://doi.org/10.1007/978-981-96-4383-7_44

Martial Arts	v/s	Self- Defense
Sport Can be pursued by enthusiasts or people eager to learn the sport	01	**Defense System & Educational** It is more of an educational, defense system designed to tackle everyday life attacks on 'the street'
Fitness-Centric It relies heavily on fitness	02	**Fitness is a byproduct** Fitness and techniques both
Combat & 'Act Now' teaching Can promote 'Aggressive', 'Strong' Persona	03	**Walk In Peace** The mantra and '101 Teaching'
Overall Development Spiritual, physical and mental development	04	**Agility & Natural Responses** It is a system based on natural responses and makes body and mind agile to unexpected situations

Fig. 1 Table of difference between self-defense and martial arts

2.1 The Difference Between Self-Defense and Martial Arts

It is common opinion that many martial arts forms are considered to be self-defense. A lack of awareness can prove to be a fatal uneducated choice and can result in the people joining different art forms that may not be able to teach the skills required to face daily life situations [7] (Fig. 1).

2.2 Teachings of Self-Defense

The figure below shows the teachings and principles that show what self-defense actually preaches contrary to the belief that self-defense is about throwing hands and legs to fight [7] (Fig. 2).

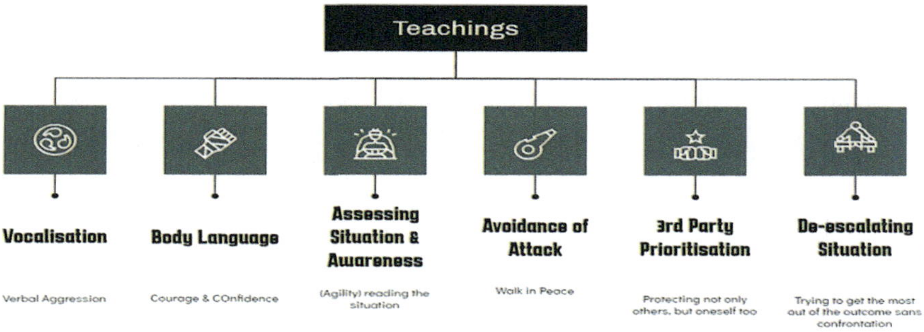

Fig. 2 Teachings of self-defense

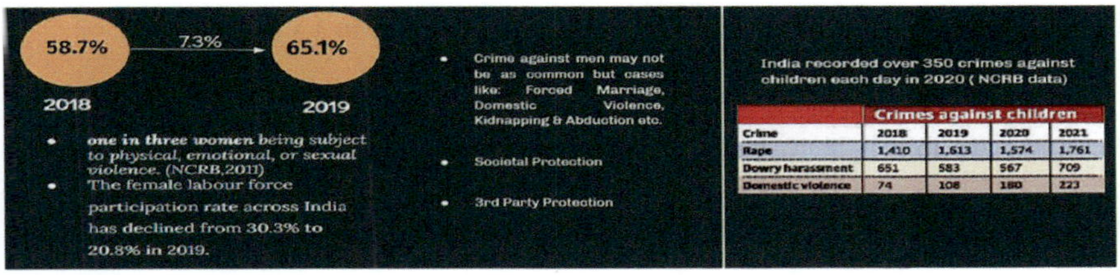

Fig. 3 Crime rates against women, men, and children (L-R) according to NCRB data of 2020

2.3 *Fight or Flight Response*

It is our body's natural response that prepares the body to "fight" or "flee" when in a situation of attack or threat. Self-defense helps in controlling this very impulse and teaches how to assess the situation and whether fighting or fleeing will be the correct response. (Prabha,R.)

2.4 *Women, Children, and Men—Everyone Should Know Self-Defense*

In the current scenario of society, self-defense is a must know not only for women, but also men and children (Fig. 3).

Although women are more commonly victims of crimes, men and children have also been increasingly falling victims to crimes. (Linehan, 2016).

2.5 *The Youth*

In India, about 1,70,426 schools got permission for training self-defense in schools in 3 years, 2015-2018. (EducationDesk, 2018.) Teaching self-defense to from the ages 15 to 21 can benefit in establishing awareness,developing strong self-discipline and developing physical skills [5].

2.6 *Existing Interventions*

The pros and cons of the existing interventions are as follows:

2.6.1 Online Learning (Free YouTube Tutorials, from Classes and Apps)

The pros are that it is cost effective as free, convenient and has a wider reach to people but also it lacks actual combat experience and may not be able to learn to make proper judgment of the moves [21].

2.6.2 Live Classes (Offline)

The pros are that it includes a proper training experience and practice with more people as a community through drills and guidance but it is costly, can be intimidating, traveling might be inconvenient and hence may not be continued for long [21].

2.6.3 Board Games/Card Games

The games are based on martial arts and are still considered as delf-defense games and although are intriguing, do not focus on educating but on storytelling and are not enough to make the users willing to join a class to learn it.

2.6.4 Video Games (Digital/AR/VR)

The current existing games promote an aggressive outlook of martial arts. While there are lesser games on self-defense, the martial arts games are not factually correct either which can lead to change in thoughts and expectations toward self-defense.

3 Methodology

The purpose is to describe the methodology used for data collection, synthesis, and analysis. This procedure helped achieve a solution according to the gaps and the problem statement.

3.1 Participants

The study, conducted in Mumbai with 112 participants from Khar and Andheri, focused on three groups: self-defense/martial arts teachers, students learning self-defense, and untrained youth. Of the participants, 87 completed Google Survey forms, and 25 took part in Empathy Interviews. These interviews confirmed issues identified in the secondary research stage and provided additional insights.

3.2 Materials and Procedure

Phase I: Research and Data collection

This phase began with mind-mapping which was followed by secondary research and establishing a target audience. After selecting a target audience, survey forms and empathy interviews were conducted to collect primary research data.

- **Secondary Research:** After an initial mind map, the potential areas of research were obtained. Desk research was then conducted which then revealed topics for Survey and Interviews and helped decide the target audience and their demographics. This also helped understand the ongoing situation of the topic (Social, Economic, and Local views and trends on the topic).
- **Primary Research:** After determining the stakeholders, the Survey and Empathy Interviews questions were framed and questionnaires were sent out to the respective audiences. These interviews helped gain new insights and pain points. Also, the mindsets and roadblocks of the stakeholders were obtained which helped understand what worked and what did not work for them to learn self-defense.

Phase II: Data Synthesis and Analysis

The data collected in the primary and secondary research was analyzed using data synthesis tools as mentioned below:

- **Personas:** It helped clarify the differences and similarities in all three target groups and helped justify their mental models and requirements better.
- **The Empathy Maps:** It helped obtain insights on their thoughts and roadblocks that prevented them from learning self-defense. Parameters like what they saw and what they heard and what they thought during their journey were considered.
- **How and Why laddering:** After grouping all pain points, thoughts, and hindrances, themes and patterns were identified which helped understand what the solution should focus on.
- **Camera study:** It helped capture the moves, strikes, and watches taught in self-defense. This was conducted at Khar and Andheri self-defense classes (KMG Global) and gave a better understanding of the scenario.

Phase III: Conceptualization

The final gaps were taken into consideration and after understanding the user's thoughts, preferences, mental models, and roadblocks, solutions were conceptualized.

- **Brainstorming:** This phase helped the research begin with brainstorming of concepts and ideas based on the gaps and problem statement.
- **Conceptualization:** The solutions that would fulfill the gaps and address the insights obtained from research were selected and sketched as concepts. One main solution was narrowed down to carry forward as the most effective solution.

- **Prototyping**: On the basis of this, Minimum Viable Product (MVP) were made to test the basic functionality and feedback in the form of low fidelity prototypes.
- **User testing:** The prototypes were user tested that helped understand which concept would aim at most problems.

4 Results

The current section talks about insights gained from each step of the methodology. The research aimed at understanding the current behaviors and mindsets of young adults living in the Urban Mumbai Areas toward self-defense. This was achieved with the tools used for research and analysis as deconstructed in the previous section of this paper.

4.1 Desk Research

The secondary research revealed that the current solutions are not motivating or nudging them. It also missed educating users on the difference between self-defense and martial arts. Most of the youth and general population did not notice the difference between self-defense and martial arts. They also have more and easier access to Martial Arts Classes in Urban and Rural Areas.

Self-defense 4 main steps to tackle any situation: *De-escalation, Avoidance, Prevention,* and the last stage should be the *Actual Fight.*

Desk research included studying crime rates and patterns over the period of time in India and case studies which revealed that:

- Although crimes against women are more common, crimes with men and children have been increasing and self-defense is a necessity for everyone
- It is better for children to learn from an early age as it is faster to learn and develop interests and skills to continue long term practice as well.

4.1.1 Parents

The parents of the target audience play a major role in influencing the choices and routine of the student. Below are the insights from the Empathy Interview:

- When the child is younger, the parents would nudge their children to pursue some martial art form, thinking it would involve self-defense and be a healthy and active sport for the child.
- When the child becomes older and the academic pressure increases with grade, parents may ask their children to leave any extracurricular activities (if taken) so as to completely prioritize academics.

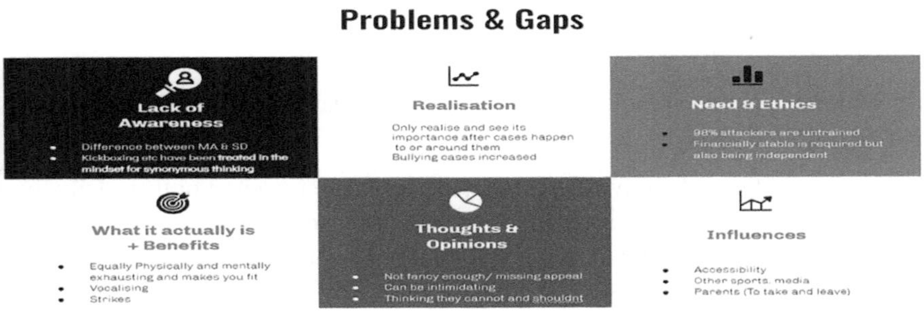

Fig. 4 Problems and gaps identified

4.1.2 Self-Defense Classes and Teachers

After interviewing students and parents, the self-defense classes and the masters were interviewed.

- The priorities and lifestyle of the students in the City and Urban areas are different (having a job, earning money, etc.) hence the need for intervention in urban areas is necessary.
- The current younger generation wants something that has more popularity as a sport, for example, Kickboxing is a sport gaining popularity especially among the youth.
- More students' request to join when a popular crime incident has happened or when an incident takes place with someone first-hand.

4.2 Gap Analysis and Direction

After evaluating insights, the secondary and primary research were mapped together and final gaps were identified. Six major gaps were identified that prevented the users from pursuing self-defense (Fig. 4).

4.3 Problem Statement

"How might we encourage and motivate young adults from the ages 15-19 to take up self-defense and remove misconceptions by making it a fun learning experience so that it becomes a must-know skill for them."

4.4 Solution Concept

The research, gaps and problem statement led to the realization of the following aims that the solution should:

- Create a nudge and motivate users to make self-defense a part of their daily lives
- Remove misconceptions regarding meaning of self-defense
- Spread awareness on the meaning and importance of self-defense
- Make self-defense interesting and fun for the users

5 Conceptualization

5.1 VR Simulation—StrikeSmart

StrikeSmart is a Cardboard VR experience that immerses users in lifelike self-defense scenarios. Divided into Learn, Apply, and Practice phases, the VR simulation teaches users about self-defense choices and distinguishes it from martial arts. Using a VR Storyboard and a Field of Vision Concept map, the experience incorporates 360° views, audio, and video effects. The high-fidelity prototype, created with Unity 3D and Fusion 360, includes Mumbai City environments designed after Camera Study and Observations. Users can choose fun environments and are encouraged to rethink self-defense, emphasizing that it doesn't require a Dojo like martial arts (Fig. 5).

Stage I—Learn

The users start with a cyberpunk-themed environment with a practice dummy in the scene and are prompted with arrows to practice "Open Palm Strikes" on the dummy. This teaches the user one type of strike and how it is performed.

Stage II—Apply

The users are given a relatable scenario of their day-to-day life on the streets of Mumbai and are asked questions to probe their decision-making and are encouraged to make observations and answer questions. They are met with a cat-calling attacker and asked to observe around for objects they can use to defend themselves and exits they can use.

Fig. 5 Environments and phases 1, 2, and 3 (Left to Right, respectively)

Fig. 6 Camera Study of Mumbai Streets at night time

Stage III—Practice

Users enter a cyberpunk-themed environment, punching focus mitts to enhance agility, mimicking real-life training. The eerie Mumbai Suburban Areas at night set the backdrop as users navigate the city on their way home from work.

5.2 *Camera Study Observations*

The environments were designed to match details after taking Camera Study and Observations for the Mumbai City Environment (Fig. 6).

5.3 *User Testing*

For user testing, 2 users were made to take the experience on the VR Cardboard Headset and 4 users were made to take an interactive experience using the software directly. All the users were from the age group 15-19 years old and belonged to the Mumbai Suburban Areas (Fig. 7).

WHAT WORKED	WHAT DID NOT WORK
Thought the probing questions helped them understand options and learn what to do next	Thought the scenario got over too quickly as they expected a longer and unfateful end to their character in the real world
The environment of the simulation was relatable to their ownlives	Expected more fancy and game-like agility practice environment
Saw the concept of self-defence in a new perspective that also made them rethink their current scenario	Expected a fancier, video game fantasy- like storyline like in movies and games
Understood some differing points of self-defence and martial arts	Consistency in the environments and backgrounds was lesser
Will perform the scanning of the environment consciously or subconsciously	Expected extremes: Too real or too scary and unreal environments
Felt the scenario was somewhat relatable to their previous experiences	
Felt a nudge to think about Self-Defense and understood it's importance	
Related the situation to their potential future situations (eg.	
Student travelling abroad) felt as if they were actually in that situation (immersed)	

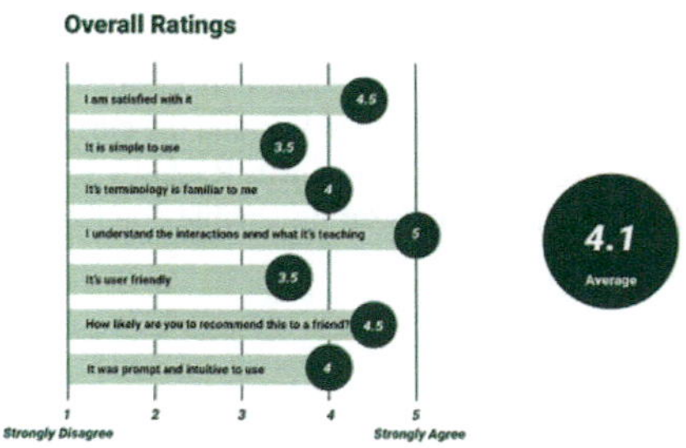

Fig.7 Table of What worked and What did not work in user testing stage along with overall and average rating of solution based on parameters

6 Discussion

Self-defense is often overlooked and misunderstood, leading to a lack of awareness about its importance. To address this, an immersive Virtual Reality (VR) solution, StrikeSmart, has been developed. This unique intervention provides users with a realistic experience, helping them understand the significance of making correct decisions in stressful situations. Unlike existing online learning sources and apps, StrikeSmart effectively engages users, with 95% expressing immediate impact and a willingness to join self-defense classes. This innovative approach contrasts with the traditional lack of interest, as demonstrated by the low 8% understanding observed in empathy interviews. The user-centered methodology, from conceptualization to prototype development, sets this solution apart in the field of self-defense education.

7 Limitations and Future Scope

This study's limitations include a simulation focused on basic-level self-defense situations, lacking extreme danger or stress. However, potential enhancements involve advancing scenarios to higher stress levels, expanding immersive details in various urban and rural settings, and customizing interactions for a more dynamic and enjoyable experience.

8 Conclusion

Self-defense education is often overlooked, commonly mistaken for martial arts. The lack of awareness regarding its importance calls for intervention, especially in early youth. Enter StrikeSmart, a Virtual Reality experience that immerses users in relatable scenarios, educating them on the distinctions between self-defense and martial arts, empowering informed choices.

Acknowledgements I extend my gratitude to NMIMS School of Design, Dr. Anirban Chowdhury, Professor Siddhesh Shirsekar, and the class of 2023 for the project opportunity and support. Special thanks to Mrs. Dipika Vaz, Ms. Rachel Vaz (Krav Maga Global Organization), and Mr. Sahil Shaikh (Parallax Labs) for their invaluable contributions.

References

1. Lalji C, Pakrashi D (2020) Self-Defense training & perception of safety among young adolescent girls: a field experiment from India. AEA Randomized Control Trials. https://doi.org/10.1257/rct.5799-1.1

2. Banks AL (2012) Developing confidence within girls through self-defense education. Strategies 25(5):26–29. https://doi.org/10.1080/08924562.2012.10592169
3. Hollander JA (2004) I can take care of myself. Violence Against Women 10(3):205–235. https://doi.org/10.1177/1077801203256202
4. Alif MN, Muhtar T, Fauzi RA (2020) Practical self-defense program for children using 4P (pray, prediction, preventive, and protective) concept: an effort to improve self-control. In: Proceedings of the 4th international conference on sport science, health, and physical education (ICSSHPE 2019). https://doi.org/10.2991/ahsr.k.200214.012
5. Potenza GM, Konukman F, Yu JH, Gümüşdağ H (2013) Teaching self- defense to middle school students in physical education. J Phys Educ, Recreat & Danc 85(1):47–50. https://doi.org/10.1080/07303084.2014.855581
6. National Crime Records Bureau. (n.d.). Retrieved from https://ncrb.gov.in/en/Crime-in-India-2020. 18 Sep 2022
7. Holland A (2021) Martial Arts vs self-defense: what is the difference and which to learn. The self defence expert.com. Retrieved from https://theselfdefenceexpert.com/martial-arts-vs-self-defense/. 18 Sep 2022
8. Vikas (2016) Self Defence Education for girls should be mandatory in Indian schools. Indian Youth. Retrieved from https://www.indianyouth.net/self-defence-education-for-girls-should-be-mandatory-in-indian-schools/. 18 Sep 2022
9. 70,426 schools got approval for self-defence training in 3 years: Upendra Kushwaha. The Indian Express. (2018, July 27). Retrieved from https://indianexpress.com/article/education/over-a-lakh-schools-get-approval-for-self-defence-training-for-girl-students-in-3-years-upendra-kushwaha-5279925/. 18 Sep 2022
10. Self-defence training for kids part of Education at govt-run primary and upper primary schools in UP. Hindustan Times. (2019) Retrieved from https://www.hindustantimes.com/education/self-defence-training-for-kids-part-of-education-at-govt-run-primary-and-upper-primary-schools-in-up/story-DAnXp2uQ1nnGqI0qbGwEkJ.html. 18 Sep 2022
11. Crime against men in India. Vidhikarya. (n.d.). Retrieved from https://www.vidhikarya.com/legal-blog/Crime-Against-Men-in-India#:~:text=There%20is%20a%20Section%20377,on%20the%20term%20'equality. 18 Sep 2022
12. Decent work for youth in India. Decent Work for Youth in India. (2012) Retrieved from https://www.ilo.org/newdelhi/info/WCMS_175936/lang--en/index.htm#:~:text=India%20has%20the%20largest%20youth,in%20the%20National%20Youth%20Policy%20
13. ABC News Network. (n.d.). ABC News. Retrieved from https://abcnews.go.com/International/blind-teenager-mumbai-fights-off-molester-defense-moves/story?id=59934261. 18 Sep 2022
14. Singh A (2021) Irked by eve-teasing, Rajasthan teenager imparts self- defense skills to other girls. The Logical Indian. Retrieved from https://thelogicalindian.com/uplifting/rajasthan-teenager-gives-self-defense-training-27827. 18 Sep 2022
15. Self-defense training should be mandatory in schools across India. Times of India Blog (2021) Retrieved from https://timesofindia.indiatimes.com/readersblog/travel/self-defense-training-should-be-mandatory-in-schools-across-india-33589/. 18 Sep 2022
16. Rita R (2020) The difference between martial arts and Combat Sports. Wado Waza Karate. Retrieved from https://www.wadowaza.it/en/news/the-difference-between-martial-arts-and-combat-sports/. 18 Sep 2022
17. Self-defense for men. Kidpower International (2022) Retrieved from https://www.kidpower.org/library/article/men-safety/. 18 Sep 2022
18. Targetstudy.com. (n.d.). Self defence—education for safety. Target Study. Retrieved from https://targetstudy.com/articles/self-defence-education-for-safety.html18 Sep 2022
19. Karen Asp MA (2021) 4 basic self-defense moves absolutely everyone should know Real Simple. Retrieved from https://www.realsimple.com/health/preventative-health/safety/4-essential-self-defense-moves-everyone-should-know. 18 Sep 2022
20. Self-defence without karate: a guide for the common man. OnManorama. (n.d.). Retrieved from https://www.onmanorama.com/lifestyle/health/2018/09/19/self-defence-without-karate-a-guide-for-the-common-man.html. 18 Sep 2022

21. Defense C O B R A S (2022) Can you really learn self-defense online? C.O.B.R.A™ Self-Defense South Africa. Retrieved from https://cobradefence.co.za/can-you-really-learn-self-defense-online/. 18 Sep 2022

Tailoring Comfort: A Comprehensive Approach to Home-Based Palliative Care for Dogs

Riya Khokhani and Arundhati Guha Thakurta

1 Introduction

In the realm of medical science, palliative care stands as an indispensable approach, one that diverges from the traditional quest for cure to emphasize the alleviation of suffering and the enhancement of the quality of life for patients grappling with life-limiting conditions. While this compassionate and patient-centered form of care is widely recognized within the realm of human medicine, its counterpart in the world of veterinary medicine, namely "veterinary palliative care," remains a relatively understudied and underappreciated facet of healthcare. This research paper delves into the significance and implications of veterinary palliative care, a therapeutic strategy that offers solace to our beloved animal companions when curative treatments may no longer provide a viable solution [1].

Veterinary palliative care comes into play when a poignant decision surfaces: the decision to halt or forgo curative treatments for our animal friends confronted with life-limiting conditions such as diabetes, chronic renal illness, or congestive heart failure. These conditions, much like their human counterparts, are relentless, often defying the possibilities of complete cure. It is imperative to understand that veterinary palliative care is distinct from managing chronic illnesses, for its primary objective is not just to extend the life of the patient but to enhance the overall quality of life in their remaining time [2].

At the core of veterinary palliative care lies the aspiration to elevate the well-being of our animal patients as they approach the inevitable end of their journey. This is achieved through the application of pain management and hospice theories and practices, a phenomenon that has gained increasing traction in recent years.

R. Khokhani · A. G. Thakurta (✉)
School of Design, NMIMS Deemed-to-Be-University, Mumbai, India
e-mail: arundhati.thakurta@nmims.edu

R. Khokhani
e-mail: riya.khokhani12@nmims.edu.in

© The Author(s), under exclusive license to Springer Nature Singapore Pte Ltd. 2025
S. Das et al. (eds.), *Human-Centered Digitalization, Volume 3*, Design Science and Innovation, https://doi.org/10.1007/978-981-96-4383-7_45

Pain, being a predominant factor in diminishing the quality of life for our canine companions, plays a pivotal role in the palliative care equation. Therefore, when embracing palliative care and hospice treatments, the effective management of pain emerges as a critical component, demanding a careful and nuanced balance between the benefits and potential drawbacks [3].

This research paper delves into the realm of veterinary palliative care, elucidating its principles, methods, and the intricate decision-making process that leads to its initiation. By exploring the driving factors behind the commencement of palliative care, such as the development or progression of life-threatening diseases and the decision to forego curative therapy, we endeavor to shed light on this compassionate discipline and underline its vital role in enhancing the final chapters of our loyal companions' lives. In doing so, we aim to promote greater awareness, understanding, and appreciation for the invaluable service that veterinary palliative care provides, both to our pets and their devoted human caregivers.

2 Literature Review

It's critical to comprehend pain well before attempting to prevent or manage it. The definition of pain offered by the International Association for the Study of Pain (IASP) offers a thorough and educational foundation. Anxiety is a multifaceted experience that involves both sensory and emotional aspects; it is not just a physical feeling. It is brought on by real or prospective tissue injury, and it may feel painful or harmful. The IASP's definition is noteworthy in part because it emphasizes how subjective pain is. Pain is an individualized experience that is impossible to quantify scientifically, in contrast to other physiological measures like blood pressure or heart rate. Accordingly, suffering cannot be minimized or rejected just because a person is unable to express it verbally.

Palliative care is the treatment of clinical symptoms of an illness without the goal of curing it. Regardless of life expectancy, the goal of palliative care is to achieve the best possible quality of life (QOL) through pain management and functional capacity restoration [5]. Animal Hospice and Palliative Care (AHPC) is a rapidly evolving field of animal health care. AHPC addresses the emotional, social, and spiritual needs of the human caregivers in preparation for the death of the animal and the subsequent grief. AHPC evolved as a philosophy of care and a system for delivery of services for seriously ill and dying patients, lagging behind but strongly influenced by the remarkable success of human hospice and palliative care [4].

Physical medicine modalities and physiotherapy are effective options for treating pain and mobility concerns in pets, complementing the pain management pyramid. Medical acupuncture, massage therapy, therapeutic laser, chiropractic adjustments, and physical therapy are among the techniques that help relieve pain, improve mobility, and promote overall well-being in animals, especially those in palliative and hospice care. Adapting the home environment, such as placing bedding and toys at lower levels or adding ramps, can also help pets with their everyday activities. Palliative treatment strategies, such as radiation therapy, may have minimal side effects that are limited to the treated area, with precautions taken to avoid complications. Understanding various sources of pain, from trauma to chronic diseases, is critical for providing proper palliative care [3].

Understanding and managing the various types of pain experienced by patients is critical in palliative and hospice care. Individuals receiving such therapy may experience a range of discomfort, from adaptive reactions to maladaptive situations. To effectively detect and manage pain, a thorough evaluation and appreciation of a patient's comprehensive healthcare history, comprising both past and present features, is required. By investigating the patient's medical history, healthcare providers can remain alert to potential sources of pain, whether acute or chronic. In the context of palliative and hospice care, trauma, surgical treatments, arthritis (particularly osteoarthritis), dental illnesses, joint problems, immune-mediated polyarthropathy, heart attacks, and respiratory ailments stand out as potential sources of pain.

Recognizing and treating these variables is critical to improving patients' quality of life during this critical stage of treatment. This comprehensive approach is consistent with the ideals supported by India's National Health Dog Portal, which emphasizes a thorough understanding of patients' health issues [2].

Palliative Care at Home

Palliative care aims to enhance the quality of life for patients with chronic serious illnesses, such as advanced cancer. It's natural for people to want to remain close to their families, especially when they have limited time left. The best care is often provided in the comfort of the patient's own home, ideally by their loved ones. Home-based palliative care offers patients a sense of security, privacy, confidentiality, and peace of mind. This type of care, delivered by a specialized team that includes doctors, nurses, social workers, volunteers, physiotherapists, and others, focuses on symptom management, health-related communication, and psychosocial support. It helps both the patient and their family better prepare for the end of life. The team regularly visits the patient's home to provide necessary medical care. With the rising costs of chronic illness treatment, maintaining care in a hospital or hospice can be financially challenging, particularly in countries like India, where many people lack financial protection. Family members play a crucial role in-home-based palliative care, receiving training from the care team to learn the basics of palliation. Integrating palliative care into routine home health services is essential [7].

Palliative care for dogs, whether delivered at home or in a clinic, has the same goals: to maximize comfort, minimize pain, and improve overall quality of life for the pet during a terminal disease or old age. Home-based care allows the dog to stay in a known and comfortable area, decreasing stress and anxiety associated with new surroundings.

For home-based care, pet parents actively participate in their pet's care by dispensing medications, providing comfort, and establishing a tight link. The care plan can be tailored to the dogs and family's specific needs and habit. It allows both the pet and the owners to receive emotional support, establishing a sense of security and friendship. Without the trip and unusual sights, sounds, and smells associated with a clinic, dogs may feel less stress. In a clinic setting, veterinary experts can provide specialized medical skills and technology for more intense monitoring and treatment. While clinical care involves medical intervention and diagnostic tools that are available immediately, allowing for quicker alterations to the treatment plan based on the dog's changing condition. Clinics can provide specialized facilities for treatments, diagnostic tests, and more advanced medical care that may not be possible in the home. There is the possibility of continuous monitoring and care in a clinic, providing fast attention to any unforeseen difficulties. A clinic setting may provide a controlled atmosphere in which the dog is isolated from potential distractions and environmental elements that may impair their well-being [6].

Opting for home-based palliative care in the final months of a pet's life is a decision that many pet parents carefully consider, and it is motivated by several important criteria that prioritize the creature's comfort and well-being during this difficult time. While many pet parents prefer home-based palliative care, it's important to recognize that the decision is highly unique and dependent on the pet's specific circumstances and needs. Certain medical interventions may necessitate the knowledge and resources of a clinic in some cases, but even in such cases, a balance can frequently be established to provide a combination of clinic and home-based care to optimize the general well-being of the pet in its final months [8].

In India, the landscape of home-based palliative care support and provisions for dogs may confront certain problems, and the situation may have changed since then. Home-based palliative care for dogs is providing compassionate and supportive care in the comfort of their own homes to pets suffering from terminal illnesses or elderly age. Recognizing the Western world's long-standing embrace of home-based palliative care support, the identification of this gap inspired the creation of the study's purpose. The purpose is to provide home-based palliative care to pet parents, with a particular emphasis on dog owners.

3 Methodology

3.1 Participants

The research comprised of 50 individuals from Mumbai, India. All the participants were interviewed at their home and clinic with their pets (dogs) and the required information was collected. To understand the perspectives of diverse participants, 40 pet parents and 10 veterinary doctors were surveyed. With interview responses, the solution scope for primary stakeholder group was narrowed down to dog owners whose dog suffering from osteoarthritis.

3.2 Data Collection and Synthesis

Secondary data was gathered from reputable sources and websites. More than thirty research papers and publications were possibly considered. This material was obtained from reliable medical sources and other health professionals. This information aided in establishing a thorough knowledge of the situation. The primary data was compared to various personas. Data from multiple personas was linked together to uncover patterns and insights. Following a thorough study, the nature of the issue in the problem region was discovered using persona-specific surveys, interviews, and extensive observation. Throughout the synthesis process, a variety of tools were used. The usage of empathy maps provided unique insights into the environments of all five personas. Affinity maps were critical for identifying trouble areas and analyzing data trends. Affinity maps were critical for identifying trouble areas and analyzing data trends. By illustrating the user's overall itinerary of the process of care giving scope at home, journey maps aided in the development of empathy. The final stage involved putting information into a how and why laddering framework in order to gain a deep knowledge of the issues that pet parents encounter.

3.3 Conceptualization

The emphasis during the brainstorming phase was on developing basic ideas based on solving challenges connected to a dog parent's goals and requirements, with a particular emphasis on emotional support and spend quality time. The main purpose was to solve these challenges using novel ways. A few prospective solutions were carefully picked from the pool of created ideas after user validation validating their validity. These thoughts served as the foundation for the creation of low-fidelity prototypes, which provided a reduced portrayal of the product's functionality and design. These prototypes provided a better understanding of how the proposed solutions would work in practice. Iterative adjustments were done based on user feedback to develop and strengthen the concepts. The minimal viable products (MVPs) developed from the low-fidelity prototypes were then shown to users in the relevant field. A brief questionnaire was created to systematically collect user feedback, with the goal of determining which components of the concepts were effective and which needed to be improved. This user validation approach was critical in refining the solutions and ensuring that they connected with the intended audience and effectively solved the stated challenges. The main concepts selected were a Palliative care kit which includes a buddy, a mat, a zine, and a harness, the second concept was an engagement toy game to help pet parents to give good palliative care to their pet who is suffering from a life-threatening disease and make them more comfortable at home while receiving it.

4 Results and Discussion

The study revealed several major gaps in the present pet care landscape, particularly in the field of palliative care. One important issue that has been noticed is a lack of information among pet owners regarding the illnesses that affect their pets and the significance of palliative treatment. This includes symptom management education, mobility help, and pain alleviation. Furthermore, the existing products are mostly intended for elderly dogs receiving palliative care, with considerable gaps in cleanliness and nutrition maintenance. Another big gap is the inadequacy of pet care education and guidance, as well as a lack of 24-hour on-call nursing coverage for emergencies and questions. In addition, the lack of comprehensive grieving support and emotional support for pet parents facing the challenges of caregiving exacerbates the existing disparities. Bridging these gaps is critical for establishing a holistic and supportive environment for both pets and caregivers on their path to palliative care.

New solutions needed to be implemented that could steadily help the pet parents to give their pets comfort and relief at home. A number of ideas and concepts were thought of that would help solve the problems that the study focused on.

4.1 Concept-1: Engagement Toy

As dogs age, their vision might degrade and they may acquire various eye disorders, just like humans. It's crucial to keep this in mind when selecting toys for older dogs, as they may not be able to gaze well enough to retrieve items or find toys. Aim was to keep the pet engaged and assure that the pet is busy playing with the toy instead of sitting. The toy has bone shaped buttons which will let the ball pop out of the hole by pressing them. This interactive design accommodates impaired vision and keeps them physically active, preventing lethargic behavior (Fig. 1).

Fig. 1 Engagement toy

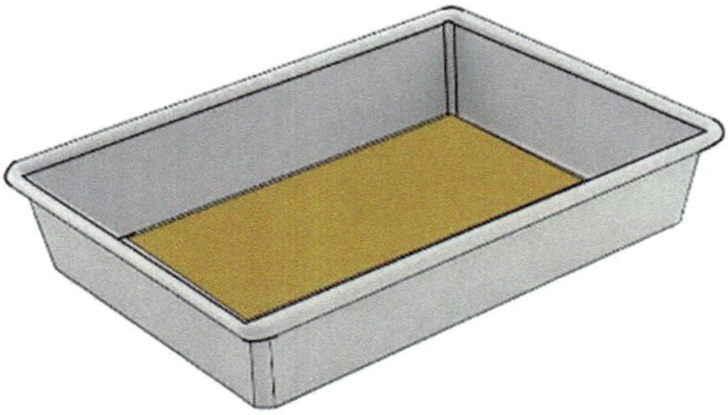

Fig. 2 Sand filled tube most of their times at home

4.2 *Concept-2: Pet Play Tub*

Dogs with this condition's brain function begins to slow down, but frequent brain training and cognitive stimulation can help them stay mentally active. So, having a tray loaded with mud is a nice way to keep them engaged, since it is good for health and brain activity. These are likewise perfect for senior dogs undergoing palliative care who are unable to run around like they used to, since it may provide them with more even engagement. The idea was to make a tub for the pet to play and relax, which will also help the pet to exercise and stay busy at (Fig. 2).

4.3 *Concept-3: Palliative Care Kit*

The palliative care kit consists of a warming mat, a harness, a manual for the pet parents to know how the kit works and lastly a buddy which helps the pet to stay active. It stands as a holistic solution catering to the unique needs of pets undergoing palliative care, fostering a more comfortable and enriching experience for both the pet and its dedicated caregivers (Fig. 3).

4.4 *Finalized Concept*

The finalized palliative care system design (Fig. 4.) is a deliberate integration of three critical components: a specific mat, a health monitoring harness, and a dedicated mobile application. These components work together to produce a complete and user-friendly strategy to improving the well-being of pets receiving palliative care in the comfort of their own homes.

The first component, the mat, is cleverly constructed to aid in the relaxing of pets during acupuncture sessions. Its distinctive design is intended to provide a soothing surface for pets, allowing them to relax quickly during therapeutic sessions. Furthermore, the mat includes thermal padding, which ensures a steady and moderate warmth that contributes to the pet's overall comfort, particularly during moments of rest. One

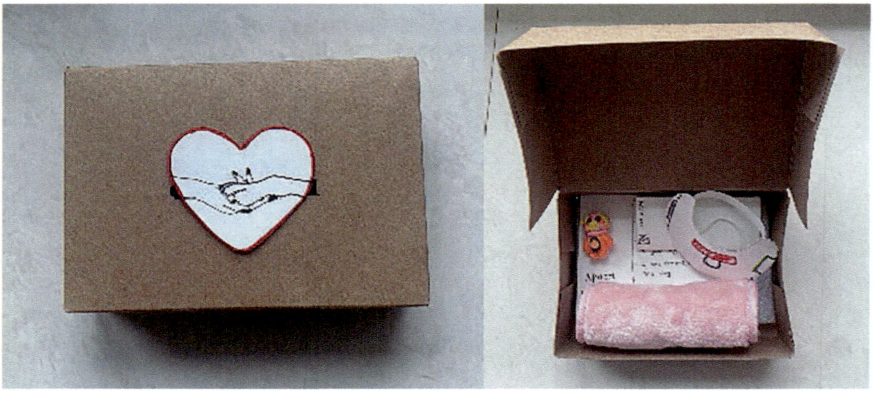

Fig. 3 Palliative care kit

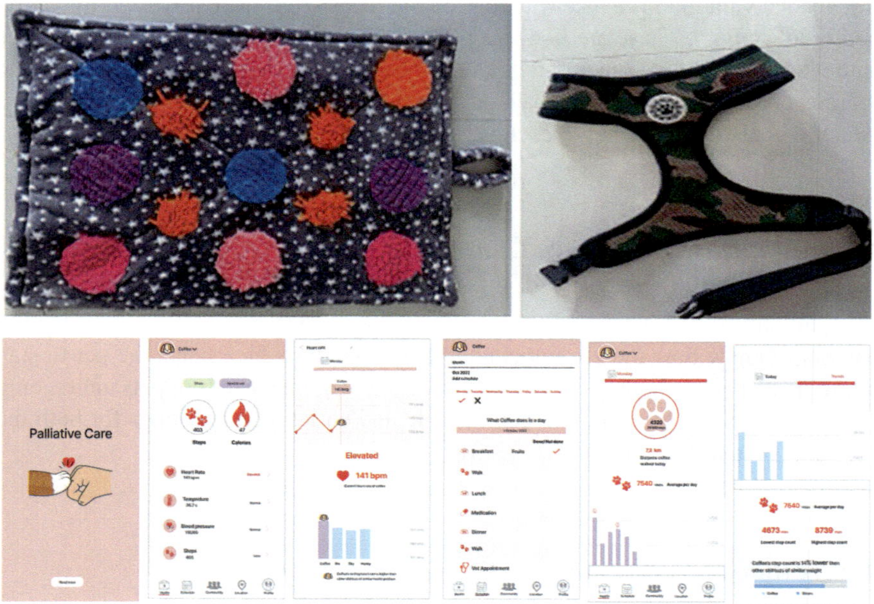

Fig. 4 Final palliative care kits for dogs at home

side of the mat includes acupuncture, which allows the pet to massage and rest. The other side features an automated warmer that keeps the cat warm and comfortable. Even if the dog isn't as lively as they once were and can't play for as long as they used to, that doesn't mean the pet parents should stop playing. Simply change the toys, limit the play hours, and adjust how intense and busy the play times are. The mat is brightly colored and has a strong odor, making it easy for the dog to find.

Because it is not only brilliantly colored, but it can also be loaded with treats, the mat is great for both young and old dogs with impaired vision.

The second component, the health monitoring harness, serves a dual goal of comfort for the pet and convenience for the owner. The harness was designed with the pet's comfort in mind, ensuring that the pet is as comfortable as possible while wearing it. It also has an integrated health monitoring system that allows pet owners to easily follow critical health data. This real-time health data provides essential insights into the pet's status, allowing for proactive treatment and early interventions when necessary. It has a temperature sensor, a blood pressure sensor, and a pulse rate sensor, all of which provide real-time data on crucial health indicators. This essential information enables pet parents and veterinarians to closely monitor the pet's physiological health, allowing for early interventions and treatment plan revisions as needed. Furthermore, the presence of a GPS device enables position monitoring, ensuring the pet's safety and permitting prompt response in the event of an emergency. The harness goes a step further by including a tiny screen that shows the collected health readings, giving both pet owners and veterinary specialists instant access to crucial information. This integrated technology not only simplifies health monitoring but also provides caregivers with the information they need to provide targeted and responsive care throughout the journey.

The dedicated smartphone application "4PAWS," which is seamlessly connected to the health monitoring harness, completes the trio. This program functions as a center for pet owners, providing a simple interface for tracking and analyzing their pet's health parameters. The software gives meaningful information and abilities required for efficient palliative care at home, going beyond conventional data collection. The app becomes a great resource for pet owners, empowering them to provide optimal care for their furry companions, whether it's advice on symptom management, recommendations for improving the pet's quality of life, or vital caregiving skills. This is an invaluable tool for pet parents navigating the difficulties of palliative care, with a variety of features precisely intended to improve the well-being of their furry companions. The app's profile function acts as a single hub for extensive pet information, allowing users to keep a thorough record of their pet's health history and preferences. The health component allows pet parents to closely monitor their pet's nutrition, symptoms, and overall well-being, allowing for proactive management. The scheduling tool allows users to efficiently manage and plan health visits, ensuring timely interventions and treatments. The community aspect provides a supportive environment by connecting pet owners and offering a platform for them to share their experiences and seek assistance. Most importantly, the GPS feature provides peace of mind by offering real-time tracking, hence improving pet safety. One outstanding aspect of the app is its capacity to ease contact with doctors; pet parents can quickly transmit health reports straight to their vets, enabling seamless collaboration in the pet's palliative care journey. In essence, "4PAWS" is a comprehensive and user-friendly platform that bridges the gap between pet parents and veterinarian treatment while also encouraging a sense of community among people caring for needy pets.

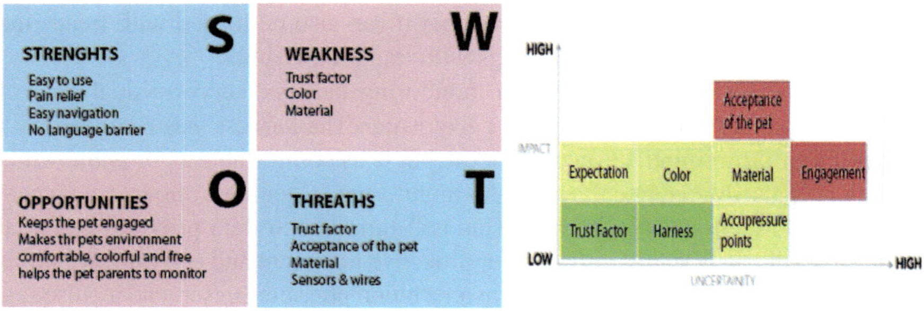

Fig. 5 SWOT analysis and risk analysis

4.5 Usability Testing

A real looking prototype of the mat and the harness was tried and tested with users. It helped in showing the engagement and the usage of the health monitoring harness which was connected to the app. The main observation made was that the patient was very comfortable and relaxed while using the mat and the app was very useful as it helped the pet parents to learn more about palliative care and it helped them schedule the pet's day. The overall evaluation score of the kit is 3.9 out of 5.

4.6 SWOT and Risk Analysis

The SWOT analysis was a great tool that helped in reviewing the solution from all aspects before it can be implemented in the real world. The figure below states all the strengths, weaknesses, opportunities, and threats that the solution holds (Fig. 5).

A risk analysis process was carried out which helped in identifying the potential problems and risks that could determine initiatives for projects. To carry out the risk analysis we first identified the likable threats and then estimated the odds that these threats will cause.

5 Conclusion

This holistic approach recognizes the specific requirements of dogs dealing with health issues and tries to improve their quality of life in the comfort of their own homes. The acupuncture carpet promotes relaxation during sessions by providing a therapeutic and peaceful sensation. The smart sensors in the harness not only assure the pet's comfort, but also allow for real-time health monitoring, allowing pet parents and doctors to make informed decisions and provide prompt interventions. The mobile app acts as a center, allowing pet owners to track and manage their pet's health, schedule appointments, and connect with veterinary specialists in real time. This integrated system not only addresses the physical components of palliative care, but it also develops a sense of connection and empowerment for pet owners, ultimately contributing to a more compassionate and customized approach to palliative care in the home setting. This palliative care home kit represents a breakthrough step in offering full assistance for dogs confronting health issues, enhancing their comfort, well-being, and the link between pets and their caretakers. These adaptable settings, which were originally created for dogs receiving palliative care for osteoarthritis, can be expanded to meet a variety of terminal illnesses and cater to a varied spectrum of pets, including cats. Adapting the acupuncture mat, specific harness, and mobile app to different conditions and animals allows a compassionate and tailored approach to palliative treatment.

References

1. Akgün KM (2017) Palliative and end-of-life care for patients with malignancy. Clin Chest Med 38(2):363–376
2. Goldberg KJ (2016) Veterinary hospice and palliative care: a comprehensive review of the literature. Vet Rec 178(15):369–374
3. Downing R (2011) Pain management for veterinary palliative care and hospice patients. Vet Clin: Small Anim Pract 41(3):531–550
4. Goldberg ME (2016) Pain management in physical rehabilitation. Vet Nurse 7(1):34–41
5. Shanan A, Shearer T (2023) What is animal hospice and palliative care? In: Hospice and palliative care for companion animals: principles and practice, pp 6–15
6. Jones K (2013) Pain management in hospice and palliative care. In: Pain management in veterinary practice, pp 431–436
7. World Health Organization (2020) Palliative Care
8. Christiansen SB, Kristensen AT, Sandøe P, Lassen J (2013) Looking after chronically iii dogs: impacts on the caregiver's life. Anthrozoös 26(4):519–533

Revamping the News Reading Culture for the Youth

Aarushi Singh and Siddhesh Shirsekar

1 Introduction

In the twentieth century, Indian newspapers played a crucial role in challenging British imperialism, fostering nationalistic sentiments, and contributing to India's independence. However, with the advent of the digital age, newspapers have struggled to engage the youth, who make up 66% of the population, due to a lack of relevant content. Over the past two decades, the rise of affordable digital services and improved connectivity has led to a significant increase in digital media content. Companies have recognized the potential of social media to captivate audiences, transforming the way news is consumed and disseminated. The shift to online media is driven by its appeal as a pastime, enabling people to share experiences and stay connected. While news reading is essential for knowledge and opinion formation, the younger generation often distrusts traditional news outlets, viewing them as politically biased. The dominance of social media as a news source, however, comes with challenges, including algorithmic influence on content and the prevalence of inaccurate and fake news. As technology evolves, there is a pressing need for a paradigm shift in news consumption methods to address the lack of engagement that discourages youth from regularly consuming news.

A. Singh · S. Shirsekar (✉)
School of Design, SVKM's Narsee Monjee Institute of Management Studies (NMIMS) Deemed-to-Be-University, Mumbai, India
e-mail: siddhesh.shirsekar@nmims.edu

A. Singh
e-mail: Aarushi.singh22@nmims.edu.in

© The Author(s), under exclusive license to Springer Nature Singapore Pte Ltd. 2025

S. Das et al. (eds.), *Human-Centered Digitalization, Volume 3*, Design Science and Innovation, https://doi.org/10.1007/978-981-96-4383-7_46

2 Literature Review

2.1 Importance of News Reading Among Youth

News provides valuable information and education on global events, covering various topics for informed discussions. Despite its importance, many young people prefer online activities over reading news, finding traditional news consumption unappealing and lacking value, especially in comparison to engaging with friends on social media. (Partner press) [1].

2.2 Newspapers in the Twenty-First-Century

The twenty-first-century newspaper industry is struggling against digital media's rise, facing a decline in readership as people shift to more convenient news sources. Once revered for their role in India's freedom movement, newspapers have lost relevance among the youth, who find them boring and overly lengthy. The younger generation seeks news that is engaging, personalized, and aligns with their interests, but traditional news companies still struggle to shed their reputation for being uninteresting and unhelpful [2–4].

2.3 Digital Effect on News Consumption

Traditional news consumption is declining globally, with 72% accessing news through smartphones, as reported by The Hindu. The twenty-first-century rise of digital media has profoundly altered daily news consumption, impacting journalism positively and negatively. Real-time information sharing empowers readers to play a more active role, and increased use of the Internet and cell phones opens up new distribution opportunities, giving readers more influence over their news preferences [5, 6].

2.4 News Consumption Through Social Media and Its Downsides

With 2.4 billion internet users, social media has become the primary source of online news, with 64.5% relying on platforms like Facebook, Twitter, YouTube, Snapchat, and Instagram. Users tend to skim articles, spending no more than 15 seconds on average, and video watch time is only 10 seconds. Social media, driven by algorithms, controls the news users see, often prioritizing sensationalized content. The prevalence of "fake news" further complicates matters, as dramatic headlines and misinformation easily spread due to a lack of fact-checking. This fosters echo chambers, reinforcing polarized thoughts and biases among like-minded groups [7–9].

2.5 *Existing Mobile News Apps and Their Limitations*

Various news apps, such as Inshorts, Times of India (TOI), Google News, and Daily Hunt, cater to daily news consumption, each with its strengths and drawbacks.

Inshorts stands out for its concise news delivery, limiting information to 60 words to save time. However, it sometimes covers vague topics that may not interest users. The app lacks motivation for daily use, relying on intriguing notifications to draw users in. TOI's app provides comprehensive news coverage but falls short in engaging the younger demographic. Its interface needs improvement, and the app tends to sensationalize news and prioritize entertainment updates, lacking features to motivate daily news consumption.

Google News aggregates content from various sources, personalizing it based on user preferences. However, it faces issues with misleading notifications, irrelevant content, and interface problems, leading to user dissatisfaction.

Daily Hunt curate's news from different publishers but lacks effective filtering, resulting in misleading headlines and misinformation.

All these apps fail to engage users sufficiently and lack motivational factors for daily news consumption. This study aims to explore new approaches to encourage youth to develop a habit of consuming news regularly, presenting information in an appealing manner that captivates their interest.

3 Methodology

The intent of this section is to provide a detailed outline on the tools and methods that were implemented for the smooth execution of the study.

3.1 *Participants*

There were in total 140 participants who took part in this research study. The study was conducted in the city of Mumbai primarily in the Andheri East locality. Due to the presence of a large number of colleges in this locality, it was easier to reach the target group which was urban middle-class youth in the age bracket of 17 to 25 years. All of the participants were students who were selected for the study by random sampling. They were either in junior college or pursuing a degree like Law, commerce, liberal arts, MBA, design and VFX animation. The educational background of the target group gave an understanding of what they look for in their news and what medium of news they consume. The target group was further divided into two groups—people who were staying away from their parental home and the people who were residing with their parents. The group division gave an idea on the habits of news consumption, various kinds of influences, accessibility to news and newspapers and their financial situation. 40 participants took part in the google survey, 20 participants sat for the empathy interview, and 80 participants took a written survey where they also validated the concept. The empathy interviews were conducted in college canteens, outside college premises, and cafes where the college students frequented the most. The insights from the google survey were validated in the empathy interviews which helped in finding out patterns and creating the questionnaire for the written survey.

3.2 Data Collection

This phase deals with collecting data through various sources and using several tools to help understand the user's thoughts and feelings. Mind maps helped in visually organizing all the data collected from the different sources and depicted a hierarchical relationship between miscellaneous variables. It also helped in narrowing the scope of research and recognizing the major stakeholders and target groups. Information was acquired during the desk research from existing sources such as articles, research papers, white papers and journals and the data collected from them supported the primary research. Post the desk research, two personas were created—students living with their parents and students living away from their parents. The personas helped in understanding the needs and motivations of users belonging to a particular group and the different problems or concerns that they have to deal with. The survey conducted during this phase gave a basic understanding of user's current news consumption habits and the empathy interviews aided in discerning the user thoughts and learning about the user's cognizance and perception on news reading.

3.3 Data Synthesis

This section synthesizes data through methods like empathy maps, color coding, and affinity mapping. Empathy maps capture user thoughts and feelings, aiding in articulating perceptions from empathy interviews and desk research. Tools organize findings based on themes, revealing patterns among participants. Prevalent themes include family, peers, work, and advertisements impacting menstruating women. Journey maps organize user tasks, providing insight into their daily activities and news consumption behavior. How and why laddering explores user needs, guiding the development of potential solutions by probing problem areas. Overall, these techniques deepen understanding and inform solutions for user-related issues.

3.4 Conceptualization

In this phase, solutions and concepts were developed to address the defined problem of youth disinterest in news consumption. Ideation and brainstorming yielded various approaches, focusing on encouraging youth engagement with news. Two potential ideas were refined into low-fidelity prototypes, showcasing basic features for quick consumer feedback. These prototypes helped validate the concept with the target group, leading to iterative improvements. The final solution emerged from user feedback, aimed at building youth interest in habitual news consumption and reshaping current consumption habits.

4 Results and Discussion

The intent of this section is to understand the key findings of the primary research and the various influences on youth which affects their news consumption pattern. The section discusses the results of the study and analyses the data collected through different research tools which further helped in narrowing down the scope of the study and focus on solving the core issue which is how to get youth interested in news.

4.1 Key Findings and Insights

4.1.1 The Monotony of Newspapers

82% of the participants interviewed did not read a newspaper at all. The participants mentioned that reading newspapers is quite boring as the newspapers are monotonous and do not present the news in an engaging and interesting way. 91% of the respondents felt that the newspapers are too lengthy which makes it unappealing as they do not know where to begin from or what to read first. 82% of the respondents mentioned that they prefer their news to be short, crisp to the point yet impactful. It was noted that all the respondents were more focused toward the aspect of convenience when it came to news consumption which the newspapers fail to provide them (Figs. 1 and 2).

4.1.2 Digital over Physical

73% of the respondents mentioned that digital is better than physical since the news is available on the go and mobile phones are handier while the other 27% mentioned that they would give physical a try provided the newspapers are available in shorter

Fig. 1 Percentage of youth reading newspaper

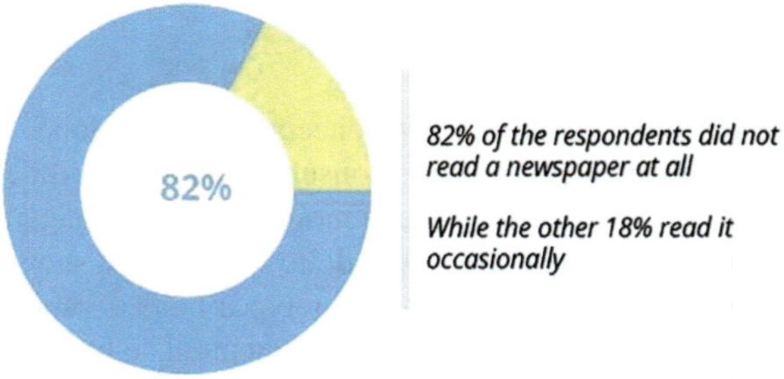

Fig. 2 Percentage of youth who use a news app

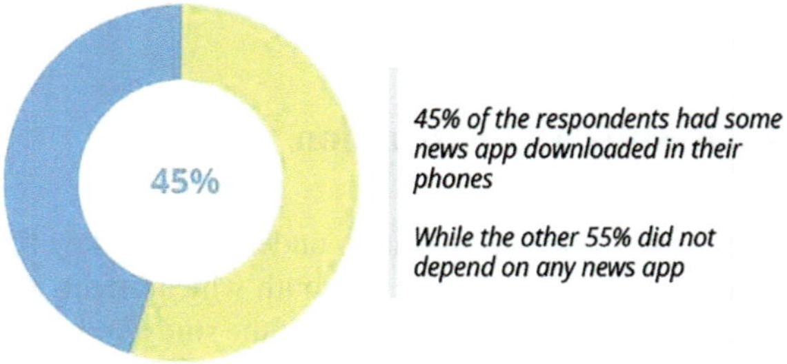

formats with more relevant news articles. 45% of the participants also had some news app downloaded on their phones as it is more convenient than reading a newspaper in terms of both accessibility and portability. Even though some participants had news apps downloaded on their phones, they rarely used it and were only interested in some news if the headline was relevant or interesting.

4.1.3 Getting Updates on Social Media

82% of respondents rely on social media for news updates, drawn by its customization aligning with individual interests. A lack of time for traditional newspaper reading is cited by 64% of respondents, emphasizing the need for easily accessible and efficient news sources. This supports findings that youth tend to stumble upon news while engaging in social media rather than intentionally seeking it. The disconnect between youth and traditional news consumption is evident, highlighting the importance of a more engaging and participatory news medium. Despite social media's appeal, there's a risk of polarization and the spread of false information, a concern often overlooked by the youth (Figs 3 and 4).

Fig. 3 Percentage of youth who depended on read news social media to get their news updates

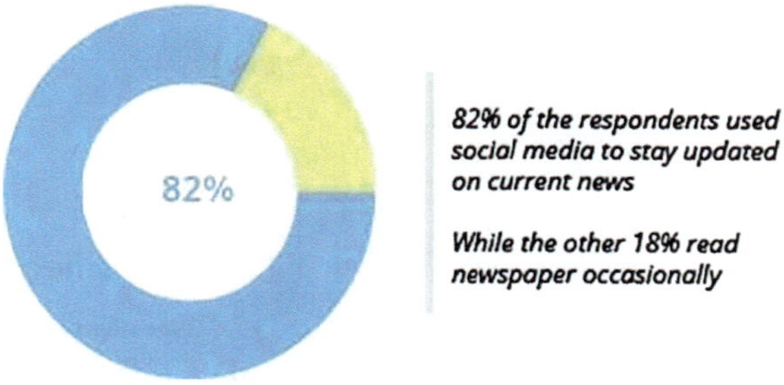

Fig. 4 Percentage of youth who don't due to time constraints

4.2 *Defining the Problem*

After finding the gaps and patterns in the data, it was determined that the intervention could be well targeted toward encouraging young people to read the news by piquing their curiosity and developing their interest in it. The youth can connect with news more easily and eventually develop an interest in it if news consumption is made more dynamic and interesting.

4.3 *Final Concept*

Newsified—Newsified is a gamified news app which makes news consumption engaging and interactive through games and quizzes. Newsified stands for news which is gamified and simplified, and the solution is focused toward building the user's interest in news and at the same time making it fun for them so that they get motivated to participate in it. It consists of newsgames which can be played individually or with others in real time and it also gives a gist of the news through short interactive comics (Fig. 5).

The proposed solution aims to cultivate a regular habit of news consumption among youth by offering an engaging and interactive platform. Through gamification, users can compete on leaderboards, earn points, and progress through levels by participating in daily news quizzes. Special rewards await those who achieve milestones, fostering continuous engagement. The application introduces a storytelling

Fig. 5 Newsified news app

element, presenting news through interactive comics, providing a unique and entertaining way to stay informed. Daily news quizzes incentivize users to accumulate points, level up, and secure a spot on the weekly leaderboard. Leagues structure the leaderboard, allowing users to advance to higher tiers based on their performance. The app also features daily news short stories presented as interactive comics, offering artists a platform for collaboration and self-promotion. Additionally, a daily "word of the day" related to recent news helps users comprehend jargon in political, business, and financial news, enhancing overall understanding (Fig. 6).

4.3.1 Features of the App

The quizzes—There would be daily news quizzes which would be related to the latest news stories. The users can participate in the quiz to collect points and move ahead in the levels. The quizzes could be played real time with other users or there would also be quizzes that could be played individually. The quiz would be in the form of an escape room game wherein the users would have to navigate through the different hints in relation to the news story and escape the room within a given time period. The player to escape the room first would be first on the leader board and the points would be distributed among the players according to their position (Fig. 7).

Points and leader board incentivize regular app use by rewarding players for quiz performance. Weekly standings are determined by accumulated points, and maintaining a top 10 position allows progression to the next league, fostering healthy competition (Fig. 8).

News stories—There would be a few news stories which would be covered in detail daily. Illustration and interactivity would help in conveying the news in an engaging manner. The illustrations would be in collaboration with artists and illustrators thus giving them a platform to showcase their work.

Fig. 6 Prototype of the app screens

Rewards—The users may get rewards that may help them in quizzes, levelling up and raise their status in the game. The rewards would also help the-layers to continue their winning streak in the game and therefore keep playing.

4.3.2 User Testing

The users liked the approach of gamifying the app since it's a great way of engaging. Users believed that engaging them through competition while also providing information would increase their interest in news. The consumers also preferred the uncommon and somewhat more distinctive news games, such as the escape room concept, because they were more entertaining. Due to missing buttons or users not understanding specific activities, some pages were difficult for users to navigate

Fig. 7 Shows leader bard, news quizzes, and badges (rewards) on completion of the quiz

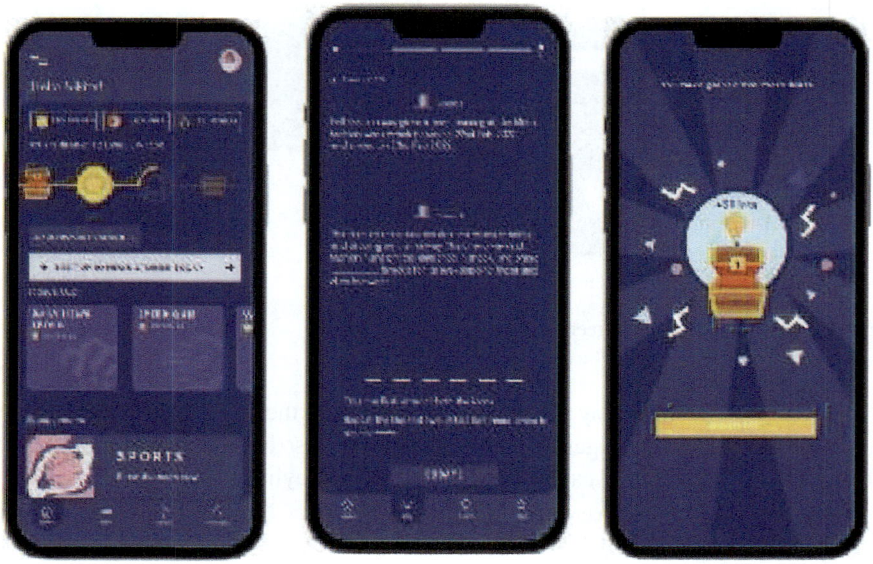

Fig. 8 Shows levels, escape room news games, and rewards on reaching milestones

functionally. The users also proposed other adjustments that could be made to the app to make it more appealing, like introducing a community page where users may follow and communicate with one another through the app. making the games more interactive so that players can even communicate with one another, like in the game "among us." The quizzes can be made more entertaining and interactive by including a plot or a narrative.

5 Conclusion

Through the research study it is quite evident that news consumption among the youth has reduced to a mere by-product of surfing on social media. News has diminished in significance and evolved into more of an attention-getting gimmick, which has eventually led to the sensationalization of some news stories. Newspapers, which were once a vital component of society, are gradually becoming irrelevant to the current generation, which makes people uninterested in reading the news. In the current technology era, it is important to stay on top of developments and accommodate young people who want news that is more interactive and engaging. The modern generation would not accept the current news consumption practices, which call for something amusing and require more of their attention. As a result, the suggested approach uses news gamification to draw users' attention to various recent news stories and events. Gamification would make information consumption more engaging and encourage competitiveness among users, which would encourage participation and lead to increased news consumption. The approach focuses on transforming how people currently consume news to something more exciting and compelling that will grab their attention and make news enjoyable.

Acknowledgements I extend my gratitude to NMIMS School of Design, Dr. Anirban Chowdhury, Professor Siddhesh Shirsekar, and the class of 2023 for the project opportunity and support.

References

1. The benefits of reading newspapers. (n.d.). Retrieved from https://www.partnerpress.be/en/blog/yet-another-blogpost-test-pager. 17 Sep 2022
2. How young people consume news and the implications for mainstream media (2022) Reuters Institute for the Study of Journalism. (n.d.). Retrieved from https://reutersinstitute.politics.ox.ac.uk/our-research/how-young-people-consume-news-and-implications-mainstream-media. 17 Sep 2022
3. How young people want their news: WARC. WARC An Ascential Company (n.d.). Retrieved from https://www.warc.com/newsandopinion/news/how-young-people-want-their-news/en-gb/42599. 17 Sep 2022
4. Adgate B (2021) Newspapers have been struggling and then came the pandemic. Forbes. Retrieved from https://www.forbes.com/sites/bradadgate/2021/08/20/newspapers-have-been-struggling-and-then-came-the-pandemic/?sh=1770bd1a12e6. 17 Sep 2022

5. Sampath G (2022) A closer look at the Reuters Institute Digital NewsReport 2022. Return to frontpage. Retrieved from https://www.thehindu.com/news/a-closer-look-at-the-reuters-institute-digital-news-report-2022/article65526579.ece. 17 Sep 2022
6. Auxier B, Arbanas J (2022) News at their fingertips: Digitaland Social Tech Power Gen Z Teens' News Consumption. Deloitte Insights. Retrieved from https://www2.deloitte.com/uk/en/insights/industry/technology/gen-z-news-consumption.html. 17 Sep 2022
7. Martin N (2021) How social media has changed how we consume news. Forbes. Retrieved from https://www.forbes.com/sites/nicolemartin1/2018/11/30/how-social-media-has-changed-how-we-consume-news/?sh=7253212c3c3c. 17 Sep 2022
8. The Echo Chamber Effect on social media PNAS (n.d.). Retrieved from https://www.pnas.org/doi/https://doi.org/10.1073/pnas.2023301118. 17 Sep 2022
9. Tian Q (2022) Impact of social media news overload on social media news avoidance and filtering: moderating effect of media literacy. Frontiers. Retrieved from https://www.frontiersin.org/articles/https://doi.org/10.3389/fpsyg.2022.862626/full. 17 Sep 2022
10. The decline in newspapers: a closer look Joseph Ahrens—Wake Review literary magazine & amp; club. (2017) Wake Review Literary Magazine & amp; Club. Retrieved from https://clubs.waketech.edu/wake-review/magazine/creative-writing/non-fiction/the-decline-in-newspapers-a-closer-look-joseph-ahrens. 17 Sep 2022
11. Online news reading behavior: from habitual reading to stumbling upon news (n.d.). Retrieved from https://www.researchgate.net/publication/227828002_Online_news_reading_behavior_From_habitual_reading_to_stumbling_upon_news. 17 Sep 2022

Analysis of Working Postures with the Work Performance of Visual Inspection Workers in the Automotive Manufacturing Industry by Using RULA and REBA

Mani Murugesan, Ramesh Babu Thangavelu, and Vimalanathan Komalanathan

1 Introduction

Visual inspection is a crucial part of the manufacturing process, identifying good and defective products before they are sent to the dispatch stage. Industrial workers often face ergonomic factors such as posture and shift work, which can impact their productivity, health, and well-being. Failure in final inspections can lead to significant losses for companies, such as rejection, customer reputation, and business loss. An ideal ergonomic workplace is essential for workers to perform their tasks efficiently. To achieve an ideal ergonomic workplace, various combinations of ergonomic factors should be considered for the operations of visual inspectors that should be done in an ideal amalgamation of ergonomic factors. An error study approach is used to identify work performance, while a posture analysis by RULA and REBA is approached to assess the posture risk of industrial workers. The final inspection is a critical feature in decision-making and involves proper training and appropriate working methods required for the process (Fig. 1).

When necessary, a magnifying lens is used during visual inspection to help spot product flaws. Visual inspectors must possess the necessary knowledge regarding probable defects. Looking for flaws during a visual inspection that fall under attribute inspection; this method is more trustworthy when considering the visual inspectors'

M. Murugesan (✉) · R. B. Thangavelu · V. Komalanathan
Department of Industrial Engineering, Chennai, Tamil Nadu, India
e-mail: Manimurugesan@gmail.com

R. B. Thangavelu
e-mail: trb@annauniv.edu

V. Komalanathan
e-mail: saivimal@annuniv.edu

© The Author(s), under exclusive license to Springer Nature Singapore Pte Ltd. 2025
S. Das et al. (eds.), *Human-Centered Digitalization, Volume 3*, Design Science and Innovation, https://doi.org/10.1007/978-981-96-4383-7_47

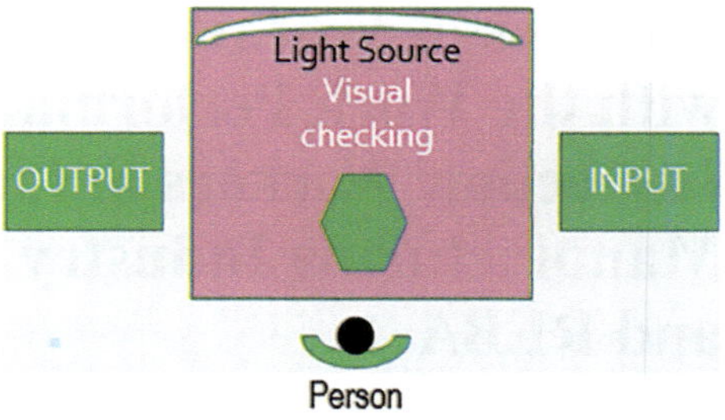

Fig. 1 Visual inspection work design

cognitive abilities. Acceptable products proceed to the final dispatch after visual inspection. To boost worker productivity in the automobile sector, ergonomic aspects are crucial in engineering since they offer answers to issues that arise during the early design stage and the evaluation of working techniques. Depending on the job done on the shop floor, a variety of ergonomic factors determine how well employees perform in the automotive industry. To determine the influence of factors on visual inspectors' performance, this study takes into account their working posture from the physical component and their shift work from the organizational element. One of the key components of visual inspectors' productivity and effectiveness is said to be their posture and work shifts. Productivity gains are largely dependent on the cognitive abilities of visual inspectors. We have thus attempted to analyze both the performance of visual inspectors by taking the posture and work shifts into consideration since research on these individual and combined characteristics can be beneficial to a variety of industry personnel.

2 Literature Review

The study indicates that major improvements in human inspection accuracy can occur [1], and that inspection accuracy may not significantly rise even at ideal levels for each variable. Practical techniques to evaluate inspection performance accuracy and a checklist to evaluate the influence of different factors on it are therefore required [2]. When making decisions without objective standards that address acceptable tolerances and fault features—often in minor or significant situations inspectors are under pressure [3]. Because of the physical demands of the workplace, the musculoskeletal system is particularly susceptible to occupational illnesses and injuries. This is why ergonomics studies humans [4]. Work posture is determined by human body organ movements, including flexion. Visual inspection is crucial in industrial settings to minimize musculoskeletal disorder (MSD) risk, especially on backbones. Studying and simulating workers is essential for identifying and evaluating potential risks. Rapid Upper Limb Assessment (RULA) [5] and Rapid Entire Body Assessment (REBA) [6] techniques are considered more versatile and less expensive than objective laboratory measures, as they allow for distinguishing disarray in various work contexts [7, 8]. RULA is applied for identifying postural discomforts in the upper part of the body, while REBA is used to establish discomforts in the complete body, based on muscular action, external loads, and grip type. RULA and REBA are considered to be risk assessment methods during the occupation and the same has been mentioned in International Standard [9] and these methods are highly cited among tools for assessing and prevention of MSDs [10]. They are incorporated to evaluate postural risk in various industries, including manufacturing, rising buildings, hospitals, oral sectors, video terminals, and waste collection activities. Work performance measurement by error study and MSDs by body parts assessment in a controlled experimental study was done. An automotive gear part selected for manual visual inspection in the manufacturing industry was selected for this study due to its widespread in many sectors and the exposure of experienced operators to safety risks [11–15].

2.1 *Inference from the Literature Review*

From the previous research, it is seen that working posture effects like musculoskeletal disorders are mostly evaluated in the field of many industries but only a few have focused on and investigated the working posture effects on the work performance of visual inspectors in visual inspection areas where workers have much more physical and mental strain due to long-standing or sitting and continuous mental work while inspecting the job. Work posture issues related to work which leads to low productivity and ill health. More research to address the gap affecting work performance which impacts the productivity and well-being of the visual inspectors' will give a huge benefit for inspectors in general, regardless of the inspection product. In this study, we have done a basic assessment of work posture evaluation for the results. This study is intense on the elements influencing the work performance of visual inspection workers in the automotive manufacturing industry, where inspection takes a crucial part in the manufacturing process and involves a huge amount of mental tasks to perform the task with a better performance. Posture and work shift factors are independently combined for evaluating the work performance of visual inspectors (Fig. 2).

3 Subjects and Methods

A field experimental study was conducted in the automotive manufacturing industry, involving 10 contract employees, who were of both genders and ranged in age from 19 to 46, all native Tamil speakers, working in the final inspection area, with allowances provided for their participation. They were all contract employees of the company working in the final inspection area as final inspectors in IP Rings Ltd. Every participant had normal or corrected vision, meaning they could all see objects clearly up

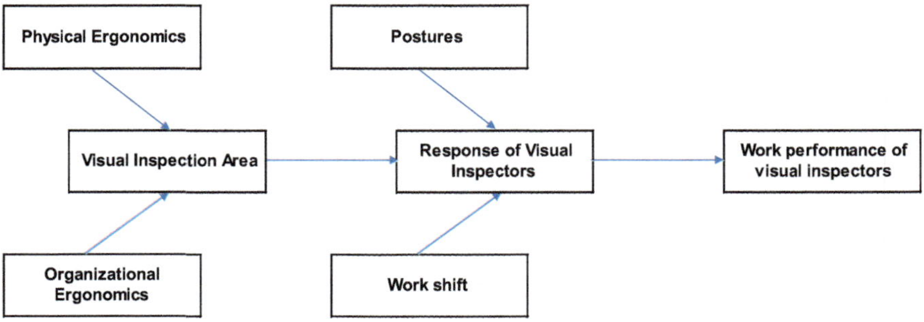

Fig. 2 Relationship between ergonomic factors on the work performance of visual inspectors

to 20 feet away from their current position and there was no evidence of color blindness. Each participant's cognitive function was normal, and none of them was using any medications associated with the mental issues. The study was conducted on the shop floor of the visual inspection area and the participants were studied individually on a single day on the shop floor at a separate inspecting table in individual team study sessions on each day with different types of postures and three types of shifts. Working posture was analyzed by scoring the assessment sheet using the video recorded during the work performed in the daily task, where the participants will inspect the sample set of jobs including OK condition and NOK condition jobs in the input. RULA and REBA Assessment sheet was utilized to assess the posture of the visual inspection task and the various factors contributing to the work performance workload, which is a simple paper and pencil sheet and it is a top-grade posture assessment method. Every experiment adhered to the ethical criteria for medical and health research involving human subjects by adhering to the Helsinki Declaration. The experimental procedure has been approved by the Department of Industrial Engineering Research Ethics Committee after assessment. Participants were advised not to take leave during their trial period, with each having a single trial day per week for 9 weeks. IP Rings Ltd approved procedures, and consent was obtained before starting. Participants were given different posture levels and shift work levels, with breaks provided after each inspection trial.

3.1 RULA

The RULA technique was introduced to evaluate a health risk present during the work linked with the upper part of the body. It was created with an emphasis on cutting, machining, clipping, inspecting, and packing in the apparel sector. The technique assesses exposure to risk factors using scoring tables and body position diagrams. These variables include external load elements such as the quantity of motions, force, static muscle effort, work postures, and continuous work duration. Workplace postures, the needless use of force or static muscular work, movement speed and precision, the frequency and length of pauses, individual factors, workplace ambient factors, and psychological variables are other elements that affect the load. Because RULA is designed without specialized equipment, investigators can conduct assessments without having to purchase extra equipment.

3.2 REBA

The REBA approach was introduced to evaluate a health risk present during the work linked with the entire part of the body. It is a simple, low-cost technique for assessing working postures that was created at Nottingham Hospital. It assesses actions oriented toward work postures in addition to identifying five danger levels. The technique is simple to use and economical; all that is needed to collect data is a pen and paper. It does, however, only permit the study of specific postures and necessitate employee approval for data gathering. Unlike previous evaluation techniques like RULA, the procedure also takes the lower extremities into account. While there is no right or wrong approach, the evaluators' circumstances and available resources determine how each method is used.

3.3 Error Study

Error study is taken as an assessment tool to identify the work performance of final inspectors in the inspection process, there are four possible human decisions in final inspection products, and normally inspectors tend to do two types of errors. Error rate is normally found by using the below formula

$$R = \frac{\text{Error identified}}{\text{Total inspected}}$$

Error study is used to observe the worker's performance by the error identified in the final inspection by the inspectors. In this research, an error study is conducted by placing the jobs on the input side of the final inspection table. This error study was conducted with each inspector with seven trials in a single day capturing all three different shifts with three various postures and each trial will have the same set of products to be inspected. In final, the error rate will be calculated by the total error identified with the total product inspected in the working tasks.

4 Experiment Procedure

The experimental study measured the error of visual inspection workers during the task in each shift during the seven trials in a single shift with all 10 participants. Factors like physical postures and shift work were evaluated. Visual inspection was carried out using a magnifying glass on a telescopic with 5 × magnification. The study involved production line parts to be inspected throughout the shift. Participants had a specialized role in identifying the defects. The study was conducted in all postures like sitting, standing, sit-stand and all three working shifts including A shift, B shift, and C shift. Observing the work performance through the error study method in the trial, working posture was also analyzed by using RULA and REBA assessment score sheets with the help of video capturing during the trial.

Visual inspectors' various working posture (Figs. 3, 4, 5).

Fig. 3 Standing posture

Fig. 4 Sitting posture

Fig. 5 Sit-stand posture

In the RULA method, there are two sections that are evaluated. In Part A, the wrists and arms are evaluated. Examine the legs, trunk, and neck in Part B. After an analysis was done using the evaluation findings for A and B, final values between 1 and 7 were obtained. In Rapid Entire Body Assessment, body posture is assessed by analyzing coupling angles, forces applied during the action for load, repetitiveness, and postural changes. It involves assembling postures into ranges and assigning scores to each position. The final value is an amalgamation of the A and B scores, with an activity score added for other static and non-static activities. The REBA score represents the REBA action level, determining whether intervention is necessary. Using the results of this method, a workstation redesign can be based on a value that indicates whether or not the worker's posture is good (Tables 1, 2, 3, 4, 5).

Minitab 14.1 was used to perform the randomized block factorial design for statistical data analysis with the 95% of confident interval level. Table 6 shows all individual and combined effects of posture and work shift to error score respectively. Figure 6 shows the residual plots of the error score.

5 Results

The outcome discloses that the combined influence of posture and work shift has a significant contribution of 09.12% on work performance. Posture has a significant effect of posture has 25.66% of contribution to the visual inspectors' work performance. Similarly, shift work has 08.65% of contribution to the work performance of visual inspectors. The optimum level of posture is the sit-stand Posture and "A" shift, both have reduced the error rate in visual inspection and increased the efficiency of performance of visual inspectors'. The impact of posture has a higher influence than shift work on work performance.

Table 1 Posture evaluation score

Method	Posture analysis		Steps														
RULA	Section	Sitting posture	1	2	3	4	5	6	7	8	9	10	11	12	13	14	15
	A	Arm and wrist analysis	1	2	3	1	3	1	0								
	B	Neck, trunk and leg analysis								4	3	2	1	3	1	0	1
REBA	Section	Standing posture	1	2	3	4	5	6	7	8	9	10	11	12	13		
	A	Neck, trunk and leg analysis								1	1	1	1	4	1		
	B	Arm and wrist analysis	2	2	2	4	0	1	1								
REBA	Section	Sit-stand posture	1	2	3	4	5	6	7	8	9	10	11	12	13		
	A	Neck, trunk and leg analysis								1	1	1	2	3	1		
	B	Arm and wrist analysis	2	2	1	3	0	3	1								

5.1 Conclusions

From the study results, posture and shift work are independently combined and have been significant in work performance. The combined effect of posture and work shift are also significantly on the performance of visual inspection workers. The result of the study is revealed by evaluating the work posture in three different types like standing, sitting and sit-stand, which shows that sit-stand posture can be adopted in visual inspection to reduce the musculoskeletal disorders in industrial sectors during the visual inspection and also to increase the productivity of visual inspectors. It concludes noteworthy association exists allying the ergonomic factors and work performance of visual inspectors'. This study investigates and finalizes that out of two ergonomic factors combined in this study, the physical factor posture impact is higher compared to the organizational factor like shift work involved.

The optimal levels selected based on minimum RPI are P3 and W1.

In this study, it is presumed that the optimum level of these two factors (P3W1 i.e., sit-stand posture and "A" work shifts) can be utilized to increase the performance of visual inspection of industrial workers and also reduce the chances of getting MSDs of visual inspectors in automotive industry.

Table 2 MSD risk score table

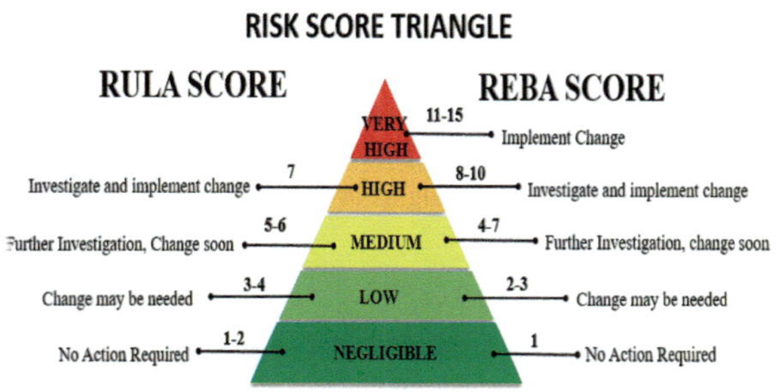

RULA & REBA MSD SCORE LEVELS		
Standing Posture	Sitting Posture	Sit-Stand Posture
6	5	3

Table 3 Errors obtained in numbers (output = error score)

Inspectors	P1			P2			P3		
	S1	S2	S3	S1	S2	S3	S1	S2	S3
Inspectors' 1	81	78	74	27	21	24	22	28	28
Inspectors 2	35	35	57	47	22	45	11	22	24
Inspectors 3	30	29	72	55	44	43	25	28	35
Inspectors 4	41	41	59	71	28	49	20	25	32
Inspectors 5	66	57	55	69	17	34	22	52	36
Inspectors 6	75	65	78	82	71	76	12	82	75
Inspectors' 7	66	56	36	28	12	62	10	24	24
Inspectors' 8	49	42	52	36	14	76	14	22	22
Inspectors' 9	52	38	71	33	11	51	18	24	32
Inspectors 10	36	29	78	41	39	58	11	23	54

6 Competing Interests

The authors proclaim that they have no conflict of interest.

Table 4 Test result for error study

Source	DF	SS	MS	F	P	Effect	%
Two-way ANOVA: error versus posture, workshift							
Posture (P)	10,073.62	2	5036.81	25.72	3.978	S	25.66
Shiftwork (S)	3396.16	2	1698.08	8.67	3.978	S	8.65
Interaction (PS)	3580.04	4	895.01	4.57	2.503	S	9.12
Block	8109.60	9	901.07	4.60	2.017	S	20.66
Error	14,098.40	72	195.81				
Total	39,257.82	89					

Table 5 Data of weights and MRPI

Trials	Factors		Response	Weights of R1	RPI
	P	S	R1		
1	1	1	531	0.0813	24.96001
2	1	2	470	0.1488	83.64513
3	1	3	632	0.1298	63.58581
4	2	1	489	0.0731	20.17373
5	2	2	279	0.0980	36.25530
6	2	3	518	0.1221	56.28204
7	3	1	165	0.0633	15.12738
8	3	2	330	0.0501	9.460011
9	3	3	362	0.0776	22.73543

Table 6 RPI scores and optimal levels

Factors	Level			Optimum level
	1	2	3	
Posture (P)	230.365	114.250	76.657	P3
Workshift (W)	159.977	163.961	97.335	W1

6.1 Author's Involvements

Mani Murugesan executed the experimental study using Error Study, RULA and REBA. Ramesh Babu Thangavelu carried through the overall supervision of the experimental study. Mani M and Vimalanathan K performed statistical analysis. All three authors read and have given acceptance to the final manuscript.

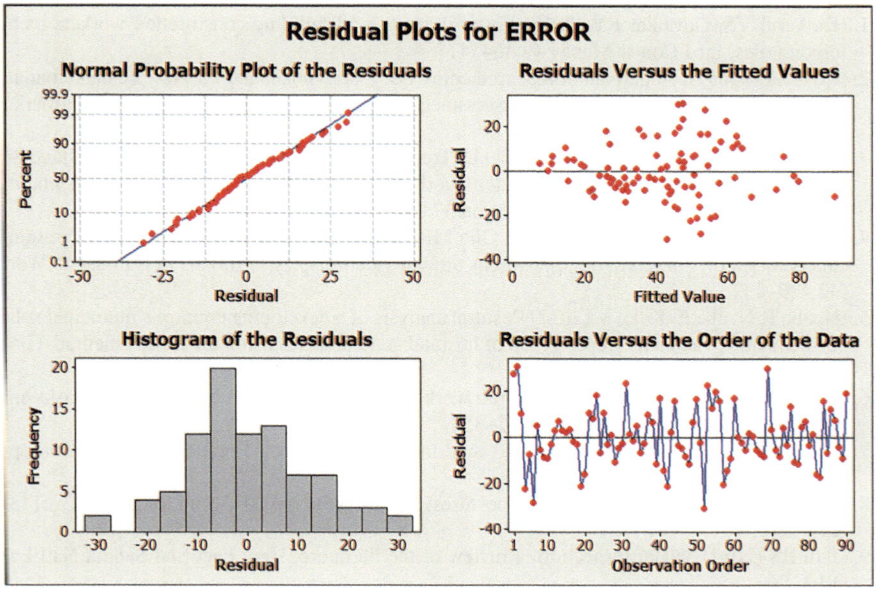

Fig. 6 Residual plots for error

Acknowledgements The author expresses gratitude to IP Rings Ltd.'s managing director, Mr. Ram Venkataramani, for his support in conducting the study in the company. Special thanks extended to the employees who helped with the study.

References

1. Schoonahd JW, Gould JD, Miller LA (1973) Studies of visual inspection. Ergonomics 16(4):365–379
2. Megaw ED (1979) The ergonomics of visual inspection. Appl Ergon 10:1–16
3. Megaw ED (1979) Factors affecting visual inspection accuracy. Appl Ergon 10(1):27–32
4. Jeffrey EF (1995) Ergonomics in the workplace. J Facil 13(4):20–27
5. McAtamney L, Nigel Corlett E (1993) RULA: a survey method for the investigation of work-related upper limb disorders. Appl Ergon 24:91–99
6. Hignett S, McAtamney L (2000) Rapid entire body assessment (REBA). Appl Ergon 31:201–205
7. David GC (2005) Ergonomic methods for assessing exposure to risk factors for work-related musculoskeletal disorders. Occup Med 55:190–199
8. Zare M, Biau S, Croq M, Roquelaure Y (2014) Development of a biomechanical method for ergonomic evaluation: comparison with observational methods. Int J Soc Manag Econ Bus Eng 8:223–227
9. ISO 11226:2000 (2000) Ergonomics—evaluation of static working postures. International Organization for Standardization, Geneva
10. Occhipinti E, Colombini D (2012) IEA/WHO toolkit for WMSDs prevention: criteria and practical tools for a step by step approach. Work 41:3937–3944

11. Kulkarni VS, Devalkar RV (2018) Postural analysis of building construction workers using ergonomics. Int J Constr Manag 19:464–471
12. Coyle A (2005) Comparison of the rapid entire body assessment and the New Zealand manual handling 'hazard control record', for assessment of manual handling hazards in the supermarket industry. Work 24:111–116
13. Janowitz IL, Gillen M, Ryan G, Rempel D, Trupin L, Swig L, Mullen K, Rugulies R, Blanc PD (2006) Measuring the physical demands of work in hospital settings: design and implementation of an ergonomics assessment. Appl Ergon 37:641–658
14. Mirmohammadi SJ, Mehrparvar AH, Olia MB, Mirmohammadi M (2012) Effects of training intervention on non-ergonomic positions among video display terminals (VDT) users. Work 42:429–433
15. Ncube F, Ncube E, Kuku V (2017) Postural analysis of a developing country's municipal solid waste handlers and a reference group of hospital general hands using the RULA method. Glob J Health Sci 9:194
16. Jung HS (2001) Establishment of overall workload assessment technique for various tasks and workplaces. Int J Ind Ergon 28(6):341–353
17. Drury CG (2002) Good practices in visual inspection. Applied Ergonomics Group Inc., pp 1–90
18. Vieira ER, Kumar S (2004) Working postures: a literature review. J Occup Rehabil 14(2):143–215
19. Judi ES (2012) Visual inspection: a review of the literature. Hum Fact Stat Sandia Natl Lab 14:1–76
20. Ranjana K, Ankit N (2022) A literature review on ergonomics, ergonomics practices, and employee performance. Quest J Manag Soc Sci 4(2):273–288
21. Mercy MA, Fitri M, Nurul ANA (2024) The influence of workstations and ergonomic work postures on employee job satisfaction in the academic library. Int J Inform Sci Manag 22(1):93–105
22. Vimalanathan K, Ramesh BT (2014) The effect of indoor office environment on the work performance, health and well-being of office workers. J Environ Health Sci Eng 12:113
23. Ansari NA, Sheikh MJ (2014) Evaluation of work posture by RULA and REBA: a case study. IOSR J Mech Civil Eng 11(4):18–23
24. Margherita MC, Ambra G, Federica C, Andrea C, Eugenio C, Stefano G (2019) Risk assessment for musculoskeletal disorders in forestry: a comparison between RULA and REBA in the manual feeding of a wood-chipper. Int J Environ Res Public Health 16(1):1–13

Identification and Prediction of Stressors Among Healthcare Workers in India During COVID-19

Ebin Mathew, K. Ratnakumar, R. Sridharan, and V. Dhanya

1 Introduction

Due to the development of a novel coronavirus that was discovered in Wuhan, China, the globe is currently dealing with a pandemic. We have to deal with uncertainty in these trying times. People will become exhausted trying to deal with unclear circumstances. To be healthy is to be in a state of complete mental, emotional, and social equilibrium, not only to be free from disease or impairment. Individuals commonly encounter tension, anxiety, and despondency during periods of uncertainty. The need for research into the causes of stress, anxiety, and depression cannot be overstated.

To preserve continuity of care and stop the pandemic outbreak from spreading, nurses and doctors in hospitals must confront a sneaky, invisible risk that is pushing the healthcare system to the limit. Due to an enormous workload, inadequate contamination protection, and the requirement to treat patients who are in severe agony and are separated from their families, medical practitioners are far more likely to experience stress and anxiety. Memory impairment, concentration difficulties, indecisiveness, diminished interest in previously enjoyed activities, overeating leading to weight gain or undereating resulting in weight loss, alongside feeling guilty, not good enough, powerless, and restless, and irritability, constitute primary clinical manifestations of depression. Suicidal ideation is also common. This case illustrates how these symptoms significantly affect essential facets of an individual's life, including learning capacity, employment prospects, and social engagement. This serves as a crucial indicator for formulating a clinical diagnosis.

Symptoms of generalized anxiety disorder (GAD) encompass anxiety, irritability, exhaustion, insomnia, gastrointestinal disturbances, panic, a feeling of

E. Mathew (✉) · K. Ratnakumar · R. Sridharan · V. Dhanya
Department of Mechanical Engineering, National Institute of Technology Calicut, Calicut, Kerala 673601, India
e-mail: ebinmathew005@gmail.com

© The Author(s), under exclusive license to Springer Nature Singapore Pte Ltd. 2025

S. Das et al. (eds.), *Human-Centered Digitalization, Volume 3*, Design Science and Innovation, https://doi.org/10.1007/978-981-96-4383-7_48

impending doom, tachycardia, diaphoresis, hyperventilation, and impaired concentration. Feeling tight or irritable, having trouble falling asleep, being run down all the time, experiencing frequent migraines, and responding too emotionally and have recurrent colds or ailments are indicative of stress.. Consequently, it may be a struggle to delineate stress, anxiety, and depression, as these conditions are often accompanied with symptoms such as sleeplessness, chest pain, fatigue, tachycardia, and impaired concentration.

A comprehensive strategy is crucial for effectively supporting healthcare personnel during the epidemic. Implementing resilience training programs can assist staff in developing coping strategies and stress management approaches, while fostering a friendly work environment through peer support systems and regular mental health assessments helps mitigate feelings of isolation and burnout. Advancing work-life equilibrium through adaptable scheduling and familial assistance, using technology for telehealth services and mental health applications, and confronting the stigma around help-seeking are essential measures. Moreover, assessing the efficacy of mental health programs via feedback systems and continuous research would enhance support strategies. Institutional and governmental policies must prioritize mental health resources and disaster preparedness strategies to guarantee that healthcare personnel receive the essential assistance required to uphold their well-being and provide high-quality care.

There are various research works which showed that the healthcare workers are experiencing stresses during the COVID pandemic which are enhanced by various sociodemographic and psychological factors.

2 Literature Review

There is a wealth of literature highlighting the challenges faced by healthcare professionals. At the first phases of the epidemic, there has been a rapid rise in the count of professionals who presented with insomnia, suspected PTSD, or psychological distress. The studies indicated that 36.0% of respondents experienced psychological distress, 30.90% may have had post-traumatic stress disorder (PTSD), and 60.9% reported insomnia. Studies have evaluated the effect of the first COVID-19 pandemic on employees' mental health. A total of 317 EMS specialists were actively sought out. The researchers examined insomnia, post-traumatic stress disorder, and psychological discomfort. At the start of the outbreak, there was a sharp rise in the number of workers who came in with mental illness, possible PTSD, or sleeplessness. Chan et al. [1] studied if there are any discernible profiles in a group of HCWs who have problems with their demography or mental health. Management should develop diverse, customized strategy approaches for each demographic group to meet distinct mental health requirements. Rois et al. [2] examined healthcare workers' mental health during the epidemic.

They employed a web-based study to collect data via an online questionnaire that was distributed over social media in a snowball effect. Questionnaires include the

PHQ-9, GAD-7, ISI, and the 10-item perceived stress scale (PSS). HCWs involved in the coronavirus disease pandemic in 2019 are exposed to a lot of traumatic or stressful situations, which has a substantial detrimental impact on their mental health. These consequences include stress-related symptoms as well as sadness, anxiety, and insomnia. Other studies also determined the neurotrophic potential of SARS-Cov-2 and its possible direct effect on mental health. They conducted a comprehensive search of the Pubmed database for studies evaluating COVID-19-related psychiatric symptoms in infected patients, healthcare employees, and non-healthcare workers. To improve treatment and mental healthcare planning, analyzing both neuropsychiatric disorders impacts and the subsequent indirect effects on mental health that resulted from this is crucial.

Tesfaye et al. [3] studied the frequency of episodes of depression and its associated variables within healthcare employees in eastern Ethiopia. Depressive symptoms were collected using a questionnaire. The data was examined using the STATA 14 software. The results were calculated using logistic regression analysis. The percentage of healthcare employees who suffered from depression was high in the study.

Pappa et al. [4] conducted a comprehensive compilation and analysis of the data that was available about the incidence of depression, anxiety, and sleeplessness within healthcare workers during the outbreak. A comprehensive search of literature databases was carried out up to the 17th of April in the year 2020. There is a significant prevalence of sadness, anxiety, and sleeplessness among those who work in the healthcare industry. Matsuishi et al. [5] examined the psychological effects of the 2009 H1N1 pandemic on healthcare staff, as well as the ways in which hospital features, gender, age, employment, and work environment affected those effects. A questionnaire with questions about sociodemographic factors was distributed. To assess the relationship between personal characteristics and multiple regression models, exploratory factor analysis and multiple regression models were performed. Workers who worked in areas where there was a high risk of infection were more nervous and tired. Liaison psychiatric services can assist decrease the outbreak's impact on hospital employees. Checked if there is a link between psychological results and physical symptoms in healthcare employees. The study included demographics, medical history, symptom prevalence, impact of event scale-revised (IES-R) and depression anxiety stress scale (DASS-21). In the case of COVID-19, there is a strong correlation between physical symptoms and psychological consequences among healthcare professionals.

3 Methodology

The current study focused on healthcare professionals who worked in hospitals in the Andhra Pradesh region that provided COVID-19 treatment. It was a cross-sectional study. We made the decision to conduct both an offline and an online survey. Telugu and English versions of a Google Forms-based online survey were created (regional language). The questionnaire was divided into five sections: sociodemographic information, patient health questionnaire (PHQ9), stress scale derived from DASS-21, general anxiety disorder (GAD7), and certain questions pertaining to COVID-19.

The replies are documented on a scale from 0 to 4, with 0 indicating Never and 4 signifying Nearly Always. There are in total 32 questions which are based on their specific parameters-depression, anxiety, and stress. Based on the responses' overall score for the above parameters are found out in a scale from normal to extremely severe.

The data was examined through MS Excel, IBM Statistical Package for Social Sciences (SPSS), and Python—Jupyter Notebook. Data visualizations have been carried out using Tableau software performing. Five different machine learning algorithms predict the stress, anxiety, and depression variables based on demographic details of the sample size population. With the best performing algorithm, the web application will be built using Flask and Python.

4 Statistical Analysis

Analysis of Moment Structure (AMOS) version 26 and the Statistical Package for the Social Sciences (SPSS) version 26 were used to do statistical analysis on the data. 23 items of the GAD, PHQ, and DASS21-STRESS questionnaires had their data validated for missing values and normality prior to analysis. Both the univariate and the multivariate levels of the distributions' normality were examined. Using Cronbach's alpha, the items' internal consistency was evaluated. We can infer from the ANOVA analysis that the demographic factors have an impact on the stress, sadness, and anxiety scores.

4.1 Exploratory Factor Analysis

The data analysis was based on the following hypotheses.

H0: There is no substantial difference that will exist between stress, anxiety, depression, and demographic characteristics among healthcare professionals during the COVID-19 pandemic.

H1: There will be significant difference between the stress, anxiety, depression, and the demographic details among medical professionals engaged in COVID-19 patient care.

To find underlying structure, EFA is utilized. It examines the linear relationship between a greater number of pertinent variables and a reduced set of unobservable variables. This is achieved by classifying variables according to how closely they correlate with one another. EFA may be carried out using factor analysis (FA). The method employed is PCA.

The correlation between the variables and the factor is known as factor loadings. A factor loading more than 0.4 is generally considered acceptable. When a variable has a strong association with several factors, it is said to be cross-loaded. Communality is the proportion of variance that a variable shares with all other variables under investigation. It speaks about the proportion of variance that may be attributed to common sources. H2 > 0.6 is considered acceptable. If h2 0.6, the result is unacceptable.

4.2 *Structural Equation Modeling*

The population covariance matrix's parameters were examined using SEM to determine how closely related they were. Determine if sample data from a theoretical model based on theory or empirical evidence is the main objective of this section.

Measured variables (observed variables), often known as items, are denoted by squares or rectangles. Here, stress, anxiety, and depression are discussed. Circular or oval shapes are used to represent factors, which are also referred to as constructs or latent variables. Stress, anxiety, and depression are the main influences in our study. Indicating a potential direct relationship between two variables is a line with just one arrow, which suggests that the variable to which the arrow points is influenced by the other. A curved, double-headed line joining two variables signifies their relationship or their shared variance. Indicating that some of each observed variable is measuring something other than the hypothesized cause, smaller ellipses reveal measurement mistakes.

AMOS used confirmatory factor analysis (CFA) to assess the measurement model. The factors for each element were tested as part of the analysis of the confirming component. There is no factor load (0.50) that is low. The general good fit of the model was assessed by model fit measurements, and the findings fell within their respective approved range [6–8]. The three-factor model (stress, anxiety, depression) produced a satisfactory fit data shown in Table 1: CMIN/df = 2.637, CFI = 0.758, TLI = 0.706, and RMSEA = 0.086. Model Fit Statistics (CMIN/df, TLI, CFI, SRMR, GFI and RMSEA) and AMOS structural equation model is shown in Fig. 1 below.

Table 1 Model fit statistics

Fit indices	Recommended values	Sources	Obtained value
P	Insignificant	Bagozzi and Yi [9]	0.000
CMIN/df	<3	<2 Ullman [8] to 5 Schumacker and Lomax [10]	2.637
TLI	>0.90	Bentler [6]	0.706
CFI	>0.90	Bentler [6]	0.758
RMSEA	<0.80	Hu and Bentler [7]	0.086

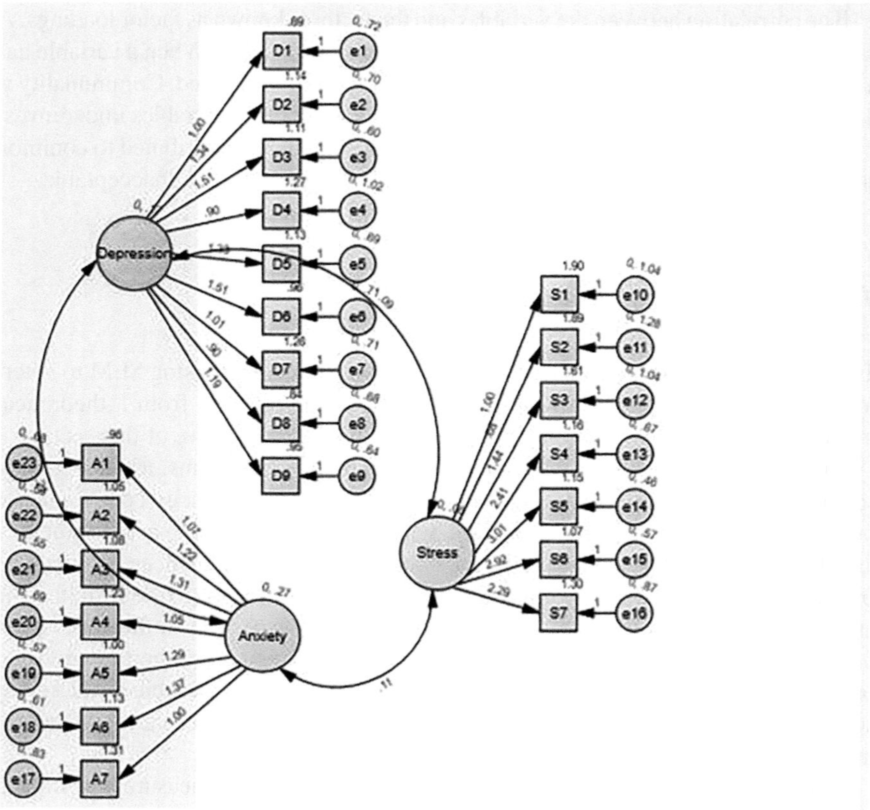

Fig. 1 Structural equation modeling

4.3 *Machine Learning Algorithms*

In order to determine which algorithm performs better with our dataset, we decided to develop a variety of different algorithms. We used five machine learning methods to do it as follows: K-nearest neighbors, SVM, decision tree, Random Forest, and XG Boost. Figure 2 below displays the degrees of accuracy for each algorithm.

The various algorithms' results for each class's the accuracy, recall, precision, error rate, specificity, and f1 results are shown in Fig. 3. XG boost achieved the highest level of accuracy for each of the three: anxiety, depression, and stress scales. The significance of a feature on demographic characteristics: The variable's significance increases with the number. Here, the importance of the role of the healthcare worker is higher, indicating that it is a significant predictor of stress, sadness, and anxiety.

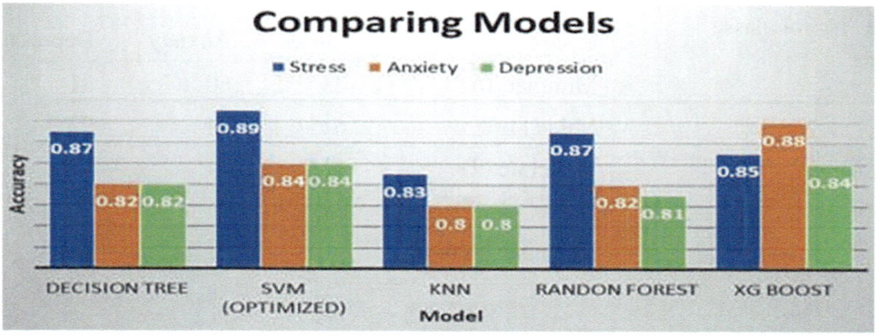

Fig. 2 Comparing accuracies of five machine learning models

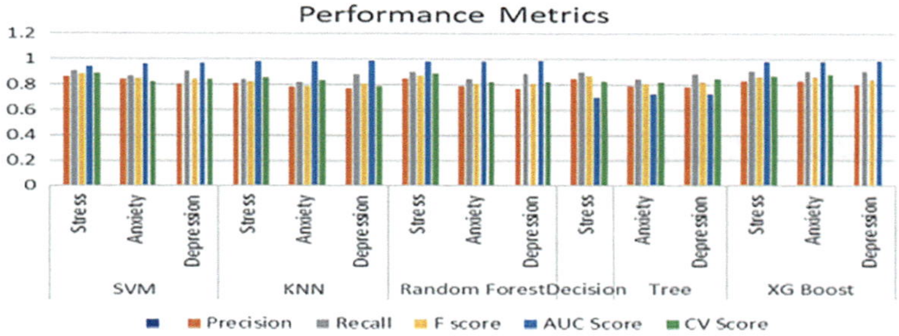

Fig. 3 Performance metrics

5 Assessment and Prediction of Stressors

Machine learning algorithms were trained to understand how to automatically categorize healthcare workers' stress, anxiety, depression levels ranging from mild to very severe, based solely on their sociodemographic profile. To forecast the stress, anxiety, and depression scores and their corresponding levels, some algorithms were also trained. Libraries used are Pandas, NumPy, and Scikit-learn. Functions used in the code are read_csv, MinMaxScaler, train_test_split, model.fit, predict, and Score.

5.1 Dealing with Imbalanced Dataset

In working with imbalanced datasets, we can use replacements to overestimate the minority class. This is called oversampling. After sampling the data, we can construct a balanced dataset for both the majority and the minority. Therefore, if both classes contain the same dataset record, the classifier may expect them to be treated equally. Unbalanced data are datasets with unequal distributions of observations in a target class, which means one class label contains a large number of observations and the other a small number. Our datasets contain uneven classes. According to Table 2,

Table 2 No. of class variables

Code	Stress	Anxiety	Depressio n
Minimal (0)	23	60	31
Mild (1)	61	88	92
Moderate (2)	112	54	62
Severe (3)	25	19	26
Extreme severe (4)	–	–	10

there were 112 answers for moderate stress and just 23 responses for minimal stress. This is equally true for stress and anxiety. Here, both the dominant and minority classes are represented. This dataset will not be evenly split, resulting in inaccurate results. We are using an upsampling technique to alleviate this issue.

5.2 SMOTE Technique

SMOTE operates by selecting an instance adjacent to the feature space, drawing a line between them, and drawing a new sample along that line. To be clearer, a random example from a minority group is initially selected. We now have 450 samples after using the aforementioned approach to upsample our data. Because there are no minority class samples, the dataset is now balanced. We can now start running the algorithms on the dataset.

5.3 Machine Learning Techniques

We chose to construct a variety of algorithms to see which one performed best with our dataset. We used five machine learning methods to do this.

They are:

1. K-Nearest Neighbor
2. SVM
3. Random Forest
4. Decision Tree
5. XGBoost.

5.4 Performance Measures

After implementing five machine learning algorithms, the metrics used for evaluating the model performance are accuracy, precision, recall, f-score, AUC and CV scores.

Table 3 Performance metrics

Classifier	Mental illness	Accuracy	Precision	Recall	F score	AUC score	CV score
SVM	Stress	0.89	0.86	0.91	0.88	0.94	0.887
	Anxiety	0.84	0.84	0.87	0.85	0.96	0.820
	Depression	0.84	0.8	0.91	0.84	0.97	0.842
KNN	Stress	0.83	0.81	0.84	0.82	0.98	0.853
	Anxiety	0.8	0.78	0.82	0.79	0.98	0.831
	Depression	0.8	0.77	0.88	0.81	0.99	0.786
Random forest	Stress	0.87	0.85	0.9	0.87	0.98	0.888
	Anxiety	0.82	0.79	0.85	0.81	0.98	0.820
	Depression	0.81	0.77	0.89	0.81	0.99	0.840
Decision tree	Stress	0.87	0.85	0.9	0.87	0.7	0.825
	Anxiety	0.82	0.79	0.85	0.81	0.73	0.822
	Depression	0.82	0.78	0.89	0.82	0.73	0.724
XG boost	Stress	0.85	0.83	0.91	0.86	0.98	0.91
	Anxiety	0.88	0.83	0.91	0.86	0.98	0.83
	Depression	0.84	0.8	0.91	0.84	0.99	0.86

All of the performance metrics for the five different models are included in Table 3 that follows.

5.5 *Confusion Matrix*

When the actual target values are compared to the predictions made by the machine learning model, the matrix produces the results. As a result, we have a comprehensive understanding of the accuracy of our categorization model as well as the many kinds of mistakes that it produces. Figure 4 below shows the confusion matrix for XGBoost algorithm

The total number of right and unsuccessful predictions is totaled and broken down per class. The program properly predicts 13 traits for minor depression but misses 5 attributes for mild depression. These five classes are associated with mild depression, but the model fails to predict them adequately. Similarly, for mild depression, the model accurately predicts 27 traits but misses two. In this manner, we can evaluate the model using the confusion matrix.

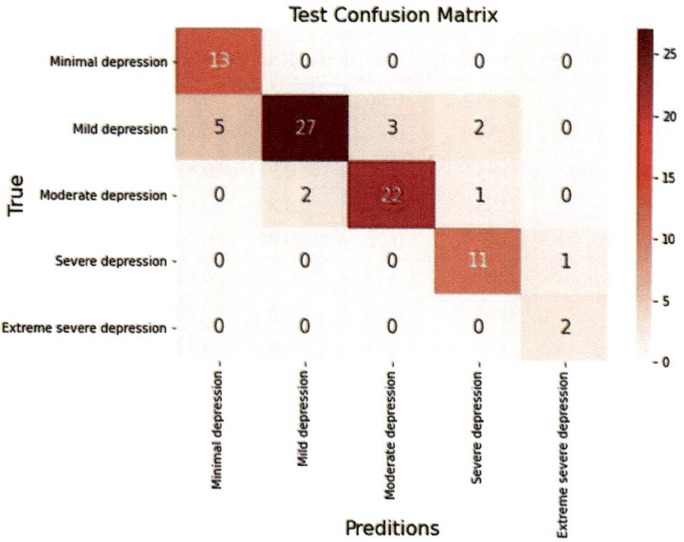

Fig. 4 Confusion matrix for XGBoost algorithm

5.6 Feature Importance

The term "feature importance" describes a set of methods for determining how much weight to give each feature while building a forecast using a predictive model. The ratings are beneficial and applicable in many contexts inside a predictive modeling scenario. The significance of characteristics in a dataset might provide insights into it. It is possible that the comparison assessments will show which characteristics are most important to the target and which ones are least important. This may be reviewed by an expert in the field and used as a basis for gathering more or different data. Feature importance of independent variables for anxiety, stress, and depression are shown in Figs. 5, 6 and 7 respectively.

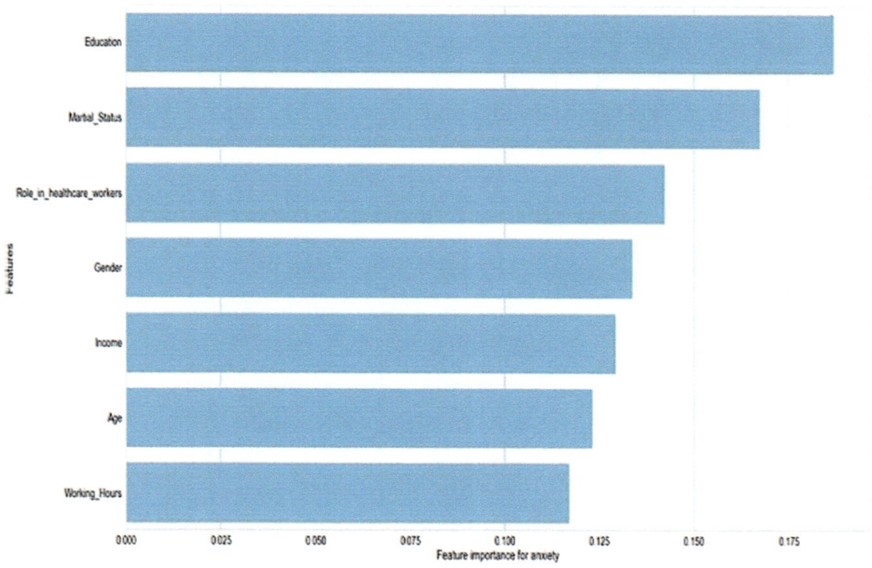

Fig. 5 Feature importance of independent variables for anxiety

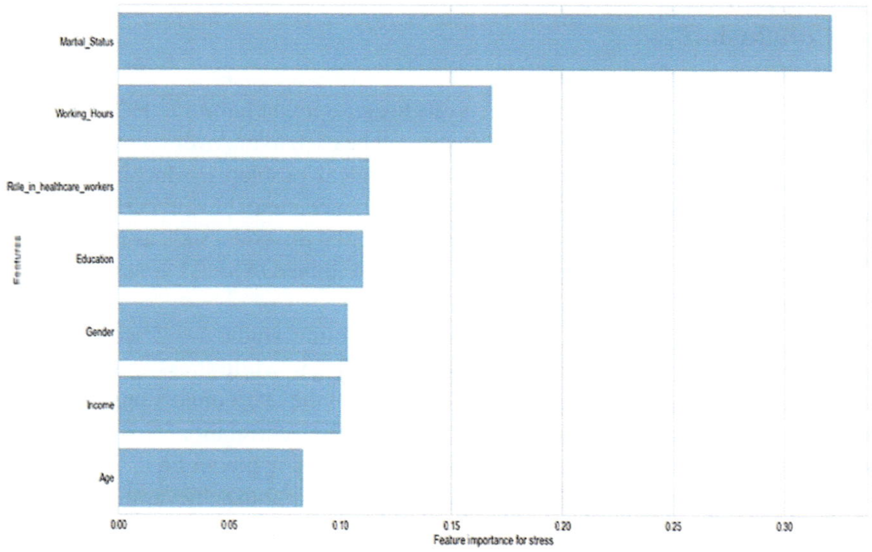

Fig. 6 Feature importance of independent variables for stress

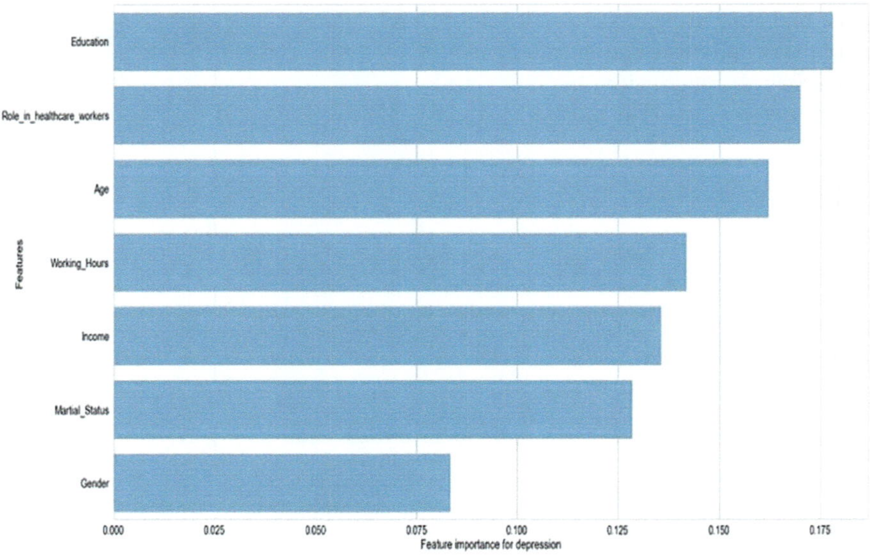

Fig. 7 Feature importance of independent variables for depression

6 Conclusion

Psychological issue is extremely dangerous because it can harm the HCW's health and social relationships. The ML framework can be used to uncover relevant components and more accurately anticipate psychological problems. Machine learning applications possess the capacity to markedly enhance mental healthcare, including the forecasting of outcomes and treatments for mental disorders such as melancholy. The questionnaire methodologies PHQ9, GAD7, and the DASS-21 stress scale have high reliability and internal consistency.

After conducting the structural equation modeling study, we discovered that depression, anxiety, and related questions are significant, while stress and associated questions are not. Furthermore, the loading factors of the interconnections are within an acceptable range. Anxiety and depression have a high correlation. Non-significant components should be eliminated to achieve a satisfactory model fit.

For depression, anxiety, and stress, the XGBoost algorithm achieves classification accuracy of 84.00, 88, and 85.00%, respectively. The KNN algorithm achieves classification accuracy of 80, 80, and 83% for depression, anxiety, and stress, respectively, which is the least among the other algorithms. Because this is an imbalanced dataset, we must consider the F1 score measure when selecting the optimum algorithm. When compared to other algorithms, XGBoost performs well in this scenario.

The speculative factors for defining their mental stress are their marital status, employment hours, and role in the healthcare sector. Similarly, education, marital status, and role in healthcare workers are preliminary characteristics for categorizing their anxiousness. Education, role in healthcare professionals, and age are preliminary characteristics for categorizing their anxiety levels.

References

1. Chan MF, Al-Adawi S, Hassan W et al (2021) Mental health outcomes among health-care workers in Oman during COVID-19: a cluster analysis. Int J Nurs Pract 14:1–9
2. Rois R, Ray M, Rahman A, Roy SK (2021) Prevalence and predicting factors of perceived stress among Bangladeshi university students using machine learning techniques. J Health Popul Nutr 40:50
3. Tesfaye Y, Agenagnew L, Anand S et al (2021) Knowledge of the community regarding mental health problems: a cross-sectional study. BMC Psychol 9:106
4. Pappa S, Barnett J, Berges I, Sakkas N (2021) Tired, worried and burned out, but still resilient: a cross-sectional study of mental health workers in the UK during the COVID-19 pandemic. Int J Environ Res Public Health 18(9):4457
5. Matsuishi A et al (2012) Psychological impact of the pandemic (H1N1) 2009 on general hospital workers in Kobe. Psych Clin Neurosci 66:353–360
6. Bentler PM (1990) Comparative fit indexes in structural models. Psychol Bull 107(2):238–246
7. Hu L-T, Bentler PM (1998) Fit indices in covariance structure modeling: sensitivity to under parameterized model misspecification. Psychol Methods 3(4):424–453
8. Ullman JB (2001) Structural equation modeling. In: Tabachnick BG, Fidell LS (eds) Using multivariate statistics. Pearson Education, Boston
9. Bagozzi RP, Yi Y (1988) On the evaluation of structural equation models. JAMS 16:74–94
10. Schumacker RE, Lomax RG (2004) A beginner's guide to structural equation modeling, 2nd edn. Lawrence Erlbaum Associates Publishers

A Readiness: Check on Human Competencies and Organization's Role for Effective Implementation of Industry 4.0

Pragya Sharma and Chandni Keswani

1 Industrial Revolution

The first industrial revolution was evident of the rise of factories, bulk production of goods, production of mechanized textile, and steam engine. The second industrial revolution marked as key development of electric engine, rise of oil and petrochemical industries, steel production, and assembly lines were prominent in bulk assembly. Third one was regarded as using technology in manufacturing, automation, and computerization. The advent of internet and World Wide Web were milestone in the late twentieth century. This was considered as digital rising. The fourth revolution, i.e., industry 4.0 is relatively different from the previous three revolutions. It is propelled by cutting-edge technologies like robotics, big data automation, Internet of Things (IoT),additive manufacturing, augmented reality, and IoS [1–7]. Industry 4.0 is a blending of physical, digital, and biological arrangements.

2 Theoretical Background

A rapid transition in industrial revolution has been noticed actively. The rapid change has extraordinarily impacted the several aspects of world's businesses. This evolution has brought research on intensity. The reason behind that is it faded the line between physical and cyber world which is known as cyber-physical world, a combination of virtual and real world which demands high potential in terms of human skills and

P. Sharma (✉) · C. Keswani
Department of Management, Shri Vaishnav Institute of Management, Indore, Madhya Pradesh, India
e-mail: pragya.sharma@svimi.org

C. Keswani
e-mail: chandani.keswani@svimi.org

© The Author(s), under exclusive license to Springer Nature Singapore Pte Ltd. 2025
S. Das et al. (eds.), *Human-Centered Digitalization, Volume 3*, Design Science and Innovation, https://doi.org/10.1007/978-981-96-4383-7_49

competencies to perform on digital, artificial intelligence, machine learning, IoT, and automated platform. Industry 4.0 has merged digital and physical domain and covered a wider area in extensive businesses. To cope up these vicissitudes, organizations are keeping abreast at the higher point of receiving, understanding, and implementing changes. Organizations must prepare themselves to act quickly and progressively to the technological rapid changes. Failing or avoidance in doing so or late response the gap will stretch and difficult to grab and rejoin the future situations. Industry 4.0 is the combination of people, machine, and things. They need to be fit together to get outcome efficiently, effectively, more optimized, value added, dynamic, and organized. Companies are expecting more than 18% higher productivity in next 5 years [8], and ~85% companies may implement industry 4.0 solutions in maximum business area. Now, we can see the changed scenario in vivid business perspectives.

Fourth industrial revolution happened in 2011 and since then organizations have been confronting wide range of challenges that demanding readiness to embrace and gear industry 4.0. The necessity is aroused due to long term change in product models and value addition [9]. Now, industries are looking at the opportunities toward implementation of industry 4.0 and seeking possibilities to improve human skills and competencies. They supposed to have a clear vision to implement industry 4.0 horizontally and vertically [10].

Technologies, human factors, and economic and social events are the future shaping factors [11]. Change in any one of them may affect the organization synchronization. Industry 4.0 has taken a fast move in various provinces of businesses and aggressively affected these future shaping factors. Therefore, organizations have started to prepare themselves with upgraded technology and essential human competencies to tackle the emerging challenges. They know that a huge growth has already been predicted if organization successfully implements industry 4.0. Organization's readiness check is depending upon to plummeting risks, capitalize production opportunities, agile, and resilience response to the unfavorable events.

3 Statements and Objectives of the Study

- To study and identify major human competencies to perform job better in the realm of industry 4.0.
- To investigate organization's role as interface in initiating and implementing industry 4.0.
- To recommend some viable suggestions to strengthen the collaborative approach of workforce and organization.

4 Research Methods

A five steps systematic literature review method [12], framing research questions, accessing research papers, selecting of the study, synthesizing, and interpreting and reporting the findings have been applied to arrive at the conclusion.

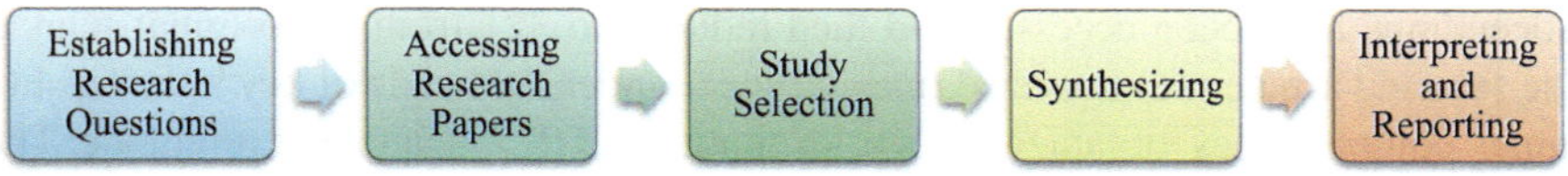

The research questions have been formed in align with the objectives:

I. **Research Questions**

- What are the major roles of industry 4.0 in the organizational performance?
- Which are the major human competencies required to implement industry 4.0 in organizations?
- What is the organization's preparedness in implementation of industry 4.0?
- What factors do justify organization's readiness for industry 4.0 implementation?

II. **Accessing Research Papers**

The criteria have been developed in this stage to find relevant research work. Three important faces: industry 4.0, human competencies, and organizations' role in industry 4.0 implementation have been identified to explore. Scopus, Web of Science, Google Scholar, Science Direct, ProQuest, IEEE Explore, National Digital Library, AIS, Emerald, Semantic Scholar, Springer, MDPI (Sci), Wiley, etc. were taken into consideration to access the research articles. The range selected for the reviewing papers in SLR method marked from 2014 to 2023. The keywords such as industry 4.0, fourth industrial revolution, smart industries, industry 4.0 implementation, human skills, abilities and competencies for industry 4.0, organization's role in industry 4.0, and organization/industry readiness. The search was over and done with progression, i.e., from basic concept of the study to the advancement and value addition of industry 4.0. To do so, the collection of the articles also completed through a systematic literature review published in peer reviewed articles [13]. The keywords come with various perspectives on search when the topic is approached and sometimes lead to the distraction in review process. To avoid this, the terms analyzed individually to maintain the authenticity of review process [14]. By this approach, a general understanding of the study has been developed. It is also ensured that the data access should maintain the novelty approach to content.

III. **Selection of Studies**

This method applied by many researchers [14, 15], in the same and different disciplines for collecting and evaluating the outcomes from multiple research works. The time span for selection of the research papers ranged from year 2014–2023. The selection range is decided with hypothetical view that it is covering the transition phases of industry 4.0 and specific human competencies requirement searched in its implementation. It is also postulated that work published during the selected period will be appropriate to quote in the study as industries especially manufacturing firms were looking forward to embrace industry 4.0. They are still in search of significant digital infrastructure and indigenous human competencies [16] to be ready to implement industry 4.0. Thousands of research article on the topic floating in digital database. It was time consuming method to get relevant research articles. Initially about 616 abstracts, articles published in various journals and conference proceedings have been accessed and then reduced to 321 from the original search. The selected articles summarized in excel sheet under the heads of key industry 4.0, human competencies, organization's role, citation, DOI, and database reference. The following inclusion and exclusion criteria were also taken into consideration:

Inclusion

- The studies published in journals, conference proceedings
- Work on industry 4.0 competencies including human skills, knowledge and abilities
- Gray literature like company research report, government policy document, case study, etc. These reports help to understand the current scenario of business market.
- Limits set to publication years, i.e., 2014–2023.
- English only language
- Full published papers
- Exact and partially related articles that justify the research questions

Exclusion

- The articles specifying industry 4.0 concept repetitively
- Articles on competencies and skills but not related to industry 4.0
- No full papers
- Research articles in other languages
- Articles not falling in the period (2014–2023)
- Technical areas such as medical, engineering

Synthesizing

To maintain the relevance of the study, the authentic digital database and quality articles have been referred. It was difficult initially to search the relevant studies as thousands of research articles available on digital database. The present study is segmented in mainly three part, i.e., importance of I4.0, human competencies, and organizational preparedness in implementing it. The studies that have applied systematic literature review methodology with the theme were the main selection criteria. During the access of the articles, it has been observed that Scopus, Web of Science (WoS), Wiley have high quality papers but in limited numbers. Whereas, Google scholar, Springer, Science Direct, MDPI were a great help in the study as good number of quality papers are available on the theme of the present study.

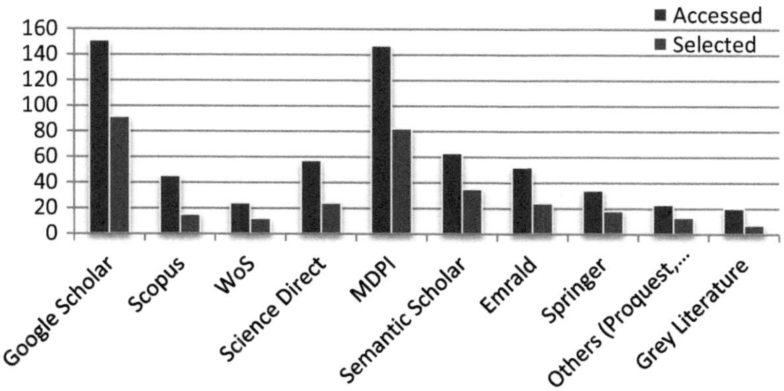

II. **Interpreting and Reporting**

Distinct conceptual research works are available on human competencies and organization's role in implementing industry 4.0. But there is dearth in the combined approach. Also, empirical studies required to be conducted especially in manufacturing industries (Table 1).

5 Discussion

Technology is upgrading day by day and becoming difficult for employees to upgrade themselves at the same pace. The humans should be well prepared with continuous learning capacity, computer vision, technological, methodological, personal, and social competencies and they still should have a room for improvement. Organizations are supposed to provide an interface between theory and practice for better competitive advantages. During review process, it has been observed that competencies may be grouped under technical and non-technical competency heads. Technical competencies are consisting of ICT, cyber-security, big data analytics, artificial intelligence, coding, computational, designing competencies, etc. [28]. Whereas non-technical skills can be elaborated as personal, social, creativity, facilitating, displaying self-efficacy, leadership, adaptability, problem solving, etc. Organizations should also have a self-check on providing resources, making strategies, developing well equipped infrastructure, organizing effective training programs on latest concepts, etc., for creating a best ecosystem in implementing industry 4.0. To maintain this ecosystem they may have different methods [29] to hire right potential to understand and run the organization in current scenario.

Table 1 SLR findings on human competencies and organizational role in industry 4.0 implementation

Human competencies	Elaboration	Author/authors
• Social, groups • Personal, and • Methodological	Imminent competencies for three professions—robotics engineers, data scientists, and food designers demanded as great industries expertise	
• Creativity-focused technology facilitating fluency (CFTF)	Such as handling complexity, rational thinking critical thought processing, predicting potentials, displaying self-efficacy and communicating skillfully	Arthur [17]
(Must have) • IT skills • Processing and analysis of data and information • Better understanding with modern equipment	Through AHP model the multicriteria issues have broken into smaller parts to get a hierarchical model More weightage given to these competencies and labeled as must have competencies	Praj et al. [18]
• Personal characteristics	Such as responsibility, loyalty, communication skills, and soft skills; high-tech industries people tend to observe the gaps in STEM and soft skills	Kveta et al. [19]
• Data • Technology and • Human literacy capabilities	Big data to improve public and business services; digital literacy for to process data and information; human literacy denotes individual character' soft skills and ability to collaborate, to adapt, and to innovate	Ellitan [20]
• Technical • Methodological • Social • Personal	With these competencies some domain specific competencies are also required such as business, engineering, and design	Hernández-de-Menéndez et al. [21]
• Data-centric and technology • Customer-centric and process • Organization and infrastructure	Interdisciplinary thinking, complex content mastery, machine interaction competence, problem solving, innovative process participation, analysis of IT business, E-commerce, management of network database, leadership skills, IT infra, and architecture	Shevyakova et al. [9]
• Manufacturing competencies	Vertical, horizontal, and end to end integration, computer machine integration	Qin et al. [22]
• ICT • Openness competencies • Being effective in autonomy	ICT has a different contribution to industry 4.0 maturity	Schumacher et al. [23]

(continued)

Table 1 (continued)

Human competencies	Elaboration	Author/authors
• System and interface competence • Creativity and methods competence for systematic idea and innovation management • Integrated product and process planning and design competence	*Required for cyber-physical (virtual, digital and real)-production system:* Cloud computing, virtual reality, data mining, IoT, RFID, sensors, and actuators Human-intelligence, senses for perception, ability to improve, learning aptitude, versatility, creativity, experience, social interaction	Jager and Ranz [24]
Organization preparedness	**Elaboration (prepare with)**	**Author/authors**
• Organization innovation in relationship with knowledge Oriented leadership • Electronic HRM	KOL-transformational, transactional, and motivational leadership to encourage in enhancing knowledge that lead to innovation E-HRM-E-HRM functions have good influence on Organizational Innovation and contributing to the industry 4.0 readiness	Rehman et al. [25]
• Technology • Behavior, and • Events (changed the world)	Technology—Big data, IoT, IIoT, AI, MI, Block chain, etc. Behavior—external (market) and internal (workplace and workforce) Events—economic, social, natural, demographical, geopolitical	Botha [11]
• Individual, and • Company specific strategies for adoption of I4.0 • Skilled employees • Organizational flexibility	How the strategies will be implemented to adopt industry 4.0 applications Skilled employees needed to implement business model and technology	Grufman et al. [26]
• Digital model techniques • Customization of products • Fostering Upskilling and reskilling culture • Measureable results	3D printing, product customization as per customer demands, motivate employees to go for upskilling and reskilling, senior people must provide measureable results	Hajoary [27]

6 Conclusion

The flood of literature gave a clear indication that industry 4.0 has nudged everywhere. Artificial intelligence, block chain, coding, machine learning, additive manufacturing, augmented reality, cyber-physical system, etc., are driving the business world and demand specific and owned human competencies, well-matched infrastructure and to maintain the pace with this competitive business world. The study helps organizations to make a thoughtful and logical blend of humans (educated and skilled), infrastructure (set of essential technologies, systems and solutions), and regulatory policies (legislative and functioning) [30]. No doubt, organizations have enormous scope to flourish in the market after successful implementation of industry 4.0. To strengthen and collaborate with industry 4.0, the workforce must be transformed. The collaborative, social [31, 32], cognitive, problem solving, organizational competencies must be combined with technical skills.

References

1. Javaid M, Haleem A, Singh RP, Suman R, Gonzalez ES (2022) Understanding the adoption of industry 4.0 technologies in improving. Environ Sustain Sustan Oper Comput 3:203–217
2. Navarro J (2022) Industrial needs in the fields of artificial intelligence. Internet Things Edge Comput Sens 5(22):4501. https://doi.org/10.3390/s2212450
3. Shet SV, Pereira V (2021) Proposed managerial competencies for industry 4.0—implications for social sustainability. Technol Forcast Soc Change 173:1080. https://doi.org/10.1016/j.techfore.2021.121080
4. Masood T, Egger J (2019) Augmented reality in support of industry 4.0: implementation challenges and success factors, robot. Comput Integr Manuf 58:181–195. https://doi.org/10.1016/j.rcim.2019.02.003
5. Antunes JG, Pinto A, Reis P (2018) Henriques, industry 4.0: a challenge of competition. Millennium J Educ Technol Health 2(6):89–97
6. Ivanov D, Dolgui A, Sokolov B, Werner F, Ivanova M (2016) A dynamic model and an algorithm for short term supply chain scheduling in the smart factory industry 4.0. Int J Prod Res 54:386–402
7. Ning H, Liu H (2015) Cyber-physical-social-thinking space based science and technology framework for the internet of things. Sci China Inform Sci 58:1–19
8. Jeevitha T, Ramya L (2018) Industry 1.0–4.0: the evolution of smart factories. Researchgate.net/publication/330336790
9. Shevyakova A, Munsh E, Arystan M, Petrenko Y (2021) Competence development for industry 4.0: qualification requirements and solutions. Insights Reg Develop 3(1):124–135
10. Rupp M, Schneckenburger M, Merkel M, Börret R, Harrison DK (2021) Industry 4.0: a technological-oriented definition based on bibliometric analysis and literature review. J Open Innov 7(68):1–20
11. Botha AP (2018) Rapidly arriving futures: future readiness for industry 4.0. South Afr J Ind Eng 29(3):148–160. https://doi.org/10.7166/29-3-2056
12. Denyer D, Tranfield D (2009) Producing a systematic review. In: Buchanan DA, Bryman A (eds) The Sage handbook of organizational research methods. Sage Publications Ltd, pp 671–689
13. Silvestri L, Forcina A, Introna V, Santolamazza A, Cesarotti V (2020) Maintenance transformation through industry 4.0 technologies: a systematic literature review. Comput Ind 123:103335
14. Oztemel E, Gursev S (2020) Literature review of industry 4.0 and related technologies. J Intell Manuf 31:127–182. https://doi.org/10.1007/s10845-018-1433-8
15. Neumann WP, Winkelhaus S, Grosse EH, Glock CH (2021) Industry 4.0 and the human factor: a system framework and analysis methodology for successful development. Int J Prod Econ 233:1–16. https://doi.org/10.1016/j.ijpe.2020.107992
16. Immerman G (2019) Industry 4.0: the history, benefits, and technologies, MachineMetrics, Industry 4.0
17. Arthur C (2020) Creativity-focused technology education in the age of industry 4.0. Creat Res J 32(2):184–191

18. Praj F, Horváthová IM, Čambál IM (2022) Employee competencies in line with industry 4.0, annual sessions of scientific papers—IMT Oradea. In: IOP conference series: materials science and engineering, p 1246. https://doi.org/10.1088/1757-899X/1256/1/012033.
19. Kveta O, Eva K, Pavel H, Ondrej V (2022) Workforce readiness for industry 4.0 from the perspective of employers: evidence from the Czech Republic. Ind Higher Educ 36(6):807–823
20. Ellitan L (2020) Competing in the era of industrial revolution 4.0 and society 5.0, entrepreneurship. Jurnal Maksipreneur Manajemen Koperasi 10(1):1–12
21. Hernández-de-Menéndez M, Morales-Menéndez R, Escobar CA, McGovern ME (2020) Competencies for industry 4.0. Int J Interact Des Manufact 14:1511–1524
22. Qin J, Liu Y, Grosvenor R (2016) A categorical framework of manufacturing for industry 4.0 and beyond. Proced CIRP 52:173–178. https://doi.org/10.1016/j.procir.2016.08.005
23. Schumacher A, Erol S, Sihn W (2016) A maturity model for assessing industry 4.0 readiness and maturity of manufacturing enterprises. Proced CIRP 52:161–166. https://doi.org/10.1016/j.procir.2016.07.040
24. Jager A, Ranz F (2014) Industry 4.0: challenges for the human factor in future production scenarios. In: Proceedings of the 4th conference on learning factories, KTH royal institute of technology Stockholm, pp 37
25. Rehman HM, Yong HN, Choong Y (2022) Facilitating the Malaysian manufacturing sector in readiness for industry 4.0: a mediating role of organization innovation. Int J Asian Bus Inform Manag 13:1–23
26. Grufman N, Lyons S, Sneiders E (2021) Exploring readiness of SMEs for industry 4.0. 146(25):54–86
27. Hajoary PK (2023) Industry 4.0 maturity and readiness: a case of a steel manufacturing organization. Proced Comput Sci 217:614–619
28. Závadský C, Olexová C (2022) Kompetencie zamestnancov pre potreby: industry 4.0 (competencies of employees for industry 4.0). Manažment V Teórii A Praxi 18:29
29. Antosz K (2016) Maintenance—identification and analysis of the competency gap. Eksploatacja i Niezawodnosc 20(3):484–494
30. MIRRI (2019) MIRRI 2030 digital transformation strategy for Slovakia. https://www.mirri.gov.sk/wp-content/uploads/2019/10/SDT-English-Version-FINAL.pdf
31. World Economic Forum (2016) The future of jobs: employment, skills and workforce strategy for the fourth industrial revolution. Global Challenge Insight Report, World Economic Forum, Geneva
32. Selamat A, Taspir SH, Puteh M, Alias RA (2017) Higher education 40: current status and readiness in meeting the fourth industrial revolution challenges. Redes Higher Educ Towards Ind 4:23–24

Effects of Using Smartphone While Standing Inside a Public Transportation with Different Positions of the Backpack on Muscle Activation and Postural Sway

Jaramier Cobrado Joibi, Jane Lee, Seoyeon Yoon, Jungyeon Park, Kim Hyunji, Woojin Park, and Myung Hwan Yun

1 Introduction

Maintaining postural stability, especially in transit environments, is critical for passenger safety. Non-collision incidents like falls due to transit maneuvers contribute to injuries, particularly for those standing [1]. Factors such as vehicle design, lack of handholds, and the actions of drivers can exacerbate the risk of falls [2]. A comprehensive approach that includes vehicle modifications, safety communication, and the use of intelligent systems is necessary to improve stability and safety for passengers [3].

Simultaneously, the prevalent use of smartphones has been linked to various musculoskeletal discomforts due to the adoption of poor postures during prolonged engagement with these devices [4–6]. Extended smartphone use can alter spinal curvature, strain muscles, and affect balance, potentially leading to increased injury risk [7, 8]. Recommendations to mitigate these risks include taking breaks and promoting safer usage habits [9, 10].

Additionally, backpack positioning significantly influences body posture, with the body compensating for the shifted center of gravity through various postural adaptations [11, 12]. Heavier and higher backpack loads can increase postural deviations and lead to muscle fatigue and pain. The placement of the load and backpack fitting are crucial in minimizing these effects, particularly in children with developing spines.

J. C. Joibi · J. Lee · S. Yoon · J. Park · W. Park · M. H. Yun (✉)
Department of Industrial Engineering, Seoul National University, Seoul 08826, South Korea
e-mail: mhy@snu.ac.kr

K. Hyunji
Department of Physical Education, Seoul National University, Seoul 08826, South Korea

© The Author(s), under exclusive license to Springer Nature Singapore Pte Ltd. 2025

S. Das et al. (eds.), *Human-Centered Digitalization, Volume 3*, Design Science and Innovation, https://doi.org/10.1007/978-981-96-4383-7_50

This paper investigates the combined effects of smartphone usage and backpack positioning on postural sway and muscle fatigue in the dynamic setting of public transit. We explore how backpack position affects neck muscle activation and postural stability. Our research aims to provide insights into ergonomic practices and inform design strategies for public transportation systems to support the health and safety of commuters.

2 Method

Figure 1 illustrates the experimental setup involving 10 healthy participants, balanced across genders and averaging 26.7 years in age. Participants were screened for musculoskeletal health, smartphone familiarity, and experience with public transit to ensure relevance to the study's focus on postural stability during transit and smartphone use.

The experiment employed a factorial design to observe the impact of backpack positioning and neck tilt on postural and muscular responses while participants engaged with smartphones. A uniform backpack, filled to represent 20% of the participant's body weight, and a standard smartphone model were used across tests to maintain consistency.

Participants underwent six testing scenarios combining different backpack placements and neck angles, standing on a force plate to capture postural stability and using EMG sensors on specific muscles to measure activity. After each scenario, participants reported their discomfort levels, providing subjective data to complement the objective measurements.

Figure 1 provides a visual overview of the posture and backpack configurations examined, helping to contextualize how smartphone use and carrying methods might influence postural sway, muscle activation, and discomfort.

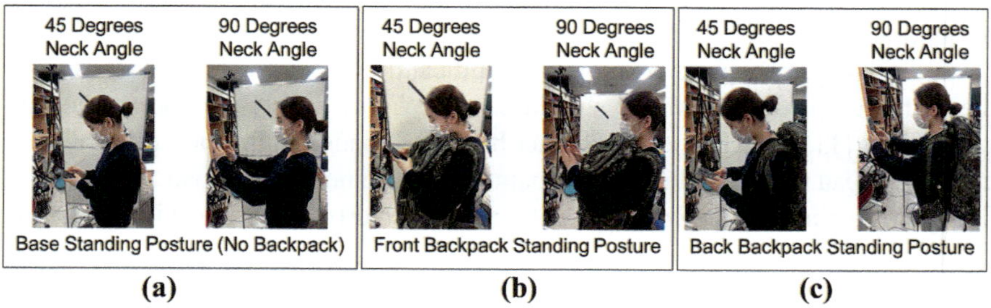

Fig. 1 Comparison of postures and backpack positions. **a** Subject standing without a backpack, viewing a smartphone at a 45° neck angle (left) and a 90° neck angle (right). **b** Subject standing with a backpack worn on the front, viewing a smartphone at a 45° neck angle (left) and a 90° neck angle (right). **c** Subject standing with a backpack worn on the back, viewing a smartphone at a 45° neck angle (left) and a 90° neck angle (right). Each posture was analyzed for postural sway, muscle activity, and perceived discomfort

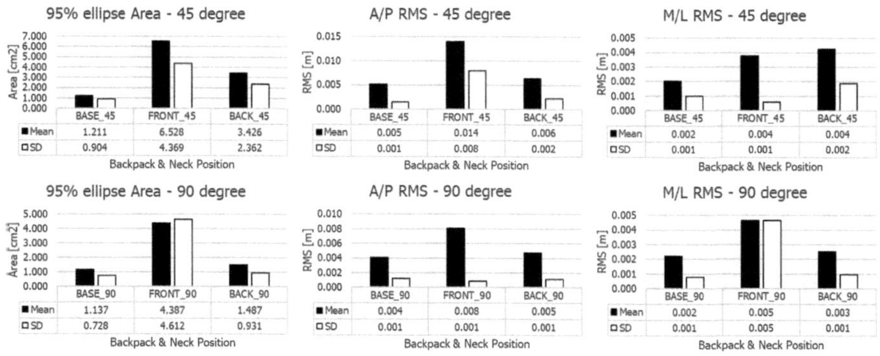

Fig. 2 Bar graphs representing postural sway metrics at different neck angles. The top row displays data for a 45° neck angle and the bottom row for a 90° neck angle. From left to right, the graphs represent the 95% ellipse area (indicating the area of postural sway), anterior/posterior root mean square (A/P RMS, indicating front-back sway), and medial/lateral root mean square (M/L RMS, indicating left–right sway). Each graph compares the mean and standard deviation (SD) values across three conditions: base standing (BASE), front backpack (FRONT), and back backpack (BACK) positions

3 Results

3.1 *Postural Sway Analysis Using a Force Plate*

The study analyzed postural sway by tracking the center of pressure (COP) using a force plate and fitting an ellipse to these points. The front backpack position induced more sway than other positions at both 45° and 90° neck angles. Statistical analysis with one-way ANOVA, considering the backpack position and neck angle, produced non-significant p-values, suggesting no substantial difference in sway between the different conditions. Both anterior/posterior and medial/lateral sway measurements supported this finding, with the front backpack position showing more sway but without statistically significant differences from other positions.

Figure 2 illustrates these results through bar graphs, contrasting postural sway metrics across base, front, and back backpack positions at varying neck angles, highlighting the consistency of sway magnitude and direction across different postures.

3.2 *Electromyography (EMG) Analysis*

The EMG analysis focused on the muscle activity of the gastrocnemius (GN), cervical paraspinal (CP), and upper trapezius (UT) muscles across different backpack positions and neck angles. The front backpack position consistently showed higher muscle activation (%MVC) across all tested muscles, indicating increased strain.

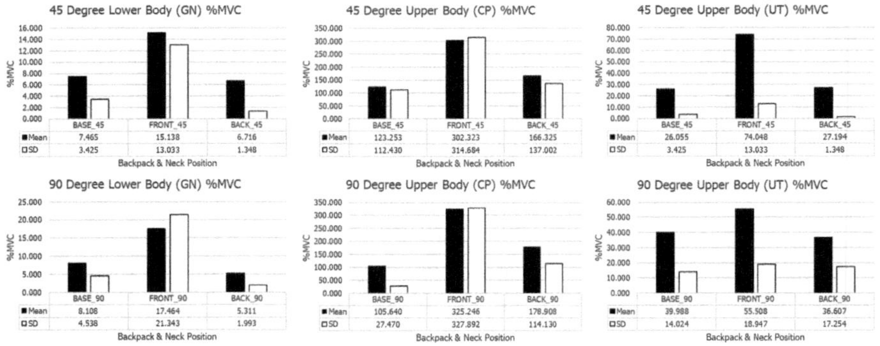

Fig. 3 Muscle activation levels by position and neck angle. Bar graphs depict the mean and standard deviation (SD) of muscle activation, expressed as percentage of maximum voluntary contraction (%MVC), across different backpack positions and neck angles. The left column shows the gastrocnemius (GN) muscle data for 45° (top) and 90° (bottom) neck angles. The middle column presents cervical paraspinal (CP) muscle data, and the right column shows upper trapezius (UT) muscle data, both at 45° (top) and 90° (bottom) neck angles. Each graph contrasts the activation levels in the base standing (BASE), front backpack (FRONT), and back backpack (BACK) positions

Statistical significance was found in the GN and CP muscles for backpack position but not for neck angle, with the front position differing significantly from the base and back positions. In contrast, UT muscle activity did not significantly differ by backpack position. The overall analysis suggested a correlation in muscle activation patterns between the upper and lower body, with some dependencies and independencies noted based on backpack position and neck angle.

Figure 3 illustrates the variation in muscle activation levels, showing the mean and standard deviation for each muscle group under different testing conditions, providing a visual representation of the data which supports the textual findings.

3.3 *Body Discomfort Survey*

Figure 4 illustrates the study's findings on body discomfort, based on participant self-reports using a scale from 0 (no discomfort/pain) to 4 (severe pain), with intermediate points for minimal, moderate, and significant discomfort/pain.

For the 45° neck position, participants reported higher discomfort levels in the front backpack position, with moderate to significant discomfort across seven areas: neck, left shoulder, left upper arm, right shoulder, right upper arm, upper back, and lower back. In contrast, the back backpack position produced similar discomfort intensity, but limited to four areas: left shoulder, right shoulder, right arm, and lower back.

At the 90° neck position, discomfort was notably lower, especially in the back backpack condition, where it was confined to the left and right upper arms. Figure 4 presents a visual comparison of these discomfort distributions, highlighting how

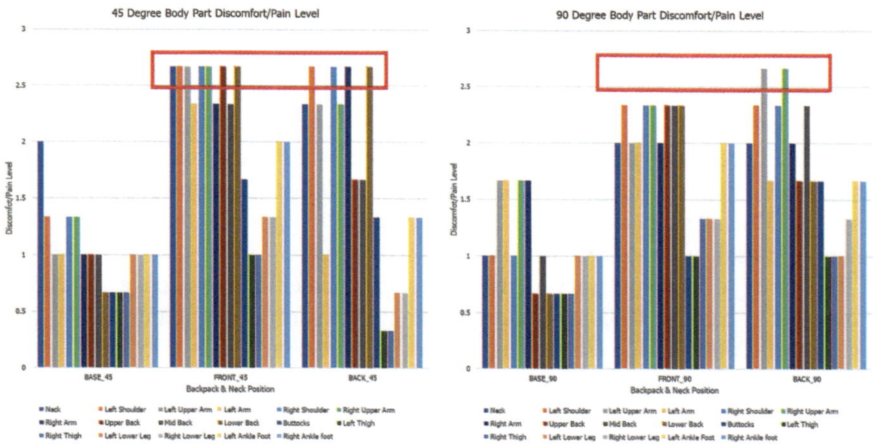

Fig. 4 Comparative body discomfort/pain levels at different neck angles and backpack positions. The cluster bar charts represent the reported discomfort/pain levels across various body parts at 45° (left chart) and 90° (right chart) neck angles. Each bar corresponds to a different backpack position (base, front, back) and is color-coded by body part, with the scale indicating the discomfort level from 0 (no discomfort) to 3 (significant discomfort/pain). Red boxes highlight the areas with the highest reported discomfort for each neck angle

neck angle and backpack position influence the location and intensity of perceived discomfort.

These outcomes highlight the variable impact of backpack position and neck angle on the participants' experience of body discomfort, with the number and intensity of discomfort areas differing notably between the front and back backpack positions, as well as across the two neck angles.

4 Discussion

The study's comprehensive results from the force plate, EMG analysis, and body discomfort survey provide valuable insights into the effects of backpack position and neck angle on postural sway, muscle activation, and body discomfort.

In the postural sway analysis using a force plate, it was observed that the front backpack position generally resulted in the most significant sway across both 45° and 90° neck angles. However, the one-way ANOVA analysis showed that the differences in postural sway for both backpack position and neck angle were not statistically significant, as *p*-values showed >0.05. This suggests that while there were observable differences in postural sway, they were not strong enough to be considered statistically significant under the conditions of this study.

The EMG analysis revealed more nuanced findings. For the gastrocnemius and cervical paraspinal muscles, the front backpack position consistently showed higher %MVC, indicating greater muscle activity. The statistical significance shown was supported by p-values <0.05 for backpack position, leading to the rejection of the null hypothesis and further analysis through post hoc tests. This suggests a notable effect of backpack position on muscle activation in these specific muscles. However, for the upper trapezius muscle, the results were not statistically significant.

Furthermore, the muscle group statistical results indicated a correlation between upper and lower body muscle groups, with significant differences observed in the one-way ANOVA and supported by post hoc analysis. This highlights the interconnected nature of muscle activation across different body regions.

The body discomfort survey results complemented the objective measures, showing that backpack position and neck angle influenced the participants' experience of discomfort. The front backpack position, particularly at a 45° neck angle, was associated with a higher number of discomfort areas, suggesting an ergonomic impact of this position.

5 Conclusion

The study provides a comprehensive analysis of the impact of backpack position and neck angle on postural sway, muscle activation, and perceived body discomfort. While the force plate analysis did not show statistically significant differences, the EMG results indicated a clear influence of backpack position on muscle activity in certain muscle groups. Additionally, the survey results revealed a perceptible impact of backpack position and neck angle on body discomfort.

These findings suggest that the design and positioning of backpacks are crucial factors in ergonomics, particularly in scenarios where individuals are required to stand for prolonged periods, such as in public transportation. While the study focused on specific conditions and muscle groups, the results underline the importance of considering ergonomic factors in the design and use of backpacks to minimize discomfort and potential musculoskeletal issues. Future research could expand on these findings by exploring a wider range of conditions, backpack designs, and population samples to further understand and mitigate the ergonomic challenges posed by backpack usage.

References

1. Silvano AP, Ohlin M (2019) Non-collision incidents on buses due to acceleration and braking manoeuvres leading to falling events among standing passengers. J Transp Health 14:100560
2. Eby DW, Kantowitz BH (2006) Human factors and ergonomics in motor vehicle transportation. In: Handbook of human factors and ergonomics, pp 1538–1569
3. Linder A, Kircher A, Vadeby A, Nygårdhs S (2007) Intelligent transport systems (ITS) in passenger cars and methods for assessment of traffic safety impact: a literature review
4. Thorburn E, Pope R, Wang S (2021) Musculoskeletal symptoms among adult smartphone and tablet device users: a retrospective study. Arch Physiother 11:1–13

5. Odole C, Olutoge DA, Awosoga O, Mbada CE, Fatoye C, Oyewole OO et al (2020) Musculoskeletal pain and postural abnormalities among smartphone-using digital natives. J Muscul Disord Treatment 6(3):65
6. Khanna V, Khanna HC (2020) Musculoskeletal disorders following excessive smartphone usage—a review article. MOJ Orthop Rheumatol 12(4):92–94
7. Shrisowmya P, Adalarasu K, Monisha-Gowri S, Aravind-Krishna T (2021) Analysis of risk factors (psychological and musculoskeletal disorders) associated with smart phone usage among indian users. International conference of the Indian society of ergonomics. Springer, Cham, pp 429–438
8. Yang SY, Chen MD, Huang YC, Lin CY, Chang JH (2017) Association between smartphone use and musculoskeletal discomfort in adolescent students. J Commun Health 42:423–430
9. Wang Y, Kim YJ, Moon KA, Park JH, Jeon HS (2023) Exploring the relation of smartphone addiction and musculoskeletal pain in the neck, trunk, and upper limbs: a cross-sectional study. Phys Therapy Korea 30(3):211–220
10. Abdullah AM, McDonald R, Jaberzadeh S (2012) The effects of backpack load and placement on postural deviation in healthy students: a systematic review. Int J Eng Res Appl 2(6):466–481
11. Janakiraman B, Ravichandran H, Demeke S, Fasika S (2017) Reported influences of backpack loads on postural deviation among school children: a systematic review. J Educ Health Promot 6:7
12. Grimmer K, Dansie B, Milanese S, Pirunsan U, Trott P (2002) Adolescent standing postural response to backpack loads: a randomised controlled experimental study. BMC Muscul Disord 3:1–10

Enhancing Defence Personnel Services Through Anthropometric-Based Design: A Focus on Soldier Safety and Efficiency

Shweta Rawat, L. Robert Varte, Inderjeet Singh, Yashmita Chaudhary, and Deepika Kakkar

1 Introduction

Anthropology (study of human beings in temporal and spatial contexts) plays a crucial role in defence personnel services through anthropometry—the measurement of human body attributes. This paper delves into the profound impact of anthropometric-based design on optimizing various products and services for defence personnel, with a specific focus on enhancing soldier safety and efficiency. Recent technological advancements, particularly three-dimensional body scanning, have significantly improved the precision of anthropometric data collection, acting as catalysts for increased accuracy in designing protective gear [16, 17, 10, 13, 21].

1.1 Background

As per situation in India, police force employs both men and women to manage emergencies like situations involving civil unrest in form of riots. The currently utilized

S. Rawat (✉) · L. R. Varte · I. Singh · Y. Chaudhary · D. Kakkar
Defence Institute of Physiology and Allied Science, DRDO, Timarpur, New Delhi 110054, India
e-mail: shwetarawat.dipas@gov.in

L. R. Varte
e-mail: robvarte.dipas@gov.in

I. Singh
e-mail: ijsingh.dipas@gov.in

Y. Chaudhary
e-mail: yashmita.dipas@gov.in

D. Kakkar
e-mail: deepikakakkar12592@gmail.com

© The Author(s), under exclusive license to Springer Nature Singapore Pte Ltd. 2025

S. Das et al. (eds.), *Human-Centered Digitalization, Volume 3*, Design Science and Innovation, https://doi.org/10.1007/978-981-96-4383-7_51

protective gears for the safety of the soldiers deployed in riot control situations are conventionally designed and offer old age solutions, with mismatched fit and poor adjustability, unapproved materials when referring to recent security specifications. Initially, women soldiers in Indian troops were required to wear ill fitted full body protector (FBP) designed and developed keeping the men troops in mind.

Instead of protecting them and aiding the female troops to give optimum performance in challenging times, the conventional FBP posed threat to their own safety and more than often hindered their mobility, making them less agile in operational environment.

The existing ensemble, based on outdated designs and lacking gender-specific considerations, presented significant challenges for female personnel. The ill-fitting male FBPs not only hindered mobility but also compromised the safety and effectiveness of female troops in riot conditions.

1.2 *Significance of Properly Fitted Protective Gear*

Full body protector (FBP) stands as a critical requirement for police and security troops, particularly in riot conditions. Properly fitting body armour is not merely a matter of comfort but holds immense significance for effectiveness of personnel in critical situations. The 2015 report from Military Medicine highlighted a crucial issue: Female soldiers in Afghanistan faced a higher rate of musculoskeletal injuries due to ill-fitting body armour compared to their male counterparts. This emphasizes critical need for well-fitted protective gear, especially for females serving in law enforcement roles [16].

1.3 *Challenges of Using Male-Centric Body Armour for Female Soldiers*

Utilization of body armour designed for men presented a myriad of problems for women soldiers in U.S. military. It significantly hindered their range of movement, affected accuracy with firearms, and posed challenges in swift entry and exit from vehicles. Perhaps more critically, poor fit created gaps that opponents could exploit during hand-to-hand combat, compromising safety and effectiveness of female soldiers on field.

Unique Challenges in Designing Female-Specific Body Protectors: Designing a female-specific body protector involves unique challenges, such as shaping curved armour plates to fit female form. This task is more complex than accommodating average male torso shape, requiring meticulous attention to anatomical nuances to ensure optimal fit and functionality.

The integration of anthropometry in the design of protective gear represents a paradigm shift in defence personnel services. Commitment to address gender-specific challenges and utilizing advanced technologies demonstrates a dedication to enhancing soldier safety and efficiency. As we navigate complexities of modern security challenges, anthropometric-based design stands as a cornerstone in providing defence personnel with tools they need for optimal performance in the field. This paper unfolds the transformative journey initiated by the Defence Institute of Physiology and Allied Sciences (DIPAS) to rectify these issues through a comprehensive anthropometric survey, ultimately resulting in the development of an ergonomically designed Female Full Body Protector (FFBP) [3].

1.4 Anticipated Outcomes

The anticipated outcome of this pioneering endeavour is a revolutionary improvement in the fit of full body protector, addressing a critical gap in operational safety and well-being for female paramilitary personnel. By meticulously analysing and validating gathered data, study aims to pave way for more effective and tailored protective gear. Overall goal is to provide safety and improve performance of women deployed in paramilitary through innovation and commitment for meeting their special needs.

2 Methodology

2.1 Anthropometric Survey and Design Process

DIPAS initiated a meticulous anthropometric survey specifically focused on female Indian troops engaged in riot control. The collected data underwent comprehensive analysis using various designing tools to create an ergonomically designed FFBP. This revolutionary approach aimed not only to rectify the existing issues but also to set a new standard for protective gear tailored to the unique anatomical characteristics of female troops [3].

The research involved an extensive anthropometric survey encompassing 325 female paramilitary troops stationed across two locations. With participants spanning an age range from 21 to 54 years, a meticulous assessment of 28 different anatomical metrics was conducted for each individual. Statistical Solution (SPSS) version 21, statistical significance and principal component analysis (PCA) were used to analyse data and identify pivotal dimensions. *K*-means clustering method was then applied to categorize population into three distinct clusters, a classification later validated through robust regression tree analysis [3].

3 Key Findings

Through this multifaceted analysis, two principal components emerged as pivotal factors: Principal Component 1, emphasizing girth dimensions, and Principal Component 2, focusing on abdomen or waist dimensions. Subsequent application of the K-means clustering method resulted in the categorization of female paramilitary personnel into three distinct clusters, each characterized by specific anatomical dimensions [19]. This insightful categorization laid the foundation for the creation of a female-specific sizing system [6, 13, 21].

Creation of Female-Specific Sizing System: The research team utilized the identified key variables to develop a female-specific sizing system for FBPs. This innovative system seeks to rectify existing disparities in protective gear by ensuring an improved and customized fit for female paramilitary personnel. The sizing system takes into account the nuanced diversity in body dimensions among female troops, addressing the critical challenge of inadequately fitting personal protective gear [6, 13, 21].

Design Philosophy: Design philosophy of FFBP revolves around not only providing maximum protection but also prioritizing user comfort and mobility. Ergonomic considerations, coupled with the use of anthropometric data, make this ensemble a pioneering solution tailored to the unique needs of Indian female paramilitary personnel. FFBP is now named as PRABLA. PRABLA is a result of meticulous design that prioritizes the unique anatomical characteristics of Indian female troops. The ensemble not only provides robust protection against projectiles and potential threats but also ensures unhindered mobility, addressing limitations faced by female troops wearing ill-fitting male FBPs in past [17, 10, 13, 21].

Key Components: PRABLA comprises several meticulously designed components, each tailored to enhance both protection and mobility:

1. **Front Shield**
    - Designed based on female anatomy/contours to accommodate user without hindrance in mobility.
2. **Back Shield**
    - Ergonomically designed with embossing and grooves to decrease impact of projectiles from rear.
3. **Shoulder Pads**
    - "Safeguards shoulder region, featuring a system for quick integration and disintegration with torso guard".
4. **Arm Guards (Lower and Upper)**
    - "Upper arm guard for upper arm region.

- Lower arm guard with full coverage and adjustability, integrated with an elbow-protecting shield for both sides of arm".

5. **Leg Protector (Lower and Upper)**

- "Thigh guard safeguards pelvic region, comes equipped with a belt for better fit".
- "Lower leg guard, associated with a knee protection shield, providing optimum protection through means of a shin guard and a calf guard, for each leg".

Versatility in Protection: One notable feature of PRABLA is its modular design, allowing troops to wear upper, lower, and peripheral protection separately based on demands of riot situation. This flexibility ensures adaptability to various operational scenarios without compromising safety [18].

Versatility in Deployment: Whether facing varying levels of threat or specific riot conditions, FFBP offers a modular approach to protection. Users can adapt ensemble based on intensity and nature of situation, enhancing both safety and operational efficiency.

4 Testing and Approval

The newly designed PRABLA underwent rigorous testing to ensure compliance with various security specifications like Anti stab, anti-impact, acid resistant, and fire resistant. Successful user trials further validated its efficacy in real-world riot control scenarios. The resolve to adhere to the latest security standards and ensuring a precise fit marked a significant departure from the issues posed by the previous ill-fitting ensembles.

5 Salient Features of the FFBP

The Female Full Body Protector (PRABLA) emerges as a vital shield for female police and security personnel, specifically tailored to address their unique anatomical considerations, especially in riot conditions. This ground-breaking protective gear is resultant of an innovative design approach, rooted in anthropometric data collected from Indian female troops. The PRABLA not only provides comprehensive protection against various projectiles and physical attacks but also prioritizes mobility and comfort, recognizing the distinct needs of female wearers [17, 10, 13, 21].

Pioneering Anthropometric Survey: The development of the FFBP is rooted in the first complete anthropometric survey of Indian female paramilitary personnel. This landmark study collected data that serves as the foundation for a revolutionary improvement in the fit of full body protectors. By meticulously analysing and validating this data, the research lays the groundwork for a female-specific sizing system tailored explicitly for the design of protective gear [6, 13, 21, 3].

Unique Anatomical Considerations: Female officers face challenges that often go unaddressed by conventional, male-specific gear. The FFBP is a response to these challenges, ensuring that protective equipment is not only effective but also considers the specific anatomical requirements of female wearers. By acknowledging these unique considerations, the FFBP sets a new standard for protective gear, emphasizing a secure fit and optimal performance.

Primary Objectives of the FFBP

1. **Comprehensive Protection**
   - Guarding against various projectiles, including stones, glass bottles, acid, and Molotov cocktails.
   - Defending against physical attacks involving sticks and chains.
2. **Mobility and Comfort**
   - Ergonomically designed to allow for unrestricted mobility, ensuring officers can respond effectively in dynamic riot control situations.
   - Prioritizing comfort to mitigate the physical strain associated with prolonged use.

The envisioned outcome of this pioneering effort is a transformative improvement in operational safety and well-being for female paramilitary personnel. By addressing the critical gap in protective gear design, the FFBP aims to optimize the performance and safety of women serving in paramilitary forces, ensuring they are equipped with gear that not only protects but empowers.

6 Conclusion

The Female Full Body Protector PRABLA stands not just as a piece of gear but as a testament to progress, inclusivity, and a commitment to the safety and well-being of female officers in the line of duty. By integrating anthropometric data and emphasizing ergonomics, this ensemble not only rectifies past challenges but sets a new standard in protective gear design. The FFBP, born out of the anthropometric survey and innovative design process, boasts several salient features [17, 10, 13, 21, 3].

It provides protection against various accidents and physical attacks that may occur during close combat situations, along with safety against potential threats and projectiles launched by rioters towards them. Designed ergonomically, using anthropometric measurements of the Indian women troops, it provides the flexibility and manoeuvrability required to manage a crisis in times of civil unrest. Available in three sizes—Small, Medium, and Large—it caters to the diverse ethnicities within the female troop contingent.

The journey from ill-fitting male FBPs to the ergonomically designed FFBP represents a significant stride in enhancing the safety and effectiveness of female troops in riot control scenarios. The anthropometric survey undertaken by DIPAS, coupled with a commitment to gender-specific design, has paved the way for a new standard in protective gear. This initiative not only rectifies existing issues but also stands as a testament to the dedication to empowering and safeguarding all members of the police and security forces, regardless of gender. The PRABLA represents a commitment to the safety, comfort, and mobility of Indian female troops, ensuring they can perform at their best in the most demanding situations. Through this research, trajectory is set for a future where protective gear is not just gender-sensitive but is a seamless extension of capabilities of every individual in service [16, 3, 7].

References

1. Aldrich W (2007) History of sizing systems and ready to wear garments. Sizing in clothing. Developing effective sizing systems for ready to wear clothing. Woodhead publishing, England, p 43
2. Boorady LM et al (2013) Exploration of firefighter bunker gear part 2: assessing the needs of the female firefighter. J Text Appar Technol Manag 8(1):14
3. British Standards Institute (1990) Body measurements of boys and girls from birth upto 16.9 years. Recommendations of body dimensions for children. BS 7231-2:1990, London
4. Carreiras H (2006) Gender and the military: women in the armed forces of western democracies
5. Centre of Diversity Expertise (2017) Defence diversity and inclusion strategy 2012–2017
6. Coltman CE, Steele JR, Spratford WA, Molloy RH (2020) Are female soldiers satisfied with the fit and function of body armor? Appl Ergon 89:1–8. https://doi.org/10.1016/j.apergo.2020.103197
7. Defence People Group (2017) Women in the ADF Report 2016–2017
8. Edwards M et al (2014) A preliminary anthropometry standard for Australian army equipment evaluation
9. Epstein Y et al (2013) Physiological employment standards IV: integration of women in combat units physiological and medical considerations. Eur J Appl Physiol
10. Gupta D, Gangadhar BR (2004) A statistical model for developing body size charts. Int Cloth Sci Technol 16(5):458–469. https://doi.org/10.1108/09556220410555641
11. Hair JF (1998) Multivariate data analysis. Prentice Hall, New Jersey, pp 87–106
12. Hsu CH, Lin HF, Wang MJJ (2006) Developing female size charts for facilitating garment production by using data mining. J Chin Inst Ind Eng 12:244. https://doi.org/10.1108/09556221211194327
13. International organisation for standardisation (2006) General requirements for establishing anthropometric databases. ISO 15535:2006. ISO, Geneva
14. International organisation for standardisation (1989) Garment construction and anthropometric surveys-body dimension. ISO 8559. ISO, Geneva
15. Park H et al (2014) Assessment of firefighters' needs for personal protective equipment. Fash Text 1(1):7
16. Rawat D, DIPAS (2019) QRs/Specs of FBP. https://www.mha.gov.in/sites/default/files/QRS_Female_06112019.pdf. Accessed 6 Aug 2024
17. Salusso CJ (2006) Improving pant fit for women 55 and older through an exploration of two pant shapes. Cloth Text Res J 24(2):147–160. https://doi.org/10.1177/0887302X0602400208

18. Simpson KM et al (2011) Effect of load mass on posture, heart rate and subjective responses of recreational female hikers to prolonged load carriage. Appl Ergon 42(2):197–205
19. Varte LR, Kakkar D, Rawat S, Singh I, Chaudhary Y, Kumar B (2021) Data mining anthropometric parameters for the design and sizing of female full body protector. Def Life Sci J 6(4):275–283
20. Winks JM (1997) Clothing sizes international standardisation: the textile institute. Redwood books, Manchester, pp 7–14
21. Yu M (2004) Human anthropometrics and sizing systems. In: Fan J, Yu W, Hunter L (eds) Clothing appearance and fit: science and technology. Woodhead Publishing Series in Textiles, Cambridge, England, pp 169–195. https://doi.org/10.1533/9781845690380.169

Quantification of Soldiers' Visual Workload During Night Operations While Viewing Through PNVS

Debojyoti Bhattacharyya, Deepti Majumdar, and Roshani Meena

1 Introduction

Human visual system is evolved to work under both day light and dark conditions to impart adequate acuity for optimal object identification by the involvement of different sets of photosensory mechanisms. However, the night/dark conditions have limitations at the domain of acquiring perceptual details of targets. Degraded visual environments under dark conditions render visual targets into faint stimulus. Consequently, the process of identification of the targets is compromised at the domain of visual acuity and contrast sensitivity, depth perception, distance and size estimation, spatial orientation, etc. [1, 2]. Exposure to dark environments is quite frequent need of military operations, where proper identification of targets is of critical importance for the success of entire mission. Therefore, human visual performance is augmented using night vision devices (NVD) for the ease of target identification under low light degraded visual environment. However, visual acuity with this augmented condition remains lower as compared to day light conditions. Therefore, cognitive workload is always supposed to be higher than any visual environments [3–5].

There are two major technologies viz., image intensification (I^2) and infrared thermal imaging (IR), widely used in the construction of the NVD [6, 7]. The present article is restricted to the I^2 technology only. For night operations I^2 technology is being frequently used and having many generations, i.e., Gen 0, Gen 1, Gen 2, Gen 3, predominantly based on the differences in the use of photocathode for image intensifier tube (IIT) [7]. The optoelectronic mechanism of the IIT is well established,

D. Bhattacharyya (✉) · D. Majumdar · R. Meena
DIPAS, DRDO, Lucknow Road, Timarpur, Delhi 110054, India
e-mail: debojyoti.dipas@gov.in

D. Majumdar
e-mail: deepti.dipas@gov.in

R. Meena
e-mail: roshanimeena.dipas@gov.in

© The Author(s), under exclusive license to Springer Nature Singapore Pte Ltd. 2025

S. Das et al. (eds.), *Human-Centered Digitalization, Volume 3*, Design Science and Innovation, https://doi.org/10.1007/978-981-96-4383-7_52

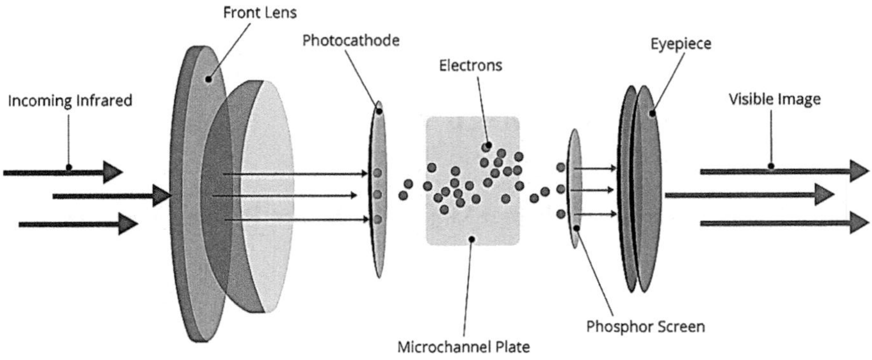

Fig. 1 Optoelectronic mechanism of the image intensification inside the tube of the night vision device

which basically collect the photons from the objects, subsequently these photons are converted into high energy photoelectrons by passing through the photocathode, then there are many folds multiplication of these electrons happened through the microchannel-plate (MCP), finally the high-density electrons hit the phosphor plates resulting in bright luminescence with the clear image formation (Fig. 1). Many of these devices do not have any source of illuminator and eventually known as the passive night vision system (PNVS).

There has been a long ancestry of research activities on the use of PNVS for military ground operations during night time in US Army [8–11]. Similarly, comprehensive research work has also been conducted in Canadian defense scenarios with PNVS emphasizing on the important aspects of night operations like aiming and shooting [12], terrain traversal by dismounted infantry soldiers [13]. Contrary to this international scenario, it has been noticed that there is a dearth of literature in the Indian scenario. Although, traces of few previous literature have been found from the perspective of aerospace medicine reasonably focusing on the aspect of visual acuity under various low light illumination condition while using different types of night vision goggle [14–16]. However, there is a dearth of empirical evidences from Indian perspective when we specifically talk about the ground operations during night conditions while using the PNVS.

The context of visual cognition is of paramount importance for human factor engineering aspect of NVD. The pivotal mechanism of visual psychophysics underlying into the process of target identification, has always brought critical challenges to the researchers during the process of scientific data collection. Although, significant volume of research work has already been conducted for target resolution while reporting the status of visual acuity, contrast sensitivity, depth perception (3D details), field of view (FOV) based either on subjective feedback or performance measure. However, in this context an objective approach like, recording of eye movement will be of keen interest. A profound neurocognitive basis of visual acuity will be revealed under various illumination conditions while looking through the PNVS. As per authors' investigation, no data is available till date on eye movement responses.

Fig. 2 300 m PNVS for INSAS Rifle

Therefore, the present study is aimed to investigate the effect of various low light illumination conditions, PNVS and distance of viewing on the changes in pupil diameter and blink responses and consequently, provide a neurocognitive basis and finally, depict suggestive measures on the present status of PNVS.

2 Methods and Materials

2.1 *Subject*

Thirty ($n = 30$) healthy soldiers with mean (SD) of age, height, weight, and experience of 35.5 (7) years, 175.4 (5.3) cm, 65.3 (3.6) kg, and 20.5 (3.2) years, respectively, from different regiments of Indian Army volunteered for the study. Participants were recruited for the study with the inclusion criteria of having no eye pathology, a normal or corrected to normal 20/20 vision, without the influence of any performance boosting drug, no history of alcohol consumption at least over a period of 24 h and adequate sleep.

2.2 *Night Vision System*

Being used most frequently, the following two PNVS systems were selected for the present investigations, viz., (I) Passive Night Vision System (300 m) for INSAS Rifle (Fig. 2) and (II) Passive Night Vision System (700 m) for Rocket Launcher (Fig. 3). Important physical characteristics of these PNVS are highlighted below in Table 1.

2.3 *Illumination Conditions*

Following three illumination conditions were selected for the present study inside the dark simulation room.

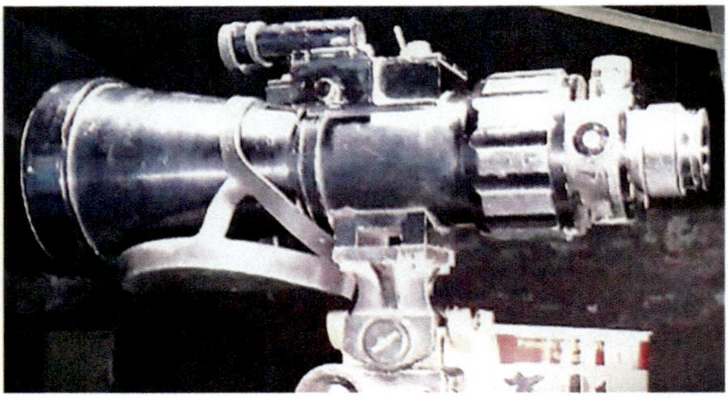

Fig. 3 700 m PNVS for rocket launcher

Table 1 Physical characteristics of PNVS

Physical properties	PNVS 300 m range	PNVS 700 m range
Generation	II generation tube	II generation tube
Magnification	4×	5.5×
Field of view (FOV)	10°	8°
Recognition range	300 m for a man size target	700 m for a vehicle
Weight	1 kg	1.7 kg

1. Star Light (SL)—10^{-3} lx
2. Half Moon Light (HML)—5×10^{-2} lx
3. Full Moon Light (FML)—10^{-1} lx

2.4 Viewing Distances

Following two viewing distances were exercised for the present study:

1. 50 m.
2. 85 m.

2.5 Experimental Protocol

Experimental protocol was first approved by the Institutional Ethics committee. Written consent was acquired from each participant. Thereafter volunteers were explained for the details of the experimental purpose and familiarized with the instruments used. Sighting systems were mounted at their respective positions of INSAS rifle (300 m) and rocket launcher (700 m), and used in the ascending range order to avoid prior recall effect. Illumination was also sequenced from lower to higher luminance levels. Participants were asked to view through the PNVS and report their visibility status verbally with in a stipulated period of visual search

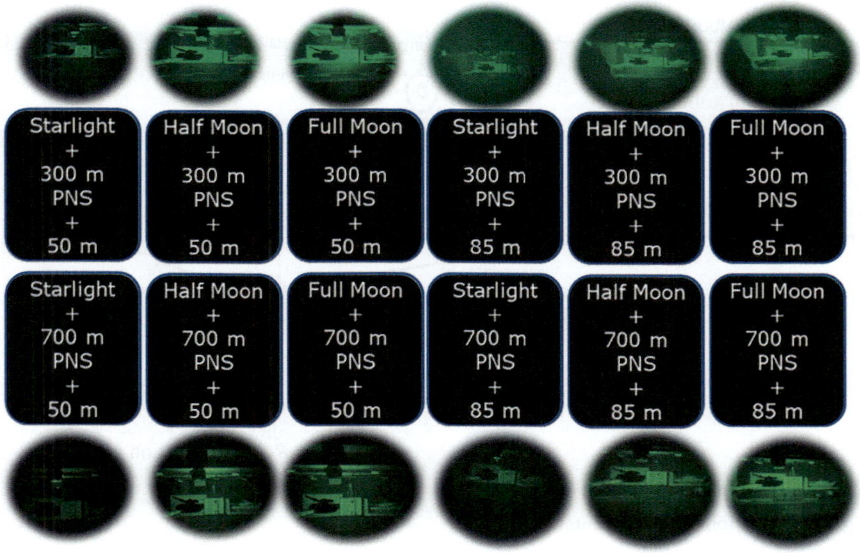

Fig. 4 Original views of the experimental stimuli through the PNVS inside dark simulation room

of 60–90 s. A combined view of illustration of screenshots from eyepiece of each system has been provided under each lighting conditions for better understanding (Fig. 4). Eye movements were recorded simultaneously by SMI ETG (Eye Tracking Glasses) with recording software iViewETG(@120 Hz). A control condition was created by viewing through the bare eyes only (without PNVS and wearing ETG—'Unaided Mesopic') with an average ambient illumination level of 400 lx from both the distances. Entire arrangement of the experimental protocol has been illustrated in Fig. 5.

2.6 Subjective Responses

While observing through the PNVS there are three essential stages for acquiring the details of the targets:

(a) *Detection*—The term implies that the observer in his search has located an object of interest in the field of view of the vision system that is being used.
(b) *Recognition*—This would mean that the observer is able to decide on the class to which the object belongs, e.g., a tank, a heavy vehicle, a group of people, and the like.
(c) *Identification*—At this level of acquisition, one should be able to indicate the type of the object, i.e., which type of tank, vehicle, or the number of people in

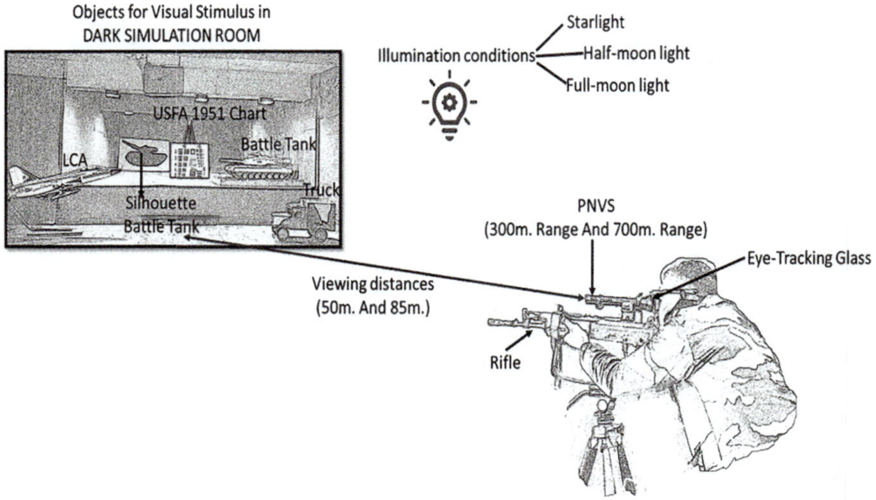

Fig. 5 Illustration of experimental protocol

a group. An important military requirement would be identification of friend or foe (IFF).

Participants were asked to register the name of the objects which they were only able to identify and based on their identifications, four different visibility zones viz., 40, 60, 80, and 100% were created, respectively (Fig. 6).

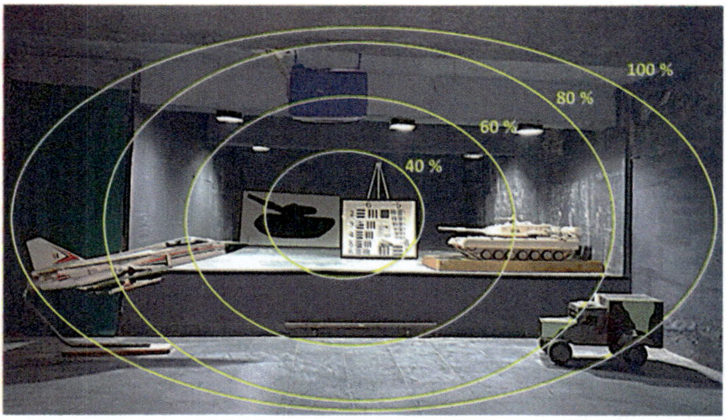

Fig. 6 Visibility zones based on various levels of identification of objects

2.7 Eye Movement

Recorded data were analyzed in BeGaze 3.7 software. For the present scenario changes in pupil diameter (mm) and total blink counts (Numbers) were selected to predict the visual workload of dark operations.

2.8 Statistical Analysis

Entire data set were treated for the normality test by Shapiro–Wilk's test, where it was observed that majority of the data were not normally distributed. Therefore, to find significant changes in the pupil diameter and blink counts, Freidman non-parametric test was applied and followed by the overall significance of the data Wilcoxon Ranked Test was also applied for Post-Hoc analysis to compare the individual pairs. Changes in dependent variables under each illumination condition for a fixed distance were compared with the control condition 'Unaided mesopic.' The Statistical Package for Social Science (SPSS) version 21 was used for statistical analysis. A value of $p < 0.05$ was considered to depict statistical significance.

3 Results

Table 2 shows the mean and SEM values of changes in pupil diameter and blink counts under various categories of independent variables. Despite exposure to various independent experimental conditions, changes in mean pupil diameter did not fluctuate over a broad range. Whereas, higher number of blink counts were observed with 700 m. PNVS. Results of Freidman test revealed significant ($p < 0.05$) changes in pupil diameter by comparing FML_300m (F), FML_700m (G), and Unaided Mesopic (C), while viewing from 50-m distance (Table 3). Post-Hoc Pairwise comparison by Wilcoxon Ranked test showed significant difference between FML_300m and Unaided Mesopic; and FML_300m and FML_700m condition. In case of viewing from 85-m distance changes in values were also found significant ($p < 0.05$) for combination of similar conditions as observed with 50-m distance (Table 3). However, pairwise comparison showed a significant change only when FML_300m and FML_700m conditions were compared. Blink count showed significant changes across all the combinations of experimental independent variables (Table 3). To find the details of the pairwise comparison followed by the overall significant please see Table 3.

Table 4 represents a percentage of visibility zones under various illumination conditions based on the successful identification of visual stimuli. It was clearly

Table 2 Descriptive statistics in the form of mean and SEM of changes in pupil diameter and blink counts under various experimental conditions

Dependent variables	Independent variables			Mean	SEM
	PNVS types	Distances (m)	Illumination conditions		
Pupil diameter (mm)	300	50	SL	4.79	0.25
			HML	4.89	0.21
			FML	5.23	0.23
		85	SL	4.31	0.34
			HML	4.33	0.34
			FML	4.85	0.28
	700	50	SL	5.31	0.10
			HML	5.03	0.21
			FML	4.76	0.13
		85	SL	4.63	0.30
			HML	4.83	0.25
			FML	4.63	0.12
	Unaided mesopic	50		4.52	0.35
		85		4.27	0.34
Blink count (numbers)	300	50	SL	9.63	2.69
			HML	9.27	2.02
			FML	9.90	2.19
		85	SL	8.83	1.77
			HML	11.87	3.09
			FML	12.70	3.16
	700	50	SL	11.97	2.85
			HML	14.93	2.91
			FML	16.37	3.28
		85	SL	13.70	2.25
			HML	14.17	2.60
			FML	15.38	2.79
	Unaided mesopic	50		16.77	2.66
		85		19.41	2.86

observed that participants' ability to scan the visibility zones as well as proper identification of the stimuli has been enhanced with the improvement of the ambient illumination conditions and with the use of PNVS with higher range of operations.

Table 3 Effects of various independent variables on changes in pupil diameter and blink count

		Pupil diameter					Blink count				
		Overall significance (freidman test)		Post_Hoc (pairwise comparison) wilcoxon ranked test			Overall significance (freidman test)		Post_Hoc (pairwise comparison) wilcoxon ranked test		
		χ^2 (df)	*P* value	A versus C	B versus C	A versus B	χ^2 (df)	*P* value	A versus C	B versus C	A versus B
50 m	A versus B versus C	0.891 (2)	0.641^{NS}				15.103 (2)	0.001*	$Z = -2.192$ $P = 0.028$	$Z = -2.883$ $P = 0.004$	$Z = -0.741$ $P = 0.459$
				D versus C	E versus C	D versus E			D versus C	E versus C	D versus E
	D versus E versus C	2.214 (2)	0.331^{NS}				10.544 (2)	0.005*	$Z = -2.467$ $P = 0.014$	$Z = -1.754$ $P = 0.079$	$Z = -2.593$ $P = 0.010$
				F versus C	G versus C	F versus G			F versus C	G versus C	F versus G
	F versus G versus C	17.315 (2)	0.000*	$Z = -2.186$ $P = 0.029$	$Z = -0.683$ $P = 0.495$	$Z = -3.436$ $P = 0.001$	12.584 (2)	0.002*	$Z = -2.672$ $P = 0.008$	$Z = -0.410$ $P = 0.682$	$Z = -2.585$ $P = 0.010$
				A versus C	B versus C	A versus B			A versus C	B versus C	A versus B
85 m	A versus B versus C	0.435 (2)	0.805^{NS}				12.268 (2)	0.002*	$Z = -3.519$ $P = 0.000$	$Z = -1.173$ $P = 0.241$	$Z = -2.541$ $P = 0.011$
				D versus C	E versus C	D versus E			D versus C	E versus C	D versus E
	D versus E versus C	1.930 (2)	0.381^{NS}				10.691 (2)	0.005*	$Z = -2.401$ $P = 0.016$	$Z = -1.558$ $P = 0.119$	$Z = -2.330$ $P = 0.020$

(continued)

Table 3 (continued)

	Pupil diameter					Blink count				
	Overall significance (freidman test)		Post_Hoc (pairwise comparison) wilcoxon ranked test			Overall significance (freidman test)		Post_Hoc (pairwise comparison) wilcoxon ranked test		
	χ^2 (df)	*P* value	A versus C	B versus C	A versus B	χ^2 (df)	*P* value	A versus C	B versus C	A versus B
			F versus C	G versus C	F versus G			F versus C	G versus C	F versus G
F versus G versus C	7.655 (2)	0.022*	$Z = -1.189$ $P = 0.234$	$Z = -0.108$ $P = 0.914$	$Z = -1.589$ $P = 0.041$	7.875 (2)	0.019*	$Z = -2.240$ $P = 0.025$	$Z = -1.128$ $P = 0.259$	$Z = -1.436$ $P = 0.151$

Note SL_300 m = A, SL_700 m = B, Unaided Mesopic = C, HML_300 m = D, HML_700 m = E, FML_300 m = F, FML_700 m = G, NS = Not significant, $*p \leq 0.05$

Table 4 Representation of visibility zones based on the target identification by the participants

Dependent variables	Independent variables			Mean
	PNVS types	Distances (m)	Illumination conditions	
Visibility zone (%)	300	50	SL	62.07
			HML	77.93
			FML	86.21
		85	SL	66.90
			HML	81.38
			FML	88.97
	700	50	SL	71.72
			HML	86.21
			FML	93.79
		85	SL	73.10
			HML	87.59
			FML	95.17
	Unaided mesopic	50		74.48
		85		73.10

4 Discussion

Present study has attempted to quantify the visual workload of military operations under dark conditions exploring all the possible effects of critical physical attributes. Results were found quite promising in predicting the cognitive demand as well as the load of the various experimental scenarios with the help of a customized methodology combining both objective and subjective approaches. In this process, the precise selection of eye movement parameters, i.e., pupil diameter and blink counts played pivotal role interpreting the results.

Pupil changes it size to achieve an optimum vision for a particular situation. One of the important strategies of optimization is to constrict the pupil for improved visual acuity and depth of field to see sharper and wider range of distances [17]. On the contrary, optimization is done to detect a faint stimulus by pupillary dilation through more sensitivity [17]. Although, previous studies [18] have already been observed the dilation of pupil diameter while performing various cognitive tasks under regular illumination conditions, emphasizing on attention [19], memory [20], and mental loading [20]. However, the present scenario is entirely different on the context of night operation, where explanation on the dilation in pupil size needs a multidimensional school of thoughts. Primarily, the challenges begin with the exposure to completely dark or very poorly illuminated conditions. Successively, this is followed by period of dark adaptation shifting the mode of operation from photopic to scotopic. Consequently, a high accuracy cone cell mediated foveal vision shifted to rod cell mediated vision. This has been considered as first stage of dilation caused by the demand of the dark environment. Second stage starts by placing the eye at the eyepiece, where images of the objects are amplified with relatively better presentation, although viewing is accomplished under monochrome (green) condition only. Therefore, the viewer face difficulties in acquiring the precise details of the stimuli, resulting in struggle for sensory processing. At this stage pupil may remain further dilated for increased effort to identify the objects. Under dark condition pupil diameter varies from 4 to 8 mm [21]. Changes in mean pupil diameter (4–5 mm) under present experimental conditions also corroborates with the literature (Table 2). Further focusing into the present raw dataset, it can also be observed that under most of the lighting conditions pupil diameter reaches up to 8 mm. Dilation of the pupil is achieved by the contraction of the dilator muscle of iris, which is regulated by sympathetic nervous system. Maintenance of the pupil diameter over a range of 4–8 mm is attributed to increased effort during visual search, negotiating with the difficulty in identification. This event in turn taxing exaggerated activities of sympathetic nervous system and sustained contraction of the dilator muscle. Consequently, participants could have experienced visual fatigue too during their search. Therefore, it can be depicted that with the existing scenario the dilation of pupil is caused by the demand or load of the tasks as well as the effort exerted in response to these demands [22–24].

The effects of a controlled condition (unaided mesopic—without PNVS and with an average ambient illumination level of 400 lx) was also explored in comparison with other experimental independent variables. Under mesopic condition both retinal cells, i.e., cone and rod, contribute to optimize the vision. The average pupil diameter values under unaided mesopic condition does not vary significantly under almost all illumination conditions, except during full moon condition from both the distances. It may be postulated that viewing distances could have played a role under other illumination conditions while there was no use of PNVS.

Changes in blink counts in present study have also paved a promising corridor for prolific discussion on visual workload. In human adults, under normal condition eye may blink ~20 times per minute [25, 26], although with a large inter-individual variability [27]. Blink counts were significantly lower during task performance with high visual load when compared to less visually demanding tasks [28]. Findings of the present study also corroborated with these previous studies. Significant reduction in blink counts were observed with lower range (300 m) of PNVS when compared with higher range PNVS. Significant increase in blinking were observed with improving illumination conditions and with 700 m range PNVS, indicating ease of search. Under mesopic condition blink counts almost restored to normal with an average searching duration of 60–90 s (Table 2). Increased visual load induced decrease in total blink count of present study could be resulted from combined effects of gradually degrading illumination condition, increasing distances of viewing and use of 300 m PNVS. This has further been established by the improved identification of the objects within different visibility zones corresponding with improving experimental conditions (Table 4).

5 Conclusions

Present study has quantified the higher demand as well as higher cognitive loading of the dark/night operation with combined approach of objective and subjective assessment. Changes in pupil diameter and blink count played very crucial role. Therefore, findings of the present study and future research attempts with improved versions of night vision systems, together will bring more clarity into the scenario for an improved human–machine interaction leading to successful identification of the targets.

Acknowledgements The authors extend their heart felt thanks to the Director, Instrument Research and Development Establishment (IRDE), DRDO, Dehradun, and all the associated staffs for providing an access to their dark simulation facilities and logistic support during the days of experiments. Authors are also immensely thankful to the participants for their continuous support toward successful data collection.

References

1. Parush A, Gauthier MS, Arseneau L, Tang D (2011) The human factors of night vision goggles: perceptual, cognitive, and physical factors. Rev Hum Fact Ergon 7(1):238–279
2. Price DR, McLean WE (1985) Aeromedical lessons learned with night vision devices. AGARD Vis Protect Enhanc 10:52
3. Melzer JE, Brozoski FT, Letowski TR, Harding TH, Rash CE (2009) Helmet-mounted displays: sensation, perception, and cognition issues. US Army Aeromed. Res. Lab., Fort Rucker, pp 805–848

4. Cuqlock-Knopp VG, Whitaker LA (1993) Spatial ability and land navigation under degraded visual conditions. Hum Fact 35(3):511–520
5. Essock EA, Sinai MJ, McCarley JS, Krebs WK, DeFord JK (1999) Perceptual ability with real-world nighttime scenes: image-intensified, infrared, and fused-color imagery. Hum Fact 41(3):438–452
6. Haque MJ, Muntjir M (2017) Night vision technology: an overview. Int J Comput Appl 167(13):8887–8937
7. Chrzanowski K (2013) Review of night vision technology. Opto-electron Rev 21(2):153–181
8. Human Engineering Lab Aberdeen Proving Ground Md. Infantry Applications of the Night Vision Goggles, AN/PVS-5.(A Pilot Study-September 1974)
9. Dyer JL, Young KM (1998) Night vision goggle research and training issues for ground forces: a literature review (DTIC No. 19980622–128). United States Army Research Institute for the Behavioral and Social Sciences, Alexandria, VA
10. Banks JH, Sternberg JJ, Cohen BJ, Debow CH (1971) Improved search techniques with passive night devices. In: Technical research report 1169. US Army Behavior and Systems Research Laboratory, Arlington, VA
11. Banks JH (1971) Effects of search area size on target acquisition with passive night vision devices, vol 2. US Army Behavior and Systems Research Laboratory
12. Angel HA (2005) Examination of the effect of night vision devices on rifle target engagement accuracy during urban operations. In: Technical reports No. CR 2005-066. Defence R&D Canada, Toronto
13. Angel HA (2004) Examination of the effect of night vision devices on dismounted terrain traverse. In: Technical report. No. CR-2004–176. Defence R&D Canada, Toronto
14. Sharma D, Tripathy NK, Raghunandan V, Sekhar BM (2021) Visual acuity through night vision goggles (NVGs): a comparative assessment between Gen 2++ and Gen 3 NVGs under different illumination conditions. Indian J Aerosp Med 65(1):17–22
15. Raghunandan V, Sekhar MB, Tripathy NK, Joshi VV (2022) Study of performance characteristics of ANVIS MK-III night-vision goggle and comparison with other NVGs for aviation usage. Indian J Aerosp Med 66(1):21–26
16. Verma SK, Joshi VV (2012) NYG research, training and operations in field: bridging the quintessential gap. Indian J Aerosp Med 56(1):29–38
17. Mathôt S (2018) Pupillometry: psychology, physiology, and function. J Cognit 1(1):132
18. van der Wel P, Van Steenbergen H (2018) Pupil dilation as an index of effort in cognitive control tasks: a review. Psychon Bull Rev 25:2005–2015
19. Laeng B, Sirois S, Gredebäck G (2012) Pupillometry: a window to the preconscious? Perspect Psychol Sci 7(1):18–27
20. Goldinger SD, Papesh MH (2012) Pupil dilation reflects the creation and retrieval of memories. Curr Direct Psychol Sci 21(2):90–95
21. Spector RH (1990) The pupils. In: Clinical methods: the history, physical, and laboratory examinations, 3rd edn
22. Hess EH, Polt JM (1964) Pupil size in relation to mental activity during simple problem-solving. Science 143(3611):1190–1192
23. Kahneman D (1973) Attention and effort, vol 1063. Prentice-Hall, Englewood Cliffs, NJ, pp 218–226
24. Kahneman D, Beatty J (1966) Pupil diameter and load on memory. Science 154(3756):1583–1585
25. Nakano T (2017) The right angular gyrus controls spontaneous eyeblink rate: a combined structural MRI and TMS study. Cortex 88:186–191
26. Thomas RJ (1994) Blinking and the release reflexes: are they clinically useful? J Am Geriat Soc 42(6):609–613
27. Nakano T (2015) Blink-related dynamic switching between internal and external orienting networks while viewing videos. Neurosci Res 96:54–58
28. Charles R, Nixon J (2017) CIEHF. In: Charles R, Wilkinson J (eds) Blink counts can differentiate between task type and load. Contemporary Ergonomics and Human Factors

Workplace Improvements with Intervention in Cashew Processing Units

Sananda Das, Mukesh Kumar Kamti, Rauf Iqbal, and Vivek Khanzode

1 Introduction

A significant tropical crop from an economic standpoint is cashew (*Anacardium occidentale* L.) [1]. Since the Indian cashew industry is export-focused, it is referred to as the nation's main source of income. More than 5 lakh people are directly and indirectly employed by it, mostly in rural areas. India is the world's greatest producer, processor, consumer, and exporter, accounting for 26.40 and 46.09% of global production, respectively. India exports cashew kernels to more than 60 nations, primarily the United States, the Netherlands, the United Kingdom, Germany, Japan, Australia, and the United Arab Emirates.

Cashew output runs from the sixth to the forty-first year following plantation. The three main cashew products are raw nuts, cashew kernels, and cashew nutshell liquid (CNSL), whereas cashew apples are typically processed and eaten locally [2].

India produced 758,410 metric tons of in-shell raw cashews in 2018, and during the projection period of 2019–2024, that production is anticipated to grow at a CAGR of 1.8%. With 5.5 lakh metric tons produced a year, India is the main cashew-growing nation in the Asia-Pacific region and the world's top producer of raw cashew nuts (RCN). Cashews are mostly farmed in India in the states of Tamil Nadu, Andhra Pradesh, Orissa, and West Bengal on the country's east coast, as well as Maharashtra, Goa, Karnataka, and Kerala along its west coast [3].

S. Das (✉) · M. K. Kamti · R. Iqbal · V. Khanzode
Indian Institute of Management Mumbai (IIM Mumbai), Mumbai 400087, India
e-mail: Sananda.das.2116008@iimmumbai.ac.in

© The Author(s), under exclusive license to Springer Nature Singapore Pte Ltd. 2025

S. Das et al. (eds.), *Human-Centered Digitalization, Volume 3*, Design Science and Innovation, https://doi.org/10.1007/978-981-96-4383-7_53

It is major in Manipur, Meghalaya, Tripura, the Andaman and Nicobar Islands, and Chhattisgarh to a minor level. In the state of Maharashtra, cashew production and productivity are at their highest levels. Maharashtra has the greatest average productivity at 1262 Kg/ha. In terms of cashew output in India, Maharashtra comes in second place, followed by Andhra Pradesh and Orissa. In order to satisfy the cashew processing industries' domestic demand, India continues to import 5.8 lakh tons of raw nuts from Africa and other nations [4]. Currently, African nations are engaged in their own cashew processing, which has limited the supply of raw cashew nuts available for processing in India [5].

Thus, in order to achieve self-sufficiency in raw cashew nut production, it is imperative to enhance domestic raw cashew nut formation by extending the areas planted with high-yielding varieties and advancing technology to boost productivity per unit area.

2 Objective and Purpose

The goal of this initiative is to

- Study the work practices being used in cashew processing units in Sindhudurg district in Maharashtra state.
- Exploring possible workplace design intervention process and improvising the interventive measures and further assessment for their work life.
- To understand the socio-economic status and morbidity disorders among cashew workers.
- To recommend a proper workplace for improving improper working postures and implement better workplace ergonomics practices for higher efficiency and productivity of the workers.

3 Methodology

The goals of this study were achieved by using the approach listed below.

1. Walk-through observational research in the real place.
2. Using a generic ergonomic checklist to pinpoint important tasks.
3. Critical job video analysis to identify ergonomic risk factors.
4. The RULA, REBA worksheet assessment for postural evaluation.
5. Statistical data analysis by using SPSS software for better results.
6. Local population anthropometric dataset generation for workstation design.
7. Designing or redesigning a workstation and tools utilizing anthropometric data from the area.
8. The creation of prototypes, laboratory testing, field testing, and deployment.

9. Physical characteristics, such as body weight, height, and BMI are measured. BMI is measured to know whether the person is obese or thin.

4 Finding and Results

1. **Finding from Field Visit**

There are three operating modes for cashew processing machines, namely (i) Automated (ii) Semi-Automated, and (iii) Minimal Automated. The Maha Cashew cluster, a special purpose vehicle (SPV) of the Maharashtra government, seeks to increase productivity and efficiency while lowering manufacturing costs. In the process, they have developed a variety of equipment and automated/semi-automated systems for processing cashews. Figure 1 displays the automated machinery present in the cluster. The process flow for the automated mode is shown in Fig. 2.

Several concerns relating to human-machine interactions were identified during the initial visit. Leamon's work system model [6], which served as a foundation for encoding workplace human-machine interactions, was utilized to pinpoint these problems [6]. The model depicts two essential interactions between a human and a machine: (a) a machine's display to a human's senses, and (b) a human's effectors—hands and feet—to a machine's controls. The model also includes two feedbacks (sometimes referred to as tactile responses): (a) machine controls to human senses and (b) machine controls to human effectors (force feedback or reaction). The interaction and feedback connected to human effectors determine the task's psychomotor skill needs, whereas the interaction and feedback related to human senses determine the task's cognitive burden.

2. **Results**

The demographic features of the participants are presented in the following table (Table 1). The physical characteristics (age, height, and weight) of the workers are

Fig. 1 Automated equipment in the cluster

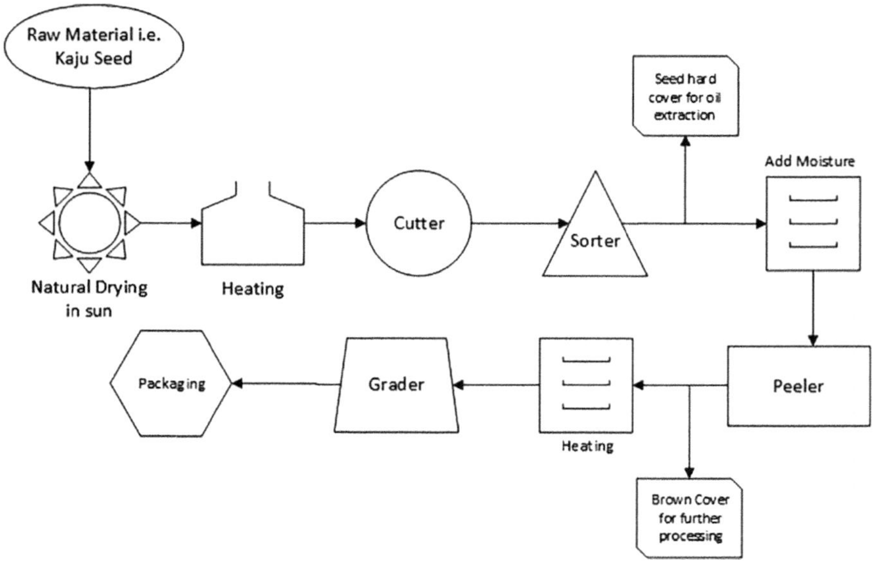

Fig. 2 Automated end-to-end process of the cashew processing unit

determined with statistical analysis by using SPSS software. We measured the physical parameters of workers which are helpful for our project. From these parameters, luckily, we found that most of the workers are in the range of normal BMI.

From our study, we found that the pain mostly felt by the workers is different. From data collection, we observed that pain is mostly felt in two body parts. Such as neck pain (31 subjects) and low back pain (30 subjects). Besides, we have shown that knee pain and ankle pain are also involved in this occupational health problem. So, there are different discomfort/pain that occurs in cashew workers. And the difference is given as a bar chart in Fig. 4.

We measured different types of hand grip strength among workers (Table 2). Then, we measured different types of pinch strength such as TIP pinch strength, KEY pinch strength, and POWER pinch strength (Table 3). Then, we evaluated the data by solving different statistical analysis such as mean, standard deviation, variance, Range by SPSS software.

Next, we performed the Pearson correlation test for the significance test. Pearson's test is the choice as it always gives the exact value. When we have two quantitative variables and want to determine whether there is a linear relationship between those variables, we use Pearson's correlation. That might be represented by a research hypothesis that one score influences the other in a particular way. A p-value can be calculated with either one or two tails.

A moderate level of association exists between the variables when the size of the correlation coefficient falls between 0.5 and 0.7. Correlation coefficients with magnitudes between 0.3 and 0.5 show low correlation between the variables.

Table 1 Physical parameter analysis through SPSS software

		Height	Weight	BMI	Age (year)
N	Valid	59	59	59	59
	Missing	1	1	1	1
Mean		143.31	49.49	24.188	35.39
Median		146.00	48.00	23.600	36.00
Mode		145	48	20.0[a]	34
Std. deviation		11.615	11.165	4.9264	9.327
Variance		134.905	124.668	24.269	87.001
Skewness		− 0.858	1.476	0.742	0.111
Std. error of skewness		0.311	0.311	0.311	0.311
Kurtosis		− 0.695	4.400	0.131	− 0.704
Std. error of kurtosis		0.613	0.613	0.613	0.613
Range		38	64	21.9	40
Minimum		122	33	16.0	18
Maximum		160	97	37.9	58
Sum		8455	2920	1427.1	2088

[a]Multiple modes exist. The smallest value is shown

Table 2 Statistical analysis from hand grip strength data

Statistics

		GS (prework-right hand)	GS (post-work-right hand)	GS (prework-left hand)	GS (post-work-left hand)
N	Valid	59	59	59	59
	Missing	0	0	0	0
Mean		14.39	14.75	12.80	12.890
Median		12.00	16.00	12.00	14.000
Mode		12	16	12	18.0
Std. deviation		5.229	5.628	5.464	5.2679
Variance		27.345	31.676	29.854	27.751
Range		24	25	30	23.5
Minimum		4	1	2	0.5
Maximum		28	26	32	24.0

GS = Grip strength

Table 3 Statistical analysis of TIP pinch strength, KEY pinch strength and power pinch strength

Descriptive statistics			
	Mean	Std. deviation	*N*
TIP pinch strength (right) kg	2.1724	0.82343	118
TIP pinch strength (left) kg	2.058	0.7801	118
KEY pinch strength (right) kg	2.614	0.9057	118
KEY pinch strength (left) kg	2.548	0.8719	118
Power pinch strength (right) kg	2.955	1.1536	118
Power pinch strength(left) kg	2.867	1.1039	118

The various working phase and working postures of the cashew workers were analyzed by postural analysis tools (RULA and REBA) with their action categories that is presented in Table 1. The posture analysis by RULA and REBA scores are indicated that the body parts were more likely to be affected for awkward bending.

5 Discussion

The entire project and all its stages are covered in the debate. The primary findings are revealed in the discussion, which also incorporates the underlying knowledge provided in the report's introduction and theoretical basis parts. As a result, several topics like those in the report's main body are briefly discussed again. The debate, however, offers a more in-depth perspective that is enhanced by the information gleaned during the project thesis compilation process.

In our project, we exposed some tables and figures as a result of this study. Specially we worked on awkward working posture problems and analyzed them. In Fig. 1, we discussed about the automated equipment in the cluster of workers in the cashew unit. In Fig. 2, we showed how the automated end-to-end process of the cashew processing unit works. In Fig. 3, we expressed the model of Leamon's work system [6], which served as a foundation for encoding workplace human-machine interactions,

The personal characteristic shows that most of the workers are female. And they are engaged with a different type of work like cutting, sieving, and grading. The physical characteristics (age, height, and weight) of the workers are determined with statistical analysis by using SPSS software shown in Table 1. From these parameters, luckily, we found that most of the workers are in the range of normal BMI. In Fig. 4, we discussed pain-related problems that occurred to awkward working postures.

In Table 2, we discussed about statistical analysis of Hand (right and left) grip strength and in Table 3, we mentioned the same with all different types of hand pinch strength for better evaluation. In Table 4, we performed the Pearson correlation test for evaluating the significance of the p-value.

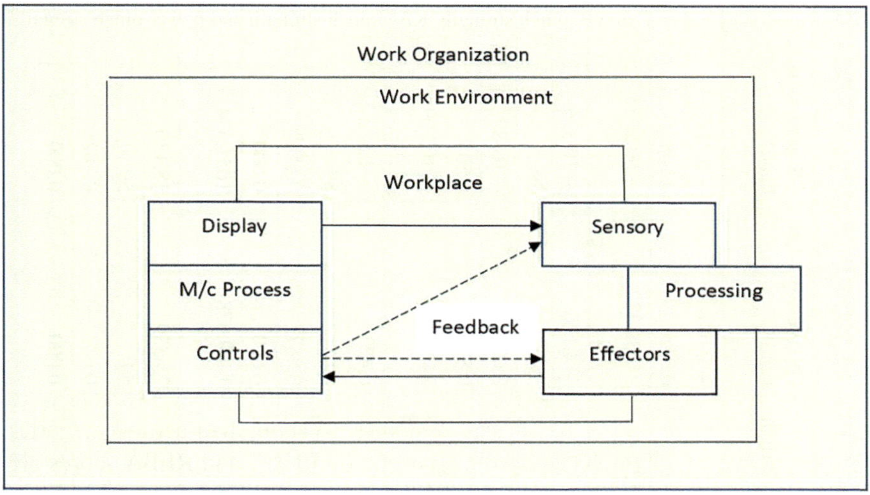

Fig. 3 Leamon's work system model [6]

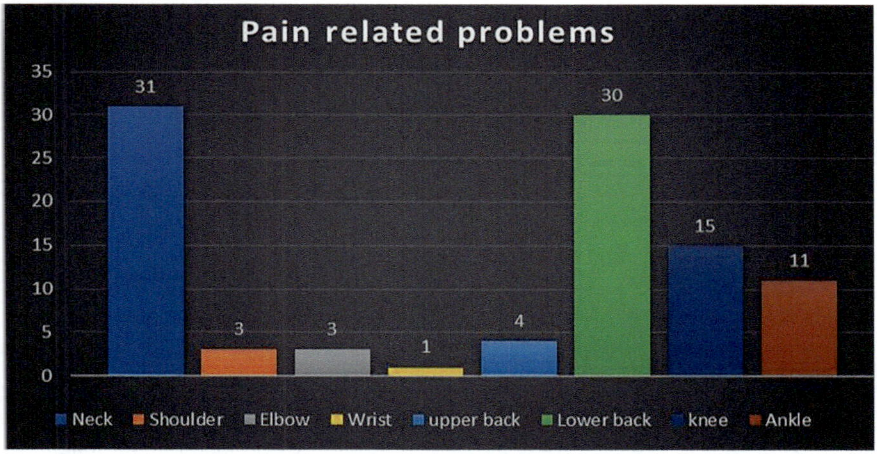

Fig. 4 Pain prevalence of different body parts among participants in cashew workplace

For posture analysis, we used the REBA and RULA assessment processes. In Table 5, some different posture shows some strain based on REBA and RULA score. In the REBA score, the action category of 4–7 indicates the working posture is at medium risk. Further investigation may be needed as change soon. Similarly, the action category of 8–10 indicates the working posture is at high risk. Further investigation is definitely needed as change soon. In our data, we found medium and high risk both in REBA score by posture analysis. In the case of RULA score, the action category of 3–4 indicates further investigation is needed and action category 5–6 indicates not acceptable posture actions should be done as soon as possible. Action

Table 4 Pearson correlation test for hand grip strength

Correlations							
		Age (year)	Experience (year)	GS (pre-work-right hand)	GS (post-work-right hand)	GS (pre-work-left hand)	GS (post-work-left hand)
Age (year)	Pearson correlation	1	0.272[a]	− 0.078	− 0.1 89	− 0.081	− 0.212
	Sig. (2-tailed)		0.037	0.558	0.152	0.542	0.108
	Sum of squares and cross-products	5046.034	569.702	− 219.966	− 575.153	− 239.322	− 602.966
	Covariance	97.001	9.822	− 3.793	− 9.916	− 4.126	− 10.396
	N	59	59	59	59	59	59
Experience (year)	Pearson correlation	0.272[a]	1	0.058	− 0.078	0.020	− 0.089
	Sig. (2-tailed)	0.037		0.661	0.559	0.880	0.503
	Sum of squares and cross-products	569.702	868.965	68.402	− 98.058	24.534	− 105.098
	Covariance	9.822	14.982	1.179	− 1.691	0.423	− 1.812
	N	59	59	59	59	59	59
GS (prework-right hand)	Pearson correlation	− 0.078	0.058	1	0.666[b]	0.887[b]	0.658[b]
	Sig. (2-talled)	0.558	0.661		0.000	0.000	0.000
	Sum of squares and cross-products	− 219.966	68.402	1586.034	1136.847	1469.678	1051.534
	Covariance	− 3.793	1.179	27.345	19.601	25.339	18.130
	N	59	59	59	59	59	59
OS (postwork-right hand)	Pearson correlation	− 0.189	− 0.078	0.666[b]	1	0.632[b]	0.882[b]
	Sig. (2-tailed)	0.152	0.559	0.000		0.000	0.000

(continued)

Table 4 (continued)

Correlations

		Age (year)	Experience (year)	GS (pre-work-right hand)	GS (post-work-right hand)	GS (pre-work-left hand)	GS (post-work-left hand)
	Sum of squares and cross-products	– 575.153	– 98.058	1136.847	1837.186	1126.949	1516.347
	Covariance	– 9.916	– 1.691	19.601	31.676	19.430	26.144
	N	59	59	59	59	59	59
GS (prework-left hand)	Pearson correlation	– 0.081	0.020	0.887[b]	0.632[b]	1	0.714[b]
	Sig. (2-talled)	0.542	0.880	0.000	0.000		0.000
	Sum of squares and cross-products	– 239.322	24.534	1469.678	1126.949	1731.559	1191.178
	Covariance	– 4.126	0.423	25.339	19.430	29.854	20.538
	N	59	59	59	59	59	59
GS (postwork-1eft hand)	Pearson correlation	– 0.212	– 0.089	0.658[b]	0.882[b]	0.714[b]	1
	Sig. (2-tailed)	0.108	0.503	0.000	0.000	0.000	
	Sum of squares and cross-products	– 602.966	– 105.098	1051.534	1516.347	1191.178	1609.534
	Covariance	– 10.396	– 1.812	18.130	26.144	20.538	27.751
	N	59	59	59	59	58	59

[a]Correlation is significant at the 0.05 level (2-talled)
[b]Correlation is significant at the 0.01 level (2-tailed)

Table 5 All the posture shows some strain based on REBA and RULA scores

Field pictures	Stick diagram of working posture	REBA score	Remark of risk	RULA score	Remark of risk	Action to be taken
	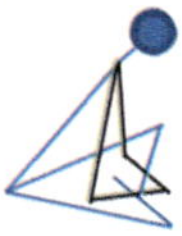	5	Medium risk	4	Change may be needed	Further investigation
		7	Medium risk	7	Not acceptable	Further investigation
		10	High risk	7	Not acceptable	Further investigation
		9	High risk	6	Not acceptable	Further investigation
		10	High risk	6	Not acceptable	Further investigation

category 7 indicates the high risk and suggested investigation and implementation change. In our study, mostly we found not acceptable postures according to the RULA assessment. So, our main focus is to form a proper workplace with better working posture.

6 Recommendation and Intervention

Considering recent technological breakthroughs, we look at the issue of human labor content in the cashew processing value chain. In the cluster, there is end-to-end automation, reducing the need for processing laborers. However, several cashew processing plants lack the scale to make use of cluster facilities. A product in over time.

Unfortunately, there are no scientifically based approaches, instruments, or methodologies that have been developed for a workplace design or human-machine interactions. These female workers consequently deal with several problems, including uncomfortable posture, repetitive motion, and imbalanced effort and force